Models for Ecological Data

Models for Ecological Data

An Introduction

James S. Clark

PRINCETON UNIVERSITY PRESS

Princeton and Oxford

Copyright © 2007 by Princeton University Press

Published by Princeton University Press, 41 William Street, Princeton, New Jersey 08540

In the United Kingdom: Princeton University Press, 3 Market Place, Woodstock,

Oxfordshire OX20 ISY

All Rights Reserved

ISBN-13: 978-0-691-12178-9

ISBN-10: 0-691-12178-8

Library of Congress Control Number: 2006930145

British Library Cataloging-in-Publication Data is available

This book has been composed in Sabon

Printed on acid-free paper. ∞

press.princeton.edu

Printed in the United States of America

1 3 5 7 9 10 8 6 4 2

Contents

Preface

THE ENVIRONMENTAL SCIENCES are witnessing a transformation in how models and data are used to draw inference and make predictions. Not more than a decade ago, ecological data were typically thought of as the products of a controlled experiment designed to test a narrow hypothesis in an abstracted setting. Statistical references used by environmental scientists contained little else. Yet data that meet the assumptions of classical statistical models remain scarce, limited by our capacity to control the environment, the temporal and spatial extent we can afford to examine, and the ability to see relevant variables. Beyond philosophical differences over, say, frequentist versus Bayes, is a more fundamental obstacle: classical methods make demands that environmental data rarely meet.

Meanwhile, environmental questions have expanded to embrace scales of space and time that could draw on suborganism-level (e.g., leaf-level) experiments to remote sensing, provided there was a framework to assimilate such mismatched data sets. Information from uncontrolled and partially observed processes has become a common basis for ad hoc inference and decision. The evolving concept of "data" extends even to such derived products as model output. Many types of information are accumulating faster than they can be assimilated in models; weather records, gene sequences, and remotely sensed imagery are examples. Environmental scientists require the capacity to draw inference based on large amounts of information, much of which cannot be shoehorned into classical models.

With the challenges have come some important new tools. The revolution in how data are used for inference and prediction is driven not only by the scale and complexity of issues confronting environmental scientists, but also by modern techniques for modeling and computation. Environmental scientists are discovering the capacity to build high-dimensional, yet coherent, models at relevant scales, and to accommodate the heterogeneous information that comes from diverse sources and was previously treated in an ad hoc fashion. Together, ambitious goals and emerging machinery have fueled growing demands on how to combine process models with data having context that can vary in space, in time, and among sample units. The approaches are technical and developing rapidly. Critical application of software requires a background that includes distribution theory and basic algorithms. This book is motivated by the need for an informal treatment of topics that could be accessible to environmental scientists who have some quantitative skills, but lack deep background in statistical theory.

In writing this book, I have tried to reach at least two audiences in the environmental sciences, including the motivated graduate student who has taken introductory courses in statistics and calculus, but has forgotten most of it, and the practicing investigator, who is faced with challenging problems. In my experience, these two audiences often possess some quantitative skills, but are stifled in their efforts to digest methods of modern inference and prediction available from the statistical literature. This introduction to concepts and methods includes material that would typically be contained in graduate-level courses in statistics and applied mathematics. The coverage has been shaped by experiences with students in my Ecological Models and Data course and in the summer institute on Ecological Forecasting at Duke University and by colleagues with whom I have collaborated and shared ideas.

In the years since the writing of this book began I have come to believe that the most difficult expectation to dispel is the one about how it should provide fodder for pithy graduate seminar discussions on the philosophy of statistics. Some readers will be disappointed not to find within these pages emphasis on the controversy common in ecological writings on Bayes. The anticipation for such a book has so dominated the communications I have received from ecologists concerning its preparation, that I will briefly explain why this book may not be what some readers expect.

The grounds for these expectations are several, but I see two factors playing an inordinate role. First is the fact that ecological discussions of Bayes have largely picked up on traditional debates in the statistics literature. These debates have been, at times, so heated (and entertaining[1],) that they have tended to overwhelm the technical developments that have breathed new life into a broad aspect of statistical practice. For the nonpractitioner (most environmental scientists fall in this category), it may be hard not to leave with the impression that "what's new" is changing philosophy.

A second and more subtle factor behind the expectation that new methods can come largely through a seminar format may be an entrenched view that ecologists can do statistics without becoming too involved in technical details. Modern Bayes demands a degree of sophistication with distribution theory that is lacking not only in the statistics courses typically available to ecologists, but also in the math courses taken at the Ph.D. level. The motivated student may gain facility with the deterministic models encountered in linear algebra and differential equations courses, while missing the stochastic models needed for modern Bayes. Even an introductory course in probability and stochastic processes will focus on theory, rather than applications that involve algorithms. In my view, the debates on philosophy can suffer from misunderstanding of some basic technical concepts.

---

[1] For example, Lange et al. (1992) responded to the claim that they were "trying to turn statisticians into Bayesians." Really, they were just trying to turn "frequentists into statisticians."

In contrast to what is often written about statistics in the ecological literature, many statisticians have recently been preoccupied with tools, rather than philosophy. The emergence of modern Bayes in so many disciplines has, to some degree, eclipsed older debates. Ironically, the excitement generated by technical developments has prompted a number of Bayesians to remind colleagues that Bayes is *more* than just machinery. Throughout, I focus on pragmatic concerns, including modeling, computation, and applications, with less emphasis on philosophy than some ecologists may expect from a treatment of environmental statistics. I introduce both classical and Bayesian frameworks, emphasizing how we might want to shift from one to the other as complexity increases. I do not dwell on philosophical divides, largely because I think most students who begin to use new tools will develop an appreciation of these issues along the way. I think that the shift to a Bayesian view of inference is so inevitable that its adoption does not really need any more arguments from people like me. In short, this book is intended for those who want to learn new techniques, including a basic introduction to statistical modeling and computation, emphasizing process models from the environmental sciences. It is a text, rather than a monograph, because the philosophical debates are available elsewhere—in the ecological literature, Hilborn and Mangel (1997) is exceptionally good—and the tools themselves are the source of current focus. A complementary volume edited by Clark and Gelfand (2006b) focuses primarily on applications and assumes a greater level of sophistication.

This effort to introduce state-of-the-art inference and prediction techniques to a nonstatistical audience is behind the nonstandard content and organization. My personal experiences with graduate students and colleagues led me to believe that a text on this material would need to accomplish several things. First, it would have to span a broader range of sophistication than is typical of most ecology and statistics references, starting at a basic level. Many students with the capability and motivation cannot get started with even the best new texts on modern Bayes, and there are now several. On the other hand, if a text starts and remains at a basic level, students fail to appreciate what it can do for them. This is certainly the conclusion many take from ecological literature on Bayes, which seems to say that Bayesians do a lot more work to get about the same results. I have been most impressed with the need for an approach that makes the connections from basic models to environmental application, with plenty of opportunity to see how to get there. Throughout, I have attempted to present material in terms that are as nontechnical and jargon-free as possible. Nonetheless, much here will challenge students entering graduate programs in environmental sciences. This text assumes that the reader is committed to investing seriously in some technical background.

So now for the nonstandard organization. A typical text with this degree of emphasis on statistics would start with probability models and other technical background. I have taken this approach in the past and found that it mostly convinces environmental students that this stuff is no fun and will take more time than they have available to them. More effective for me has been the treatment of the technical tools as progressive goals to be assimilated incrementally. Rather than starting with the hardest topics, I provide extended appendixes and encourage students to skim them repeatedly. Even the background distribution models can be most accessible when they come in the form of computer applications that are backed up by application.

The first eight chapters can be the basis for a self-contained graduate course, such as my Ecological Models and Data at Duke University. The first two chapters are intended to move rapidly through some motivation, basic concepts, and process models, providing an introduction only. Clearly, there are whole texts devoted to this material. In Chapter 2, my emphasis is on the broad outline of how one might think about process models, with a general introduction to applications and behavior. In my course, I move quickly to inference, which begins in Chapter 3. This is the point where students start to feel more engaged and make connections to data. By the end of my course (Chapters 1 through 8 of this book), many students progress to the point of writing their own Gibbs samplers. Throughout, I attempt to fill in many of the steps that seem obvious to the practitioner, but can baffle the nonstatistician. I emphasize individual goals for the course, with computer labs leading directly to student independent projects. The computer lab manual that accompanies this text, *Statistical Computation for Environmental Scientists in R*, contains substantial cross-referencing with this text. The student projects assure that students apply the tools and move at their own pace.

Chapters 8 through 10 are more advanced. Here I assume additional sophistication and include fewer of the technical details. These chapters provide an overview of the complexity that can be addressed with modern techniques. For some topics (e.g., spatio-temporal models), I have resorted almost exclusively to primary statistics literature. Even with advanced material, I have attempted to explain the basic modeling concepts and how one can begin to develop algorithms for computation. These chapters represent a point of departure for some of the recent excellent texts in this field.

Finally, I have to include a few remarks on literature coverage. This is intended as a text, not a review of either statistics or ecological modeling. I do not attempt to reference all relevant papers. Indeed, the many examples overemphasize work from my own lab for the simple reason that I have worked extensively with these data sets and models and can readily summarize them in this text. Environmental data sets are inevitably complex, far more so than is generally acknowledged in publications. I have tried to emphasize tools to deal with these issues, best done with examples I know well. The coverage also diminishes after 2003. The draft of this book was essentially completed by early 2004. I have included some references through 2006, but these were primarily already known to me before they were published. Thus, the treatment of publications in 2004 and 2005 is necessarily more cursory than that of earlier works.

Acknowledgments

I have benefited from the interactions with many students and ecological colleagues and from the indulgence of several superb statisticians, who have tolerated my meddling in their discipline. Foremost, I have to thank lab members, present and past, who have reviewed drafts of the text, some of them repeatedly. These include Brian Beckage, Kendrick Brown, Phil Camill, Mike Dietze, Michelle Hersh, Janneke HilleRisLambers, Ines Ibanez, Shannon LaDeau, Jason Lynch, Jason McLachlan, Jackie Mohan, Mike Wolosin, and Pete Wyckoff. I owe much to students in my Ecological Models and Data course, who have provided fodder, in the form of interesting modeling problems, while serving as experimental subjects for material in this text and the accompanying lab manual. I thank Peter Morin and Fred Adler for feedback on earlier drafts of the full text. An anonymous reviewer made many valuable suggestions.

Of my statistical colleagues, I must especially thank Alan Gelfand, Montse Fuentes, Michael Lavine, and Chris Wikle. Alan has been a continuing inspiration with his support and friendship. I have abused Chris Wikle with more than one draft of the entire text, and he has been incredibly generous with his time and feedback. I have had the benefit of significant feedback from Michael Lavine over the years and some very constructive collaborations. Montse Fuentes has been supportive through the latter stages of this project. Kate Calder and Chris Pajorek provided very insightful thoughts on presentation of the material in Chapter 10. At various times along the way I have benefited from the additional inputs from Dave Higdon and Peter Mueller. Additional feedback came from Brad Carlin and Tony Ives.

Finally, Chantal Reid has provided valuable criticism and support throughout.

Part I Introduction

The first two chapters lay out goals and challenges and introduce basic elements of models used for ecological data. In Chapter 1 I discuss issues that arise in the development and application of models for environmental inference and prediction. In Chapter 2 I apply some examples from population dynamics as the basis for introducing the different types of models used for ecological processes. These process models are combined with data and parameter models in Parts II, III, and IV.

1 Models in Context

1.1 Complexity and Obscurity in Nature and in Models

This book deals with the use of data and models that can enhance understanding and contribute to prediction. These two goals are complementary. Both involve inference, and model analyses can take the form of predictive distributions. For environmental scientists, the challenge stems from the fact that natural and managed systems are high-dimensional, meaning that many interacting forces are at work (Levin 1998; Clark 2005; Clark and Gelfand 2006a). Much of nature is unmeasurable, unobservable, or both. Much cannot be manipulated. Faced with obscure, complex, and uncontrolled processes, environmental scientists have long recognized the need for abstraction (Schaffer 1981; Caswell 1988). Theoreticians and experimentalists attempt to extract the important relationships from nature so they can be studied in a controlled setting. Ecologists write models with only a few variables and parameters. They design experiments with only a few treatments.

The need to simplify on both the theoretical and experimental sides leaves a gap that can isolate those who analyze ecological models from those who collect and draw inference from data. This gap makes it difficult to test theory with data and to model data in appropriate ways (e.g., Oreskes et al. 1994). The goal of this book is to describe methods that can help to bridge the gap, starting from concepts that underlie traditional process and statistical models, and moving toward modern techniques that allow for deeper integration. This introductory chapter starts with some background and motivation.

1.1.1 Why Ecological Models Are Low-Dimensional

Attempts to abstract key features of a process are an important component of all scientific disciplines. From the conceptual (theory) side, this abstraction is accomplished with process models that contain few variables and parameters. High-dimensional process models are intractable; without simplified models, they cannot be analyzed to yield transparent relationships. Complex process models are difficult to apply beyond the context for which they were developed. If we can abstract the important elements and develop a simple model thereof, analysis might allow us to understand how the process behaves and why.

The simplification needed to describe systems mathematically often requires assumptions that cannot apply to ecological data. Theorists may speak in terms of processes that apply everywhere, in a general sense, but nowhere in particular. In the light of the complexity mismatch between theory and the real world, it is not surprising that mathematical models are often viewed as irrelevant (e.g., Simberloff 1980). In over a decade of teaching mathematical models I do not need one hand to count the number of times that a basic model described in ecological textbooks has been directly applied to a student data set. When it comes to models, irrelevance can be the price of tractability.

1.1.2 Why Statistical Inference Is Low-Dimensional

Traditionally, the statistical analysis needed for inference and prediction was possible only for data collected under a rigid design. Here again, simplification is achieved through use of model systems. Statistical models simplify with assumptions such as each observation is independent of all others; uncertainty is typically allowed only for a response variable; the variables must be observable, in the sense that we can assume that values assigned to predictor variables represent truth. To meet these assumptions, experiments rely on strict design.

By contrast, ecological data are typically complex and interrelated in space and time (Ellner and Turchin 1995; Scheffer and Carpenter 2003). Broad-scale and long-term interactions with superimposed high frequency and difficult-to-measure fluctuations are pervasive. Only small parts of a process readily submit to experimental control and typically only in terms of highly abstracted experimental designs (Carpenter 1996; Brown 1995; Skelly 2002; Hastings 2004; Clark and LaDeau 2005). The small amount of variance in data that is explained by ecological models reflects the high dimensionality of nature, the many interacting processes that affect observations. The fact that simple models typically account for a small fraction of the total variance, even in experimentally controlled settings, leads to obvious questions: Are the important factors included in the analysis? Do the assumptions of independence and uncertainty confined to the response variable impact inference? Can the experimental results be extrapolated to nature?

Because many important processes cannot be studied in controlled experiments, there is a tendency to overlook model assumptions and apply statistical models where assumptions are violated. Spatio-temporal aspects of data and most sources of uncertainty tend to be ignored. Whether nature is abstracted to the point where simple model assumptions can be satisfied or assumptions are violated in the analysis of natural settings, the barriers between data and theory can be large.

The contrast between the high dimensionality of natural systems versus the simplicity that is manageable in experiments and models explains part of the historic gulf between ecological models and data. Throughout, I suggest that simple process models can be powerful. A goal of this book is to demonstrate that this gulf can often be bridged, but only if process models can remain simple enough to be tractable, and data/parameter models can be sufficiently sophisticated to allow for the high dimensionality of nature. Rather than design away natural influences, I emphasize inference based on relevant scales and settings; the focus is on bringing models to nature, rather than nature to models.

1.2 Making the Connections: Data, Inference, and Decision

The challenges of inference, prediction, and decision making faced by ecologists are shared by many disciplines. I use two examples to highlight some of the specific issues. To emphasize the generality of both the challenges and the emerging tools needed to address them, I draw the first example from medicine. The second example comes from community ecology.

1.2.1 Example: Soft Data, Hard Decisions

Most information does not come from experiments designed by statisticians, and most decisions are subjective. Consider the example of a standard treatment of kidney stones with extracorporeal shockwave lithostripsy (Madigan and York 1995). The stone is located on a real-time ultrasound image by an operator who will focus shockwaves capable of disintegrating it. If the image is of high quality (I), there is an increased chance that it will be properly identified and disintegrated (D). If disintegrated, there is increased chance of clearance through the urinary tract (C). The decision framework involves interpretation of data with a flow outlined in Figure 1.1a. Note that the medical practitioner is concerned with the probability of successful clearance (C) given image quality (I), or $\Pr\{C|I\}$. Between these two events is the probability of disintegration (D). The decision is subjective, in the sense that, faced with the same evidence and choices, two practitioners could arrive at different decisions.

The basic elements of this problem are familiar to environmental scientists, managers, and policy makers. For instance, information is limited, but it can accumulate with time and experience. To many scientists, these data

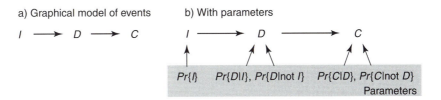

FIGURE 1.1. A graphical model of the kidney stone example (modified from Madigan and York 1995). In (*a*) is the graphical model of events that the image is of high quality (*I*), the stone is disintegrated (*D*), and the remnants are cleared through the urinary tract (*C*). To calculate Pr{C|I} we require five parameters (*b*). By making them stochastic (giving them priors), we can write simple rules for updating knowledge based on experience with new patients.

seem soft. The decision must be based on inadequate knowledge. We may not have the luxury of putting off a decision until hard data can be collected from, say, a series of controlled experiments. However, we would value a means for updating knowledge that can result in better decisions, that is, an adaptive management framework.

How can the practitioner use the model to learn from accumulated experience? The answer is, in large part, technical and a principal motivation of this book. But I provide a partial answer here. The parameters that influence decisions include the probability of C given that disintegration did or did not occur ($\Pr\{C|D\}$ and $\Pr\{C|\text{not } D\}$,[1] respectively), the probability of D given that image quality was good or bad ($\Pr\{D|I\}$ and $\Pr\{D|\text{not } I\}$, respectively), and the probability that the image was of good quality ($\Pr\{I\}$) (Figure 1.1b). The values of these five parameters determine probability of success. Clearly, the more we know about the values of these parameters, the more informed the decision. If we treat these parameters as being fixed values, there is no opportunity for learning. To allow for updating knowledge, the parameters are taken to be random. This random character of parameters allows for regular updating, with current understanding being further refined by each new experience (observation). The posterior knowledge taken from each patient becomes the prior for the next.

Whether the goal is increased understanding (as in, say, inference), prediction, or decision, the model graph provides a road map that facilitates not only modeling, but also computation. It emphasizes the importance of conditional probability, represented by arrows connecting events that are directly linked. In this particular instance, it describes a decision process that involves uncertainty that can be reduced though a prior-update-posterior cycle.

I use graphs like Figure 1.1 to represent models throughout this book. Early in the text, I use the convenient structure that involves data, process, and parameter submodels. This hierachical framework for submodels serves to decompose potentially complex problems into simple elements. By the end of this book I extend this framework to the more general notion of models as networks.

1.2.2 Example: Ecological Model Meets Data

A second example illustrates how the graphical framework of Figure 1.1 extends to ecological processes and brings in some of the challenges that confront ecologists attempting to integrate models and data. Ecological models predict that differences in how species respond to limiting resources can determine whether they can coexist (Levins 1979; Armstrong and McGehee 1980; Tilman 1982; Pacala and Silander 1990; Grover 1991; Pacala and Rees 1998; Murdoch et al. 1998). Is one species better able to exploit a limiting resource than another species? Does the advantage depend on how much resource is present? To evaluate the role of limiting resources, ecologists gather data describing growth responses at different resource levels.

---

[1] The notation | means "given that." Thus, $\Pr\{C|D\}$ means "the probability of event C given that event D has occurred."

Figure 1.2 shows data on seedling response (height growth) to a resource (light). The degree of scatter in Figure 1.2a and 1.2b is not unusual (Kobe 1999; Finzi and Canham 2000). Growth rate data are obtained from measurements of seedling height, together with estimates of light that penetrates the forest canopy. Full sunlight has a value of one, and complete darkness has a value of zero. Although the raw data do not show obvious differences between the two species, the fitted model says that the differences are highly significant, with 95 percent confidence intervals of each assigning near zero probability to the other (Figure 1.2c). It might seem paradoxical that these broadly overlapping data clouds are represented by significantly different models, yet this common situation is rarely mentioned. In fact, it is central to the interpretation of models.

Here is the standard analysis that led to this result. A model might include parameters that describe a minimum light requirement, a half-saturation constant, and a maximum rate, or asymptote. We could write the model as

$$y_{ijt} = \mu(x_{jt};\theta) + \varepsilon_{ijt}$$
$$\varepsilon_{ijt} \sim N(0,\sigma^2)$$

where x_{jt} and y_{ijt} are the predictor and response variables (light and growth) for the ith individual at location j at time t, and μ is the saturating function of light availability x. θ represents the three parameters that determine μ, representing (1)

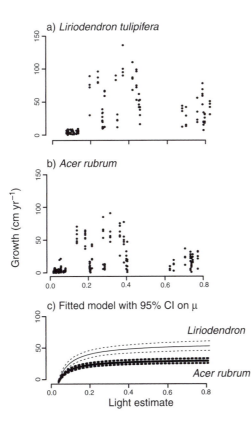

FIGURE 1.2. Growth rate responses to light availability. (*a*) and (*b*) show data obtained as measurements of seedling height growth and light availability. (*c*) shows a traditional model fitted to these data, the solid line being the estimate of μ. The dashed lines are 95 percent confidence intervals for μ. From Clark et al. (2003).

the minimum light requirement, (2) a half-saturation constant, and (3) the asymptotic growth rate. This saturating function does not fully describe the relationship between x and y—there is residual scatter. To accommodate this scatter, this function is embedded within a stochastic shell, which might allow for the fact that the model cannot be correct, and growth rates are not precisely known. Together, deterministic and stochastic elements are termed a *likelihood function*. This model says that y is assumed to have a normal distribution with a mean determined by the function μ and residual variance σ^2. This is the fitted model of Figure 1.2c.

The saturating function μ runs through the data cloud and is described by estimates of three parameters θ, which are represented by error distributions in Figure 1.3. The spread of these error distributions represents the level of confidence in parameter estimates, which depends, in turn, on the number of observations, the scatter, and the model. A 95 percent confidence interval can be viewed as the central 95 percent of the error distribution (although this is not its precise definition, as discussed in Chapter 5). The errors in these estimates are asymptotically zero in the sense that, as sample size increases, the confidence intervals (spread of error distributions) decrease. Hereafter, when I speak of asymptotics, I mean it in this sense of sample size. A predictive interval for the function μ is obtained by propagating the error in parameters θ to error in μ (dashed lines in Figure 1.3).

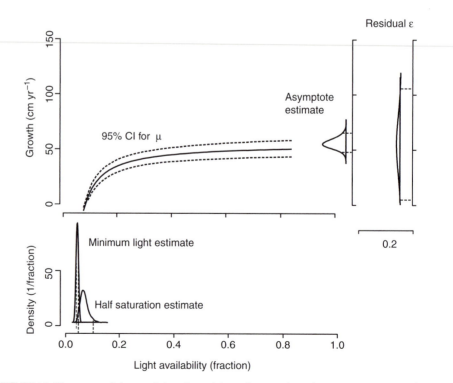

FIGURE 1.3. Elements of the traditional model used to analyze data in Figure 1.1. There are error distributions for three parameters and residual stochasticity in the error term ϵ (From Clark et al. 2003).

Obviously, the 95 percent confidence interval on $\mu(x)$ does not bound 95 percent of the data. The uncertainty in parameter estimates, which is used to construct the envelope for μ, is relatively small in comparison with variability in the data. The bulk of the variation is associated with the error term ε_{ijt}. This scatter is represented by a density shown on the far right-hand side of Figure 1.3. We could calculate a predictive interval for y that incorporates this scatter (it integrates ε) (Feiburg and Ellner 2003; Clark 2003), but that is rarely done. We would do this to predict as-yet unobserved growth response data y. But if this scatter is viewed as noise or error, we might not have much use for this predictive interval—it adds clutter, but not insight. Moreover, if we thought the scatter captured by ε_{ijt} was anything other than error, it might be difficult to justify this model in the first place.

So what is this leftover scatter? I began by saying that the scatter sopped up by ε_{ijt} might be associated with observation errors or model misspecification. But measurements of seedling height can be off by a centimeter. They are not off by a meter. So observation error is not the explanation for the broad scatter in Figure 1.2. If the deterministic part of the model μ is inadequate, we might increase its complexity in terms of a more flexible form or by including additional covariates that could explain the scatter. In either case, we require more parameters. In fact, ecologists have studied seedling growth many times, and measurements of more variables often do not explain much additional scatter. In other words, we often cannot account for this variability by increasing the complexity of the deterministic part of the model $\mu(x)$ (Clark et al. 2003a). If the scatter is not observation error, and we cannot accommodate it by incorporating more deterministic complexity, traditional methods do not leave many options.

In fact, there are many ways in which stochasticity might stand in for unobservable aspects of this relationship. For example, the light data x might be variable or imprecisely known. This brings an additional source of stochasticity and is sometimes termed an error-in-variables problem. I distinguish between the light seen by a plant at location j, x_j, and the observation of it using the notation $x_j^{(obs)}$. To allow that observations depend on the true light level, with uncertainty, I include a density for light observations, $x_j^{(obs)} \sim p(x_j, \phi)$, where ϕ represents any parameters that enter the model for observations. We now have the model

$$y_{ijt} = \mu(x_{jt}; \theta) + \varepsilon_{ijt}$$

$$x_{jt}^{obs} \sim p(x_j, \phi) \qquad \text{uncertainty in } x$$

$$\varepsilon_{ijt} \sim N(0, \sigma^2) \qquad \text{error in } y$$

As in Section 1.2.1, we can represent this model with a graph. The basic model that ignores uncertainty in light is represented by Figure 1.4a. If implemented in a Bayesian framework (Chapter 4), I could refer to this as "simple Bayes." The stochastic element in this graph is represented by the connection between σ and y, indicating the error in y. In part b of this figure, there is an additional source of stochasticity associated with observations of x. The graph has increased in complexity to accommodate this additional relationship.

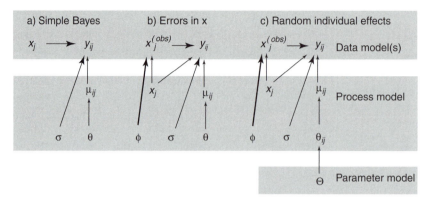

FIGURE 1.4. Four models that can be used to model growth response to light. Complexity increases from a simple model with error in y (a), to error in both variables (b), to variability among individuals (c). Modified from Clark (2005).

But there is still much more going on in Figure 1.2. We expect that individuals will have different responses to light, depending on many factors that cannot be measured. If individual responses result from unmeasured or unmeasurable factors, then we can include random effects in the model. In this case, there is a parameter vector that applies to each individual θ_{ij}. These individual parameters are drawn from a distribution having parameters Θ. This model is:

$$y_{ijt} = \mu(x_{jt};\theta_{ij}) + \varepsilon_{ijt}$$
$$x_{jt}^{(obs)} \sim p(x_j,\phi) \qquad \text{error in } x$$
$$\theta_{ij} \sim p(\Theta) \qquad \text{random individual effects}$$
$$\varepsilon_{ijt} \sim N(0,\sigma^2) \qquad \text{error in } y$$

There are now $n+k$ new parameters that must be estimated, one θ_{ij} for each individual, and k parameters in θ, describing population heterogeneity.

The model is getting complex, and the list of potentially important influences is still not exhaustive. For example, we might include random effects associated with location, fixed or random effects for years, and there might be autocorrelation between years or with distance. We might have additional sources of information that are not yet accommodated by the model. Already the model is beginning to look like a network of elements, some of which are known and some unknown. This network perspective is readily depicted with graphs. The modules labeled as "Data," "Process," and "Parameter" on the right-hand side of Figure 1.4 help organize relationships.

With these basic principles in mind, I return to the traditional analysis, which superficially appears to have clarified a relationship involving broad scatter with a tight relationship between resource and growth (Figure 1.2). So where did the scatter go? And did we lose anything important? Using models available in standard software, there is a deterministic function $\mu(x)$ buried in noise. The function $\mu(x)$ is like the mean response, and it is surrounded by error that is independent of $\mu(x)$. We now have the answer to the first question: the scatter was sopped up by a stochastic shell.

The second question is more difficult to answer. The insights used to extend the simple model came from recognition that observations of height growth are imprecise, light observations or treatments are variable and imprecisely known, individuals have different genotypes, and their responses to light depend on other factors that vary in time and space. Some of these factors can vary on such fine spatial and temporal scales that we could not hope to measure them. Because there are response times associated with many variables, we would have a hard time even deciding on a scale to measure them. If all of these factors contribute to the scatter, then the problem is high-dimensional.

How do the assumptions of the statistical model used to fit Figure 1.2c compare with the factors we identified on the basis of ecological insight? One could argue that at least one of these factors, observation error, is consistent with the statistical model (the deterministic-response-in-a-stochastic-shell approach). This would be the case if there were a deterministic response that applies to all individuals, and it was obscured by errors that result from sampling. The two might be independent if meter stick error does not depend on tree height. But the data suggest that this assumption does not apply here. The measurements could be off by perhaps a centimeter, but not by a meter.

Once we move beyond simple observation error, the deterministic-response-in-a-stochastic-shell model becomes less plausible. The other factors are not well described by this model, because their effects depend on the function $\mu(x)$ itself. Of course, we could make the stochasticity depend on $\mu(x)$ in a specific way by, say, modeling log values. Where such specific transformations can provide a quick fix, use them. For a number of reasons, this quick fix will not suffice here. If the estimates of the light level at which a plant grows are only approximate, then x is stochastic, and, in turn, $\mu(x)$ is stochastic; it is a function of stochastic x. If individuals have different responses to light due to genotype or factors that vary, then there is a function $\mu(x)$ for each individual i.

Is this simply nit-picking? How much of a difference can it make? Would we not fit roughly the same model, regardless of seemingly arcane decisions about what causes the scatter? Can these details affect the inference on the process of interest, example, competition and coexistence? The answer is, "it depends." We will confront this issue repeatedly. For now simply consider a few more points. First, the range of variability in growth described by the fitted function $\mu(x)$ in Figure 1.2c is a small fraction of the total (Figure 1.2a, b). If pressed to identify the dominant term in our model, we would have to say that it is the part we have relegated to the noise or error term, not to the signal $\mu(x)$.

The first point leads to a second. Inference is based on a model that differs substantially from our view of the relationship. If the scatter were modest we might ignore these factors, but how do we justify it here? If we are satisfied with sweeping under the rug the dominant part of the relationship, is there any point to statistics? Do we have any more confidence in a predictive interval constructed from the traditional model than we would in a line drawn through the scatter by eye? It is worth mentioning that the prevailing confidence in ecological predictions is not great, in part because it is not clear how they relate to underlying processes and to the data used to fit them. The traditional view that the scatter is error underlies the interpretation that one species grows faster at all light levels (Figure 1.5a). In light of the assumptions, can we be sure that tulip poplar outcompetes red maple?

If we trust the data and we trust the theory (e.g., growth is a saturating function of light), then the statistical analysis is the weak link. Off-the-shelf software does not provide flexible options for most ecological data sets. In other words, it may be a bad idea to let default assumptions from canned software determine everything beyond the basic process model.

As preface to techniques covered in this book, consider what happens if we allow for the sources of variability that are consistent with ecological insight (uncertainties in observations, light, and the growth response $\mu(x)$). An analysis that admits these considerations suggests broad overlap (Figure 1.5b). It is true that we could also construct broadly overlapping prediction intervals for the classical approach in Figure 1.5a. This interval could be constructed by putting back the scatter (the stochastic shell) that we threw away to produce Figure 1.5a. I discuss this in later sections of the book. This is generally not done in ecological analyses, because modeling begins from the premise that everything other than $\mu(x)$ constitutes error and is independent of $\mu(x)$. Moreover, the predictive interval obtained by mixing in ε_{ijt} is not the one we obtain under the assumption that many factors contribute stochasticity, and many of them come through $\mu(x)$ itself.

Throughout this book I focus on the weak interface, the critical connection that is needed to evaluate relationships and to predict. Modern approaches use structure to allow for complexity and stochastic elements to stand in for uncertainty. Getting this connection right helps make theory relevant, and it allows us to exploit data fully.

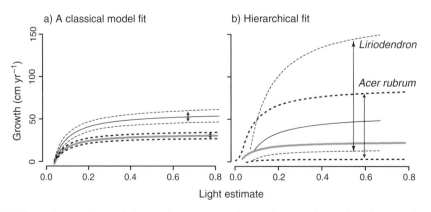

FIGURE 1.5. Comparison of the predictive interval on μ from a classical analysis, as has been standard practice in ecological analyses, with the predictive interval on y from the model in Figure 1.4c.

1.3 Two Elements of Models: Known and Unknown

I emphasize two elements of models, including variables and relationships that are known, either because they can be seen or we can assume them to be true, and those that are unknown, because they are obscure. The former constitutes much of traditional modeling in ecology. Unknowns are treated stochastically. Section 1.3.1 gives a brief overview of process models, followed in Section 1.3.2 by some aspects of unknown model elements, which are taken up in stochastic components.

1.3.1 A Point of Departure: Ecological Process Models

Environmental scientists do not have the luxury of continuing to limit attention to simple models. Low-dimensional models and designed experiments will continue to contribute insight, but the growing demand for relevant inference and prediction calls for the added capacity to address complex interactions. Modeling often begins with a process component to describe how things work. We may seek to determine whether or not relationships exist (e.g., hypothesis tests), to quantify relationships, and to make predictions. Where possible, the modeling strategy will focus on simple process models and, in many cases, allow for complexity on the stochastic side. Still, many processes have multiple interacting elements. Here I mention some of the types of environmental questions that entail integration of data and models, some of which are necessarily high-dimensional. Not all of these examples are included in this book, but they are all amenable to the approaches considered here.

Populations—At a time of intense interest in protecting rare species, the awareness that many nonendangered species undergo extreme fluctuations in density is of great interest. A broad range of questions and processes is addressed using population models and data. What places a population at risk? Will changes in the demography or age structure of a population have consequences in the future? What are the ecological implications of this structure? What kinds of constraints and trade-offs gave rise to observed life history schedules?

Ecologists use models to explore relationships involving individual behavior and dynamics of populations. For example, reproductive episodes can result in oscillations if there are strong feedback effects of density on population growth rate. Although evidence for chaos in population dynamics is weak (Hassell et al. 1976; Ellner and Turchin 1995), feedback involving resources or natural enemies can result in complex dynamics (Elton 1924; Saucy 1994; Ostfeld and Jones 1996; Earn et al. 2000). Oscillations can bring a population seemingly close to extinction, only to be followed by rebound to high abundances. Such fluctuations can occur repeatedly. They can be periodic, and, thus, are potentially predictable. Important strides in the 1970s (May 1974) initiated fertile research that has expanded to include feedbacks and environmental variation (e.g., Bjørnstad et al. 2001). Age represents one of several kinds of structure in populations. It can have several classes of effects, depending on its contribution to time delays and response to perturbations (Lande et al. 2002; Coulson et al. 2001). The inherent spatial variability in nature (e.g., soils and topography) can explain much pattern in populations, but spatial pattern can also result from processes endogenous to populations themselves. Spatial coupling can result from environmental variation (the Moran effect) (Grenfell et al. 2000), dispersal (Kelly and Sork 2002), or other biotic interactions (Ostfeld and Jones 1996).

The challenges faced by scientists and managers come from the fact that populations are highly variable and subject to many influences, and demographic processes are hard to see. The basic process models used to infer population dynamics are reviewed in a number of recent theoretical ecology texts (Chapter 2). Spatio-temporal variability in demographic rates can complicate models (Wikle 2003a; Clark et al. 2003a) and require many years of data to estimate parameters (Clark et al. 1999b; Coulson et al. 2001). Many of the variables that ecologists would like to infer or predict, such as demographic rates for endangered species and extinction time, occur at low density. Because different processes may dominate at low density, including Allee effects, we cannot simply extrapolate parameter estimates obtained from populations studied at higher densities. Spatio-temporal processes are inherently high-dimensional, in the sense that variables not only interact with one another, but they do so in ways that depend on when and where they occur (Legendre 1993a; Ranta et al. 1995). Natural ecosystems are characterized by both continuous and discrete variation in space (Hanski 1998). Parameter uncertainty may have a large impact on predictions of extinction risk (Ludwig 1999; Ellner and Feiberg 2003). Ecologists have long struggled to characterize land cover in ways that are tractable, yet relevant for population dynamics (Lande 1988; Franklin et al. 2000).

Ecological Communities—Models are used to understand trophic interactions among species (Hutchinson 1959; MacArthur 1972; Tilman 1982, 1988). For example, why are there so many species? Why are some abundant and others rare? Trophic interactions and environmental controls contribute to these patterns. Interactions among species on the same and on different trophic levels must somehow explain much of the pattern (or lack thereof) observed in nature. Why don't the best competitors for the few resources that appear to limit growth drive all others to extinction (Tilman 1988)? How do species interact by way of shared natural enemies (Holt 1977) or other types of indirect effects (Wootton 1993)?

Early ecological models were deterministic, but incorporation of temporal stochasticity has become increasingly popular. Ecologists early suspected that variability in time could have some obvious as well as some mysterious implications for fundamental ecological processes (e.g., Elton 1924; Hutchinson 1959). Fluctuations in the environment coupled with temporal integration by organisms (e.g., long life, dormancy) can contribute to coexistence of species, and it may have evolutionary consequences (Chesson 2000). I consider different approaches for accommodating temporal heterogeneity in Chapter 9.

Spatial relationships complicate modeling, including movement and re-source patchiness. Recent theoretical models have increasingly focused on how predictions of spatial models differ from those of nonspatial ones (e.g., Lewis 1997; Neuhauser 2001; Bolker et al. 2003). A rich literature is developing that emphasizes problems related to spatial covariance and scale (Levin 1992). Theory has facilitated understanding of how aggregation of individuals in space affects interactions such as competition and predation. New statistical approaches allow us to consider processes where spatial relationships are not spatially coherent and change over time (Chapter 10).

Recent models have contributed to our understanding of how disease can affect population dynamics. For example, models have proved invaluable for exploring the spread of AIDS (Anderson and May 1991), foot-and-mouth (Keeling et al. 2001), and the temporal trends in measles (Bjørnstad et al. 2002) and whooping cough (Rohani et al. 1999). Examples of spatio-temporal analyses based on epidemiological data arise in Chapters 9 and 10.

Ecosystem Function—Biogeochemistry involves models and data at a range of spatial and temporal scales. While the 7 percent perturbation of carbon exchange between plants and the atmosphere has been enough to awaken global concerns of climate change, it is one of many interacting cycles that are absorbing the dramatic alterations by humans (Schlesinger 2004). Freshwater shortages, eutrophication from mobilized reactive N and P, acidified precipitation from mobilized S, and pollution of rivers and coastal oceans are just a few of the now-recognized transformations in the chemical environmental upon which life depends. Fertilizer applied in the Upper Midwest impacts fisheries in the Gulf of Mexico (Galloway 2004). Growing awareness that important human perturbations transcend the periodic table (e.g., recent concerns for Pb, Fe, Cl, and B, to name a few) underscores the need to understand linkages. Biology is central to such linkages. Stochiometric relationships demanded by organisms mean that change in the supply of one constituent can cascade through many others (Elser and Sterner 2002). Stochiometric relationships put biology in the driver's seat, as a regulator and a place where nonlinear feedbacks can reside, while simultaneously lending a degree of predictability to element interactions.

With a pending global energy crisis and its broad impact through climate, carbon will keep the attention of biogeochemists for some time (Jackson and Schlesinger 2004). Environmental scientists must anticipate not only the future impact of continuing human perturbations, but they also must weigh in on potential engineering fixes, such as the large-scale N fertilization of soils and Fe fertilization of oceans to stimulate uptake of atmospheric CO_2.

The modeling challenges are great, involving physical processes in soils, waters, and the atmosphere at a range of scales. I mention several continental-scale analyses involving air quality in Chapter 10.

Biodiversity Feedbacks on Ecosystems—Through integration of data and models, the melding of the classically distinct disciplines of community and eco-system ecology has shown that nutrient addition can lead to reduced diversity (Wedin and Tilman 1996), while diversity can play an important role in nutrient supply (Vitousek et al. 1996) and retention (Tilman et al. 1996). Reorganization of food webs comes with changes in nutrient loading that propagate through primary producers to herbivores (Rosenzweig 1971; Carpenter et al. 2001). Nonindigenous species not only respond to land use changes, native biodiversity, and climate change, but they can also change native ecosystems, altering fire regimes (D'Antonio et al. 2000), promoting spread of infectious disease (some of which are themselves caused by non-indigenous pathogens) (Daszak et al. 2000), and affecting nutrient cycling (Vitousek et al. 1996). How does the loss or addition of species affect ecosystem functioning? Is there a minimal number of species needed to maintain ecosystem function in a given setting? Is redundancy needed as a buffer against change (e.g., Loreau et al. 2001)? Does variability at the level of individual populations propagate broadly? Or are species with similar or complementary function and niche requirements roughly equivalent (Doak et al. 1998; Reich et al. 2003)?

The many scale-dependent issues that arise in the context of ecosystem function can be modeled using the graphical framework outlined in Section 1.2. This framework is applied throughout this book.

Human Dimensions—Models and data are used to provide guidance on potential impacts of climate and land-cover change. The pace of contemporary climate change threatens reorganization of food webs, as species outrun their natural enemies and hosts, forming new networks of interaction, while others are left behind. Anthropogenic climate change has already affected many populations, but to what degree, and how has it propagated through communities? Lags, transient effects, and poorly understood changes in phenology are already combining to produce unanticipated effects on migratory birds, pollinators, and plant reproduction (Parmesan and Yohe 2003). Migration potential of plants became a research priority when general circulation models (GCMs) of the atmosphere suggested that contemporary migration rates would need to exceed apparent dispersal capacities of many species (Pitelka et al. 1997; Clark et al. 1998, 2003b). A growing number of examples from contemporary invasions demonstrate the extensive impacts that can follow from a small change in species richness (Mack et al. 2000; Callaway and Manon 2006). Ecologists are increasingly applying models to help understand which species, communities, ecosystems, and habitats are susceptible to climate change impacts.

Models and data are central to CO_2 fertilization study. Where can we expect CO_2-enhanced growth responses, for which species, and how will changing competitive relationships affect diversity (Hattenschwiller and Korner 1996; Mohan et al. 2005)? How do climate and CO_2-induced shifts in the length and timing of the growing season affect species distributions and interactions, and ultimately ecosystem-level processes (Farnsworth et al. 1996; DeLucia et al. 1999; Parmisan and Yohe 2003; Root et al. 2003; Ibáñez et al. 2006)? How do the combined effects of climate, CO_2, and habitat fragmentation affect migration potential?

These examples, drawn from a range of ecological concerns, have been traditionally modeled deterministically, in the sense that unknowns are ignored or treated in unrealistic ways. There is a growing number of important exceptions that constitute many of the examples used in this book. Because process models do not include everything that goes on in nature, we need to consider what to do with the things that are left out. As suggested in Section 1.2, these leftovers are treated stochastically.

1.3.2 Stochasticity and a Structure for Complexity

The foregoing processes can be complex and obscure, and involve the types of unknowns described in Section 1.2.2. As discussed for that simple example, statistical inference is used to integrate data with the phenomena that are formulated as process models. That integration involves stochasticity. I began this chapter by saying that a statistical model may be as simple as a process model wrapped in stochasticity, a place to park the scatter. In fact, stochasticity is not confined to places where data enter the model. In Section 1.2.2 it stands in for unknowns, such as differences among individuals that cannot be related to specific variables or processes that are only partially understood. As background and motivation for the methods that follow, I consider the concept of stochasticity a bit further.

In this book, I use the term *stochasticity* to refer to things that are unknown. Although we could speak of a stochastic process, it can be most productive to think of processes themselves as deterministic. *Models* are stochastic, because we cannot measure or specify everything that is relevant. We choose what to treat deterministically and what must be treated stochastically. This pragmatic position is encountered in traditional mathematical treatments (e.g., Taylor and Karlin 1994) and in computer sciences applications such as machine learning (Mitchell 1997; Pearl 2002), and it is unavoidable in modern statistics (e.g., Dawid 2004). Nonetheless, this view of stochasticity may appear to conflict with terminology in the ecological literature, in that it does not suggest that processes are themselves stochastic. When pressed, those who argue that, say, population dynamics are inherently stochastic, may appeal to Heisenberg's uncertainty principle, which says that we cannot simultaneously know a particle's position and momentum. The uncertainty it describes has no demonstrated (or hypothesized) relevance at the level of observable phenomena. There is no obvious answer to the question of whether processes are really stochastic, but there is a practical approach—deterministic relationships are used for relationships that are known; stochasticity stands in for the unknowns. For example, Palmer et al. (2005) discuss the classic Lorenz (1963) attractor, a type of chaotic behavior, produced by a set of three coupled differential equations. Standard model selection arguments (Chapter 6) might tell us to truncate the model to two equations, because the third equation accounts for only 4 percent of the total variance. It is well known that this truncation completely changes the behaviour—it is not even chaotic. So what is the statistician to do with so little information on one-third of the state variables? One answer is a stochastic autoregressive term, which recovers the chaotic behavior, including the Lorenz attractor with the two-equation truncated version. In other words, judicious application of stochasticity stands in for additional complexity, which may be unknown, unobserved, or both. For an ecological example, births and deaths are modeled as demographic stochasticity in models that treat each individual as a black box; we do not care to model the physiology of individual bodies. But a physiologist might do precisely that and would not view birth and death as stochastic, but rather as explicit processes about which much is known. We could have a philosophical debate about this; thus, I simply adopt this perspective on pragmatic grounds, several of which are outlined here:

1. It helps clarify the role of stochasticity in models—it stands in for things that are unknown.
2. It helps clarify the trade-off between determinism and stochasticity in models. If we could (or desired to) know everything, we would have a high-dimensional, deterministic model. If little is known, we may require a sophisticated treatment of stochasticity. Or we might choose a stochastic representation simply to reduce the complexity of a model.
3. It emphasizes that the terms *noise* and *error* are not always the best way to think about the role of stochasticity. In many cases, variability emanates from many unidentified sources, including from the process itself and from the context of the process. Sources of variability may be unknown, yet still demand careful treatment.

For pragmatic reasons I take the view that there are real processes behind the widely ranging growth rates in Figure 1.2. Stochastic representation is used when we do not know the underlying processes, they cannot be measured, or we do not know how to describe them in a deterministic model. The many components may operate on different scales. Simply collecting them all in one stochastic shell seems to agree with the concept of noise, but the processes they represent should impact the model in diverse ways. Figure 1.4c showed three different ways in which stochasticity stands in for uncertainty in x, in y, and in θ. This example illustrates the pragmatic treatment of stochasticity that is essential to modern data modeling. I began with a deterministic process model that might be motivated by previous understanding (saturating light-growth response). I then identified aspects of the problem that are unknown, including the resource, the differences among individuals, and the process model itself. Stochasticity is introduced for each of these unknowns.

Modern statistical methods provide a framework for complexity that allows for the comprehensive treatment of unknowns that contribute to Figure 1.2d. Assuming the wrong structure for stochastic elements might be just as naïve as an inappropriate function for the deterministic process. The basic approach involves a decomposition of complex problems into those that control the process of interest, those that generate data, and the underlying parameters. This structure admits influences that may impinge in diverse ways, without necessarily requiring that they can be observed or even identified.

The basic structure I follow throughout is outlined in Figure 1.6. This particular decomposition comes from Berliner (1996) and has been used in several recent ecological examples (e.g., Wikle 2003a; Clark et al. 2003b, 2004). Like

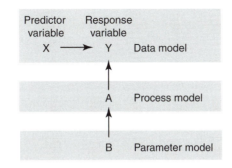

FIGURE 1.6. A graphical structure for an inferential model. Arrows indicate direct relationships that are specified explicitly as conditional distributions.

Figures 1.1 and 1.4, it is a graphical representation that provides a road map for the model. Attention may initially focus on the process level, where we allow for stochasticity in terms of process error or model misspecification. The connection to data accommodates sampling and the stochasticity that results from observation errors, sampling, missing data, and so forth. The parameter model can accommodate structure or relationships among sample units that might result from location, aggregation of individuals or traits, and so on.

By decomposing the problem, we can work directly with a process model that may be simple (and general), while allowing for the complexity that applies to a specific system (parameter models can absorb structure among variables) and data set or experimental design (data model) at other levels. This structure also admits data assimilation, meta-analysis, prediction, and decision analysis. To emphasize this connection between models and data it can be useful to think of inference and forward simulation as two sides of the same coin, one being the inverse of the other.

1.4 Learning with Models: Hypotheses and Quantification

Broadly speaking, the principal products of an analysis consist of (1) quantifying relationships and states (estimates and confidence intervals), (2) testing hypotheses (including model selection), (3) prediction, and (4) decision. Of these four activities, ecological analysis has tended to focus on hypothesis testing. I will not dwell on philosophy, but my inclination to soft-peddle hypothesis testing throughout this book needs some explanation. I discuss hypothesis testing many times, but it will often not be a prime motivation.

Before discussing why I place less emphasis on hypothesis testing than is customary, it is important to first say that hypotheses are critical. Any hope of success requires experiments and observations that are motivated by a clear set of questions with careful thought as to how the range of potential outcomes can lead to progress in understanding. A focus on multiple hypotheses, as opposed to a null hypothesis (e.g., Hilborn and Mangel 1997), can sharpen the approach. The lack of consensus on how to test hypotheses and to interpret them plays into a broader misunderstanding of what models are and how they can be effectively exploited. I summarize some of the issues that arise with hypothesis testing followed by general considerations concerning the inevitable subjectivity of statistics.

1.4.1 Hypothesis Tests

Consider the (null) hypothesis H_0: $\theta = q$. We collect data and obtain an estimate of θ. Can we make a probability statement about the result? Something about our confidence that reality is different from or the same as q?

Three approaches to hypothesis testing are reviewed by Berger (2003). Fisher significance testing involves a test statistic S that increases in value the farther the estimate is from the hypothesized value q. The larger the value of S, the greater the evidence against the null hypothesis. A P value is associated with S, but it is not the error probability (i.e., the probability of being wrong). It is calculated as the area in the tail of the distribution of S, the tail being everything more extreme than S. If treated as an error probability (as is often done), the evidence against the null hypothesis is overstated. So P values are not error probabilities (they violate the frequentist principle—see below); they are based on values of S more extreme than actually observed.

Neyman-Pearson hypothesis testing has an explicit alternative hypothesis and appeals to the frequentist principle. This principle can be stated in different ways. Berger's (2003) practical version says that if we could repeat the experiment many times, then the calculated Type I and Type II error probabilities (falsely reject the null, falsely accept the null) would agree with the fraction of Type I and Type II errors from the repeated experiments. Whereas the Fisher alternative is vague (H_0: $\theta = q$, H_1: $\theta \neq q$), the Neyman-Pearson alternative is explicit (H_0: $\theta = q_0$, H_1: $\theta = q_1$). As with a Fisher hypothesis test, there is a test statistic S that increases in value with evidence against the null. The method requires a predefined critical value of S for accepting versus rejecting H_0. The outcome does not discriminate between outcomes (values of S) that may be barely significant versus very significant, because the critical value (e.g., $\alpha = 0.05$) only has meaning if it is designated in advance. Unlike the Fisher method, we do not conclude that S had a P value of, say, 0.03, unless this happened to be the preselected critical value. S is either less than the critical value or not, and the conclusion is the same regardless of whether S exceeds the critical value by a little or by a lot. In other words, data sets that lend very different weights to the null hypothesis in the Fisher sense might lead to the same inference in a Neyman-Pearson sense.

Jeffreys hypothesis testing also involves an explicit alternative hypothesis. The null hypothesis can be accepted if it results in a better fit (a greater likelihood) with a probability determined under the assumption that prior probabilities for null and alternative $= \frac{1}{2}$. One criticism of this approach has been the need to specify a prior probability (Fisher 1935).

The point of this summary is to emphasize that thoughtful experts do not agree on how to test a hypothesis. And the challenges mount as we move to high-dimensional hierarchical models, where the concepts of sample size and number of parameters do not have obvious interpretations (Gelfand and Ghosh 1998; Spiegelhalter et al. 2002). The desire to assign a probability statement to a study is a laudable goal. It is often unrealistic in more than a highly qualified fashion. Ecologists can take sides in the debates and hope to maintain the "illusion of objectivity" (Berger and Berry 1988; Clark 2005). This may be the only option for those wishing to justify heavy reliance on hypothesis tests. At best, hypothesis tests are guidelines.

Even where we can agree on a hypothesis test, designs based on rejecting the null may not yield much. A study deemed unpublishable, because it fails to reject the null, may have been vacuous from the start. If showing an effect is viewed as the only informative outcome, there probably is no informative outcome. Studies founded on questions like "Does X affect Y?" (H_0: $\theta = q$) rarely provide much guidance, regardless of the P value. Such approaches often devolve to sample size (Spiegelhalter et al. 2003). Alternatively, if designed to parameterize a relationship, as opposed to rejecting a null hypothesis, that relationship may be the important product of the study.

1.4.2 The "Illusion of Objectivity"

The foregoing concerns bear on broader issues regarding what models represent, how they relate to theories and hypotheses, how they are used, and what can be inferred when they are combined with data. Models area not truth, just tools.

I will not advocate a distinction between scientific versus statistical models or between empirical versus mechanistic models. Such terms have had a place in the context of classical data modeling, because theory and data can be difficult to combine. For instance, it is common to fit data to a model like that shown in Figure 1.2, and then to use estimates that fit in a more complex model, for example, one that is dynamic with additional assumptions about other variables, perhaps even at different spatial and temporal scales. More desirable would be to fit the data directly to the model that will ultimately be used for prediction (Pacala et al. 1996). This direct connection is more desirable, because predictive intervals depend on the relationships that are lost when models are fitted piecemeal. The full model is harder to link directly with data, because it is complex.

By admitting complexity, modern methods allow for more direct connections between theory and data. The emerging techniques increasingly make distinctions between scientific and statistical models unnecessary. A motivation for setting such terms aside is the sense they can foster that there is a correct model, a best type of data, and an objective design; that is, there should be some formal and objective statistical test that will tell us what is best and correct. Models differ in complexity for many reasons, including not only the complexity of the process, but also how much we can know about it. Models are caricatures of reality, constructed to facilitate learning. In a world where data are accumulating at unprecedented rates, far more rapidly than we can assimilate them in models (e.g., remote sensing, climate variables, molecular data), we need ways to combine the information coming from multiple sources. Much of this book addresses formal structures for integrating multiple data sources within complementary models.

In most cases, there will be more than one way to model the same process. How well the model fits a data set is one of several considerations in deciding among candidate models. In several sections, I discuss model selection as a basis for helping to identify models that can be of most use. I follow the standard practice of parsimony: increasing complexity in a model requires justification. Still, I place less emphasis on model selection than is typically done within a classical approach focused on rejecting a null hypothesis. This diminished emphasis on model selection stems not only from the view that there is rarely a correct model, but also on several characteristics of ecological phenomena. Ecological processes are inherently spatio-temporal, and the best model can vary from place to place and change over time (e.g., West and Harrison 1997). The frequentist concepts based on the idealized notion of resampling an identical population often does not directly translate to environmental data sets. This is inevitable in high-dimensional systems that are subject to processes that operate at a range of scales. This view can shift the emphasis from that of identifying the correct model to one of identifying models that can be useful in different settings and the need to consider model uncertainty as integral to inference. In some cases it can motivate combining models or model averaging.

Finally, formal statistics should not trump all other considerations, especially when it comes to model selection. The fit to data is not the sole basis for model selection. General circulation models of the atmosphere (GCMs) were constructed to embrace certain physical relationships. Although weather and climate models can benefit from better integration with data (e.g., Bengtsson et al. 2003), a hypothesis-testing framework was not the basis for their formulation and does not play a role in continuing model development.

1.4.3 Quantifying Relationships

Unlike hypothesis tests, confidence envelopes are less controversial than many ecologists think. Confidence envelopes estimated for a given model by different methods and stemming from different perspectives often yield similar results. The processes under investigation are often known to exist. There may be no point in rejecting the existence of an effect. There may be value in quantifying it. For example, the effect of interest may contribute a relatively small fraction of the variance to a data set, because there are other large effects. If so, model selection (a hypothesis test) may reject the added complexity of a model containing the variable of interest. This should not necessarily deter us from examining those variables. Lange et al. (1992) found slight improvement in the fit of a change-point model of treatment effects on CD4 lymphocytes, a surrogate for AIDs infection, in patients with advanced HIV infection (models of this type are described in Chapter 9). The more complex model that included the variable of interest was still valuable, despite the fact that it did not explain nearly as much of the total variance as did other factors. Clark et al. (2004) found that temporal effects on tree fecundity overwhelmed individual effects, yet individual differences were still large. The more complex model that contained individual effects allowed for quantification of individual effects, *in spite of* large stochasticity standing in for other effects. A posterior density centered near zero might suggest use of a simpler model, yet this parameter might be retained simply as insurance against effects on estimates of other parameters.

1.4.4 Learning from Models

Learning is progressive. It entails continual updating or assimilating information as it accumulates (Section 1.2.1). Models have a central role in this process. This process might be represented as

$$\text{updated knowledge} = f(\text{data, previous knowledge})$$

Bayesian methods do this formally as

$$posterior \propto likelihood \times prior$$

Learning can occur sequentially, with the posterior or predictive distribution that derives from information available before serving as a prior to be combined with data that comes available in the future (West and Harrison 1997). Whether or not we do this formally, the process of successively updating understanding as information accumulates is standard practice. The Bayesian framework for learning plays a large role in this book. The likelihood in this expression admits data. For processes that evolve over time, we might term this the *update/forecast cycle*, with the goal being to ingest new information continually.

1.5 Estimation versus Forward Simulation

Model analysis and prediction are ideally based on the same process model as that used for estimation (Figure 1.7). This connection is critical if we are to construct prediction envelopes that accurately reflect uncertainty in data and models. For a number of reasons, few ecological analyses proceed this way. Predictions are typically made from models that are not the same as those used to obtain parameter estimates. And ecological predictions typically come from process models only after discarding the stochasticity (Clark 2003). It is not necessarily wrong to analyze models that are not fitted to data. But arbitrary treatment of uncertainty can make predictions misleading or even meaningless. This book is partly motivated by the fact that there are fewer obstacles in the way of integrating theory and data than there were just a decade ago. New approaches provide a natural framework for data assimilation and for decision (Figure 1.7).

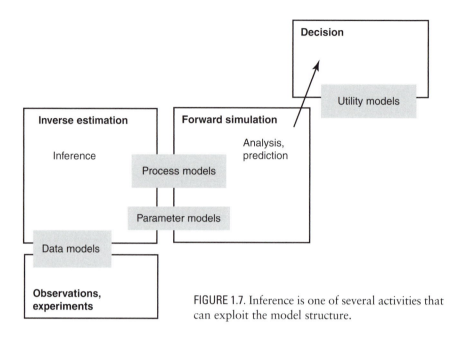

FIGURE 1.7. Inference is one of several activities that can exploit the model structure.

1.6 Statistical Pragmatism

No practicing ecologist is ignorant of the debates over frequentist and Bayesian approaches (e.g., Dixon and Ellison 1996). Already, philosophy has come up in reference to stochasticity (Section 1.3.2) and hypothesis testing (Section 1.4), but it does not play a large role in this book. In light of the philosophical emphasis of ecological writings about Bayes, I should provide some justification for my largely nonphilosophical treatment in this text. Here I summarize several issues that I have written about elsewhere (Clark 2005).

First, the points of controversy have traditionally revolved around *frequency* versus *subjective* concepts of probability (Chapters 4 and 5). The different concepts are especially important when one tries to make sense of classical hypothesis testing, but they also arise within the Bayesian community, in the context of how strictly prior specification must represent current understanding. Ecological writings on Bayes have focused on philosophical differences, advising that an analysis begins with one's philosophical stance, which then prescribes a classical versus a Bayesian approach. This prescription follows past debates in the statistical literature that have become less common and less polarized in recent years. I have expressed my view that the dissipation of such debates and the emergence of modern Bayes in applied fields have less to do with philosophy than pragmatism (Clark 2005). In fact, the philosophy remains important, being central to the challenge of doing objective science using the subjective tools of statistics (e.g., Berger and Berry 1988; Dawid 2004). The expansion of modern Bayes owes much to developing machinery.

Bayesians can now take on high-dimensional problems that were not accessible in the past.

In my view, the persistent focus on philosophy in ecological writings has become counterproductive (Clark 2005). It should not be news to any scientist that strong priors can affect estimates. Weak priors generally do not. The nonpractitioner must wonder at the standard examples provided in many ecological demonstrations of Bayes. Typically a simple Bayesian analysis requires far more work to arrive at essentially the same confidence envelope that could have been obtained with standard software in a classical framework. Given that most ecologists have limited philosophical baggage in this respect, why complicate the analysis? Moreover, the focus on philosophical differences can confuse the issues. For example, although Bayesians refer to parameters as "random," and frequentists do not, in both cases, parameters are fixed. Going further, both approaches view parameters as uncertain. Ecologists have long confused references to random parameters with the idea that the underlying value of a parameter fluctuates. Classical confidence intervals and Bayesian credible intervals both express uncertainty about the underlying true value, which is unknown.

I say more about the underlying assumptions of classical and Bayes models as we begin to implement them. For now, I simply forewarn that both are important in this book, with applications being pragmatic rather than philosophical. By pragmatic, I mean that I avoid unnecessary complexity. The initial models I discuss in Chapter 3 are classical. Bayesian methods require evaluation of integrals, analytically or through simulation. For simple problems this can mean that Bayes requires more effort. I also discuss Bayes for relatively simple problems, but it becomes the dominant approach in later chapters to allow for external sources of information and uncertainty. As models become more complex, the level of difficulty associated with classical and Bayes can reverse. Not only will the Bayesian framework bring flexibility to address complex problems (even where classical methods are still an option), Bayesian approaches can be *easier*, facilitated by new computational techniques.

A pragmatic approach need not be controversial. As suggested above, primary focus on quantification, rather than hypothesis testing, means that the products of a classical and Bayes analysis (e.g., confidence and credible intervals) may often be similar (e.g., Cousins 1995; Clark and Lavine 2001). Controversy can still arise over the role of priors in a Bayesian framework. I will say that informative priors are a good idea, when external information is available. Noninformative priors are useful when information is not available, or when we want to isolate the contribution of a particular data set. Those just testing the Bayesian waters may be motivated by machinery. Those who initially balk at informed priors may come to recognize each new data set as a way of updating what is already known. Priors can be an efficient way to introduce partially known relationships, in contrast to the traditional practice of assuming such relationships are completely known. I return to some of these issues in the final chapter.

Even as models become complex and (in this case) Bayesian, many of the tools will be familiar from classical statistics. Bayesian models will have embedded within them many traditional structures. The power of modern Bayes comes from the capacity to integrate traditional techniques within high-dimensional models.

Although I embrace both frequentist and Bayesian approaches, there is much that I avoid. In both cases, there is an established framework that underpins the analysis. If I say that I have constructed a confidence envelope based on Fisher Information or a Bayesian posterior, you will have a good idea of what it means (Chapter 5). Although challenging problems can foster excursions into ad hockery (a much bigger temptation for the inflexible classical methods), the formalism of these two approaches is well established.

In this regard, the expanding ecological literature on fitting models with ad hoc approaches that are neither frequentist nor Bayesian is avoided in this book. Many such approaches are highly creative. But there can be numerous pitfalls in terms of inference, prediction, and communication to others. The goal of inference is a probability statement about unobservables—quantities we wish to estimate. A probability statement relies on a framework that is consistent throughout. Ad hockery tends to break down the connection between confidence envelopes and the data used to construct them. Moreover, ad hockery is often unnecessary. When things get complicated, we will go Bayesian. It admits complexity within a consistent, reproducible framework. Although there remains much to understand in the Bayesian world (e.g., selection among hierarchical models in Chapter 8 and complex spatio-temporal models in Chapter 10), there is rapid progress and it builds on a firm foundation. Because we have the Bayesian alternative, there are many recent papers on how to run lines through data points that will not be covered here.

Finally, although I cover basic concepts in classical statistics, I make no effort at comprehensive coverage of classical designs and hypothesis tests. Classical statistics may seem to burden the scientist with a different design and test for each new application. There is substantial jargon that continues to find its way into ecological statistics books, much of which is largely historical and rarely encountered in the modern statistics literature. I focus on general approaches and attempt to minimize jargon.

2 Model Elements: Application to Population Growth

IN THIS CHAPTER I INTRODUCE some of the basic types of process models used in ecology. Population growth provides an example, but the principal goal is to summarize basic features of models that are formulated to describe a process. These concepts include model state, linear versus nonlinear models (and how these terms are used in mathematics and in statistics), and structure in models.

Because the focus of this book is data-model connections, my exposition of standard ecological models differs, in some respects, from those contained in most texts. Population models are the basis for some excellent recent ecology texts, including Hastings (1997), Kot (2001), Caswell (2001), Morris and Doak (2003), and Lande et al. (2003). These references are highly recommended for ecologists who study populations, as they contain more analysis of models than is included here. I provide a condensed treatment of this material, but emphasize aspects of models that will be important as we begin to incorporate data in Part II. For ecologists who do not study population-level phenomena, I recommend this section for general background on model development.

2.1 A Model and Data Example

In the early part of this century Raymond Pearl (1925) used a simple sigmoid (logistic) curve, which he sometimes referred to as the *law of population growth*, to predict that the U.S. population would reach an upper limit of 197 million shortly after the year AD 2000 (Figure 2.1a) (Pearl 1925; Kingsland 1985). The basis for this prediction was U.S. Census Bureau data, from which he obtained decadal population statistics from AD 1790 to 1910, and the model in which he placed great faith—so much faith that he reported the inflection point of the curve to lie at April Fools Day, AD 1914. Note from Figure 2.1a that population sizes after 1910 are based on extrapolation.

The future has come to pass, and the modern census data do not look like the predictions of the early twentieth century (Figure 2.1b). History provides two lessons that this and subsequent chapters explore: (1) models can be importantly wrong (they may fit well for the wrong reasons), and (2) model evaluation requires estimates of uncertainty.

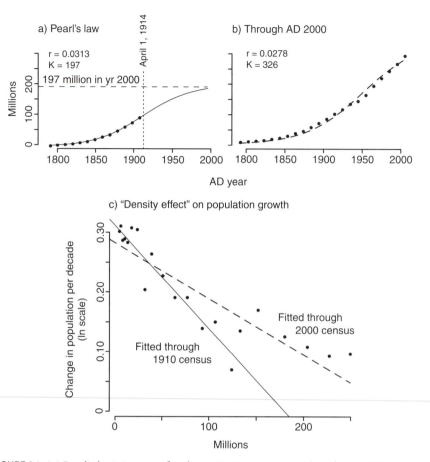

FIGURE 2.1. (a) Pearl's logistic curve fitted to U.S. Census Bureau data from 1790 through 1910. (b) A comparison with the longer series available in AD 2000. (c) The model assumption that population growth declines with density fitted to two subsets of the full census data. The parameters r and K describe per capita growth rate and the carrying capacity, respectively. Census data from www.census.gov/population/censusdata/table-16.pdf.

What did Pearl's model assume? One of the oldest ideas about population growth is that it cannot continue indefinitely (Darwin 1859; Lack 1954). The rate of growth will decline as population size increases, because growth will eventually become limited by resources, natural enemies, and so forth. If density exceeds some carrying capacity, growth rates can be negative, and density will either stabilize or decline.

Model construction begins with an assumption of how density affects growth rate. This relationship is complex and poorly understood. For instance, fertility might decline, mortality might increase, both in ways that vary in time and space. Lacking knowledge of these details, ecologists commonly use an assumption that the per capita rate of growth declines as a constant fraction of population size. The *per capita rate* of increase is the proportionate rate (Appendix B). As a rough approximation, this assumption of constant decline might be reasonable, regardless of whether density influences fertility or mortality,

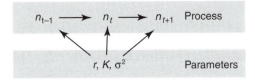

FIGURE 2.2. A graph of the standard logistic model. The process steps forward as governed by two process parameters (r and k) and process error, having variance σ^2.

because both are per capita rates. Let n represent density. This linear relationship is written as

$$\begin{array}{c} \text{proportionate} \\ \text{change in } n \end{array} = \beta_0 + \beta_1 n$$

There is a potential rate of increase β_0 that applies when density is low (n is close to zero). The slope parameter β_1 is negative and describes the strength of population regulation. This relationship between population size and growth is shown in Figure 2.1c as estimated for examples in 2.1a and 2.1b. This is the logistic model of population growth that was used by Pearl. It is intended to represent feedback of density on population growth. This model could be viewed as having a process that is governed by parameters that must be estimated from data (Figure 2.2).[1] The parameters β_0 and β_1 can be expressed in terms of Pearl's r and K (see below).

The assumptions seem plausible, so why did the model mislead? The logistic model of population growth fits human census data remarkably well (Figure 2.1a). This close fit can inspire confidence that it represents the correct model for population growth. We might quarrel with Pearl's extrapolation from pre-1910 data on uncertainty grounds—had he attempted to determine the uncertainty that could have been quantified as part of the model fit, the subsequent outcome might have been within, say, a standard derivation or so of model predictions. We could then argue about whether the prediction was close enough to be useful (model adequacy). We could have this debate whether or not we agree that this is the most realistic model of population growth (model selection).

So there is a second issue that concerns whether the model captures the process in a reasonable way. Obviously, demographers applying the same model today could reach an equally confident, albeit substantially different, answer (Figure 2.1b). After all, this tight fit is the outcome that fosters confidence and justifies detailed model analysis and projection.

The close fits in Figure 2.1 belie inaccurate description of the process. For example, much of the U.S. population increase resulted from immigration, a process that affects model behavior in a different way than fertility of the resident population. In 1850 (the first year that the U.S. Census attempted to gather nativity data), nearly 10 percent of the censused population was foreign-born. By 1910 (Pearl's data set), foreign-born residents reached nearly 15 percent (www.census.gov/population/www/documentation/twps0029/ tab01.html). In this example, extrapolation was impressively inaccurate, in part, because

---

[1] This is not precisely the model fitted by pearl.

immigration might respond to extrinsic factors to a much greater degree than does fertility. Immigration came in waves, for reasons that had little to do with U.S. population size. This weak link between population size and immigration means that it is not well described as a per capita rate. Whereas no model would be precisely correct, this one has serious flaws.

Environmental scientists embrace a broad range of questions, consider many types of data, and employ a variety of models. To get started, I devote this chapter to the deterministic elements that are typically included in process models, such as this one for population growth. Some ecologists think of this as the theory or process part of the problem. I include only the most basic tools needed for derivation and analysis, which appear primarily in appendixes. These process models provide fodder for the data modeling that begins in Chapter 3.

2.2 Model State and Time

In this section I cover some background concepts, terms, and notation. First I consider the distinction between *linear* and *nonlinear* models, followed by state variables and structure.

2.2.1 When Is a Model Linear?

There is a simple way to determine whether a model is nonlinear: the second derivative of a linear model is equal to zero. Because this book deals with models that come from both ecological and statistical traditions, there are different terminologies that can create confusion; in this case, we must agree on which derivative. Statisticians apply this distinction with respect to the *parameters* of a model. Consider the equation

$$f(n) = \beta n^2$$

There is a state variable n and a parameter β. (I say more about state variables in a moment.) In a statistical context, this model is termed *linear*, because it is *linear* in β. In other words, $\dfrac{d^2 f}{d\beta^2} = 0^2$. For a mathematician, the model $f(n) = \beta n^2$ might be termed *nonlinear*, because

$$\frac{d^2 f}{dn^2} \neq 0$$

In other words, mathematicians tend to think in terms of state variables.

Like mathematicians, ecologists tend to think in terms of state variables and often use the terms *density-independent* and *density-dependent*, respectively, for linear and nonlinear models in the state variable population density

---

[2] Differentiate once with respect to β to obtain $df/d\beta = n^2$, their differentiate again to obtain zero.

(e.g., May 1981). Some ecologists will find it confusing to see polynomials referred to as "linear models." Recall that statisticians come from a tradition of thinking about parameters. Of course, models can be nonlinear both in parameters and in state variables, such as $f(n) = \dfrac{n}{n + \beta}$ and $f(n) = 1 - e^{-\beta n}$. I retain this confusion of terms, because it is entrenched. Throughout, I define how the terms are used in specific cases. Except when used in a statistical context, I use the terms *linear* and *nonlinear* with respect to the state variable, such as concentration of a nutrient, population size or density, and so forth.

A second source of confusion comes from the fact that population ecologists often apply the terms *linear* and *nonlinear* to the effect of density on the per capita growth rate, $1/n \, dn/dt$. This terminology is fine, so long as it is qualified by context. However, a model that is linear in per capita growth rate is not a linear differential equation.

In human demography and population biology, linear models (in terms of the state variable density or abundance) are typically used to describe current population growth. Standard human population growth statistics, such as "country X is growing at a rate of Y percent per year," are examples. They are rarely used to predict long-term population trajectories (i.e., over generations), because they cannot be extrapolated with confidence. Many factors that might be ignored in the short term have long-term or cumulative effects. These regulatory factors can directly or indirectly involve density or age structure. Nonlinear models, when used to describe population growth, are termed density-dependent, because the per capita rate of growth depends on the state variable density. Such models describe the phenomenon of population regulation or the tendency for population density to approach some stationary distribution having a mean, a variance, and so forth (May 1973; Dennis and Taper 1994; Turchin 1995). Nonlinear models are also used when the rate of a reaction depends on the concentration of substrate.

2.2.2 Time as a Continuous Variable

Many ecological models are dynamic, with time as an explicit variable. Time can be treated as discrete or continuous. Continuous time models are used to describe processes where change occurs continuously and are written as rate equations. For change in density $n(t)$, we write a differential equation as dn/dt.

Differential equations can be linear or nonlinear. The equation

$$f(n) \equiv \frac{dn}{dt} = rn$$

is a linear differential equation, because the second derivative with respect to the state variable n is

$$\frac{d^2}{dn^2}f(n) = 0$$

This terminology may seem confusing at first, because the solution to this particular linear differential equation is exponential (nonlinear) in the continuous variable *time*. In population biology, parameter r is sometimes termed the "intrinsic rate of increase." It is the instantaneous contribution of each individual to the population through its birth rate and mortality risk, which is easy to see if we express this rate on a per capita basis,

$$\frac{1}{n}\frac{dn}{dt} = \frac{d(\ln n)}{dt} = r$$

It has the solution

$$n(t + dt) = n(t)\, e^{r \cdot dt} \tag{2.1}$$

where dt is a time increment (Appendix B). I use the notation $n(t)$ to denote time as a continuous variable, to distinguish it from the notation for discrete time n_t in Section 2.2.3. Short-term rates of population change are often reported as annual rates of population growth and presented as percents. These statistics are calculated as $100\, r$. To determine r based on knowledge of $n(t)$ and $n(t + dt)$, take logs of equation 2.1

$$\ln n(t + dt) = \ln n(t) + rdt$$

and rearrange,

$$r = \frac{\ln n(t + dt) - \ln n(t)}{dt} \tag{2.2}$$

Population increase is described by $r > 0$ and vice versa. Figure 2.3 shows an application of this model, as it was used to describe the twentieth-century increase in the nesting population of black noddies (*Anous minutus*) on Heron Island, Great Barrier Reef (Ogden 1993). The rapid increase, estimated as $r = 0.081$ (8.1 percent per year), cannot continue indefinitely, yet it appears to fit the existing data well.

Nonlinear differential equations are commonly used in ecology. If the environment has a limited capacity to support new individuals, the per capita rate must decline as population density increases. Pearl used the simplest assumption of a straight line (left side of Figure 2.4). This equation could be written as

$$\frac{1}{n}\frac{dn}{dt} = \beta_0 + \beta_1 n \tag{2.3}$$

for two parameters, an intercept β_0 and a slope β_1. In this logistic model, the first parameter is positive so that growth rate is positive at low density. The second parameter is negative, reflecting the negative effect of density on growth rate (Figure 2.4a). Ecologists often write this as

$$\frac{1}{n}\frac{dn}{dt} = r\left(1 - \frac{n}{K}\right)$$

with parameters β_0 being equivalent to r and β_1 equivalent to $-r/K$. This is the "continuous logistic" or (formerly) the "Law of Population Growth" of

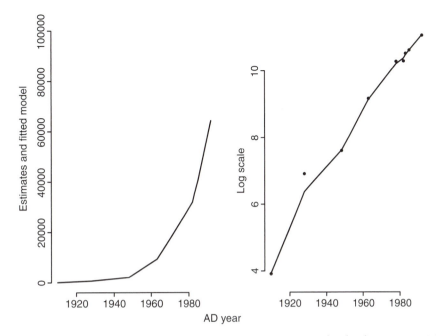

FIGURE 2.3. The black noddy population increase on Heron Island. The exponential model was fitted to this data set, with allowance for process error and observation errors (Chapter 9).

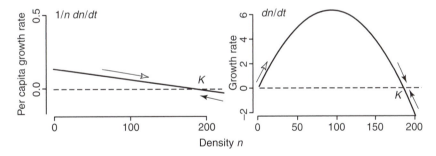

FIGURE 2.4. A linear effect of density on per capita growth rate, $1/n\ dn/dt$, from Equation 2.3, with parameters transformed to r and K. Although the effect of density on per capita growth rate is linear (*left*), this is a nonlinear differential equation, sometimes called the logistic equation for population growth. *Right* is the same model, plotted as simple population growth, dn/dt, rather than per capita growth. Parameters are taken from the fitted model for the BPF moose data (Fig. 2.5). Note that the carrying capacity estimate would be slightly below 200 individuals.

Pearl (Figure 2.1). The logistic model of population growth is nonlinear, because the second derivative of dn/dt with respect to n does not equal zero. This is clear from the "hump" in Figure 2.4b. Nonlinear models usually cannot be solved (the logistic model is an exception—Appendix B).

2.2.3 Discrete Time

Discrete time models describe changes that occur at discrete intervals in time, such as a population of organisms where births and deaths occur at discrete intervals. These may be periodic, example, annual, but they need not be equally spaced in time—Chapter 9 provides examples of unequally spaced intervals. A discrete time model consists of an expression describing change from one time increment to the next. If the time increment is, say, one unit, then there is an expression for $n_0, n_1, n_2, \ldots$. The time increment might be one generation of duration T, in which case there would be an expression for $n_0, n_T, n_{2T}, \ldots$. Discrete time process models are termed *difference equations*.

As an example, the simplest population model expresses density at the next time step as the sum of individuals present at the current time plus new individuals added by births minus those lost to mortality, $n_t = n_{t-1} + B_{t-1} - D_{t-1}$. Let $B_t = n_t b$ and $D_t = n_t \rho$, with the two new parameters being per capita rates. Then

$$
\begin{aligned}
n_t &= n_{t-1}(1 + b - \rho) \\
&= n_{t-1}\lambda
\end{aligned}
\tag{2.4}
$$

where the composite parameter $\lambda = 1 + b - \rho$ is sometimes termed the *finite rate of increase*. This *first-order linear difference equation* has a solution that we can discover by iteration (Appendix B). This difference equation $n_t = f(n_{t-1}) = n_{t-1}\lambda$ is linear in terms of n, because $d^2 f/dn^2 = 0$. In population biology, it is often written in terms of change in log density, $\ln n_t = \ln n_{t-1} + \ln \lambda$, to express change in proportionate terms, that is, on a per capita basis. The per capita rate of change is $\ln \lambda$, and is often represented by the quantity $r \equiv \ln \lambda$.

Nonlinear difference equation models are common in the environmental sciences. An example from population ecology refers to factors that might maintain a population within some range of densities, similar to the continuous logistic model of the previous section. Per capita birth rates, death rates, or both may decline as density increases due to diminishing resources, crowding, and so on. A conventional model that describes a decline in growth rate as density approaches some carrying capacity K is written as

$$
n_t = n_{t-1}\lambda^{\left(1 - \frac{n_{t-1}}{K}\right)}
$$

As density approaches K, the exponent approaches zero, and density does not change. If density is greater than K, then the exponent is negative, and density declines. Ecologists usually write this model in terms of the parameter r, or $\lambda \equiv e^r$, to obtain an equivalent discrete Ricker (1954) model

$$
n_t = n_{t-1}\exp\left[r\left(1 - \frac{n_{t-1}}{K}\right)\right]
\tag{2.5}
$$

(Appendix B). This is a nonlinear difference equation, because $d^2 f/dn^2 \neq 0$. An example is shown in Figure 2.5b, where interannual changes in moose density are modeled with an underlying process described by Equation 2.5.

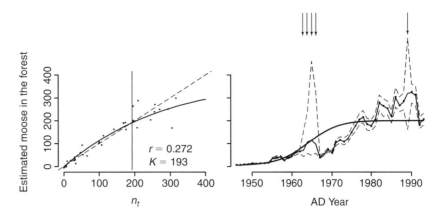

FIGURE 2.5. Models fitted to the BPF moose data (using methods of Chapter 9) compared with the deterministic version of the logistic model (smooth line). Left is the lag-one plot with observations and the fitted model (solid curve). This plot could be used to construct a cobweb diagram (Appendix B) that would show the constant solution to be stable: from any initial density we would converge to $n^* = 193$. This convergence comes from the curved relationship in the figure, *left* (Section B.5.4). *Right*, is the time plot. The vertical arrows indicate years of missing data. Dashed lines bound the 95 percent credible interval, indicating the degree of uncertainty as to the true population size. The uncertainty reflected in both graphs raises doubts about the existence of this stable solution $n^* = K$. From Clark and Bjørnstad (2004).

Discrete time models are often applied to processes that are continuous, either for convenience or for sampling considerations. In the latter case, it is important to recognize that discrete and continuous time models do not always exhibit the same behavior. Most ecological processes are not precisely continuous in time, yet difficult to characterize through a simple discretization of time. There can be some advantages to difference equations, particularly when data are concerned. It is becoming increasingly common to work directly with discretized versions of continuous time models with allowance for the error associated with discretization. Due to the focus on data, this book mostly discusses discrete time models. It is still important to understand continuous models, because environmental processes are often analyzed in continuous time. For example, the continuous formulation

$$\frac{dn}{dt} = rn\left(1 - \frac{n}{K}\right)$$

can be discretized to

$$n_{t+dt} \approx n_t + dn \quad \text{current density plus increment in } n$$

$$= n_t + \left[\left(\frac{dn}{dt}\right)dt + \varepsilon_t\right] \quad \text{increment in } n \text{ is linear}$$
$$\text{approximation plus error}$$

$$= n_t + rn\left(1 - \frac{n}{K}\right)dt + \varepsilon_t \quad \text{substitute for } dn/dt$$

To follow this transition from continuous to discrete time, recall from introductory calculus that dn is defined as the small increment in n that occurs with a small time increment, dt. After review of Appendix A, you should recognize this as the Taylor series for n_{t+dt}, with ε_t representing error that comes from ignoring higher-order terms. The discrete and continuous versions will behave similarly, provided that the time increment dt is not too large and nonlinearities are not too large. The process error term ε_t accommodates the error that comes from the fact that I did not integrate the rate equation, but, rather, simply scaled the rate of change, dn/dt, by the time increment dt. This error will often be small relative to other sources of process error, meaning that $rn(1 - n/K)$ is a crude approximation of the growth process. When discretized in this manner, the error associated with ε_t increases with the time elapsed between t and $t + dt$. Difference equations can be unstable due to the coupling of nonlinear dynamics with a time lag dt (May 1974). It is important to note, however, that the instability that can result from using a difference equation to approximate a nonlinear differential equation is less of an issue for data modeling than it is for simulation. With data modeling, we are not solving equations, but rather estimating parameters. We want to allow for process errors, regardless of their source. Scaling of process error is discussed in Chapter 9.

If generations overlap, dynamics depend on some underlying interactions that are not explicitly included in the models considered thus far. The difference equation $n_t = f(n_{t-1})$ says that density at time t is fully determined by density one time step ago. A model for growth that depends on population size in the more distant past could allow for change in age distribution (e.g., Nisbet and Gurney 1982). It is also possible that the population interacts with other species that are not explicitly included in the model (Schaffer 1981). Although unknown, it is possible to allow for such effects deterministically, by including delayed feedback. In Section 9.9, I mention models of the form

$$n_t = f(n_{t-1}, \ldots, n_{t-p})$$

This model says that density at time t directly depends not only on density at $t - 1$, but also at times back to $t - p$. In this case, p is the *order* of the model.

2.2.4 Discrete States and Structure

In addition to continuous versus discrete time, models have continuous versus discrete state variables. In the context of population growth, *state* refers to population size (number of individuals) or density (number per area or volume). The state variable could also refer to the concentration of a nutrient, to the volume or mass of moisture in soil, to the land cover classification of a county, or to the number of infections or deaths in a defined area. Confusion can arise when there are multiple state variables, because the terms *continuous state* versus *discrete state* are sometimes used to refer to the quantity within a given class or category or to the way in which the categories themselves are defined (Table 2.1). The differential equation and difference equation

TABLE 2.1.
Treatment of Time, State, and Structure in Some Common Process Models

Process Model	Time		State		Structure		
	Continuous	Discrete	Continuous	Discrete	Continuous	Discrete	Stochastic
Differential equation	x		x				
Difference equation		x	x				
Partial differential equation	x		x		x		
System of differential equations	x		x			x	
System of difference equations/matrix models		x	x			x	
Discrete state Markov process/cellular automata		x		x		x	
Birth/death process (continuous time Markov process)	x			x			
Integro-difference/ integral projection		x	x		x		
Random effects*		x	x	x	x		x

* Random effects arise in the context of data modeling, discussed in Chapter 9.

models discussed in the previous section contain a single state variable (density) that is continuous, in the sense that density can assume any values greater than or equal to zero (although the zero class is still typically viewed as discrete and requires special consideration—more on this later). Alternatively, a model for the number of births or deaths could be modeled with discrete numbers, the state variable assuming integer values 0, 1, 2, ...

A *structured model* indicates discrete classes or categories. A structured model can be written as (1) multiple attributes for a state variable or (2) multiple state variables. In the latter case, each of the classes is viewed as a different state variable and represented by a separate equation. Together the equations constitute a system of differential (continuous time) or difference (discrete time) equations. In the former case, attributes may be represented as continuously varying (e.g., age) or as discrete classes (e.g., classes for juveniles and adults). Either way, there is structure (e.g., age or stage) associated with the single state variable density.

If the stages are discrete, the full system of equations may be represented with a matrix having elements that describe transitions among states. This is standard practice when models are linear in state variables. Deterministic matrix models have become extremely popular in population ecology, in part because analysis is accomplished with linear algebra (most such models are linear; see the thorough treatment of Caswell 2001). For example, the demography of a population could be represented as a collection of demographic rates organized in a table or matrix. Matrix models are discrete in time and in structure—all individuals have a specific class or stage to which they belong. However, within a stage, density is a continuous variable. A transition matrix contains demographic rates that are the basis for projecting stage structure forward in time. I take a few moments here to outline some of their key features.

Classes or stages are defined based on age (age-structured model) or other attributes (stage-structured model). Age-structured models can be viewed as a special case of stage-structured models. They are often used when age information is available, having the advantage that transitions among classes (i.e., ages) can occur at the same rate as the passage of time. Individuals who survive advance one age increment in one time increment. There is no uncertainty associated with timing of transition. The U.S. Census Bureau uses age structure, because it is available for a large segment of the U.S. human population. The transition matrix based on age is often called a *Leslie matrix* (Leslie 1945).

Stage structure is typically used when age is uncertain and when stages provide a useful summary of demographic rates. Perennial plants may have a number of stages, including mature individuals, rosettes, and seeds in the soil

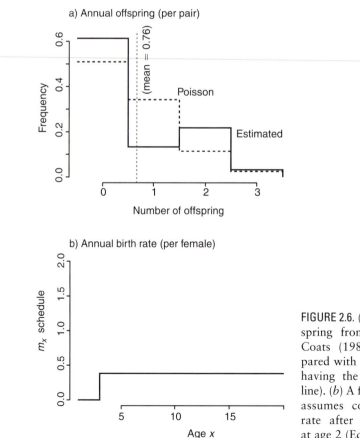

FIGURE 2.6. (*a*) Distribution of offspring from Barrowclough and Coats (1985) (solid line) compared with a Poisson distribution having the same mean (dashed line). (*b*) A fecundity schedule that assumes constant reproductive rate after reaching maturation at age 2 (Equation 2.6).

that cannot be aged accurately. Invertebrates pass through a discrete set of metamorphic stages that often are readily identifiable. Because transitions among stage classes are more variable than are those among age classes, models must accommodate additional considerations. Analysis involves more work, and estimates of transition rates tend to have greater uncertainty.

A classic example for the northern spotted owl (NSO) (*Strix occidentalis*) includes a fecundity schedule m_x, consisting of transitions related to births and deaths. A simple birth schedule includes a maturation age of $x = 2$ years and an average production of $b = 0.382$ offspring per year thereafter. This schedule could be written as

$$m_x = \begin{cases} 0 & x < 2 \\ b & x \geq 2 \end{cases}$$ 2.6

(Lande 1987, 1988; McKelvey et al. 1993) (Figure 2.6b). Deaths are modeled using age-specific survival parameters, often interpreted as the probability that an individual in age class x will survive the interval from, say, t to $t + 1$. If age classes and time are measured in the same units, then survival is sometimes estimated as the fraction of individuals in age class x that survive during the age interval $(x, x + 1)$ relative to the fraction present at the beginning of the age x,

$$s_x = 1 - \rho_x = \frac{n_{x+1}}{n_x}$$

where n_x is the density of individuals in age class x and ρ_x is interpreted as the probability of mortality. The *survivor function* can be obtained from the age-specific survival parameters as

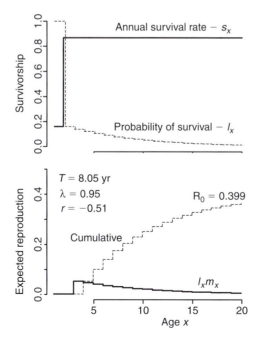

FIGURE 2.7 Life history schedules in this example for NSO. (*a*) Survival the first year is low, resulting in a large initial decline in the survivorship curve l_x, followed by geometric decline thereafter. (*b*) At the time of birth, the expected reproduction from an age x individual, m_x, must be scaled by its probability of surviving to age x, that is $l_x m_x$. Net reproductive rate R_0 is this expected rate summed over all age classes (Equation E.1).

$$l_x = \prod_{i=1}^{x-1} s_i = s_1 \times s_2 \times \cdots \times s_{x-1} \qquad 2.7$$

and interpreted as the probability of surviving to age x, obtained as the joint probability of surviving all previous age intervals (Figure 2.7a).[3] NSOs have high juvenile (first year) mortality. McKelvey et al. (1993) use a low juvenile survival rate of $s_1 = 0.159$ and a higher rate of $s = 0.868$ for subadults and adults. The plot of survivorship shows the large decline during the first year and slower decline thereafter (Figure 2.7a).

Analysis of population growth rate, population structure, and contributions of demographic rates to population growth rate can be accomplished directly from these equations (Appendix E) or using matrix algebra (Appendixes C, E). For the latter, demographic rates are assembled in a matrix $\mathbf{A}$, having elements a_{ij}, describing the contribution of stage j to stage i. Here is a matrix for the NSO model just discussed,

$$\mathbf{A} = \begin{bmatrix} 0 & 0 & b \\ s_1 & 0 & 0 \\ 0 & s & s \end{bmatrix} \qquad 2.8$$

The contribution of stage j to the population at the next time increment is determined by the elements in j^{th} column of $\mathbf{A}$. Individuals in stage 1 can only survive to stage 2 ($a_{21} = s_1$) or die, with probability $1 - s_1$. Individuals in stage 3 not only can survive in stage 3 ($a_{33} = s$), but they can also produce offspring ($a_{13} = b$). Several examples in this book involve matrix population models (Sections 9.16, E.3). If a population grows linearly, and it possesses a stable stage structure, then the growth rate is obtained as the dominant eigenvalue of $\mathbf{A}$ (Appendix E).

Instead of discrete classes, there may be a continuous range of states, with the rate taken to be expectations for a random variable. The states may be represented by a continuous function describing how this expectation varies with size or age. For example, fecundity, natality, or age-specific fertility schedules can be represented by smooth functions of age or size (Figure 2.8). For the continuous age variable a, define expected fecundity $m(a) =$ expected offspring per unit time per age-a female per time. Because a is continuous, per capita births during the age interval $(a, a + da)$ are approximately $m(a)da$ for some age interval da. Time and age are often measured on the same units, with $da = dt$, although this need not be the case. This expectation is a continuous function of time and of state. As in Figure 2.8, such functions are often expressed in terms of size, say, $m(x)$ for diameter x, rather than age a. To obtain the age schedule, we need an additional piece of information, the growth rate dx/da,

---

[3] The indexing of classes needs careful thought (e.g., Cawell 2001). I adopt the following convention. Define the first age class to be $x = 1$. The individual enters this age class at birth. At this point, the survivor function is $l_1 = 1$. At the end of age class 1, the survivor function is $l_2 = l_1 \times s_2 = s_1$. This is the probability of surviving to the beginning of age class 2. If the survival rate is the constant rate x_1 then $l_x = s^{x-1}$.

$$\frac{dm}{da} = \frac{dm}{dx}\frac{dx}{da}$$

(recall the chain rule). Continuous survival functions are summarized in Section E.6, with inference discussed in Sections 3.8 and 8.2.

In continuous time, structure can be included as systems of *ordinary differential equations* (ODEs) and *partial differential equations* (PDEs). A system of coupled ODEs describes rates of change among discrete stages. There is a rate equation for each state, and equations are coupled by terms that express the continuous transitions among these discrete states. Ecological models are often posed as coupled ODEs.

A PDE represents how the condition of the state variable changes continuously with time, in this case represented by the index of age and time on the state variable n,

$$\frac{\partial n(a, t)}{\partial t}$$

Models involve an additional equation for boundaries (the *boundary condition*). For an age-structured model, there is a boundary at age zero, which can be a function describing fecundity. Finally, there will a function for the initial condition. Unlike a simple differential equation, where the initial condition is a constant, with a PDE, the initial condition is a function of (in this case) the variable a, describing the age structure at time $t = 0$. PDEs apply to many types of structure, including size, physiological state, and location. Metz and Diekmann (1986) provide an overview. Parameterization can be challenging (Wood 2001), and stochasticity is typically ignored or limited to simple Gaussian variability. Some basic concepts are included in Section E.7.4. They appear in Chapter 10, where we consider diffusion.

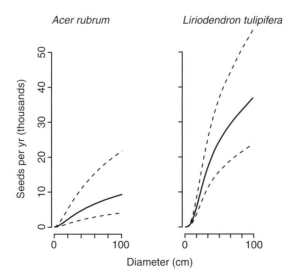

FIGURE 2.8 Fecundity schedules for two tree species treated as a continuous function of tree size. These are schedules for expected fecundity. Actual fecundity shows large fluctuations (chapter 10). From Clark et. al. (2004).

Integral projection models are discrete in time, but continuous in state (Easterling et al. 2000). The term *integro-difference* refers to the integrated effects of all stages at the current time on the density of stages at the next time. This combination provides an alternative to matrix projection models for cases where stages are not clearly differentiated, but shift more-or-less continuously. Over a discrete time increment, the population structure, described by a continuous variable x, changes as

$$n_{t+1}(x) = \int_0^\infty k(x|y)n_t(y)dy$$

where the kernel $k(x|y)$ represents transitions from state y to state x. The kernel summarizes transitions that result from fecundity, survival, and growth. Although not yet widely used, integral projection models could be more broadly applied (Section E.8), because they do not assume a single transition parameter. I do not explicitly consider this approach here, because Bayesian approaches in Chapter 9 can be used to address a wide range of deterministic and stochastic components of growth rates (Section 9.17.4). A brief introduction to stochasticity in models follows.

2.3 Stochasticity for the Unknown

Continuous time models can accommodate stochasticity (Bartlett 1960; Dennis et al. 1991; Lande et al. 2003), but they are typically treated in a deterministic fashion. In ecology, deterministic models are used to model expected values of a variable, for example, change in the mean density of a population. In such cases, the population *size* is assumed to be large. *Density*, being the number of individuals per unit area or volume, is a real number that results from division. Thus, a large population does not imply that density is large, because density could be defined over an arbitrary area or volume. Rather, both the population *density* and *size* must be sufficiently large that we can ignore the stochasticity represented by individual births and deaths. Large population size does not assure that such stochasticity can be ignored, because individual births and deaths can be important if density is effectively low. Moreover, variability among individuals can and, perhaps typically, does have impact even when populations are large. I devote substantial attention to this in later chapters.

Models where the state variable is represented by a discrete number of items or events involve probabilities of transition that result from, say, births and deaths. Unlike continuous state population models, which can be stochastic, but are typically treated deterministically, models for discrete numbers or events are invariably stochastic. The fluctuations that result from the fact that births and deaths are necessarily discrete events are termed *demographic stochasticity* (May 1973; Engen et al. 1998), although the same term has been applied to variability among individuals in terms of fitness (Lande et al. 2003). Because this book focuses on data-model connections, I deal with discrete counts more often than is typical in traditional ecological models. A connection between discrete and continuous models is obtained with a deterministic, continuous model for an underlying process, such as natality or risk, superimposed with a stochastic, discrete model for the events themselves, such as births and deaths, for observations thereof, such as counts, or both. An example is shown in Figure 2.6, where the fecundity schedule m_x is a deterministic relationship (Figure 2.6b) for the mean number of births, from which actual births are taken to be stochastic (Figure 2.6a).

Several types of stochastic models are used for ecological processes. A *Markov chain* is a process where there is a probability associated with transitions among states, and each transition probability depends solely on the current state. In population models, transitions result from birth, growth, and death. A model for discrete numbers of individuals in each state could involve the same matrix that was described for the deterministic model in Equation 2.5, but now transitions occur stochastically, with each column of the matrix being one element of a multinomial parameter vector. Such models are taken up in Chapter 9.

There are many combinations of discrete versus continuous time, state, and structure (Table 2.1), and stochasticity can enter in a number of ways. For example, population models that are continuous in time, but have a discrete number of individuals in a state are termed *birth-death processes*. Let $p_n(t)$ be the distribution of densities n at time t, where $n = 0, 1, 2, \ldots$. A rate equation dp_n/dt describes the instantaneous rate of change in the probability of state n. Because time is continuous in this formulation, development of the model entails writing down instantaneous probabilities of transition from one state to a neighboring state. A birth-death process is a *continuous time Markov chain*. Birth-death processes are sometimes implemented in discrete time. If several births and deaths might occur from one time increment to the next, there are a potentially large number of transition possibilities, each with an associated probability. This model is sometimes termed a *branching process*. A matrix can be used to define probabilities of transition from a given state to any other state. Rather than defining a large number of transition probabilities among all pairs of possible states (population sizes), we might focus on a random variable representing the number of new individuals that enter or leave a population during a given time step (Kot 2001; Caswell 2001). I do not dwell on continuous time birth-death models, because they have not been well developed for data modeling, which is our focus, and potential for analysis is limited. Solutions are available for some simple linear models, and there are useful results for some nonlinear models (Nisbet and Gurney 1982; Ricciardi 1986). Overviews of such models include Bartlett (1960) and Kot (2001).

A *random walk* is a Markov process where there are nonzero probabilities of moving to either of the adjacent states, there is zero probability of all other transitions in a single time increment, and the transition probability does not depend on the state variable itself. Models of random walks arise in a number of contexts, often where the state variable is location rather than population size. A biased random walk has unequal probabilities of moving to adjacent states, with the bias sometimes being termed *advection*.

In addition to some of these standard models that regularly appear in the ecological literature, there is an additional class of models that is increasingly applied to data, termed *random effects* models (Table 2.1), which come from the statistical literature (Clark 2003; Clark et al. 2003a). In this case, the structure is defined stochastically to allow for the differences among sample units that will be estimated from data, as opposed to being defined in advance. The variability among seedlings in terms of growth response to light is an example (Section 1.2.2).

2.4 Additional Background on Process Models

The general classes of models mentioned in this chapter are thoroughly discussed in a number of recent texts on ecological theory applied at the population level (e.g., May 1973; Edelstein-Keshet 1988; Murray 1989; Hastings 1997; Kot 2001; Caswell 2001; Case 2000) and ecosystem level (DeAngelis 1992; Agren and Bosatta 1996). Some basic elements of analysis are provided in the form of appendixes for basic process models (Appendix B), matrix manipulations (Appendix C), and life history calculation (Appendix E). This material is placed in appendixes, because it is referenced in several sections of the book and is intended for use on an as-needed basis.

Part II Elements of Inference

In this section I move directly to inference, the practice of estimating unknowns based on models and data. I begin with the likelihood, introduce inference by maximum likelihood, and then mention the method of moments. The remaining chapters of Part II take up confidence envelopes and the basics of model selection.

3 Point Estimation: Maximum Likelihood and the Method of Moments

3.1 Introduction

The basic models from Chapter 2 are used to describe a process. The process might be viewed as the core of a stochastic model that brings in sources of uncertainty and additional variability associated with the collection of data. I now consider connections between models and data.

Statistical inference involves constructing and evaluating models in the context of data. Inference can be motivated by a desire to understand the process that generated the data, to make predictions, and/or to inform a decision-making process. The process can involve comparing models, evaluating plausible parameter sets, and assigning goodness of fit. Probability statements are made in light of data, which might come from several sources and accumulate over time. Decisions involve combining the uncertainties estimated from the analysis with perceived benefits and consequences (gains, losses, risks) of alternative decisions.

There are two common frameworks for formal statistical inference that have broad application. I refer to these as classical and Bayes. It is important to understand both. Likelihood is fundamental to both frameworks, so I begin here. I then take up maximum likelihood for point estimation. A *point estimate* is the value of a parameter that finds most support in the data set. After introducing the concepts using examples based on survival, I consider applications to population growth and some standard approaches to survival analysis. At the end of the chapter, I discuss the method of moments and some general considerations about sampling distributions. This is a good time to begin familiarizing yourself with the material in Appendix D.

3.2 Likelihood

Likelihood is the probability of a data set given that the model for those data is deemed to be correct (Fisher 1934). The *likelihood principle* says that a model (with parameter values) is more likely than another if it is the one for which the data are more probable. This principle provides a basis for saying which models explain a data set better than others, that is, model selection. The easiest way to grasp this concept is with an example.

3.2.1 An Exponential Model

To estimate mortality rate from an experiment, begin by specifying a model. Initially I assume that there is one observation, the life span of a plant. If mortality rate is continuous and constant, say, ρ per unit time, then the probability that death occurs at an age between a and $a + da$ can be represented by the notation $\Pr\{a < a_1 < (a + da)\}$ (Section E.6). This notation represents the probability that the observed time of death, a_i, occurs after age a and before age $a + da$. The subscript i could indicate that this is the time of death for the ith individual. This is a joint probability of two events, (1) the plant is still alive at a, and (2) the plant dies before $a + da$. If these two events are independent, then the joint probability is their product:

$$\Pr\{a < a_i < (a + da)\} = \Pr\left\{\begin{array}{c}\text{die now given that}\\\text{plant is still alive}\end{array}\right\} \times \Pr\left\{\begin{array}{c}\text{plant is}\\\text{still alive}\end{array}\right\}$$

$$\approx h(a)da \times l(a)$$

The factor $h(a)$ is the age-specific risk per unit time, and it is scaled by the duration of the increment da to obtain the (approximate) probability of death between a and $a + da$ (Figure 3.1). The last factor $l(a)$ is the survival function, or the probability of surviving until age a (Equation E.18). If the age-specific risk is a constant $h(a) = \rho$, then $l(a) = e^{-\rho a}$, and we have

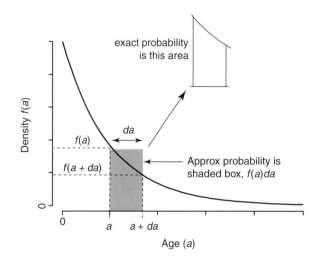

FIGURE 3.1. The relationship between probability of death in an age interval from a to $a + da$ (extracted at *upper right*) and the approximation $f(a)da$ (shaded). As the interval da becomes small, the two areas converge.

$$\Pr\{a < a_i < (a + da)\} \approx \rho da \times e^{-\rho a}$$
$$\approx f(a)da$$

These relationships are reviewed in Section E.6. The density of deaths $f(a)$ describes how the probability of death changes with age a (solid line in Figure 3.1). The term *independent and identically distributed*, or i.i.d., is used to indicate random variates that are drawn independently from a distribution. For example, the notation $y_i \overset{i.i.d}{\sim} N(\mu, \sigma^2)$ indicates that the random variates y are independently drawn from a normal distribution with the same parameter values. The i.i.d. designation is often omitted, where context makes it obvious. In the present context, I could write this assumption as $a_i \overset{i.i.d.}{\sim} Exp(\rho)$.

Multiplying the factor $h(a)$ (and thus, $f(a)$) by the factor da makes this density approximately equal to a probability (recall that density has units of a^{-1}, and da has units of a).[1] The exponential density applies when there is constant probability of an event.

The error in the approximation is the difference in area between the shaded box and the area under the curve in Figure 3.1. However, for purposes of parameter estimation, we can ignore the factor da, because this constant will not change the location or shape of the likelihood function I am about to write. But it will be important to remember that likelihood functions are based on probability. So, ignoring da, the model is the exponential density

$$f(a) = \rho e^{-\rho a} \qquad\qquad 3.1$$

This is the likelihood function.

3.2.2 An Observation

Now suppose I observe that a plant dies on the tenth day. If I make observations daily, then da = one day, and the probability of observing death on the tenth day is approximately

$$f(10) \approx \rho e^{-\rho \times 10}$$

From this single observation, what would be the best estimate of the daily mortality rate ρ? The maximum likelihood (ML) procedure for obtaining a point estimate can be summarized in two steps:

1. Write a likelihood function, describing the likelihood of the observation assuming the model to be correct.

---

[1] A precise probability would be obtained by integrating over elapsed time $(a, a + da)$, so this is an estimate that is approximately true when da is small.

2. Find the value of the parameter that maximizes this likelihood. This
    step involves differentiation.

For this single observation intuition might tell us that the answer should be $\frac{1}{10}$.
Here are these two steps for the exponential model:

Step 1. The likelihood is the probability of the observation under the
assumption that the model is true. For the exponential density (Equation 3.1)
that likelihood is

$$L(\text{data;model}) = L(a;\rho) = \rho e^{-\rho a} \qquad\qquad 3.2$$

This function is the likelihood of observing death at age a under the
hypothesis that ρ assumes a particular value. It is equivalent to the density $f(a)$.
I use the different notation $L(a;\rho)$ to emphasize that we are now thinking in
terms of a function of ρ, as opposed to a density of a. The best value of ρ is
the one that maximizes this likelihood. The likelihood function for $a_i = 10$ is
shown in Figure 3.2a.

Step 2. The ML estimate of ρ occurs at a critical point of L. To find this
critical point, set the derivative equal to zero. To simplify, first use the convention of taking logs. The critical point of $\ln L$ coincides with the critical
point of L, and the derivatives of products are more complicated than are
those of sums (joint probabilities involve products). So first, take the log of
Equation 3.2,

$$\ln L = \ln \rho - \rho a$$

and differentiate,

$$\frac{\partial \ln L}{\partial \rho} = \frac{1}{\rho} - a$$

Now set this equal to zero and solve for ρ,

$$\rho_{ML} = \frac{1}{a}$$

So the best guess at the value of ρ after a single observed death at $a_i = 10$ days
is

$$\rho_{ML} = \frac{1}{10 \text{ day}} = 0.1 \text{ day}^{-1}$$

The maximum, indicated by the dashed line in Figure 3.2a, is the most
likely value of ρ. Because log likelihoods are usually negative (the log of a
value between zero and one), they are often plotted as the negative $\ln L$, which
is positive (lower panels of Figure 3.2). Having said this, do not be aghast if
you obtain a positive log likelihood. This occurs if the density upon which
the likelihood function is based has a value greater than one. This can occur

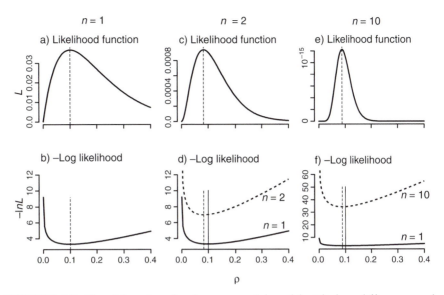

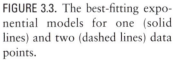

FIGURE 3.2. Likelihood functions for the exponential model with three different sample sizes. Note the different scales on the vertical axes.

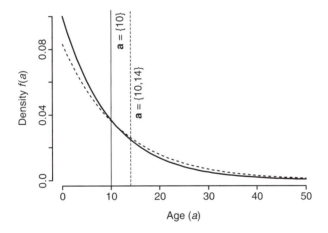

FIGURE 3.3. The best-fitting exponential models for one (solid lines) and two (dashed lines) data points.

when the likelihood is written as a density, although not for a probability (Appendix D).

The best estimate of the model $f(a)$ is shown in Figure 3.3. The single observation is represented by the vertical dashed line at age = 10 days. It lies near the center of this density; if it did not, we would suspect an error in the calculation. Confidence in an estimate based on one observation is low. I discuss how to place confidence intervals on this estimate in Chapter 5. For now, consider how the estimate changes with additional observations.

3.2.3 Additional Data

Now suppose that I observe a second plant that dies at day 14. The two observations are now deaths at days 10 and 14. How does a second observation influence the estimate of ρ? Recall the two-step procedure. The first step is to write a likelihood function for the two observations. The likelihood, or sampling distribution, is still assumed to be exponential, so the form of the likelihood function does not change. But the data have changed. If the ages of death are independent, then the probability of observing these two life spans, a_1 and a_2, is simply the product of observing each individually. As a probability density, write this relationship as $f(a_1)f(a_2)$. As a likelihood function, I use the L notation,

$$L(a_1, a_2; \rho) = L(a_1; \rho) \times L(a_2; \rho) = \rho e^{-\rho a_1} \rho e^{-\rho a_2}$$

$$= \rho^2 e^{-\rho(a_1 + a_2)}$$

Proceed as before, by taking logs,

$$\ln L = 2 \ln \rho - \rho(a_1 + a_2)$$

Step 2 (find the maximum) is completed by differentiating

$$\frac{\partial \ln L}{\partial \rho} = \frac{2}{\rho} - (a_1 + a_2)$$

Set this derivative equal to zero and solve for ρ,

$$\rho_{ML} = \frac{2}{a_1 + a_2}$$

Compare this result with that obtained for a single observation; with two observations the estimate is the reciprocal of the mean life span. The peak of the likelihood function has shifted from the previous value of 0.1 to the new value of 0.083 (Figure 3.2c, d). The additional observation means we have more information about the parameter, so the likelihood function has greater curvature near the maximum likelihood estimate (MLE) (Figure 3.1). The fitted density of mortality ages reflects the new observation (Figure 3.3).

3.2.4 A Data Set

The extension to larger data sets is straightforward. For n observations, Step 1 involves taking the product over the likelihood of each observation,

$$L(\mathbf{a}; \rho) = \prod_{i=1}^{n} \rho e^{-\rho a_i} = \rho^n \exp\left[-\rho \sum_{i=1}^{n} a_i\right] \qquad recall: a^x a^y = a^{x+y} \quad 3.3$$

Step 2 involves taking logs,

$$\ln L = n \ln \rho - \rho \sum_{i=1}^{n} a_i$$

and differentiating with respect to ρ

$$\frac{\partial \ln L}{\partial \rho} = \frac{n}{\rho} - \sum_{i=1}^{n} a_i \qquad recall: d\ln x/dx = 1/x$$

The ML estimate lies at the critical point,

$$\rho_{ML} = \frac{1}{\bar{a}} \qquad\qquad 3.4$$

where

$$\bar{a} = \frac{1}{n} \sum_{i=1}^{n} a_i$$

is the arithmetic average of the life spans. For a sample of ten observed deaths $\mathbf{a} = [10, 14, 15, 22, 24, 1, 7, 1, 11, 9]$ the MLE is 0.0877 day^{-1}. The likelihood function now has far more curvature than that for a single observation (Figure 3.2e, f), reflecting the greater amount of information that went into the estimate.

The likelihood function has thus far provided a tool for finding the "best" parameter estimate (in the ML sense). The changes in curvature with increasing data suggest that the likelihood function can also help define a confidence interval. Before taking up this topic, I provide a bit more practice with likelihoods using a different sampling distribution.

3.3 A Binomial Model

So far I have discussed a model to estimate mortality rate by observing when individuals died. Suppose instead I only have information at the end of an experiment, after many individuals have already died. I may not know when they died, but I can still estimate ρ. I begin by considering the parameter describing the probability of surviving to the end of the experiment, represented by parameter θ. This parameter is best estimated as the number remaining alive y divided by the number of initial individuals n. If the probability is the same for all individuals, this model is binomial (Section F.1.1). I bother to demonstrate that y/n is the ML estimate of the probability of survival θ, because the binomial is the basis for more complex problems that frequently arise.

Step 1. The likelihood of observing y survivors out of n initial plants depends on the parameter θ,

$$L(y;\theta) \propto \theta^y (1-\theta)^{n-y} \qquad\qquad 3.5$$

This is the binomial distribution without the normalization constant (Section F.1.1). I have ignored the binomial coefficient, because it drops out when I differentiate $\ln L$ with respect to θ. Although n is technically a parameter in the binomial model, I assume here that it is fixed by the experimental design. In other words, I will not estimate it in this example, because I know precisely how many plants were studied. We only estimate unknowns.

Step 2. I now proceed as with the exponential model. Taking logs,

$$\ln L = y \ln \theta + (n - y) \ln(1-\theta)$$

and differentiating gives

$$\frac{\partial \ln L}{\partial \theta} = \frac{y}{\theta} - \frac{n - y}{1 - \theta}$$

Set this equal to zero and solve for θ,

$$\theta_{ML} = \frac{y}{n}$$

The ML estimate of the probability of surviving is just the fraction that survive, in agreement with intuition.

3.4 Combining the Binomial and Exponential

Ecological data commonly involve complex models, because sampling distributions depend on several sources of variability that may be beyond an investigator's control. As a first introduction to some of this complexity I consider a model based on a combination of the two models discussed above. It actually falls under the heading of survivor analysis, a topic I revisit later in this chapter.

Consider again the mortality experiment. After twenty days I found that only six of thirty original plants remain alive. I use the binomial together with the survivor function to estimate ρ based only on the data collected at day 20.

Before writing a likelihood for this model, recall the distribution and survivor functions for the exponential density from Appendix E. The *distribution function* $F(a)$ is the probability that death will occur on or before age a, and it is determined by integrating the density $f(a)$ over $(0, a)$,

$$F(a) \equiv \Pr\{a_i < a\} = 1-e^{-\rho a} \qquad 3.6$$

(Section E.6.1). Recall that the probability of surviving at least until age a is the *survivor function* and is obtained by integrating $f(a)$ over all ages $> a$, because the probability of not dying before a is the complement of dying sometime after a,

$$l(a) = \int_{a}^{\infty} f(u)du = e^{-\rho a} \qquad 3.7$$

Note that $l(a) = 1 - F(a)$. Here is the likelihood function for the combined model.

Step 1. Let T be the time at which the experiment ends. Then the probability of surviving is binomial (Equation 3.3), with the parameter θ being equivalent to the survivor function 3.5. In other words, the parameter θ of the binomial in Equation 3.3 is the probability of surviving until the end of the experiment. For the exponential model, $\theta \equiv l(T) = e^{-\rho T}$. Making this substitution, the binomial Equation 3.5 becomes

$$L(y; \rho) \propto \theta^k (1-\theta)^{n-k} = l(T; \rho)^y (1 - l(T; \rho))^{n-y} \qquad 3.8$$

or

$$\ln L = y \ln l(T; \rho) + (n-y) \ln (1 - l(T; \rho))$$

Substituting for the exponential survival function

$$l(T; \rho) = e^{-\rho T}$$

results in

$$\ln L = -y\rho T + (n-y) \ln (1 - e^{-\rho T})$$

Step 2. The derivative is

$$\frac{\partial \ln L}{\partial \rho} = -yT + (n - y) \frac{Te^{-\rho T}}{1 - e^{-\rho T}}$$

Set this equal to zero and rearrange. At the critical point,

$$\rho_{ML} = \frac{\ln \left(n/y \right)}{T}$$

Satisfy yourself that this is correct by the substitution

$$y/n = \theta_{ML} = e^{-\rho T}$$

that is, the survivor function of Equation 3.7. So, for the example problem,

$$\rho_{ML} = \frac{\ln(30/6)}{20} = 0.080 \text{day}^{-1}$$

Thus, using the assumption that age-specific mortality risk is a constant value ρ, we were able to estimate that risk based only on knowing how many individuals survived twenty days. That estimate depended on assumptions that all individuals are identical, that both mortality and observations thereof were independent across all individuals, and that the mortality risk was constant.

3.5 Maximum Likelihood Estimates for the Normal Distribution

Maximum likelihood can be used with multiple parameters. The following example applies the method of maximum likelihood to the normal distribution, a widely used distribution having two parameters, a mean and a variance.

Example 3.1. A normal distribution
Estimate the mean μ and variance σ^2 parameters of a normal distribution from a sample of size n.

Step 1. The likelihood for this data set is

$$L(\mathbf{y};\mu,\sigma^2) = \prod_{i=1}^{n} N(y_i|\mu,\sigma^2)$$

$$= \prod_{i=1}^{n} \frac{1}{\sigma\sqrt{2\pi}} \exp\left[-\frac{(y_i - \mu)^2}{2\sigma^2}\right]$$

$$= \frac{1}{\sigma^n(2\pi)^{n/2}} \exp\left[-\frac{1}{2\sigma^2}\sum_{i=1}^{n}(y_i - \mu)^2\right]$$

Step 2. Taking logs yields

$$\ln L = -n\ln\sigma - \frac{n}{2}\ln(2\pi) - \frac{1}{2\sigma^2}\sum_{i=1}^{n}(y_i - \mu)^2$$

The MLE for μ is found by differentiating with respect to this parameter,

$$\frac{\partial \ln L}{\partial \mu} = \frac{1}{\sigma^2}\sum_{i=1}^{n}(y_i - \mu) \propto \sum_{i=1}^{n}y_i - n\mu$$

Setting this equal to zero and solving for μ, we obtain

$$\mu_{ML} = \bar{y}$$

that is, the sample mean. Repeat the process for σ,

$$\frac{\partial \ln L}{\partial \sigma} = -\frac{n}{\sigma} + \frac{1}{\sigma^3}\sum_{i=1}^{n}(y_i - \mu)^2$$

Set this equal to zero and solve,

$$\sigma^2_{ML} = \frac{1}{n}\sum_{i=1}^{n}(y_i - \bar{y})^2$$

In this last step, I have plugged in the ML estimate for μ. The MLE for the variance is the average squared deviation from the mean or the sample variance.

It is worth mentioning at this point that the MLE for μ is independant of the MLE of σ^2. But the converse is not true.

Note that the two parameters of the normal distribution are the first parametric moment, the mean μ, and the second central parametric moment, the variance σ^2. The ML estimates (MLEs) of these parameters are the corresponding empirical moments (Appendix D.3). It turns out that the estimate of the ML variance parameter is biased. A corrected version is obtained using restricted maximum likelihood, which places $n - 1$, rather than n, in the denominator (e.g., McCulloch and Searle 2001). This is the restricted maximum likelihood (REML) estimate of the variance parameter.

There are at least two reasons why the normal distribution is important. First, it is often justified as a sampling distribution, because the sums of random variables that are not normal are (usually) normal. This property is expressed in the Central Limit Theorem. Second, we can solve for many quantities that are useful for inference and prediction. The normal distribution appears throughout this book.

3.6 Population Growth

Now apply maximum likelihood to population growth models introduced in Chapter 2. Population growth rates are often calculated from samples taken over time. The linear model from Chapter 2 (linear in state variable n) is

$$n_t = n_{t-1}\lambda = n_{t-1}e^r$$

There are many considerations that affect how population growth rate is estimated. For example, I might allow for variability in the process that is not fully accommodated by the deterministic model. This variability is not the same as observation error, which I take up in Chapter 9. For now, I simply allow for process error, also termed *model misspecification*. It represents changes in densities that are not accounted for by the simple deterministic model.

Here are two simple models that provide additional practice with maximum likelihood. In each case I include an error distribution that does not admit negative population densities. The first is a lognormal density, which can be used when population size is large enough to ignore demographic stochasticity. The second is Poisson, which can be appropriate when population size is small. These are continuous and discrete state models, respectively, both unstructured (Chapter 2).

3.6.1 Lognormal Error Distribution

If the population density is large, then the *process error* distribution is frequently taken to be lognormal. This distribution assumes that the state variable n_t is continuous. The easiest way to model this process involves modeling *log density* with a normal likelihood. The exponential model with error could be written as

$$n_{t+1} = n_t e^{r+\varepsilon_t} \qquad\qquad 3.9$$

Taking logs, this exponential model in time becomes linear both in time and in the parameter r

$$\ln n_{t+1} = \ln n_t + r + \varepsilon_t$$

or

$$x_t = x_{t-1} + r + \varepsilon_t$$

where $x_t = \ln n_t$. I have included an error term ε_t, that allows for the fact that the exponential model does not explain everything about population growth. This is process error.

To estimate the parameter r, I begin with a likelihood function. Start with a single observation. Because I assume that the error has a lognormal distribution, and I am working with log values, the term ε_t has a normal distribution with mean 0 and variance σ^2. I express this density with the notation

$$\varepsilon_t \sim N(0, \sigma^2)$$

Because the mean and variance of the normal distribution are independent,[1] I could add this random variate ε_t to the deterministic part of the model, $x_{t-1} + r$, thus, simply shifting the mean to write the density for x_t,

$$x_t \sim N(x_{t-1} + r, \sigma^2)$$

This says that the tth value of x is expected to have a mean value that is greater than the previous value by an amount r, with variance that comes from the error term ε_t. The likelihood for this particular value of x is

$$L(x_t; x_{t-1}, r, \sigma^2) = N(x_t | x_{t-1} + r, \sigma^2)$$

For a time series of values the likelihood function is

$$L(\mathbf{x}; x_0, r, \sigma^2) = \prod_{t=1}^{T} N(x_t | x_{t-1} + r, \sigma^2) \qquad 3.10$$

Note that the product series begins with the observation indexed $t = 1$. For now, I will not model the initial value x_0, but take it to be a known constant. The likelihood consists of densities for each observation conditioned on the preceding one; I discuss this further in Chapter 9. Given the assumptions this model is

$$L(\mathbf{x}; x_0, r, \sigma^2) = \prod_{t=1}^{T} N(x_t | x_{t-1} + r, \sigma^2)$$

$$= \left(\frac{1}{\sqrt{2\pi}\sigma} \right)^T \exp\left[-\frac{1}{2\sigma^2} \sum_{t=1}^{T} (x_t - x_{t-1} - r)^2 \right]$$

---

[1] The variance is independent of the mean, but the *estimate* of the variance is not independent of the *estimate* of the mean (Section 3.5).

Taking logs and ignoring terms that do not contain parameters I obtain

$$\ln L = -T\ln \sigma - \frac{1}{2\sigma^2}\sum_{t=1}^{T}(x_t - x_{t-1} - r)$$

The MLE for the growth parameter requires the derivative

$$\frac{\partial \ln L}{\partial r} = \frac{1}{\sigma^2}\sum_{t=1}^{T}(x_t - x_{t-1} - r)$$

Setting this equal to zero and rearranging, I obtain

$$r_{ML} = \frac{1}{T}\sum_{t=1}^{T}(x_t - x_{t-1}) = \frac{x_T - x_0}{T} \qquad \text{(all but first and last terms cancel)}$$

$$= \frac{\ln n_T - \ln n_0}{T} \qquad\qquad 3.11$$

Thus, the MLE for growth rate is the total change in log density divided by elapsed time. This estimate of per capita growth rate agrees with Equation 2.2. The estimate for the variance is

$$\sigma_{ML}^2 = \frac{1}{T}\sum_{t=1}^{T}(x_t - x_{t-1} - r)^2$$

Compare this result with that obtained in Example 3.1.

3.6.2 Poisson Distribution

Individuals are discrete, so at some level, dynamics will be modeled with stochastic additions and losses through births and deaths (Bartlett 1960). Models tend to ignore this source of stochasticity when population sizes are large, but it dominates when a population is small. Ecologists refer to this as demographic stochasticity. To model it requires a discrete state model (Section 2.2.4). The Poisson distribution is a common distribution for discrete variables (Section F.1.2). If each individual lives a single time step and begets a Poisson number of new individuals per time unit with mean b (e.g., Figure 2.6), we have the model

$$n_t|n_{t-1} \sim \sum_{i=1}^{n_{t-1}} Pois(b) = Pois(n_{t-1}b) \qquad\qquad 3.12$$

In this model, each individual contributes a random number of offspring drawn from the distribution $Pois(b)$. The summation in Equation 3.12 provides the total offspring produced by all n_{t-1} individuals. This is a branching process (Section 2.3). The last step follows from a peculiar property of the

Poisson distribution, which says that $n \, Pois(b) = Pois(nb)$ (Section F.1.2). The likelihood for the full data set is

$$L(\mathbf{n};n_0,b) = \prod_{t=1}^{T} Pois(n_t \,|\, n_{t-1}b)$$

$$= \prod_{t=1}^{T} \frac{[n_{t-1}b]^{n_t} e^{-n_{t-1}b}}{n_t!}$$

As in the last example, I have not modeled n_0. Now take logs

$$\ln L = \sum_{t=1}^{T} [n_t(\ln n_{t-1} + \ln b) - n_{t-1}b - \ln(n_t!)]$$

and differentiate with respect to b,

$$\frac{\partial \ln L}{\partial b} = \frac{1}{b}\sum_{t=1}^{T} n_t - \sum_{t=1}^{T} n_{t-1}$$

Set this equal to zero and solve for b to obtain

$$b_{ML} = \left. \sum_{t=1}^{T} n_t \middle/ \sum_{t=1}^{T} n_{t-1} \right. \tag{3.13}$$

This is the maximum likelihood estimate (MLE) of the growth parameter.

To estimate growth rates in these two examples, I used the standard assumption that observations are independent. This assumption is implicit when the likelihood for the full data set is taken to be the product of the likelihood for each observation. The assumption of independence is seriously violated in the case of population growth, because observations involve sequential observations that are related to one another. This violation does not affect the point estimates of parameters (in this case), but it does keep us from estimating uncertainty, which would be expressed in the form of confidence or credible intervals for parameters or predictive intervals for future values of density. I revisit population growth in Chapter 9, examining how to reduce this interdependence by replacing population density with change in population density and by exploiting conditional independence (from a Bayesian perspective).

3.7 Application: Fecundity

LaDeau and Clark (2001) obtained maturation and fecundity data from counts of cones on pine trees subjected to ambient and elevated CO_2 (Figure 3.4). Their model took into account the fact that trees might be mature or not

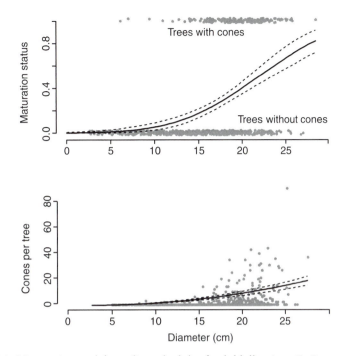

FIGURE 3.4. Maturation and fecundity schedules for loblolly pines (LaDeau and Clark 2001). The fitted model for this example is shown in the lower panel. An extension of this model is taken up in Example 5.6.

(reproductive status) to estimate conditional cone production for trees that are mature. Ignore reproductive status for now. Because seed production is often assumed roughly proportional to crown area (e.g., Ribbens et al. 1994; Clark et al. 1998), the fecundity schedule looks like this:

$$g(x_i;a) = ax_i^2,$$

where x_i is the diameter of the ith tree and a is a fitted parameter. Further assume that cone production per tree y_i follows a Poisson distribution. We wish to estimate the parameter a.

The likelihood function is the joint probability for all trees,

$$L(\mathbf{y};\mathbf{x},a) = \prod_{i=1}^{n} Pois\left(y_i \,\middle|\, g(x_i;a)\right)$$

This likelihood function says that each y_i has a Poisson distribution, the mean of which depends on the tree's diameter x_i by way of a function g. Writing out the Poisson distributions and substituting for g yields the likelihood

$$L = \prod_{i=1}^{n} \frac{[g(x_i;a)]^{y_i} e^{-g(x_i;a)}}{y_i!} = \prod_{i=1}^{n} \frac{[ax_i^2]^{y_i} e^{-ax_i^2}}{y_i!}$$

Taking logs,

$$\ln L = \sum_{i=1}^{n} [y_i(\ln a + 2\ln x_i) - ax_i^2 - \ln(y_i!)]$$

and differentiating with respect to a gives

$$\frac{\partial \ln L}{\partial a} = \frac{1}{a}\sum_{i=1}^{n} y_i - \sum_{i=1}^{n} x_i^2$$

Now set this equal to zero and solve for a:

$$a_{ML} = \left. \sum_{i=1}^{n} y_i \middle/ \sum_{i=1}^{n} x_i^2 \right. = \frac{\bar{y}}{\overline{x^2}}$$

In other words, the parameter estimate is simply the mean cone production divided by the mean squared diameter. This model is not precisely the same as that shown in Figure 3.4b. The full analysis used to construct Figure 3.4 is given in Example 5.6.

3.8 Survival Analysis Using Maximum Likelihood

In Section 3.3 I discussed use of maximum likelihood to estimate survival in terms of (1) time until death and (2) the number of subjects that survive to the end of an experiment. Most survival experiments involve both of these possibilities; for individuals that die, we might know when death occurred. For those that survive, we do not know how long they would have survived had the experiment not been terminated. These represent two different, albeit related, contributions to the likelihood. Analysis of survival data is an especially important and well-developed topic, being used not only in clinical studies, but also in industry, where it is known as failure time analysis. It can provide inference on a range of factors that affect the survivor functions outlined in somewhat more detail in Appendix E.6.1. I stay with this topic a bit longer, because the extensions illustrate some of the general ways in which maximum likelihood can be tailored to more complex issues. This will not only help with the fundamentals behind the many statistical packages available for survival analysis, but it will also demonstrate how to develop algorithms for the cases where existing software will not suffice.

3.8.1 The Likelihood for an Exponential Model

Survival analysis is based on the binomial sampling distribution, introduced earlier in this chapter, because each individual has one of two possible states, alive or dead. It could be viewed as a special type of binomial involving time, because mortality involves passing from the one state to the other. Inference comes from a model describing when this transition occurs. The relationship is easiest to see starting from the binomial likelihood for a sample of y dead out of n total individuals,

$$L(y;\theta) \propto \theta^y (1 - \theta)^{n-y} \qquad 3.14$$

where θ is the probability of death. (It does not matter whether I define θ to be probability of death, in which case y is the number that die, or the probability of survival, in which case y is the number that survive.) Survival analysis builds on this binomial, by admitting that different individuals may die at different times. Recall that a distribution of times of death $f(a)$ is related to a survivor function as

$$l(a) = \int_a^\infty f(x)dx$$

If $f(a)$ is continuous (as in this case), then the probability of dying on the interval $(a + da)$ is approximately $f(a)da$ (Section 3.2).

In most experiments or observational studies, not all individuals die. Those that survive to the end of the experiment and those that meet their end by means other than death are said to be *censored*. We distinguish between those that die and those that are censored, because censored individuals contribute differently to the likelihood. Designate as a_i the age when an individual dies, and as c_i the censoring time of one that survives. Each individual that dies contributes to the likelihood a factor $f(a_i)da$. Each censored individual contributes $l(c_i)$. I now modify likelihood 3.14 to allow for this temporal information:

Step 1. To write a likelihood function for survival data, begin by replacing θ and $(1 - \theta)^{n-y}$ with products over all individuals (because each individual has a different time of death):

$$L(y;\theta) = \prod_{i=1}^{y} \theta_i \times \prod_{i=1}^{n-y} (1 - \theta_i)$$

Note that if all values θ_i are the same, then this collapses to Equation 3.14. Now substitute $f(a_i)$ for θ_i, because $f(a)da$ is the probability for death at age a, and da will drop out when I differentiate $\ln L$. Then substitute $l(c_i)$ for $(1 - \theta_i)$, because $l(c)$ is the probability of not dying before age c. The likelihood is now

$$L(\mathbf{a,c};\theta) = \prod_{i=1}^{y} f(a_i;\phi) \times \prod_{i=1}^{n-y} l(c_i;\phi) \qquad 3.15$$

where ϕ represents parameters in the model. In the following example, I extend the exponential example at the beginning of this chapter as a full survival analysis.

Example 3.2. I would like to estimate a survival function for seedlings, assuming constant risk of mortality, that is, an exponential survival function. Ten seedlings are planted, and censuses are taken at daily intervals for three months. There are $y = 6$ individuals that die at ages $a_1, \ldots, a_6$ out of a total of $n = 10$ individuals. To determine biomass, I remove $k = 2$ seedlings, one each after the first and second months, or $c = [30, 60]$. These individuals are censored at the time of harvest—I know they survived at least that long, but I do not know when they would have died had they not been sacrificed. These individuals have censoring times $c_1 = 30$ $c_2 = 60$. The remaining $n - y - k = 2$ seedlings survive to the end of the experiment, which terminates at time $T = 90$ days. The data are shown in Table 3.1.

To estimate the mortality rate (the exponential parameter) ρ write a likelihood as the product of the contributions from y individuals that die, k individuals that are censored by harvest, and $n - y - k$ individuals that survive to the end of the experiment,

$$L(\mathbf{a},\mathbf{c};\rho) = \prod_{i=1}^{y} f(a_i|\rho) \times \prod_{i=1}^{k} l(c_i|\rho) \times [l(T|\rho)]^{n-y-k}$$

Note that each of the three elements of the likelihood describes the probability somewhat differently, depending on the fate of each individual and when that fate occurred. The last two elements refer to censored individuals. For an exponential density (constant mortality risk) this likelihood is

$$L(\mathbf{a},\mathbf{c};\rho) = \prod_{i=1}^{y} \rho e^{-\rho a_i} \times \prod_{i=1}^{k} e^{-\rho c_i} \times e^{-\rho T(n-y-k)}$$

TABLE 3.1.
A Survival Data Set

Individual	Fate	Contribution to Likelihood*
A	Died on day 2	$a_1 = 2$
B	Died on day 10	$a_2 = 10$
C	Died on day 7	$a_3 = 7$
D	Died on day 5	$a_4 = 5$
E	Harvested on day 30	$c_1 = 30$
F	Died on day 45	$a_5 = 45$
G	Harvested on day 60	$c_2 = 60$
H	Died on day 50	$a_6 = 50$
I	Survived to end of experiment	$c_3 = 90$
J	Survived to end of experiment	$c_4 = 90$
		Sum = 389

* a = age of death; c = age at harvest.

Taking logs, obtain

$$\ln L(\mathbf{a},\mathbf{c};\rho) = y\ln\rho - \rho\sum_{i=1}^{y}a_i - \rho\sum_{i=1}^{k}c_i - \rho T(n - y - k)$$

Differentiating with respect to ρ and setting the resultant function equal to zero gives the ML estimate of ρ,

$$\rho_{ML} = \frac{y}{\displaystyle\sum_{i=1}^{y}a_i + \sum_{i=1}^{k}c_i + T(n - y - k)}$$

$$= \frac{\text{number that died}}{\text{total times to death or censoring}}$$

You should convince yourself that the denominator is just the sum of the total time at risk. For this data set, then, the estimate of mortality rate is

$$\frac{6}{389} = 0.0154 \text{ day}^{-1}$$

The approach taken in Example 3.2 is similar to that used by Kobe (1999) to estimate seedling survival at different light levels. In that case, the survivor function was assumed to depend on light availability, which was parameterized from planted seedlings in the forest understory. He used different forms of the light relationship for different species (Figure 3.5).

3.8.2 Nonparametric Life Table Estimate of Survivorship

The foregoing method assumes that I have some functional form in mind for $l(a)$ (and thus for the age-specific risk $h(a)$ and for the density of deaths $f(a)$) for the exponential model $h(a) = \rho$. I return to parametric models when I consider covariates in Chapter 8. For now, I introduce the nonparametric approach. A nonparametric method is useful when parametric assumptions seem unwarranted. Rather than select a functional form that might apply to the full survivor function, I can simply fit a survival probability for the observed times of death a_i. Recall from the consideration of life tables (Sections 2.2.4 and E.1) that a discrete survivor function can be written as

$$l_a = \prod_{a_i < a}(1 - \rho_i)$$

where ρ_i is mortality probability associated with time of death at time a_i (see Equation 2.7).

Now consider a likelihood for mortality of a sample of individuals that die at ages $a_i, i = 1, \ldots, n$ for census intervals j of duration dx_j. In this case, the binomial likelihood can be cast in terms of the mortality rate for an age class,

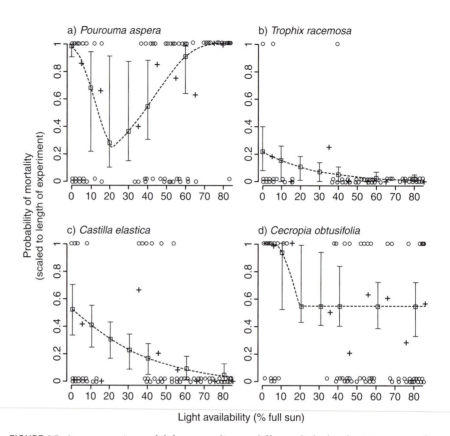

FIGURE 3.5. A parametric model for mortality at different light levels. Mean mortality rate for individual light levels, binned for each 10 percent increment, is indicated by +. From Kobe (1999).

the probability of death, $\rho_j dx_j$, and the probability of survival, $1 - \rho_j dx_j$. If n_j is the number of individuals alive in age class j, d_j of which die, the likelihood for age class j is

$$L(d_j;\rho_j) = (\rho_j dx_j)^{d_j}(1 - \rho_j dx_j)^{n_j - d_j}$$

Taking logs and setting the derivative equal to zero gives the ML estimate for ρ_i,

$$\rho_i = \frac{1}{dx_j} \cdot \frac{d_j}{n_j}$$

that is, the fraction of those entering the age class that die per unit time.

For now, assume that the times of death are known to occur at intervals of equal and sufficiently short duration that we can forget about dx. There still may be ties, with more than one event occurring at the same time that need special consideration. Modify the bookkeeping somewhat, using a_i to indicate

TABLE 3.2.
A Life Table Developed from Table 3.1

Ordered Mortality or Censor Times (a_i)	Number (n_i)	Fate	d_i	ρ_i	$1 - \frac{d_i}{n_i}$	Product Limit Estimate of l_a	Parametric Estimate of l_a Exponential Model
0	10		1			1	1
2	9		1	0.100	0.900	0.900	0.970
5	8		1	0.111	0.889	0.800	0.926
7	7		1	0.125	0.875	0.700	0.898
10	6		1	0.143	0.857	0.600	0.857
30	5	censored	0	0	1	0.600	0.630
45	4		1	0.200	0.800	0.480	0.500
50	3		1	0.250	0.750	0.360	0.462
60	2	censored	0	0	1	0.360	0.396
90	2	censored	0	0	1	0.360	0.250

the time of an event, d_i to represent the number of events at a_i, and n_i for the number of potential events. Then the *product limit estimator* (Cox and Oakes 1984) of the survivor function is

$$l_a = \prod_{a_i < a}\left(1 - \frac{d_i}{n_i}\right)$$

where I simply replaced the parameters ρ_i with their ML estimates.[2] To calculate the product limit estimator of survival, first sort the individuals according to their times of death or censoring. This allows us to readily see how many individuals are at risk through time. Table 3.2 contains the sorted data set for Example 3.3. Note how the censored individuals affect the product limit estimate of the survivor function. The age-specific mortality estimates ρ_i are zero for all age classes not explicitly listed in the table. As a result, the product-limit estimator "steps down" at each time a death occurs.

Figure 3.6 shows nonparametric survivor functions for tree seedlings grown in the forest understory for five years. Many survived the entire period ($c_i = T = 5$ yr), some were harvested before the end of the experiment ($c_i < 5$ yr), and some died before the end of the study ($a_i < 5$ yr). Note the large number of censored seedlings during the experiment. Because of this censoring, we could not have estimated the survivor functions simply by dividing the number remaining in a given year by the initial number. The two species have very different survivor functions due to their tolerances of low light conditions in the understory. *Pinus taeda* colonizes high-light environments. *Cercus canadensis* is most common in shaded sites.

---

[2] Life tables are used to organize data for these calculations. They are discussed in Appendix E.

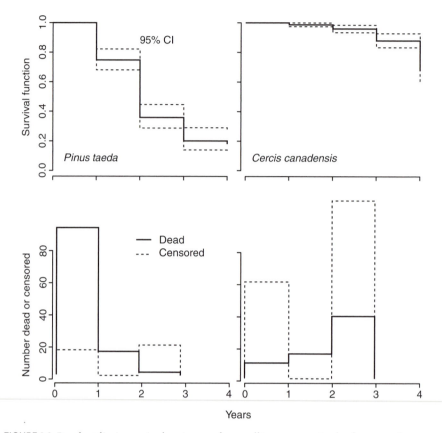

FIGURE 3.6. Product limit survival estimates for seedlings grown in the forest understory. Confidence intervals (dashed lines in upper panels) are considered in Example 5.12. From Mohan et al. (2006).

3.9 Design Matrixes

A vector of covariates for an observation, such as the length-p vector $\mathbf{x}_i$ of Box 3.1, can be arranged into a matrix of n samples by p covariates, $\mathbf{X}$. This *design matrix* contains the independent variables for a model that typically involves multiplication by a vector of parameters. If the model contains an intercept parameter, then the first column of $\mathbf{X}_i$ contains all ones. For a model having only a slope and intercept, the design matrix is

$$\mathbf{X} = \begin{bmatrix} 1 & x_i \\ \vdots & \vdots \\ 1 & x_n \end{bmatrix}$$

Where experiments involve application of a treatment, or not, the corresponding column of $\mathbf{X}$ has zeros (e.g., control treatment) and ones. If there are different

Box 3.1 Semi-parametric models for covariates

There are times when we want to understand how differences among individuals affect survival. For the nonparametric approach, we can compare survival for two populations that might differ in terms of a treatment, species, gender, and so on. We cannot assess how continuously varying factors influence survival. For that we need a parametric model. For survival models, these terms have specific connotations. Here are the terms used for different types of survival analysis:

1. *Nonparametric*—the product limit estimator. Used to compare different survivor functions with no assumptions concerning shape of the survival function or of covariate effects (see above).
2. *Semi-parametric*—a parametric form for covariates, but not for the survivor function. I discuss the Cox proportional hazards model in this section.
3. *Parametric*—a parametric form for the survivor function and for covariates. I use the accelerated time model for covariates in Chapter 8.

Here I consider the semi-parametric proportional hazards model for estimating covariate effects.

Consider a length p row vector of covariates $\mathbf{x}_i = [x_{i1}, \ldots, x_{ip}]$ that apply to the ith individual. These might represent treatments to which the ith individual is exposed, properties of the individual itself, or factors affecting it. The associated parameter vector is $\beta = [\beta_1, \ldots, \beta_p]^T$. These two vectors are multiplied to obtain a linear equation. For two covariates this yields

$$\mathbf{x}_i\beta = [x_{i1} \quad x_{i2}]\begin{bmatrix}\beta_1 \\ \beta_2\end{bmatrix} = \beta_1 x_{i1} + \beta_2 x_{i2}$$

In this example, covariate x_1 represents log height growth of a seedling (vigorous individuals are more likely to survive) and x_2 represents light availability. The additional subscript i refers to the value of the covariate for observation i.

The *proportional hazards* model assumes that there is a base age-specific risk, call it $h_0(a)$, that is accelerated by the effects of these covariates. A convenient way to formulate the problem is

$$h(a_i|\mathbf{x}_i) = h_0(a_i)e^{\mathbf{x}_i\beta}$$

The exponential form means that when the covariates are zero, we recover the base hazard. Positive effects in the exponent accelerate life ($h(a) > h_0(a)$) (shorten time to death), and negative effects prolong it ($h(a) < h_0(a)$). In other words, we expect a negative parameter estimate for a variable or treatment that prolongs life, and vice versa. The base survivor function is

$$l_0(a) = \exp\left[-\int_0^a h_0(u)du\right]$$

(Continued on following page)

so the proportional hazards survivorship is

$$l(a_i|\mathbf{x}_i) = \exp\left[-\int_0^a h_0(u)e^{\mathbf{x}_i\beta}\,du\right] = \exp\left[-e^{\mathbf{x}_i\beta}\int_0^a h_0(u)du\right]$$

$$= [l_0(a)]^{\exp(\mathbf{x}_i\beta)}$$

To write a likelihood for the proportional hazards model, I make use of some relationships from Section E.6.1. From Equation E.19 recall that the density for death at age a is the product of age-specific risk and the survivor function,

$$f(a) = h(a)l(a)$$

Making the appropriate substitutions in Equation 3.15 yields the likelihood

$$L(\mathbf{a},\mathbf{c};\phi) = \prod_{i=1}^{y} h(a_i|\mathbf{x}_i\beta)l(a_i|\mathbf{x}_i\beta) \times \prod_{i=1}^{n-y} l(c_i|\mathbf{x}_i\beta)$$

I have taken the product for the survivor function over all n individuals (censored or not) and the risk function only over the individuals that actually died. This can be written equivalently as

$$L(\mathbf{a},\mathbf{c};\phi) = \prod_{i=1}^{y} h(a_i|\mathbf{x}_i\beta) \times \prod_{i=1}^{n} l(T_i|\mathbf{x}_i\beta)$$

where T_i is simply the time of the observed event, either a_i or c_i.

Due to widespread use of survival analysis in many disciplines, particularly in biomedicine, there is standard software available for a variety of designs (e.g., Klein and Moeschberger 1997). The Cox proportional hazards model makes use of a partial likelihood, which is a variant on the ML theme (e.g., Cox and Oakes 1984).

Figure 3.7 shows estimates of years until half of the seedlings shown in Figure 3.6 would be expected to die, accounting for differing light availability and seedling

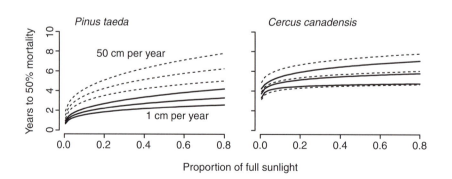

FIGURE 3.7. The expected number of years until 50 percent of transplanted seedlings die at different light levels and growth rates. Dashed lines are for individuals increasing in height at 50 centimeter per year. Solid lines are for individuals growing one centimeter per year. From Mohan et al. (2006).

growth rates. This smooth relationship represents the parametric part of this semi-parametric model. The vector **x** consists of log height growth (cm yr^{-1}) and log light availability (fraction of full sunlight). The parameter vectors are $\beta = [-0.864,\ -1.585]^T$ for *Pinus* and $\beta = [-0.168, -1.231]^T$ for *Cercus*. Negative values indicate that both of these factors are associated with longer life. The more shade-tolerant *Cercus* is predicted to survive longer than *Pinus* individuals that are growing slowly (Figure 3.7). *Cercus* survival is relatively unresponsive to light availability. It is important to interpret the results appropriately. Although fast-growing *Pinus* trees are predicted to survive several years, provided that they are growing rapidly, *Pinus* rarely grows rapidly at low light levels. Thus, we would tend to interpret the solid line at low light and the dashed line at high light.

levels of a treatment, there may be a different column of zeros and ones for each level of the treatment. In this case, treatments are handled as factors. Alternatively, if levels are viewed as differing in quantity of a particular covariate, there can be a single column containing the level of the covariate that applies to observation *i*. There will be many examples of design matrixes in this book.

3.10 Numerical Methods for MLE

Solutions for MLEs are often unavailable. This occurs when, say, we differentiate the log likelihood and find that a parameter of interest cannot be isolated on one side of the equation. If there is no closed-form solution for a parameter, then numerical methods are used to locate the parameter values where the likelihood surface reaches its maximum. There are many software options for maximizing a function (see lab manual). Numerical approaches can work well with a relatively small number of parameters. As models become more complex, it can be difficult to identify the global maximum in a sea of local maxima or along ridges in the likelihood surface (e.g., Figure 5.9). There are limits as to the complexity of the model that can be fitted with any approach, including ML. A different limitation comes when the raw data are inaccessible. When a computer or the data are not available, the method of moments, or *moment matching*, can sometimes be used to obtain parameter estimates.

3.11 Moment Matching

Moment matching involves estimating parameters by equating them with sample (empirical) moments. In some cases (normal, Poisson), the MLEs of parameters *are* sample moments, so methods are equivalent (empirical moments are discussed in Appendix D). The Poisson parameter is the sample mean. The two parameters for the normal distribution are termed the *mean* and *variance*, because the mean parameter is estimated as the sample mean, and the variance parameter is estimated as (approximately) the sample variance (Example 3.1).

There are at least two situations in which the method of moments may provide a more practical means for parameter estimation than maximum likelihood. First, solving for the MLEs can, at times, be difficult. For example, the gamma density (Section F.3.3) has a likelihood that involves a factorial (gamma function) for one of the parameters, which can only be differentiated numerically. Without a computer one can still estimate parameters by calculating the mean and variance of the data and using these moments to calculate the two parameters of the gamma density. From Section F.3.3 we know the mean and variance of the gamma density to be

$$E[x] \equiv \bar{x} = \frac{\alpha}{\beta}$$

and

$$\mathrm{var}[x] = \frac{\alpha}{\beta^2}$$

Substitute for the moments and solve these two equations in two unknowns to obtain

$$\alpha = \frac{\bar{x}^2}{\mathrm{var}[x]}$$

$$\beta = \frac{\bar{x}}{\mathrm{var}[x]}$$

This method requires the same number of sample moments as parameters to be estimated. The single parameter of the Poisson requires only the first sample moment, the sample mean. For this example of a gamma density (two parameters), we needed the mean and variance. Of course, with access to the computer, we could estimate the gamma parameters using maximum likelihood, because the computer could evaluate the gamma function in the likelihood.

A second application of the method of moments comes when data are unavailable, but moments are reported. Without the data, we cannot evaluate the likelihood function. But we can use the moments to approximate a functional form. For example, the estimate for the binomial parameter θ may be reported together with its standard error. In a Bayesian context (Chapter 5), we might approximate the distribution of θ with a beta density (Section F.3.5) using moments. Let the variance in parameter θ be $\mathrm{var}[\theta] = se_\theta^2$ and use the relationships from Box D.1 to equate moments for the beta parameters,

$$E[\theta] = \frac{\alpha}{\alpha + \beta}$$

$$\mathrm{var}[\theta] = \frac{\alpha\beta}{(\alpha + \beta)^2(\alpha + \beta + 1)}$$

By substitution both parameters can be expressed in terms of the mean and variance,

$$\alpha = E[\theta]\left[\frac{E[\theta](1 - E[\theta])}{\text{var}[\theta]} - 1\right]$$

and

$$\beta = (1 - E[\theta])\left[\frac{E[\theta](1 - E[\theta])}{\text{var}[\theta]} - 1\right]$$

Thus, the point estimate of θ, together with its standard error, provides a basis for approximating the parameter distribution.

Example 3.3. Point counts of orange-crowned warblers in nonriparian Ponderosa pine stands yielded the frequencies in Table 3.3. Compare parameter estimates for the negative binomial distribution obtained by maximum likelihood and the method of moments.

Maximum Likelihood—The likelihood for the negative binomial is obtained for data set **y** using the distribution function from Appendix F. Take the product over $n = 152$ sample locations, each location having count y_i,

$$L(y\,|\,\alpha,\beta) = \prod_{i=1}^{n} \frac{\Gamma(y_i + \alpha)\beta^{\alpha}}{\Gamma(y_i + 1)\Gamma(\alpha)(1 + \beta)^{y_i+\alpha}}$$

$$= \frac{\beta^{n\alpha}}{\Gamma(\alpha)^n} \times \prod_{i=1}^{n} \frac{\Gamma(y_i + \alpha)}{\Gamma(y_i + 1)(1 + \beta)^{y_i+\alpha}}$$

Taking logs gives the log likelihood

$$\ln L(y\,|\,\alpha,\beta) = n(\alpha \ln \beta - \ln \Gamma(\alpha)) + \sum_{i=1}^{n}[\ln \Gamma(y_i + \alpha)$$

$$- \ln \Gamma(y_i + 1) - (y_i + \alpha)\ln(1 + \beta)]$$

This must be solved numerically (see lab manual), because it involves gamma functions of parameters we wish to estimate. The numerically obtained MLEs are

$$\alpha = 0.431$$
$$\beta = 3.640$$

and $-\ln L$ is 57.16 (Figure 3.8).

Method of Moments—The data have a mean and a variance of 0.1184 and 0.1448, respectively. From moments provided in Section F.1.3 solve for parameters as

$$\alpha = \frac{\bar{x}^2}{\text{var}[x] - \bar{x}} = 0.531$$

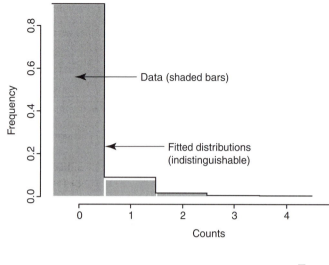

FIGURE 3.8. Negative binomial distributions fitted to bird count data by ML and moment matching yield identical fits, albeit with somewhat different parameter values.

$$\beta = \frac{\overline{x}}{\text{var}[x] - \overline{x}} = 4.485$$

These values appear different from the MLEs, but the $-\ln L = 57.18$ obtained using these estimates indicates a fit that is almost identical (see lab manual). In other words, these estimates do not differ importantly from the MLEs. How do we know that this difference in likelihoods is small? This question is addressed in Chapter 5. For now, you might consult Figure 2.6 of the lab manual for a contour plot of this likelihood surface, which indicates the correlation between these two parameters.

3.12 Common Sampling Distributions and Dispersion

Likelihoods are based on sampling distributions. The examples considered thus far have included several types. The most common sampling distributions are (1) normal and lognormal for continuous state variables, (2) binomial for binary responses, (3) Poisson for counts, and (4) multinomial for nominal or ordinal classes. Also frequently used is the Bernoulli, which is a binomial with sample size $n = 1$. Because the lognormal is usually implemented as a normal distribution on log values (exceptions to this arise several times in this text), I consider that together with the normal distribution. This leaves the normal, binomial, Poisson, and multinomial to serve as the workhorse sampling distributions. When you construct a likelihood you will usually turn to one of these distributions. All have overdispersed counterparts. A fifth distribution, the exponential, usually arises in the context of time, as discussed in this chapter. In this section I mention additional considerations about dispersion with respect to the four most common sampling distributions.

Although each of the standard sampling distributions can represent some idealized situations, data are commonly more dispersed than those idealized situations. For example, continuous data may have fatter tails than does the

TABLE 3.3.
Point Counts of Orange-crowned Warblers.

Count	Number of point count locations
0	137
1	12
2	3

Source: From White and Bennett 1997

best-fitting normal distribution. Such distributions make up for the fat tail by having less in the shoulders. They are said to be overdispersed. The normal distribution is a logical choice for continuous variables, where the sampling distribution is roughly symmetric. It is motivated by the Central Limit Theorem, which says that the sum of random variables that might have non-Gaussian distributions usually tends to normality as the sample size increases. An overdispersed counterpart for the normal distribution is the Student's t distribution, which results from mixing the normal sampling distribution with an inverse gamma distribution for the variance σ^2. Consider the normal distribution, written this way:

$$x \mid \mu, \sigma^2 \sim N(\mu, \sigma^2)$$

We could think of a random variable x as having a normal distribution that depends on the values of μ and σ^2. The variability (spread) is determined by σ^2. The larger the value of σ^2, the larger the variability. Regardless of the value of σ^2, the distribution is still normal. Among other things, this means that the tails of the density fall off at a rate of $\exp((x - \mu)/\sigma)^2$.

By "mixing" I mean that σ^2 might vary *within the same sample*. If there is random variation in σ^2, the distribution of x is termed a *mixture*. That mixture is obtained by integrating over the variability in σ^2, that is,

$$P(x/\mu) = \int N(x/\mu, \sigma^2)_p\, (\sigma^2) d\sigma^2$$

This marginal density $P(x/\mu)$ now has tails that are fatter than $\exp((x-\mu)/\sigma^*)^2$. The overdispersed Student's t assigns higher probability to values of x far from the mean and close to the mean; it has less in the shoulders. I discuss mixtures further in Appendix D and in Chapter 5, where I consider conjugate pairs (the normal likelihood with inverse gamma on the variance parameter is an example).

The integral for this mixture is simulated as follows. Suppose the distribution for the variance is inverse gamma. We can draw samples from this inverse gamma density and then draw a sample from the normal distribution. In other words, we do this:

$$\sigma_i^2 \sim IG(a, b)$$

$$x_i \sim N(\mu, \sigma_i^2)$$

To construct a frequency distribution, repeat this process, saving the estimates of x_i.

Overdispersion applies to other distributions. Data obtained from a process that admits only two outcomes often have an empirical variance that is larger than the binomial sampling distribution we might fit to those data. The binomial distribution, describing the number of successes in n trials, has an overdispersed counterpart in the beta-binomial. The beta-binomial arises when the binomial probability θ is drawn from a beta distribution. The beta-binomial is a mixture.

For large n and small θ, the binomial can be approximated by a Poisson distribution with mean parameter $\lambda \approx n\theta$. Example 3.3 demonstrates that count data may be better described by the negative binomial than by the Poisson. The negative binomial distribution is the overdispersed counterpart for the Poisson. It is a mixture that results from a gamma-distributed Poisson parameter λ. The overdispersed multinomial is achieved with a Dirichlet mixing distribution for parameters.

In Example 3.3, the problem of overdispersed data was addressed by using a sampling distribution having a greater dispersion (negative binomial) than the standard one (Poisson). An alternative approach is to treat the process that produces this excess variability more explicitly at a lower stage in model. The sampling distribution is viewed as the data model, which might be conditionally normal, binomial, Poisson, or multinomial but *marginally* overdispersed. Process models at a lower stage may include not only stochastic elements, but also deterministic relationships that involve covariates. When I consider sampling-based approaches for determining a Bayesian posterior density in Chapter 8, I commonly work with standard sampling distributions at the data model stage and allow for additional variability at the process stage.

3.13 Assumptions and Next Steps

Maximum likelihood is a widely used method for obtaining estimators. It is relatively easy for simple models, such as those that apply to simple experimental designs. Point estimates involve derivatives, which are usually available. Confidence intervals may come from one of several approaches (Chapter 5).

Nonetheless, there are some potentially restrictive assumptions associated with maximum likelihood and some limitations to its application. In practice, it typically requires that we can write a model in a single stage. It can be difficult to construct models for cases where samples are not independent. If there is stochasticity that goes beyond the sampling distribution, we usually need to marginalize in some way, such that we can derive estimates. This marginalization requires integrals, which will often be unavailable for applications of interest.

In the next chapter I consider Bayes, a second common approach. Bayes combines the likelihood with priors for parameters, a structure that admits information that comes from additional sources. Confidence envelopes for both approaches follow in Chapter 5.

4 Elements of the Bayesian Approach

> Not so many years ago, "bayesian statistics" was frequently
> viewed as the antithesis of "frequentist statistics," and the feeling
> among some was that their favorite of these two theories would
> eventually triumph as the other failed ignominiously. It now is
> apparent that this will not happen. There is much evidence that
> we are currently in the midst of a productive amalgamation of
> the two schools of statistics.
>
> —Brown (2000, p. 1278)

In many respects, science is a Bayesian activity. Inference and decision inevitably involve synthesis. A body of knowledge develops from many individuals working in different settings. The challenges for inference and prediction include integrating a heterogeneous body of evidence that may bear on the question at hand and the capacity to admit multiple hypotheses within the context of a comprehensive analysis. Progress comes when new information is placed in the context of what is already known, including previous studies of others, experiments, nonexperimental observations, and theory. Scientific debate typically focuses on the different ways in which each of us frames new evidence in the context of our broader understanding of a process. A classical view does not easily permit this integration of evidence from multiple sources. Bayesian analysis provides a formal mechanism for synthesis.

Several decades ago it was easy to find statisticians who disliked the Bayesian approach, in part, because they viewed the integration of information external to the experiment at hand as too subjective for science. There is growing appreciation that statistics is inevitably subjective (Berger and Berry 1988). Model construction and selection involve many decisions that cannot be resolved by appeal to objectivity. Such decisions must be made in the context of limited knowledge. They involve different types of (and often, apparently, conflicting) evidence. Carefully designed experiments may represent only a fraction of the full weight of the evidence. As we learn more, our understanding needs to be updated, but in a manner that can be clearly communicated and is consistent with probability models.

Bayesian analysis formalizes the process of integration in a way that allows different views to be compared within a common framework. For data integration and complex models, it is frequently the only practical option. The more complex the problem, the greater can be the advantage provided by a Bayesian structure. And the more apparent becomes the inevitable subjectivity that is part of any thoughtful analysis. In this chapter I introduce the basic elements of a Bayesian analysis. I draw some additional comparisons with classical approaches when it comes to confidence envelopes in Chapter 5 and model assessment in Chapter 6. Chapter 7 takes up computational considerations.

4.1 The Bayesian Approach

As with many classical treatments the Bayesian approach is based on the likelihood. Bayes includes a second element, the *prior* density for parameters. The prior does two things for us. First, the prior makes it possible to calculate a posterior parameter density. Second, the prior provides an entry for information that is external to the data set(s) at hand.

In Box 4.1 I summarize Bayes' example of posterior probability. The experiment involved casting a second billiard ball repeatedly onto a pool table and inferring the location of a first ball based on the number of times the second ball landed to the right of the first ball. The location of the first ball is a random variable, in the sense that it is uncertain, having a distribution of its own. Before the experiment begins, the distribution for the first ball is uniform: it could land anywhere between the two sides of the table. This first density can be thought of as the prior, the description of the ball's location before the experiment. The distribution (our view of where the first ball lies) is changed by the experiment, because the data update our understanding of its location; the more times we throw the second ball, the greater is our knowledge of the first ball's location. Although the density assigned to the first ball is uniform *prior* to obtaining data, the density is beta (and more peaked) after obtaining data. The latter is the *posterior* distribution.

In a binomial experiment such as this, the parameter describing the probability of an event can be likened to Bayes' first billiard ball. Information from the prior and the likelihood are combined using Bayes' theorem to solve for the posterior density of the parameter conditioned on the data. The result is a posterior density

$$p(\theta|y) = \frac{p(y|\theta)p(\theta)}{p(y)}$$

$$= \frac{p(y|\theta)p(\theta)}{\int_{-\infty}^{\infty} p(y|\theta)p(\theta)d\theta} \qquad 4.1$$

(Box 4.1) or

$$\text{posterior} = \frac{\text{likelihood} \times \text{prior}}{\int\limits_{-\infty}^{\infty} (\text{likelihood} \times \text{prior})d\theta}$$

On the left-hand side is the posterior, the probability of the model (represented by parameter θ) given the data y. The numerator on the right-hand side is the product of the likelihood (the probability of data given the model) and the prior density of θ. The denominator is the probability of the data, which requires integration. Because the denominator is a constant (it does not depend on θ), we often work with a proportional form of Bayes' theorem,

$$p(\theta|y) \propto p(y|\theta)p(\theta)$$

The importance of the prior is now clear. Calculation of the posterior density requires it (Equation 4.1). And it provides a formal method to assimilate information. In Box 4.1, the prior was flat, because we expect that the table is flat. If we saw a book under one leg of the table, a prior that incorporates that information might have led to a more informative analysis.

These equations summarize a perspective that is basic and can be related with an anecdote on Bayesian logic. You submit the first paper from your thesis to the journal *Ecology*. You understand from others that the acceptance rate θ is 30 percent. This information is the basis for your prior on θ, which might be a density on (0,1) with most weight near 0.3. Your paper is accepted. This is the likelihood for observation y. What do you conclude? The sample size is $n = 1$. From this Bernouilli trial, you obtain data $y = 1$. Based on this observation, the estimate of acceptance rate is 100 percent. But you do not conclude that the acceptance rate is 100 percent. You might decide that the rate is higher than 30 percent. Or not. It depends on how much faith you place in the prior estimate of 30 percent and on how you view your observation. Alternatively, you might poll $n = 1000$ ecologists and find that 20 percent of the submitted papers had been accepted. Some of you who would not change your opinion based on $n = 1$ might change your opinion based on $n = 1000$.

In the foregoing example, there is a balance between evidence contributed by the experiment at hand and by evidence that is external to that experiment. That balance depends not only on the amount of evidence (e.g., n), but also on the perceived relevance and validity of the external evidence. Ecologists raised in the tradition of rejecting null hypotheses may find this subjective balance unsettling. Bayesians would point out that any scientist weighs the evidence from different sources, whether or not it is done as part of a formal analysis. Bayes provides a formal method that is reproducible and can accommodate a range of views.

Because we can have varying degrees of insight as to the prior density $p(\theta)$, there can be disagreement regarding its specification and the probability statements that arise from the Bayesian approach. In the past, some of the criticism leveled at the Bayesian framework focused on the seeming arbitrary selection of priors, perhaps for ease of analysis (*conjugacy* in Chapter 5), but not much else. Fair or not, modern computational methods covered in this book have largely neutralized this criticism; there is great flexibility in modern Bayes.

Box 4.1. Inverse probability from Reverend Thomas Bayes

Basic elements of joint probability are illustrated by the example from the Reverend Thomas Bayes, published two years after his death in 1763. The problem involves two random events. We wish to infer something about one event based on the outcome of the second. The analysis is relevant to the problem of inferring the value of a parameter (an unknown, and therefore, stochastic quantity) on the basis of sampling. This example illustrates concepts to which we will return many times. I follow Karl Pearson's (1920) use of the analogy to a billiard table. Suppose we have two billiard balls. We toss the first ball onto the table and allow it to settle. We then toss the second ball and determine whether it settles to the right or to the left of the first ball. Assume that we cannot see the first ball, but we somehow know whether the second ball lies to the right of it (Bayes made this up). By repeatedly tossing the second ball and recording right or left, we might develop insight into the location of the first ball. Our goal is to determine the location of the first ball. Our repeated tosses of ball 2 represent data, and the location of the first ball could be thought of as an underlying parameter. Bayes derived some basic probability rules that address this situation.

For simplicity assume that the width of the table spans the range (A, B) in Figure 4.1. The first ball settles at location o. Let θ be the relative distance that the first ball settles from the right side of the table. θ is ratio Ao/AB. Let y represent the number of times the second ball settles to the right of the first ball in n trials. At the outset, both θ and y must be treated as random (in the subjective probability sense). Represent their *joint probability* density as $p(\theta, y)$, the probability that the

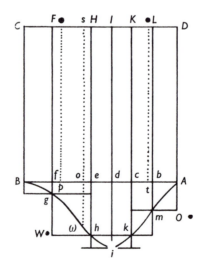

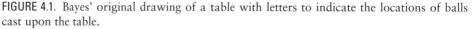

FIGURE 4.1. Bayes' original drawing of a table with letters to indicate the locations of balls cast upon the table.

first ball comes to rest on some interval on the table and the second ball settles to the right of the first. The probability density $p(\theta,y)$ is related to *conditional densities* using Equation D.22:

$$p(\theta,y) = p(\theta|y)p(y) = p(y|\theta)p(\theta)$$

The conditional density $p(\theta|y)$ is the density of θ given that we obtain a data set, represented by y. The conditional density $p(y|\theta)$ is the probability of a data set y given that θ assumes a particular value. Together these relations say that the joint density is the product of the density of one variable conditioned on the second having some value multiplied by the probability that the second variable assumes that value.

The *marginal density* of y can be obtained by knowing the marginal density of θ and the conditional density. By integrating over the effect of θ on we obtain the marginal density from Equation D.24,

$$p(y) = \int_0^1 p(y|\theta)p(\theta)d\theta$$

We now have a means for determining the location of the first ball, that is, estimating the parameter θ. Rearranging we can obtain Bayes' rule,

$$p(\theta|y) = \frac{p(\theta,y)}{p(y)} = \frac{p(y|\theta)p(\theta)}{\int_0^1 p(y|\theta)p(\theta)d\theta}$$

Thus, Bayes expressed the probability of θ conditional on the data y as a ratio of the joint probability of data and θ. By expressing the joint probability in terms of the conditional probabilities, he arrived at an expression involving things he could measure. We examine each of those things in turn.

First, consider the prior density of θ. This density is uniform, because the first ball settles anywhere on (A,B) with equal probability, which I equate here with $[0,1]$,

$$p(\theta) = 1 \quad 0 \leq \theta \leq 1$$

or

$$\theta \sim Unif(0,1)$$

Convince yourself that this is a uniform density on $(0,1)$ (i.e., it integrates to one) (Section F.3.2).

(Continued on following page)

There are two fates for the second ball: it can settle to the right of the first ball, with probability θ, or it can settle to the left, with probability $1-\theta$. The binomial applies here,

$$p(y|\theta) = \binom{n}{y}\theta^y(1-\theta)^{n-y}$$

This function is a likelihood. Substituting into Bayes' rule we have

$$p(\theta|y) = \frac{\binom{n}{y}\theta^y(1-\theta)^{n-y}\cdot 1}{\binom{n}{y}\int\limits_0^1 \theta^y(1-\theta)^{n-y}d\theta}$$

This looks complicated, but, with a few relationships from Box D.1, it collapses to something simple. Cancel the binomial coefficients, and recognize the integral as a beta function. We then have a beta density,

$$p(\theta|y) = \frac{\theta^y(1-\theta)^{n-y}}{B(y+1,n-y+1)}$$
$$= Beta(\theta|y+1,n-y+1)$$

This density describes the location of the first ball conditioned on the data set. We expect that the more data we collect, the better we should be able to locate the position of the first ball. The most probable location of the first ball coincides with the mode of this distribution where

$$\frac{d}{d\theta}p(\theta|y) = 0$$

which is, of course, y/n. Finding the probability that θ lies to the left of an arbitrary location (the distribution function of θ) is more troublesome, involving an incomplete beta integral. The problem eluded Bayes, but Laplace made progress using an expansion about the mode of the distribution (Stigler 1986). Figure 4.2 demonstrates how understanding of θ is refined with increasing sample size.

It is worth noting that, before the first ball is thrown, all outcomes (values of y from zero to n) are equally likely. This can be demonstrated with the marginal distribution of y,

$$p(y) = \int\limits_0^1 p(y|\theta)p(\theta)d\theta$$
$$= \int\limits_0^1 \binom{n}{y}\theta^y(1-\theta)^{n-y}\cdot 1 d\theta = \binom{n}{y}B(y+1,n-y+1)$$

Now rewrite the beta function in terms of gamma functions (Box D.1),

$$= \binom{n}{y} \frac{\Gamma(y+1)\Gamma(n-y+1)}{\Gamma(n+2)}$$

Recalling that $\Gamma(n + 2) = (n + 1)\Gamma(n + 1)$, and $\Gamma(n + 1) = n!$, this simplifies to

$$= \binom{n}{y} \frac{y!(n-y)!}{(n+1)n!} = \frac{\binom{n}{y}}{(n+1)\binom{n}{y}}$$

$$= \frac{1}{n+1}$$

Thus, despite the fact that we are tempted to view the data as a binomial process, it is only binomial when we assume that there is, indeed, some underlying, fixed probability θ that the event will occur. If instead, all values of θ are equally likely (which is true before the first ball is thrown), the process is not at all binomial, but uniform, with each of the $0, \ldots, n$ possible outcomes being equally likely.

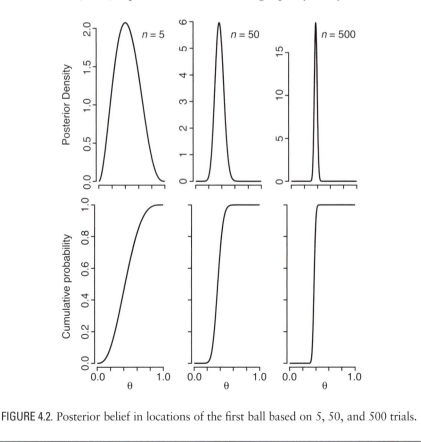

FIGURE 4.2. Posterior belief in locations of the first ball based on 5, 50, and 500 trials.

4.2 The Normal Distribution

I have changed notation for the likelihood from $L(y;\theta)$ in Chapter 3 to $p(y|\theta)$ in Section 4.1. These two different ways of referring to the likelihood do not reflect a change in how they are constructed. Rather, the change in notation reminds us that we treat the classical likelihood as a function of parameter(s) θ. The likelihood begins life as a density or a distribution. But in the classical framework, we immediately shift to this view of it as a function, which reminds us that there is no justification for integrating over θ. In the Bayesian framework, we never really forget this treatment of y and θ as jointly distributed random variables. The notation reflects a different view of the likelihood that comes from the fact that the estimate of θ is treated as a random variable, but only in the subjective probability sense. The stochastic representation of θ does not mean that θ can fluctuate, that is, assume different values at different times and places. It is assumed to have a constant fixed value, with stochasticity used to represent our degree of belief in the different values it might assume.

4.2.1 The Mean with Known Variance

As with classical statistics, the normal distribution plays a central role and provides a point of departure. It has two parameters, but there can be reasons to view one or the other as being known or fixed. For example, the variance might be fixed if we are using an instrument with known measurement error. In this case, we might condition on a fixed variance and draw inference on the mean of n measurements. Alternatively, the mean might be fixed. For example, the mean parameter of the Gaussian distribution could be viewed as the source location of particles that subsequently diffuse away from that source. In this case, inference focuses on the variance, which represents how far particles travel. A more general motivation for considering conditional relationships will follow when we take up more complicated models. I consider each of the possibilities in turn.

4.2.2 An Observation

The likelihood for an observation y for a normal sampling distribution is

$$p(y|\mu) \propto \exp\left[-\frac{(y - \mu)^2}{2\sigma^2}\right] \qquad 4.2$$

As before, I have excluded factors that do not contain the parameter of interest, μ. Assuming a fixed variance σ^2, I choose a prior density for the mean μ that is also a normal density,

$$p(\mu) = N(\mu|\mu_0, \tau^2)$$

$$\propto \exp\left[-\frac{(\mu - \mu_0)^2}{2\tau^2}\right] \qquad 4.3$$

This prior density has a mean of μ_0, which represents my best understanding of the mean before observing the data set at hand, and τ^2, a variance parameter that expresses prior confidence in the prior estimate μ_0. This two distributions, prior and likelihood, are drawn for specific values in Figure 4.2. Note that a small value of τ^2 would produce a prior distribution that is concentrated near the estimate μ_0. I would use such a prior in cases where I have a high degree of confidence that the mean is close to μ_0. This confidence would come from information external to the data y. The information from y enters through the likelihood. I do not want the same information to enter twice.

A large value of τ^2 would make the prior distribution essentially flat, in which case the prior estimate of the mean would have little impact on the analysis. For this reason, $\frac{1}{\tau^2}$ is often termed the *prior precision*, indicating the level of precision assigned to the prior estimate of μ. To see this, calculate the posterior for μ, which is proportional to

$$p(\mu|y) \propto p(y|\mu)p(\mu)$$

$$\propto \exp\left[-\frac{(y-\mu)^2}{2\sigma^2}\right]\exp\left[-\frac{(\mu-\mu_0)^2}{2\tau^2}\right]$$

$$= \exp\left[-\frac{1}{2}\left(\frac{(y-\mu)^2}{\sigma^2} + \frac{(\mu-\mu_0)^2}{\tau^2}\right)\right] \qquad 4.4$$

Here I simply multiplied likelihood and prior, both normal, dropping all coefficients that do not contain μ. Some unexciting algebra (e.g., Berger 1985, p. 127) leads to a posterior normal distribution. There are some shortcuts for finding parameter estimates for normal distributions that will be especially handy when we take up more complex models. Here I describe one.

Because this posterior distribution is normal, it will have the form

$$p(\mu|y) \propto \exp\left[-\frac{(\mu-V\nu)^2}{2V}\right], \qquad 4.5$$

where $V\nu$ and V are the posterior mean and variance, respectively. Check Equation 4.2 to convince yourself that $V\nu$ is the mean and V is the variance. I use $V\nu$ to represent the mean simply for convenience. This will be apparent in a moment.

To obtain the posterior distribution, we need to put Equation 4.5 in a form that allows us to identify ν and V. The method is termed *completing the square*. The algebra used to arrive at the full solution by this method is tedious and unenlightening. I call this shortcut *incompleting the square*. First, expand the exponent in Equation 4.5, ignoring for now the leading coefficient $-\frac{1}{2}$,

$$\frac{\mu^2 - 2\mu V\nu + V^2\nu^2}{V} = \mu^2 V^{-1} - 2\mu\nu + V\nu^2$$

The posterior variance is found in the coefficient of μ^2 in the first term. The coefficient ν occurs in the second term multiplied by -2μ. Said another way, the two coefficients V and v can be obtained from just the first two terms as

$$\mu^2 V^{-1} - 2\mu v + C \qquad\qquad 4.6$$

where C is a constant, because it does not contain μ. The goal is to write Equation 4.4 in the form of Equation 4.6. When we do this, we will have the posterior mean and variance.

Expand the two terms in Equation 4.4,

$$\frac{y^2 - 2\mu y + \mu^2}{\sigma^2} + \frac{\mu^2 - 2\mu\mu_0 + \mu_0^2}{\tau^2}$$

Because we want the normal distribution for μ, collect terms in μ,

$$\mu^2\left(\frac{1}{\sigma^2} + \frac{1}{\tau^2}\right) - 2\mu\left(\frac{y}{\sigma^2} + \frac{\mu_0}{\tau^2}\right) + \frac{y^2}{\sigma^2} + \frac{\mu_0^2}{\tau^2}.$$

Now identify the posterior variance from the first term,

$$V^{-1} = \frac{1}{\sigma^2} + \frac{1}{\tau^2} \qquad\qquad 4.7$$

The coefficient v is obtained from the second term,

$$v = \frac{y}{\sigma^2} + \frac{\mu_0}{\tau^2} \qquad\qquad 4.8$$

Thus, Equation 4.4 can be written as

$$N(\mu | V v, V) = N\left(\mu \left| \left(\frac{y}{\sigma^2} + \frac{\mu_0}{\tau^2}\right) \middle/ \left(\frac{1}{\sigma^2} + \frac{1}{\tau^2}\right), \left(\frac{1}{\sigma^2} + \frac{1}{\tau^2}\right)^{-1}\right.\right)$$

Note that the posterior mean is a weighted average of prior mean and data. The weights are the *precisions* of the two means, the precisions being the reciprocals of the variances. The posterior variance is the harmonic mean of the two variances. So the likelihood and prior are given equal weight. I call this *incompleting the square*, because I stopped as soon as I identified the coefficients V and v from the first two terms. From now on, I will incomplete the square.

4.2.3 A Data Set

For a data set of n observations the approach is the same. I continue to assume that σ^2 is fixed. The likelihood is now

$$p(y|\mu) \propto \prod_{i=1}^{n} \exp\left[-\frac{(y_i - \mu)^2}{2\sigma^2}\right] = \exp\left[-\frac{1}{2\sigma^2}\sum_{i=1}^{n}(y_i - \mu)^2\right]$$

and the posterior is proportional to

$$p(\mu|y) \propto \exp\left[-\frac{1}{2}\left(\frac{\sum_{i=1}^{n}(y-\mu)^2}{\sigma^2} + \frac{(\mu-\mu_0)^2}{\tau^2}\right)\right]$$

Incompleting the square, I obtain posterior parameter values that are weighted by the sample size:

$$\nu = \frac{n}{\sigma^2}\bar{y} + \frac{\mu_0}{\tau^2} \qquad\qquad 4.9$$

$$V^{-1} = \frac{n}{\sigma^2} + \frac{1}{\tau^2} \qquad\qquad 4.10$$

In other words, the larger the data set (large n), the greater the dominance of the data over the prior and the smaller the posterior variance. From Equations 4.9 and 4.10 come the asymptotic relationships (as n becomes large) for mean and variance,

$$\lim_{n\to\infty} V\nu \to \bar{y}$$

$$\lim_{n\to\infty} V\nu \to \frac{\sigma^2}{n}$$

Thus, with large sample size the posterior distribution of μ is approximated as

$$p(\mu|y, \sigma^2) \approx N\left(\mu\Big|\bar{y}, \frac{\sigma^2}{n}\right) \qquad\qquad 4.11$$

This is a normal distribution centered on the sample mean with a standard deviation equal to the classical standard error (Chapter 5). Again, the weight of the prior depends on the relative magnitude of terms in Equation 4.10 being equivalent to τ^{-2}. Prior and data will have similar weights when the prior precision is roughly equivalent to n/σ^2.

In the foregoing example, the likelihood and prior for μ were both Gaussian, and they led to a Gaussian posterior. Specific combinations of likelihood and prior that lead to a posterior having the same form as the prior are termed *conjugate* (Appendix G). I return to this topic shortly.

Example 4.1. The weight of the prior
The prior density for the example in Figure 4.3 is

$$p(\mu) = N(\mu|90, 2)$$

The prior precision is $\frac{1}{2} = 0.5$. The two data sets have sample size $n = 5$ and $n = 20$. In both cases, the sample means and variances are roughly 100 and 10, respectively. To assess the weight of the prior, I compare the prior precision of 0.5 to the sample size divided by the variance (n/σ^2 in Equation 4.11). In the case of $n = 5$ this ratio is approximately equal to the prior precision, both being about 0.5.

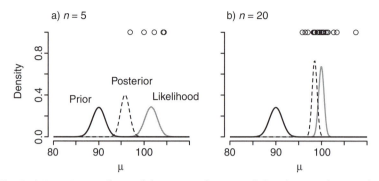

FIGURE. 4.3. A Bayesian analysis of the mean of a normal distribution showing how the posterior balance the input of the prior and the likelihood. Sample values (x_i) are shown as dots above. Curves are plotted against values of parameter μ.

This equal contribution from prior and likelihood explains why the posterior in Figure 4.3a splits the difference between the two. In Figure 4.3b, the ratio is $n/\sigma^2 = 2$. This places more weight on the likelihood than on the prior, because $0.2 > 0.05$. The posterior in Figure 4.3b shows this stronger influence of data. If the sample size were $n = 100$, we would find no discernible effect of the prior.

At this point one might ask whether the density 4.11 is of much use. The parameter μ lies to the left of the bar, and y and σ^2 are to the right of the bar. In other words this is the posterior density for the mean μ conditioned on a known variance—σ^2 is written to the right of the vertical bar on the left-hand side to indicate that it is known. This is consistent with the stated objective of determining the posterior for the mean, when the variance is known. Usually σ^2 will also be unknown, in which case we need to estimate both mean and variance from the data. In other words, it would be more useful to know the joint distribution of the two unknown parameter values, $p(\mu, \sigma^2|y)$, and the marginal posterior of the mean that integrates the uncertainty in σ^2. When the density for the variance is inverse gamma (following section), this integral happens to be a Student's t density,

$$p(\mu|y) = \int p(\mu, \sigma^2|y)d\sigma^2$$
$$= t_{n-1}\left(\mu \middle| \bar{y}, \frac{\sigma^2}{n-1}\right)$$

Often such integrals are not available. It turns out that sampling-based approaches of Chapter 8 for simulating a posterior distribution rely on simple conditional relationships, such as that for the mean in Equation 4.11. So this is an important tool that will be especially useful when we get to computation. Before proceeding further, I consider calculations for the variance conditioned on a known mean.

4.2.4 The Variance of a Normal Distribution

The mean of a normal distribution might be known, and we wish to determine the posterior variance. A prior commonly used for the variance is inverse gamma, because it is conjugate with the normal. This example of a prior inverse gamma density for the variance demonstrates this relationship. Here is the prior:

$$IG(\sigma^2|\alpha, \beta) \propto (\sigma^2)^{-(\alpha+1)}\exp\left[-\frac{\beta}{\sigma^2}\right]$$

Combined with the normal likelihood this yields

$$p(\sigma^2|y,\mu) \propto p(y|\mu,\sigma^2)p(\sigma^2)$$

$$= \overbrace{(\sigma^2)^{-n/2}\exp\left[-\frac{1}{2\sigma^2}\sum_{i=1}^{n}(y_i - \mu)^2\right]}^{likelihood} \times \overbrace{(\sigma^2)^{-(\alpha+1)}\exp\left[-\frac{\beta}{\sigma^2}\right]}^{prior}$$

Now collect coefficients containing σ^2 to obtain a new inverse gamma density,

$$= (\sigma^2)^{-(\alpha+n/2+1)}\exp\left[-\frac{1}{\sigma^2}\left(\beta + \frac{1}{2}\sum_{i=1}^{n}(y_i - \mu)^2\right)\right]$$

$$= IG\left(\sigma^2\middle|\alpha + n/2, \beta + 1/2\sum_{i=1}^{n}(y_i - \mu)^2\right)$$

Thus, conditioned on a specific value of the mean, the posterior variance is also inverse gamma. As before, the shape of the distribution depends on both the prior and the likelihood. The inverse gamma distribution is sharply peaked when its parameter values are large. Both parameter values are sums of a prior parameter and terms that increase with sample size. As seen in the previous example, the prior has impact when prior parameter values are small relative to the amount of data.

Example 4.2. Dispersal studies can involve a Gaussian dispersal kernel for the scatter of seeds about a parent plant (e.g., Clark et al. 1998). In this case, the mean of the distribution is known (at distance zero), and we seek to estimate the variance. A dispersal kernel is two-dimensional. For a normal dispersal kernel with no directional bias, we have

$$p(x,y|\sigma^2) = \frac{1}{2\pi\sigma^2}\exp\left[-\frac{1}{2\sigma^2}\sum_{i=1}^{n}(x_i^2 + y_i^2)\right]$$

Note that the normalization constant for this bivariate density has units of distance^{-2}. To simplify, replace the Cartesian coordinates with distance from source,

$$r^2 = x^2 + y^2$$

(see Clark et al. 1999b for details). The probablity that a seed travels distance r is obtained with an arcwise integration over arc angle ϕ,

$$p(r|\sigma^2) = \frac{1}{2\pi\sigma^2} \oint_{2\pi} \exp\left[-\frac{r^2}{2\sigma^2}\right] d\phi = \frac{r}{\sigma^2} \exp\left[-\frac{r}{2\sigma^2}\right]$$

If you are unfamiliar with this type of integration, it can be found in a standard calculus text where polar coordinates are discussed. It is not important for you to understand polar coordinates (or arcwise integration) for this example—it is solved for you in the following equation. For n released seeds the likelihood, where x and y have been replaced with distance r, is

$$p(\mathbf{r}|\sigma^2) = \prod_{i=1}^{n} \frac{r_i}{\sigma^2} \exp\left[-\frac{r_i^2}{2\sigma^2}\right]$$

You should convince yourself that the maximum likelihood estimate for the dispersal parameter is

$$\sigma^2_{ML} = \frac{\overline{r^2}}{2}$$

that is, half the mean squared dispersal distance. The posterior estimate of the dispersal variance with an inverse gamma prior is

$$p(\sigma^2|\mathbf{r}) \propto p(\mathbf{r}|\sigma^2)p(\sigma^2)$$

$$\propto (\sigma^2)^{-n} \exp\left[-\frac{1}{2\sigma^2}\sum_{i=1}^{n} r_i^2\right](\sigma^2)^{-(\alpha+1)} \exp\left[-\frac{\beta}{\sigma^2}\right]$$

$$\propto (\sigma^2)^{-(\alpha+n+1)} \exp\left[-\frac{1}{\sigma^2}\left(\beta + \frac{1}{2}\sum_{i=1}^{n} r_i^2\right)\right]$$

$$= IG\left(\sigma^2\bigg|\alpha + n, \beta + n\frac{\overline{r^2}}{2}\right)$$

This posterior has the most probable value (mode) at

$$\frac{\beta + n\dfrac{\overline{r^2}}{2}}{\alpha + n + 1}$$

which tends to the ML estimate with large n.

Results involving the normal distributions will arise repeatedly throughout this book.

4.3 Subjective Probability and the Role of the Prior

This is a good time to clarify two broad concepts of probability that contribute to the differences between classical and Bayes. These differences have been implicit throughout this chapter, and they are explicit in the following chapter on confidence and credible intervals.

In the frequentist view, probability statements imply the frequency of outcomes for repeated experiments. In Chapter 5 I discuss a classical confidence interval—a 95 percent confidence interval is expected to bound the true parameter value in 95 percent of repeated experiments. In other words, if we could repeat the experiment 100 times, the true value should lie within about 95 percent of the estimated confidence intervals.

In the Bayesian framework, we use the notion of *subjective probability*. Consider the prediction that an asteroid may strike the earth in your lifetime with probability P. Your lifetime will not occur 100 times. It will occur once. The event will occur, or not. Here the concept of probability in a frequency sense does not apply. Yet we use the concept of subjective probability on a regular basis. A subjective probability can be used to communicate uncertainty about an event, even if it will not be repeated. It integrates our knowledge about the process, our confidence in those making the prediction, and so forth. We have more confidence if the prediction comes from an astronomer than we would if it came from an astrologist. We do not blindly accept the information, but rather combine it with previous knowledge.

A Bayesian analysis provides a simple structure for updating prior knowledge conditioned on new information. The posterior is this update. If we have a large amount of information that is external to the data at hand, the prior might have substantial weight. Different people have different priors—not all individuals are privy to the same information, and not all would weight it in the same way. The value of the Bayesian structure comes with the capacity to admit those differences. An undesirable alternative would be to introduce them post hoc in some informal fashion.

Noninformative priors are used when the investigator knows nothing or when it is deemed desirable to isolate the contribution from a specific data set. You may encounter some disagreement here. A strict subjectivist might say that one always knows something or they would not be analyzing the problem in the first place. Moreover, a case can be made that substantial effort is needed to ensure that the prior density accurately reflects current knowledge for all values of a parameter. I suggest pragmatism (Chapter 1). If the focus is a specific data set, the prior can usually be made inconsequential (noninformative). If the prior is believed to have more influence than desired, see what happens when it is made flatter. If specific ranges of a parameter value and especially critical for inference, prediction, or decision, focus attention on getting the prior right for those values.

In practice, one frequently encounters analyses that use *improper* priors, so flat that they cannot be integrated. In many cases this is fine, because a posterior based on such priors often *can* be integrated. But it requires some care to ensure that the posterior is proper. In other words, the posterior must have a finite integral (it is normalized to be a proper density). Sometimes the prior is flat, but proper. This prior is uniform on an interval $a < \theta < b$ with density $(b - a)^{-1}$, being noninformative over a range of values, yet proper. Again, care must be taken with such vague priors to ensure that the posterior is not highly sensitive to the arbitrary limits placed on the prior. If the interval (a,b) is chosen to simply cover the likelihood, then one is effectively using the data twice. This concern can likewise apply to nonuniform vague priors (e.g., Berger et al. 2001). When a model is hierarchical (Chapter 9), it may not be possible to apply a noninformative prior. For example, a variance on random effects may not converge unless some information enters through the prior. In general, it is good to be proper, because, with large models, there are fewer chances for identifiability problems (Chapter 7). Yet, for hierarchical models, these priors must have some rationale—some of them will affect the posterior.

A posterior may be *robust* to different priors. The sensitivity of the posterior to the prior assumptions can be explored by changing them. We often want to know how the prior and likelihood contribute to the posterior. The prior and likelihood might be compared with the posterior to determine if they strongly disagree. If prior and posterior are well displaced from one another, it might be worth rethinking the problem—it is possible that the process does not operate as anticipated, that prior and data apply to different situations, or both.

If this seems uncomfortably subjective, consider the alternative. In practice, when it is difficult to draw inference on multiple parameters, parameter values are usually fixed. When this is done, there is no possibility of comparing prior and likelihood. The transparent subjectivity of Bayes can be unsettling at first, because, in a classical setting, many aspects of data modeling that need careful thought are not apparent to the investigator, fostering the "illusion of objectivity" (Berger and Berry 1988). The following chapter provides concrete examples to illustrate the role of the prior and how classical and Bayes approaches compare.

5 Confidence Envelopes and Prediction Intervals

During the 1980s, concerns over the viability of northern spotted owl (NSO) *(Strix occidentalis)* populations in the Pacific Northwest focused attention, controversy, and even litigation on confidence intervals. NSO populations are sparsely distributed in and around old-growth Douglas fir (*Pseudotsuga menziesii*) forests from southern British Columbia through northern California (Gutierrez et al. 1985). Widespread logging led to declining populations, a fact that did not escape notice of those seeking to preserve old-growth forests. The Forest Service and other agencies are required to protect the owl. If it were determined that populations were not replacing themselves ($\lambda < 1$), then a case could be made to curtail additional logging. Industry and conservation interests represent competing pressures to avoid over- or underestimating the area needed to protect owls (Harrison et al. 1993). High stakes made for scientific scrutiny of model selection, data quantity and quality, inference, and decision (Swedlow 2003).

Ecologists prepared testimony on the minimal amount of old-growth forest needed to minimize extinction risk. Much of the effort involved estimates of the growth rate parameter λ (Chapter 2), which can be calculated from demographic data on fecundity, growth, and mortality (Section E.2). Populations are structured in various ways, and heterogeneity associated with age, size, and location affects dynamics. Various models make accommodation for areas lost to logging (Lande 1987, 1988; Anderson and Burnham 1990; Lamberson et al. 1992). Some are framed as simple demographic models, others as matrixes (Equation 2.8), and still others as spatially explicit landscape simulations with varying degrees of complexity (Doak 1989; Lande 1988; McKelvey et al. 1993). Variability within populations and sampling considerations can make it difficult to obtain sufficient data on all life history stages (Franklin et al. 2000; Cam et al. 2002; Clark et al. 2005). Not surprisingly, parameter estimates can be uncertain, and they vary from place to place and from study to study. Calculations of population growth rate should reflect the uncertainty in parameter estimates that make up a matrix such as Equation 2.8.

Figure 5.1 depicts some of the different estimates of population growth rate λ—drawn using estimates of the growth rate estimates and standard errors—from

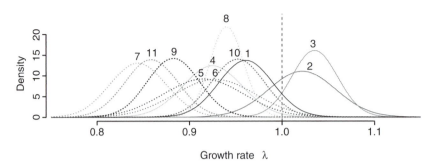

FIGURE 5.1. Estimates of population growth rate with propagated parameter error for NSO from a number of published studies. Values of $\lambda < 1$ imply population decline (from Clark 2003, orginal sources cited there in).

studies available at the time of the NSO litigation and some more recent. As with any type of estimate, the analyses represented here start with an assumption that there is a true value that can never be known precisely. The spread of each density reflects underlying uncertainty in demographic parameter estimates. We can learn more about the true growth rate by obtaining more observations. Increasing the sample size has the effect of making this density more peaked, thus narrowing the range of plausible estimates. From this figure it is easy to see why it would be difficult to say whether or not the population was in decline. The degree of uncertainty suggested by these different estimates is not unusual for environmental problems. In the case of NSO there were high economic stakes that prompted extensive and long-term efforts to estimate demographic rates from field data. Such motivation and investment is unusual.

In this chapter I discuss how estimates of uncertainty are derived and what they stand for. Although scientists can disagree about hypothesis testing and use of prior distributions (Chapter 1), there is general agreement on the need for some type of confidence envelope. There are different philosophies that underlie the confidence envelopes that derive from classical and Bayesian methods, and it is important to recognize the assumptions behind the confidence interval that is reported. For example, although densities in Figure 5.1 are based on parameter estimates obtained from classical analyses, my depiction of them is intuitive and Bayesian, and not really consistent with assumptions behind the analyses themselves. Moreover, Figure 5.1 might seem at odds with the statement that increased sampling increases confidence in the estimate. Rather than building support for a particular estimate of λ, we might say that repeated studies of NSO produced more, not better, estimates of λ; they do not seem to be in agreement. The eleven distributions differ because they come from different data sets (e.g., on populations from different locations, times, on one or both genders), some are based on more data than others, and some are based on different models. These differences are inevitable, because in nature (in contrast to a tightly controlled experiment), we are never resampling from the same idealized population.

All of the approaches discussed here begin with the likelihood function. In a classical framework, confidence intervals are obtained by extending concepts introduced in Chapter 3. In the Bayesian framework, the credible interval comes directly from the posterior density of Chapter 4. In Sections 5.1 and 5.2, I introduce concepts and calculations behind classical and Bayesian confidence envelopes, respectively, using examples based on familiar sampling distributions. I then consider multiple parameters (Section 5.3), with regression as a specific example (Section 5.4). Subsequent sections include predictive intervals and discussion of how parameter error affects model inference in a broad sense. This background leads to model assessment in Chapter 6.

5.1 Classical Interval Estimation

In Chapter 3 the maximum likelihood estimate (MLE) of a parameter was defined as the value finding most support from the data. But how much better is the MLE than some other value? The MLE for a rate parameter ρ estimated from a single observation (death at ten days) did not inspire confidence, because $n = 1$ (Section 3.2). As data accumulate, confidence in the MLE should increase. The shape of the likelihood function contains additional information about a parameter estimate (Fisher 1959); increasing curvature in $-\ln L$ with accumulation of data describes increasing confidence in the ML estimate (Figure 3.1).

A *frequentist confidence interval* is interpreted as the fraction of intervals calculated from a large number of data sets generated by the same process that would include the true parameter value. The true (but unknown) parameter value is viewed as being fixed and the data as being random. The term *frequentist confidence interval* is used to distinguish this interval, which is based on the idea of coverage, from a Bayesian confidence interval or credible interval, which comes from integrating a likelihood times prior as though it were a parameter density (Chapter 4). In a frequentist framework we do not speak of the probability that the parameter assumes a particular value, because the parameter is assumed to be a fixed constant, and degree of belief (the subjective probability of Bayes—Section 4.3) is not the basis for its calculation. Because the data are viewed as random, the confidence interval is random—like the point estimate, the confidence interval is calculated from random data.

Recall from Chapter 4 that, although Bayesians speak of the parameter as random, both of these approaches assume some underlying true value for the parameter that is fixed, and both produce estimates that are, in different senses, random. The Bayesian approach uses a probability model to quantify uncertainty. The frequentist approach calculates an estimate from data that are viewed as random. Again, these concepts of randomness apply to the *estimate*, not to the parameter itself. In both cases, the confidence limits represent uncertainty as to the true value. In this sense, Bayesian and classical approaches are not as different as is often portrayed in the ecological literature. But depiction of estimates as densities in Figure 5.1 is clearly a Bayesian view. In Section 5.2, I compare different types of confidence envelopes—a term I use to include both confidence and credible intervals—but focus first on a classical notion of a confidence interval.

5.1.1 A Confidence Interval Learned in Statistics Class

A brief example outlines the frequentist notion of a confidence interval. Consider a random variate x_i drawn from a normal distribution,

$$x_i \sim N(\mu, \sigma^2)$$

with parameters μ and σ^2. For the moment, suppose the variance parameter σ^2 is known, as would be the case if x_i were a measurement obtained on an instrument used previously. Based on this one observation, uncertainty as to the true value of μ is represented by the density drawn in Figure 5.2a. This density is centered on the sample value x_i, because x_i is the best (ML) estimate of μ (recall Example 3.1). To see this, consider the normal density centered on the value μ,

$$p(x_i | \mu, \sigma^2) = \frac{1}{\sqrt{2\pi}\sigma} \exp\left[-\frac{(x_i - \mu)^2}{2\sigma^2} \right] \qquad 5.1$$

The "most likely" value of μ is that which yields the highest probability of the data, that is, the ML estimate μ_{ML}. To demonstrate that x_i is the ML estimate of μ, recall the log likelihood (including only terms involving μ),

$$\ln L(x_i | \mu, \sigma^2) \propto - (x_i - \mu)^2 / 2\sigma^2$$

(Section 3.5). Differentiating, setting the derivative equal to zero, and solving for μ gives the value μ_{ML}, which is equal to x_i. Because x_i is the only observation, a density centered at x_i is more likely than is a density centered on any other value of x.

To construct a confidence interval for, say, $\alpha = 0.1$ (i.e., a 90 percent confidence interval) draw two other densities, one for the upper confidence limit μ_l (Figure 5.2b) and one for the lower μ_u (Figure 5.2c). The upper confidence limit comes from a density that has an area to the right of the estimate $\mu_{ML} = x_i$ equal to $\alpha/2$ or 0.05 (Figure 5.2b). The mean of this density is the lower confidence limit μ_l, which is selected such that the density $N(\mu_l, \sigma^2)$ has area $\alpha/2 = $ of 0.05 to the right of the estimate $\mu_{ML} = x_i$,

$$\int_{x_i}^{\infty} N(x; | \mu_l, \sigma^2) dx = \alpha/2$$

The lower confidence limit comes from a second density having an area of 0.05 to the left of x_i, that is, $N(\mu_u, \sigma^2)$,

$$\int_{-\infty}^{x_i} N(x | \mu_u, \sigma^2) dx = \alpha/2$$

(Figure 5.2c). The interpretation reflects the view that the confidence interval is random and μ is fixed. With each new sample there is a new estimate for the

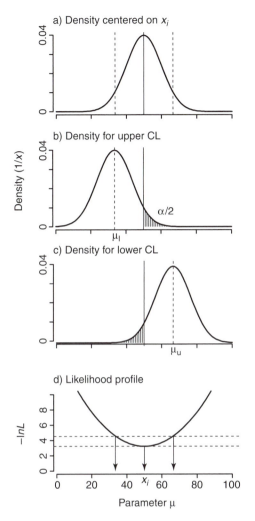

FIGURE 5.2. The relationships among densities used to construct confidence intervals and the likelihood function. The solid vertical line indicates MLE. The dashed lines in *d* denote the confidence interval, the upper line being 1.31 units above the lower line (Section 5.1.3).

confidence interval. The concept of *coverage* means that $100(1 - \alpha)\%$ of confidence intervals calculated from similar experiments should contain μ.

The foregoing method follows a classical definition of a confidence interval. But it is not the way we usually construct confidence intervals, and it is not the way we usually think about them. Instead of two distributions in Figure 5.2b and 5.2c, we tend to think of one distribution centered on the best estimate, with the confidence interval defined by the areas in the two tails (Figure 5.2a). In fact, a strict frequentist viewpoint does not permit this approach, because, except in the normal case, it does not yield the classical confidence interval. Moreover, nothing in the frequentist setting justifies integrating a likelihood function, because it is not a density. We continue to think of the problem in terms of this density centered on the MLE, because (1) for the normal case, the integral of the tails in Figure 5.2a yields the same answer as that obtained in Figure 5.2b, c; (2) for non-normal cases having large sample sizes, it converges to the same answer (there are some qualifications to this); and (3) because of (1) and (2), our statistical training is dominated by normal sampling distributions. But this method is not correct for asymmetric sampling distributions with small sample sizes, as shown in Box 5.1.

The classical confidence interval can be calculated or approximated in several ways. I introduce the topic with a review of the standard error. Three methods for constructing confidence intervals that follow all involve the likelihood function. The first technique, the *likelihood profile*, focuses on likelihood shape. I follow with *Fisher Information*, which summarizes that shape in terms of its curvature at the MLE. Next is a numerical method, the *bootstrap*, which yields a frequency distribution of estimates that has the shape of a likelihood function. All of the methods approximate the frequentist idea of coverage (the true parameter value lies within a fraction $1 - \alpha$ of the confidence intervals that would be constructed from similar experiments). These methods only *seem* akin to the intuitive notion of a distribution centered on the MLE itself (Figure 5.2a), with the confidence interval determined by integrating the tails. A fourth method, integrating the likelihood, only makes sense in the context of Bayes' theorem, assuming that the prior is so flat as to be ignored. The concept illustrated by Figure 5.1 is Bayesian in this regard. Last, I describe a Bayesian *credible interval*, which comes from mixing the likelihood with the prior in Section 5.2.

Box 5.1. Confidence intervals for the Poisson

The confidence intervals for the Poisson parameter θ are not symmetric about the MLE. This example is adapted from Cousins (1995). First, just a few remarks about the Poisson distribution. The Poisson distribution is used to describe the number of events to occur in a prescribed area (a spatial Poisson process) or during an interval of time. Unlike the binomial, which describes the number of events to occur from a total of n potential, the Poisson has no such upper bound. We use a Poisson in the case where n is so large and the probability of any one event is so low that n can be ignored.

The probability of y events is described by the Poisson distribution

$$p(y|\theta) = Pois(y|\theta) = \frac{\theta^y e^{-\theta}}{y!}$$

where the mean number of events is the parameter θ.

White and Bennett (1997) used the Poisson to describe samples of birds (Example 3.3). Consider an observation of $y = 2$ birds. The MLE of the Poisson parameter is $\theta_{\mathrm{ML}} = 2$ (Figure 5.2a). A classical confidence interval is defined by the two values of θ between which the true value of θ is expected to lie in $100(1 - \alpha)$ percent of all

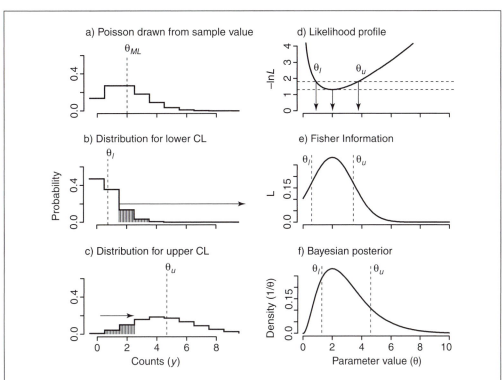

FIGURE 5.3. Confidence intervals for the estimate of a Poisson mean parameter θ.

trials. The lower limit is found by locating the distribution that has area $\alpha/2$ at and above $\theta_{ML} = 2$. The mean of this distribution is the lower confidence interval, θ_l

$$\sum_{j=y}^{\infty} p(j|\theta_l) = \alpha/2$$

where $y = 2$. This summation is highlighted in Figure 5.2b. The upper limit is the mean of the Poisson distribution having area up to and including θ_{ML},

$$\sum_{j=0}^{y} p(j|\theta_u) = \alpha/2$$

(Figure 5.2c). Note that $y = 2$ is part of the summation for both lower and upper confidence intervals. The 68 percent confidence interval shown in Figure 5.2 is (0.755, 4.67).

Because y is discrete and θ is continuous, there is no clear analogy that could tie this Poisson example to the normal case depicted in Figure 5.1. Because of its discrete nature, we generally do not think about finding an area $\frac{\alpha}{2}$ in one of the tails of Figure 5.2a.

(*Continued on following page*)

TABLE 5.1.
Confidence Intervals for the Poisson Distribution at One Standard Error from the Point Estimate 68 percent

Quantiles	$\theta_l = 0.16$ $\theta_l,\ 16\%$	$\theta_u = 0.84$ $\theta_u,\ 84\%$
Classical	0.72	4.62
Likelihood profile	0.91	3.75
Fisher Information	0.59	3.41
Bayesian Posterior (flat prior)	1.37	4.62

We shift our attention from the distribution function of y to the likelihood function (a function of θ)—the likelihood is a function of θ rather than y. This happens to be equivalent to the gamma density,

$$Pois(y|\theta) = Gam(\theta|y + 1,1) = \frac{\theta^y e^{-\theta}}{\Gamma(y + 1)} = \frac{\theta^y e^{-\theta}}{y!}$$

(see Appendix F for the gamma density). Some quantiles from this density are given in Table 5.1.

From Table 5.1 it is evident that, unlike the Gaussian case, the integrated likelihood function (designated "Bayesian posterior (flat prior)" in Table 5.1) yields a different confidence interval from the classical one. The upper limits happen to be the same due to an unusual quality of the Poisson distribution, namely,

$$\int_{\theta_u}^{\infty} Gam(\theta|y + 1,1)d\theta = \sum_{j=0}^{y} Pois(j|\theta_u)$$

(The finite series on the right-hand side is used to calculate the incomplete gamma function, which arises on the left-hand side; Gradshteyn and Ryzhik 1980.) The lower confidence intervals are quite different; although 68 percent of classical confidence intervals would admit a true value of $\theta = 1$ (the classical $\theta_l = 0.72$), this value falls well below the confidence interval obtained from the integrated likelihood (Bayesian posterior).

The remaining confidence envelopes listed in Table 5.1 are calculated by methods we discuss shortly. Here I simply point out that each is different, and it is important to understand the differences.

5.1.2 The Standard Error as Basis for a Confidence Interval

The standard error expresses uncertainty in an estimate. This example concerns the standard error for a mean parameter of a normal distribution μ, which is estimated by taking the average over n samples, that is, the sample mean

$$\bar{x} = \frac{1}{n}\sum_{i=1}^{n} x_i \qquad\qquad 5.2$$

This is the maximum likelihood estimate of μ (Example 3.1). Confidence in this estimate increases with sample size n. The standard error summarizes support from the data for a particular value. It is the square root of the variance divided by the sample size,

$$se_\mu = \sqrt{\frac{\text{var}[x]}{n}}$$

5.3

where

$$\text{var}[x] = \frac{1}{n-1}\sum_{i=1}^{n}(x_i - \bar{x})^2$$

5.4

is the sample variance and the estimate of the true variance. (The $n - 1$ in the denominator corrects for bias in the variance estimate for small n.) This is the standard error of the estimate of μ. From the formula for the standard error it is clear that the width of the confidence interval must decrease with increasing sample size.

Recall from introductory statistics that 68 percent of the intervals spanning one standard error on either side of the mean estimate are expected to contain the true value. Ninety-five percent of the intervals spanning 1.96 standard errors of the mean estimate should contain the true value. This is equivalent to saying that the frequency distribution of parameter estimates we would obtain from repeating the experiment can be approximated by a normal distribution with mean $\bar{x}$ and standard deviation equal to the standard error,

$$\mu \sim N(\bar{x}, se_{\bar{x}}^2)$$

Another way of writing this is

$$\frac{\mu - \bar{x}}{se_\mu} \sim N(0,1)$$

which is shifted to have mean zero and rescaled in units of standard errors.

Figure 5.4a shows a random sample of $n = 5$ observations drawn from the normal density with mean μ and variance σ^2. The sample mean $\bar{x}$ lies to the right of the true population mean μ, but not by much. So with just five observations we have a reasonable estimate of the mean. Ninety-five percent of the confidence intervals obtained from samples of size $n = 5$, calculated as 1.96 standard errors of sample means, should contain the true estimate μ. Figure 5.4c shows the two densities used to estimate confidence intervals. Because the likelihood is Gaussian, these two densities yield the same answer we would obtain if we simply integrated the tails of the density $N(\bar{x}, se_{\bar{x}}^2)$ in Figure 5.4b, that is, one centered on the sample estimate $\bar{x}$. A sample of size $n = 50$ results in an estimate of μ that is much closer to the true value (Figure 5.4d). The increased information available from a sample of size $n = 50$ over one of size $n = 5$ is reflected in a narrow 95 percent confidence interval (Figure 5.4f).

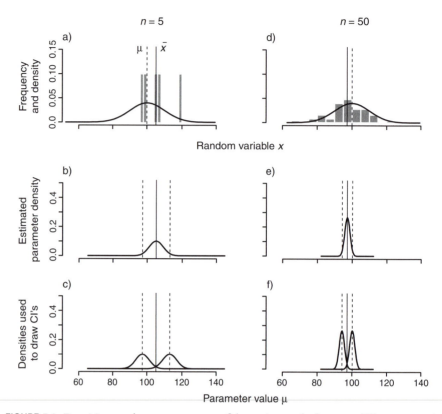

FIGURE 5.4. Densities used to construct confidence intervals for two different sample sizes. Vertical solid lines represent sample means. The vertical dashed lines in *a* and *d* show the parameter μ used to generate data. Vertical dashed lines in *b*, *c*, *e*, and *f* bound 95 percent confidence intervals.

Figure 5.4 further illustrates the random character of a classical confidence interval. Any given sample from the same population would produce a different confidence interval, referenced by the sample estimate $\bar{x}$ and its standard error. Thus, a confidence interval can be calculated from a standard error. When the likelihood is not normal, the confidence interval is approximate (Section 5.1.4).

5.1.3 Likelihood Profile

As discussed in the previous section, confidence intervals are the basis for probability statements about parameter estimates. An absolute probability cannot be assigned to a particular parameter estimate or to a model. We can think of the probability of a particular value of a parameter only in the context of some other value or model. The likelihood function is used for such comparisons. Confidence intervals can be derived using methods that involve ratios of likelihoods using a Likelihood Ratio Test (LRT). This approach is also used for model selection in Chapter 6.

Let the vector $\mathbf{x} = [x_1, \ldots, x_n]$ represent n observations and θ be one or more (a vector of) parameters about which we wish to draw inference. Ideas about θ can take the form of hypotheses (Chapter 1). Because a classical approach does not yield a direct probability statement about θ, hypothesis testing is evolutionary, through progressive competitions with alternative views about θ. By eliminating the alternatives, one at a time, the idea is that confidence will accumulate in support of a particular hypothesis. A probability statement is made in the context of two competing hypotheses, for example, a "null" (say, $\theta = \theta_0$) and an alternative, such as $\theta \neq \theta_0$ or $\theta = \theta_1$ (Chapter 1). Standard tests arise in the context of classical methods for relating likelihoods of two hypotheses based on the view that both values of θ are fixed and the data are random. The probability involves two likelihoods, both of which assume the respective hypothesis to be true. I begin with a sample mean.

Suppose we wish to make a probability statement about the value of μ finding most support from a data set, that is, μ_{ML}. The alternative hypothesis concerns a rival value of μ, call it μ_0. The test statistic for this comparison involves a likelihood ratio (LR),

$$LR = \frac{L(\mathbf{x};\mu_0)}{L(\mathbf{x};\mu_{ML})}$$

This is the ratio of two likelihoods taken over the same data $\mathbf{x}$, but assuming different values of μ. The likelihood ratio can be evaluated for any two values of μ. When the denominator is taken at the MLE (as in this case), the LR is termed the *normed likelihood function* (Lindsey 1999),

$$R(\mu_0;\mathbf{x}) \equiv \frac{L(\mathbf{x};\mu_0)}{L(\mathbf{x};\mu_{ML})}, \qquad 5.5$$

and has range [0, 1]. The test statistic is sometimes called a *deviance*,

$$D = -2\ln R$$

and is distributed as χ^2 with, in this case, one degree of freedom, because there is only one parameter at issue.

Consider again the likelihood for the mean of the normal distribution. From Equation 5.5 write the likelihood ratio

$$R(\mu_0;\mathbf{x}) = \frac{L(\mathbf{x};\mu_0)}{L(\mathbf{x};\mu_{ML})} = \frac{(2\pi)^{-\pi/2}\sigma_0^{-n}\exp\left[-\frac{1}{2\sigma_0^2}\sum_{i=1}^{n}(x_i - \mu_0)^2\right]}{(2\pi)^{-\pi/2}\sigma_{ML}^{-n}\exp\left[-\frac{1}{2\sigma_{ML}^2}\sum_{i=1}^{n}(x_i - \mu_{ML})^2\right]} \qquad 5.6$$

This simplifies with substitution for the two values of σ that would be estimated for each value of μ. From Section 3.5, the MLE for σ^2 is

$$\sigma_{ML}^2 = \frac{1}{n}\sum_{i=1}^{n}(x_i - \mu_{ML})^2$$

that is, the mean squared deviation from the MLE of the mean. For the alternative value of μ, the estimate of σ^2 is

$$\sigma_0^2 = \frac{1}{n}\sum_{i=1}^{n}(x_i - \mu_0)^2$$

Then the summations in both exponents of Equation 5.6 are equal to $-n\sigma^2$ for the two values of σ Plugging in these two values gives the normed likelihood

$$R(\mu_0;x) = \frac{L(x;\mu_0)}{L(x;\mu_{ML})} = \frac{\sigma_0^{-n}\exp\left[\dfrac{n\sigma_0^2}{2\sigma_0^2}\right]}{\sigma_{ML}^{-n}\exp\left[-\dfrac{n\sigma_{ML}^2}{2\sigma_{ML}^2}\right]} = \left(\frac{\sigma_{ML}}{\sigma_0}\right)^n \qquad 5.7$$

and deviance

$$D = -2\ln R = 2n(\ln\sigma_0 - \ln\sigma_{ML})$$

Now, if μ_0 is taken to be the value of μ where the ln likelihood function is $\frac{1}{2}$ unit below the value at the maximum, then $D = 1$, and the probability of D (χ^2 with 1 df) is 0.68. Confidence intervals of 90 percent and 95 percent have $-\ln L$ values of 1.31 and 1.92 below the ML, respectively. These quantiles can be obtained from standard software packages. In other words, the likelihood ratio test provides a confidence interval that is identical to that found by the integral method in Figure 5.2b and 5.2c. Figure 5.2d shows $-\ln L$ with a 90 percent confidence interval drawn as a dashed line 1.31 above the minimum. Because the sampling distribution is normal, the log likelihood (Equation 5.1) is a parabola and symmetric about the minimum at μ_{ML}.

Example 5.1. The MLE for exponential parameter ρ in the example from Section 3.2 was $\rho_{ML} = 0.083$ day^{-1} (for a sample of mortality for ages 10 and 14 days). Use the likelihood ratio and deviance to determine the confidence interval.

For the mortality example of Section 3.2, the MLE for ρ was

$$\rho_{ML} = \frac{1}{\bar{a}}$$

The likelihood ratio (normed likelihood) is

$$R(\rho_0;a) = \frac{L(a;\rho_0)}{L(a;\rho_{ML})} = \frac{\rho_0^n\exp[-\rho_0 n\bar{a}]}{\rho_{ML}^n\exp[-\rho_{ML}n\bar{a}]}$$

Substituting for the ML estimate and rearranging gives the normed likelihood

$$= (\rho_0\bar{a})^n\exp[n(1 - \rho_0\bar{a})]$$

and the deviance

$$D = -2\ln[R(\rho_0;\mathbf{a})] = 2n[\rho_0\bar{a} - \ln(\rho_0\bar{a})-1]$$

(Figure 5.5). For this example, $n = 2$, $D = 2.12$, and the P value for D (from χ^2 with 1 df) is 0.145.

Unlike the normal example, the confidence interval for the exponential Example 5.1 is not symmetric (Figure 5.5). The relationship between likelihood and deviance is apparent from a plot of the normed likelihood function R, where the likelihood in the denominator is taken to be the MLE (Figure 5.5a). The normed likelihood thus has a maximum value of one at the MLE and declines on either side. At an α level of, say 0.05, the confidence interval can be constructed by finding the values of D that yield $(1 - \alpha) = 0.95$ probability from the χ^2 test. At the MLE, $R = 1$, $D = 0$, and $P = 1$, because the

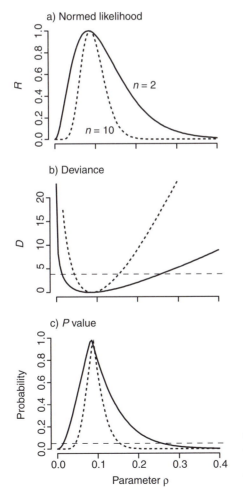

FIGURE 5.5. Likelihood ratio test for the exponential example with two sample sizes (Example 5.1). The horizontal dashed lines in b and c intersect curves at the 95 percent confidence interval.

ratio is taken for two equivalent values finding equal support from the data. The profile itself is constructed by successively calculating R for a range of parameter values ρ_0 in the neighborhood of the MLE. Deviances increase on either side of the MLE, whereas the associated P values decline. The normed likelihood shows that, for $n = 2$ (solid curve in Figure 5.5a), ρ values < 0.03 and > 0.15 are less than half as well supported by the data as the MLE at $\rho = 0.083$. For $\alpha = 0.05$, we reject the hypothesis that ρ lies outside the interval bounded by the horizontal dashed line at $P = 0.05$ (Figure 5.5c) or by deviances that lie above 3.84 (Figure 5.5b), the χ^2 value that yields $P = 0.05$ with one degree of freedom.

The likelihood profile is based on the distance below the maximum likelihood, and not on integrating the likelihood (recall that integrating the likelihood requires a Bayesian view). The likelihood ratio provides an economical means for comparing parameter values, because all coefficients that do not contribute insight about a specific comparison drop out. The normal example collapsed to a ratio of squared deviations, because other coefficients were redundant between the two models being compared. We can ignore some coefficients provided the comparison involves the same functional form. Model comparisons that involve different functional forms can still be done using likelihoods, but all coefficients must be retained (Chapter 6).

5.1.4 Approximate Confidence Intervals from Fisher Information

The likelihood profile makes use of the likelihood function for the information it provides about a parameter. The broad likelihood function in Figure 3.1a indicates that the ML of $\rho = 0.1$ is not much more likely than would be a value of, say, 0.2. Recall that this likelihood function is based on a single observation. Figure 3.1e, based on $n = 10$, indicates that a value of 0.2 is highly unlikely.

The decline in the normed likelihood function on either side of the MLE can be summarized by the width (curvature) of the function at the MLE (Fisher 1959). *Fisher Information* uses the curvature of the ln likelihood function to estimate a variance for the parameter error distribution, which might be obtained as a frequency of estimates acquired by repeating the experiment. The curvature is actually a quadratic approximation (second derivative) of the log likelihood function in the neighborhood of the MLE. If this curvature is slight, then the data do not contain much information about the parameter and the standard error is large, and vice versa. The quadratic is exact for a normal sampling distribution, because the log likelihood is quadratic, but it can be a poor approximation for asymmetric likelihoods when sample sizes are small.

A Single Parameter—Fisher Information is the expected value of the width of the likelihood function, which is inversely related to the curvature. First, write the Taylor expansion of $\ln L$ about the MLE,

$$\ln L(\theta) = \sum_{k=0}^{\infty} \frac{(\theta - \theta_{ML})^k}{k!} \times \frac{d^k \ln L}{d\theta^k}\bigg|_{\theta_{ML}}$$

(Appendix A). Including up through quadratic terms this series is

$$\ln L(\theta) = \ln L(\theta_{ML}) + (\theta - \theta_{ML})\frac{d \ln L(\theta_{ML})}{d\theta} + \frac{(\theta - \theta_{ML})^2}{2}\frac{d^2 \ln L(\theta_{ML})}{d\theta^2} + \cdots$$

The second term disappears, because the derivative of $\ln L$ at the MLE is zero[1]. To simplify notation, let

$$I = -\frac{d^2 \ln L(\theta)}{d\theta^2}\bigg|_{\theta_{ML}} \qquad\qquad 5.8$$

Then

$$\ln L(\theta) = \ln L(\theta_{ML}) - \frac{I(\theta - \theta_{ML})^2}{2}$$

or, equivalently,

$$\ln L(\theta) - \ln L(\theta_{ML}) = -\frac{I(\theta - \theta_{ML})^2}{2}$$

The left-hand side of this expression is just the log of the normed likelihood, that is,

$$R(\theta) = \exp[\ln L(\theta) - \ln L(\theta_{ML})] = \exp\left[-\frac{I(\theta - \theta_{ML})^2}{2}\right]$$

This has the same shape as (is proportional to) a normal density having I^1 as the variance parameter. This approximation is sometimes termed the *marginal likelihood*. Calculating the standard error by this method involves two steps:

1. Determine the curvature of $-\ln L$ near the MLE. For one parameter, Fisher Information is

$$I(\theta) = -\frac{\partial^2 \ln L}{\partial\theta^2}\bigg|_{\theta_{ML}}$$

2. Estimate the standard error as

$$se_\theta = \frac{1}{\sqrt{I}}$$

I is known as the observed information, because it is obtained by plugging in the MLE obtained from the data.

---

[1] If we take expectations for both sides of this equation, giving expected $\ln L$ on the left-hand side, then the second term still disappears ($E(\theta - \theta_{ML}) = 0$).

Example 5.2. Fisher Information is exact for the mean of a normal sampling distribution. To find the standard error of the mean, take the second derivative of the likelihood function with respect to μ,

$$\frac{\partial^2}{\partial \mu^2} \ln L(\mathbf{x};\mu,\sigma^2) = -\frac{n}{\sigma^2} \qquad\qquad 5.9$$

Then Fisher Information is

$$I = \frac{n}{\sigma^2}$$

with the standard error

$$se_\mu = \frac{1}{\sqrt{I}} = \frac{\sigma}{\sqrt{n}}$$

For a normal sampling distribution, the MLE together with its standard error suggests that we could draw the likelihood using the standard error from Fisher Information,

$$\theta \sim N(\theta_{ML}, se_\theta^2)$$

as was done for growth rates in Figure 5.1. Neglecting the normalizing constant of the normal distribution, the approximate normed likelihood function from Equation 5.6 is

$$R(\theta) \approx \exp\left[-\frac{(\theta - \theta_{ML})^2}{2se_\theta^2}\right]$$

which has the value of one at the maximum located at θ_{ML}. Thus, the normed likelihood obtained in the previous section is approximately proportional to the normal density having a variance derived from Fisher Information. A comparison of these functions is shown in Example 5.3.

Example 5.3. Estimate the standard error for the exponential model of Example 5.1 using Fisher Information.
The likelihood function for the exponential

$$\ln L = n \ln \rho - \rho \sum_{i=1}^{n} a_i$$

yields Fisher Information

$$-\frac{\partial^2 \ln L}{\partial \rho^2}\bigg|_{\lambda_{ML}} = \frac{n}{\rho_{ML}^2}$$

Thus, the standard error is

$$se_\rho = \frac{1}{\sqrt{I}} = \frac{\rho_{ML}}{\sqrt{n}} = \frac{1}{\bar{a}\sqrt{n}}$$

Note that this is equivalent to the square root of the variance (the variance for the exponential is $\frac{1}{\rho^2}$ —see Appendix F) divided by the square root of the sample size as was obtained in Example 5.2. For the example with $n = 2$ (dashed line in Figure 5.6a), the normal distribution drawn with this standard error of 0.0589 implies substantial probability of impossible negative values. However, with n still as small as 10 (Figure 5.6c) the normal approximation with a standard error of 0.0277 is much tighter, it does not imply appreciable probability of negative values, and it is similar to the exact normed likelihood. One standard error of the mean is shown in Figure 5.6b and 5.6d.

The normal approximation for the exponential likelihood (dashed lines in Figure 5.6a, c) is obtained by plugging in estimates of the mean and standard error,

$$R(\rho) \approx \exp\left[-\frac{(\rho - \rho_{ML})^2}{2se_\rho^2}\right] = \exp\left[-\frac{n}{2}(\rho\bar{a} - 1)^2\right]$$

Figure 5.6 illustrates how the normal approximation improves with increasing sample size, being not too bad in the neighborhood of the MLE when $n = 10$, but poor for $n = 2$.

Example 5.4. Estimate the standard error for the binomial parameter θ for $y = 1$ success out of $n = 5$ trials.
Fisher Information for the binomial is

$$I = -\frac{y}{\theta_{ML}^2} - \frac{n - y}{(1 - \theta_{ML})^2}$$

Substituting for $y = n\theta_{ML}$ gives the standard error

$$se_\theta = \sqrt{\frac{\theta_{ML}(1 - \theta_{ML})}{n}}$$

The normal approximation is poor for $n = 5$, showing substantial probability of negative values, but good for $n = 50$ (dashed lines in Figure 5.7).

For the binomial model, the normal approximation to the normed likelihood is

$$\lim_{n \to \infty} R(\theta) \propto N(\theta_{ML}, se_\theta^2).$$

The approximate normed likelihood function is proportional to

$$R(\theta) \approx \exp\left[-\frac{n(\theta - \theta_{ML})^2}{2\theta_{ML}(1 - \theta_{ML})}\right] = \exp\left[-\frac{n(n\theta - y)^2}{2y(n - y)}\right]$$

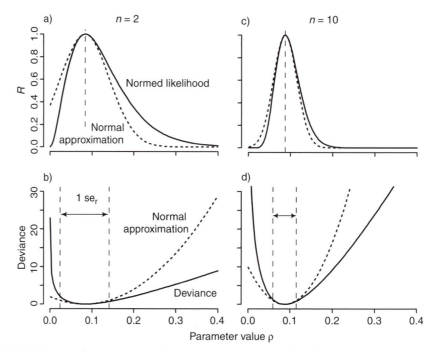

FIGURE 5.6. Confidence intervals for the exponential model and two sample sizes using the normal approximation with Fisher Information (dashed line) and the likelihood profile (solid line) or normed likelihood. In *b* and *d* are shown the deviance and the approximation obtained using Fisher Information. See Example 5.3.

This is the normal approximation, because I substituted point estimate and standard error for the normal mean and standard deviation, respectively. This is compared with actual normed likelihood for the binomial

$$R(\theta) = \left(\frac{\theta}{\theta_{ML}}\right)^y \left(\frac{1-\theta}{1-\theta_{ML}}\right)^{n-y} = \left(\frac{n\theta}{y}\right)^y \left(\frac{n(1-\theta)}{n-y}\right)^{n-y}$$

in Figure 5.7. Again, sample size has a large effect on the approximation.

Confidence intervals are often taken from a normal approximation with the mean and variance taken as the MLE and squared standard error, respectively. Thus, a 95 percent confidence interval would be approximately $\theta_{ML} \pm 1.96 se_\theta$. These confidence intervals assume that repeated experiments would yield a normal distribution of parameter estimates, which can often be a reasonable assumption for large samples.

Information for Multiple Parameters—Fisher Information is readily determined for models containing multiple parameters. For p parameters there is a p by p symmetric information matrix **I** consisting of elements,

$$I_{ij} = -\frac{\partial^2 \ln L}{\partial \theta_i \partial \theta_j}\bigg|_{\theta_{ML}}$$

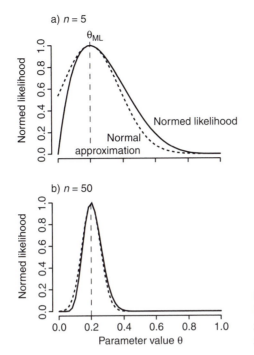

FIGURE 5.7. Normed likelihood for the binomial parameter (solid line) and normal approximation using Fisher Information (dashed line).

$\mathbf{I}^{-1}$ is the covariance matrix for parameters, with standard errors represented by the square root of the diagonal of this matrix. Analogous to the one-parameter case, for p parameters the asymptotic approximation is

$$\theta \sim N_p(\boldsymbol{\theta}_{ML}, \mathbf{I}^{-1})$$

where $\boldsymbol{\theta}_{ML}$ is the vector of MLEs.

5.1.5 Bootstrap

The bootstrap is a numerical method that can be used to obtain confidence intervals (Efron and Tibshirani 1993). The basic methodology can be extended for propagation of error for purposes of prediction, sometimes termed *Monte Carlo simulation*. Unlike other methods, bootstrapping does not require fancy math. If a point estimate is available, so is the bootstrap. There are two types of bootstraps to consider, *nonparametric* and *parametric*.

***Standard Errors and Confidence Intervals by Nonparametric Bootstrap*—** The nonparametric bootstrap produces estimates based on resampling the data. The resampling procedure involves sampling from the data with replacement. Suppose the sample consists of n observations. The method involves drawing a sample of size n from the data set. This means that once an observation has been included in the sample, it can still be sampled again. This procedure simulates the frequentist concept of obtaining estimates from repeated similar experiments. It substitutes resampling of one data set for repeated experiments.

The following five steps can be used to estimate standard errors and confidence intervals simultaneously. The extension to multiple parameters is straightforward. This recipe assumes a sample of size n for which we can estimate the parameter θ using an estimator function (e.g., a likelihood function).

1. Draw with replacement a resample of size n from the original sample.
2. Estimate θ from this resample. Let $\hat{\theta}_b$ represent this b^{th} estimate of θ.
3. Repeat this procedure B times. For a standard error estimate, B might be as low as 50. For a 95 percent confidence interval, B might be more like 2000. (Smaller α values require larger samples.) There are now B estimates of the parameter, one for each bootstrap resample.
4. Estimate the parameter standard error as the standard deviation of the B replicates,

$$ se_\theta = \sqrt{\frac{\sum_{b=1}^{B}(\hat{\theta}_b - \bar{\theta})^2}{B - 1}} $$

where

$$ \bar{\theta} = \frac{1}{B}\sum_{b=1}^{B}\hat{\theta}_b $$

is the mean of B estimates.
5. Estimate the confidence interval as the interior $100(1 - \alpha)\%$ of the bootstrapped estimates.

Example 5.5. Recall the example of ten mortality ages assumed to have an exponential distribution:

Plant:	1	2	3	4	5	6	7	8	9	10	ρ_{ML}
Days:	10	14	15	22	24	1	7	1	11	9	0.0877

The MLE (shown to the right of the sample values), is the reciprocal of the mean age of death. Below are three bootstrap resamples from this data set. Note that some ages are sampled more than once. In the first resample ($b = 1$), age 14 appears five times. Other ages do not appear at all. The same resample does not contain the ages 1, 15, 22, or 24, which are present in the data set. Here are three bootstrap resamples:

b	1	2	3	4	5	6	7	8	9	10	ρ_b
1	14	9	14	7	9	11	10	14	14	14	0.0862
2	10	9	14	14	14	15	7	22	1	11	0.0855
3	1	1	22	15	15	1	1	7	9	1	0.137

The MLE for each of the resamples is shown to the right. The full distribution of 2000 bootstrapped estimates is shown in Figure 5.8a. The standard error of 0.0214 is the standard deviation of this distribution. As expected from the shape of the likelihood function, this distribution is skewed to the right. The 95 percent confidence interval bounds the interior 1900 estimates (95 percent of 2000). The vertical dashed lines in Figure 5.8a delineate the tails, each containing 2.5 percent of the bootstrap estimates (i.e., fifty estimates). This 95 percent confidence interval is (0.0613, 0.143).

Example 5.6. Simultaneous fecundity and maturation schedules in pines
In Section 3.7 I considered fecundity schedules in pines assuming that all trees were of the same status (capable of producing cones at any size). LaDeau and Clark (2001) accommodated the fact that the probability of reproductive maturity increased with tree size. A given tree is reproductive with probability $\theta(x_i)$, which increases as a cumulative normal (probit) function,

$$\theta(x_i) = \Phi(x_i; \mu, \sigma^2)$$

where $\phi(x_i; \mu, \sigma^2) = \int_{-\infty}^{x_i} N(x|\mu, \sigma^2) dx.$

Given that a tree is mature, it produces cones according to the conditional fecundity schedule

$$g(x_i) = a x_i^2$$

The likelihood function contains total probabilities of observing y_i cones, given that the tree could be in either of two states,

$$L(y_i; x_i, \mu, \sigma^2, a) = \begin{cases} 1 - \theta(x_i) + \theta(x_i) e^{-g(x_i)} & y_i = 0 \\ \theta(x_i) \dfrac{g(x_i)^{y_i} e^{-g(x_i)}}{y_i!} & y_i > 0 \end{cases}$$

The first line applies to trees where no cones are observed. This can occur if the tree is immature, with probability $1 - \theta$, or if the tree is mature, with probability θ, and no cones are observed, with probability given by the zero class of the Poisson, e^{-g}. The second line applies to observations of at least one cone. In this event, the tree must be mature, so we have the joint probability of being in a mature state θ and of producing $y_i > 0$ cones, taken to be Poisson. MLEs for the three parameters can be estimated by optimization techniques described in the lab manual. Confidence intervals and propagated errors can be obtained by a bootstrap. Each bootstrap resample contains n trees and is used to estimate a_b, μ_b, and σ_b. The two functions $\theta_b(x)$ (Figure 3.3a) and $g_b(x)$ (Figure 3.3b) are calculated and saved for determination of bootstrapped confidence intervals (dashed lines in Figure 3.3).

Parametric Bootstrap—The foregoing approach is termed *nonparametric*, because the empirical (frequency) distribution of data is resampled, each time refitting parameters. For the *parametric bootstrap*, we assume a parametric form for the data, fit parameters, and then resample from the distribution described by those parameters to bootstrap a distribution of estimates.

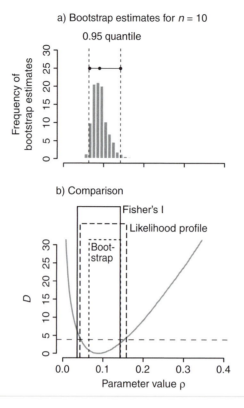

FIGURE 5.8. Confidence intervals obtained by three different methods for the exponential of Example 5.5. In *a* is a histogram of 2000 bootstrapped estimates. In *b* is deviance from the likelihood profile method, with dashed lines outlining confidence intervals obtained by other methods described in this chapter.

In Section 3.7, the parameter *a* describing increase in fecundity with tree size was estimated using the likelihood

$$L(\mathbf{y};\mathbf{x},a) = \prod_{i=1}^{n} Pois(y_i|g(x_i;a))$$

where x_i is tree diameter, y_i is the cone count, and $g(\ )$ is the allometric function describing increase in cones with tree size. In Example 5.6 I extended the model to allow for maturation and bootstrapped estimates of three parameters. This was a nonparametric bootstrap, because I resampled pairs of (x_i,y_i) directly from the data set.

A parametric bootstrap would begin with an estimate of $\hat{a}$, such as the MLE, draw a sample of cone counts of size *n* from

$$\mathbf{y}^{(b)} \sim Pois(g(\mathbf{x};\hat{a}))$$

and estimate a new value $a^{(b)}$. The collection of $a^{(b)}$ estimates would be treated as previously.

Why bother with a parametric bootstrap, when we can sample directly from the empirical data? In some cases, data may be sparse, yet there is good reason to believe in a particular sampling distribution. The Poisson distribution is fully defined by one parameter, and that parameter can be estimated rather confidently from relatively small data sets. If we are sure that we are dealing with a Poisson process, we may decide to estimate a confidence interval by sampling from the parametric distribution described by that estimate.

More important, there will be cases where resampling from the data does not duplicate the data-generating mechanism assumed by the model. Time series data are a good example. I consider this approach in Section 9.5 for time series and in Section 10.3 for a biogeographic example.

5.2 Bayesian Credible Intervals

The *Bayesian credible interval* can be obtained directly by integrating the posterior. This posterior is like the density in Figure 5.2a, and so returns us to the way we typically think about confidence intervals. For a 95 percent credible interval, we seek the upper and lower interval limits such that

$$0.95 = \int_{\theta_i}^{\theta_u} p(\theta|y)d\theta$$

or, if the parameter θ assumes discrete values,

$$0.95 \leq \sum_{\theta_i \leq \theta \leq \theta_u} p(\theta|y)$$

This intuitive definition represents the subjective probability that the true value of θ lies between θ_1 and θ_u. For discrete θ, this interval includes θ_i and θ_u. This differs from the frequentist concept of coverage, which invokes the hypothetical idea of repeating the experiment many times and determining the fraction of confidence intervals that contain θ. Of course, the values of θ_1 and θ_u that satisfy these equalities are not unique, and we would like the interval to be as narrow as possible. The *highest posterior density* (HPD) is defined by a horizontal line drawn through the density. This line intersects the posterior density in two places, thus defining two tails. A line drawn such that the fraction of the density in the tails is $1 - \alpha$ (a 95 percent credible interval would have $\alpha = 0.05$) defines the HPD. The areas in the two tails can differ by this method. Much easier to compute is the *equal-tail* interval, where each tail has the fraction $\frac{\alpha}{2}$ of the density, and is hereafter used throughout this book. The two are equivalent for symmetric, unimodal densities. Otherwise, the equal-tail interval is slightly wider.

As outlined in Box 4.1, Bayes' original example involved binomial trials with n trial and y successes. The goal was to estimate a parameter θ describing the probability that a trial is successful. The posterior density for this parameter was

$$p(\theta|y, n) \sim Beta(\theta|y + 1, n - y + 1)$$

Here is an example that uses the same model with application to different variables.

Example 5.7. Credible interval for a binomial parameter
This example uses prior information with the likelihood to illustrate a credible interval for the binomial parameter θ. Suppose there are four trees, one of which dies during a one-year interval. From this limited sample size I could use maximum likelihood to estimate a mortality rate (probability of death per year). The likelihood is based on the binomial,

$$p(y|\theta) = \binom{n}{y}\theta^y(1 - \theta)^{n-y} = \frac{4!}{1!3!}\theta^1(1 - \theta)^3$$

Taking logs and setting the derivative with respect to θ equal to zero yields the ML estimate of $\theta = \frac{1}{4}$. This is an estimate of risk for an interval of one year. I cannot make probability statements about this estimate, except with respect to some other estimate. In other words, I could calculate the likelihood of the data under the assumption that, say, $\theta = \frac{1}{3}$ and then assign a probability using a likelihood ratio test. Such statements are relative.

Now suppose I wanted to use the knowledge I have from a previous observation that one of two trees had died under comparable circumstances during the previous year. Because circumstances are similar, I might combine them to improve the overall knowledge of mortality rates. Bayes provides a way of combining this information.

I assume the prior density of θ to be a beta density with parameters determined by knowledge gained from the previous information, a total sample of $n_0 = 2$ trees with one death ($y_0 = 1$). This beta prior is

$$p(\theta) = \frac{\theta^{y_0-1}(1 - \theta)^{n_0 - y_0 - 1}}{B(y_0, n_0 - y_0)}$$

The mean of this density is the ratio of dead to live, $d_0/n_0 = 0.5$.

Before going further, I can speculate on the answer. Together there are six trees, two of which died. I therefore expect that the posterior estimate should have a mean value near $\frac{1}{3}$. Note that this value falls between the mean of the prior ($\frac{1}{2}$) and the data ($\frac{1}{4}$).

With both a likelihood for the data set and a prior density for θ. I apply Bayes' theorem to estimate the posterior density,

$$p(\theta|y, y_0) = \frac{\theta(1 - \theta)^3 \times 1}{\displaystyle\int_0^1 \theta(1 - \theta)^3 \times 1d\theta} = \frac{\theta(1 - \theta)^3}{B(2,4)}$$

$$= \frac{\Gamma(6)\theta(1 - \theta)^3}{\Gamma(2)\Gamma(4)} = 20\,\theta(1 - \theta)^3$$

(beta and gamma functions are described in Box D.1). This posterior density is beta(2,4) and has a mean value of $\frac{1}{3}$, which we can obtain by integration (calculating the first moment—Appendix D),

$$E[\theta|y,y_0] = \int_0^1 \theta p(\theta|y, y_0)d\theta = \frac{1}{B(2,4)}\int_0^1 \theta^2(1 - \theta)^3 \, d\theta$$

$$= \frac{B(3,4)}{B(2,4)} = \frac{\Gamma(3)\Gamma(4)}{\Gamma(3 + 4)} \times \frac{\Gamma(2 + 4)}{\Gamma(2)\Gamma(4)}$$

$$= \frac{2!3!5!}{6!1!3!} = \frac{1}{3}$$

that is, the total that died over the total number observed. The mode (most probable value) is still $\frac{1}{4}$, because the posterior amounts to a simple integration of the likelihood using a flat prior. In general terms, this result is

$$E[\theta|y, y_0] = \frac{y_0 + y}{n_0 + n}$$

A Bayesian credible interval is obtained by integrating the posterior,

$$\int_0^A p(\theta|d)d\theta = \int_{\rho_u}^1 p(\theta|d)d\theta = \alpha/2$$

For the beta density, these values are obtained numerically. The central 68 percent of this density is contained in the interval (0.147, 0.523) and the central 95 percent is contained in (0.0527, 0.716).

Example 5.7 involves a prior based on historical data. There may be a series of experiments, sometimes even on the same subjects (e.g., carcinogenicity studies), results of which can enter by way of the prior. The weight of the prior and likelihood might be completely determined by respective sample sizes (e.g., Example 5.7), but they need not be. Typically, the prior would be accorded less weight than the current study. Ibrahim and Chen (2000) explore the so-called power prior, whereby the likelihood for the historical data is raised to a power between zero (flat) and one (equal weight). This approach effectively changes the prior sample size.

In the foregoing example, the credible interval had much to do with the weight of data and prior. I considered this relationship for a normal likelihood and prior in Example 4.1. Here it arises in the context of a binomial likelihood with a beta prior. How would the credible interval change if I knew in advance that $y_0 = 10$ from a total $n_0 = 20$ trees, rather than one of two? How about $y_0 = 10$ of $n_0 = 40$? As seen in Chapter 4, the answer depends on the balance of input between prior and data. I extend this example to consider the relative weights of prior and data, much like the normal-normal example in Example 4.1. This exploration is facilitated by the concept of conjugacy.

The example above involves a binomial likelihood

$$Bin(y|n,\theta) \propto \theta^y(1-\theta)^{n-y}$$

Note the beta prior density for the parameter θ had the same form for coefficients involving θ,

$$Beta(\theta|y_0,n_0) \propto \theta^{y_0-1}(1-\theta)^{n_0-y_0-1}$$

When this is the case, the posterior has the same form as the prior (here, a *beta* density),

$$Beta(\theta|y_0 + y, n_0 - y_0 + n - y) \propto \theta^{y_0+y-1}(1-\theta)^{n_0-y_0+n-y-1}$$

This combination of prior and likelihood that yields a posterior of the same form is termed a *conjugate pair*. Conjugate pairs are useful due to their tractability; they can be readily updated as new data accumulate. In this case, simply sum the observations to update the posterior beta density.

For conjugate prior-likelihood pairs, the parameters of the posterior often involve sums that reflect the weights of prior and data. The beta-binomial pair has posterior parameters consisting of the sum of the total observed trees $n_0 + n$ and the total dead $y_0 + y$. If there are many more observations contributing to the prior density, then $n_0 \gg n$, and the prior dominates the posterior. Alternatively, the data dominate when $n \gg n_0$.

Example 5.8. For a Poisson example in Box 5.1, the integrated likelihood did not produce a classical confidence interval. For a single observation, this integrated likelihood was a gamma distribution with parameters $(y + 1, 1)$. Find the integrated likelihood for a sample of n observations.

For multiple observations, the Poisson likelihood is

$$p(\mathbf{y}|\theta) = \prod_{i=1}^{n} \frac{\theta^{y_i}e^{-\theta}}{y_i!}$$

The integral needed for the denominator of Equation 4.1 is

$$\int_0^\infty p(\mathbf{y}|\theta)p(\theta)d\theta = \int_0^\infty \prod_{i=1}^{n} \frac{\theta^{y_i}e^{-\theta}}{y_i!}d\theta = \frac{1}{\prod_i y_i!}\int_0^\infty \theta^{\Sigma_i y_i}e^{-n\theta}d\theta$$

Using the substitution $u = n\theta$ the integral simplifies to a gamma function,

$$= \frac{\int_0^\infty u^{\Sigma_i y_i}e^{-u}du}{n^{1+\Sigma_i y_i}\prod_i y_i!} = \frac{\Gamma(1 + \Sigma_i y_i)}{n^{1+\Sigma_i y_i}\prod_i y_i!}$$

Combined with the numerator, this yields the gamma posterior density

$$p(\theta|\mathbf{y}) = Gam\left(\theta\middle|1 + \sum_{i=1}^{n} y_i, n\right)$$

Thus, the earlier solution was the special case of $n = 1$.

Example 5.9. Now assume a gamma prior for the Poisson parameter θ. The two parameters of the gamma density allow us to change the weight of the prior. Here is the gamma prior density:

$$p(\theta) = Gam(\theta|\alpha,\beta) = \frac{\beta^{\alpha}\theta^{\alpha-1}e^{-\beta\theta}}{\Gamma(\alpha)}$$

The integration constant (the denominator) for the posterior is solved using the substitution $u = \theta(n + \beta)$,

$$\int_0^\infty p(\mathbf{y}|\theta)p(\theta)d\theta = \int_0^\infty Pois(y|\theta)p(\theta)d\theta$$

$$= \frac{\beta^{\alpha}}{\Gamma(\alpha)\prod_i y_i!} \int_0^\infty \theta^{\alpha-1+\Sigma_i y_i}\, e^{-\theta(n+\beta)}d\theta$$

$$= \frac{\beta^{\alpha}\Gamma(\alpha + \Sigma_i y_i)}{\Gamma(\alpha)(n + \beta)^{\alpha+\Sigma_i y_i}\prod_i y_i!}$$

Together with the numerator, we have a new gamma posterior,

$$p(\theta|\mathbf{y}) = Gam\left(\theta\Big|\alpha + \sum_{i=1}^n y_i, \beta + n\right) \qquad 5.10$$

Thus, the integration for a Poisson likelihood in the last example is simply a special case of a gamma prior as $\alpha \to 1$ and $\beta \to 0$. The posterior has the same form as the prior, that is, both are gamma densities.

   To be precise, we could extend the notation to emphasize that the posterior is conditioned not only on the data, but also on the fixed constants that define the prior. In other words, instead of $p(\theta|\mathbf{y})$, we would write $p(\theta|\mathbf{y},\alpha,\beta)$. In fact, we are really conditioning on many unstated things, so it is often reasonable to simplify notation. But it is important to recall this conditioning on fixed constants is implicit is the simplified notation $p(\theta|\mathbf{y})$.

   Note the conjugacy represented by this gamma-Poisson pair. The Poisson likelihood is proportional to

$$Pois(\mathbf{y}|\theta) \propto \theta^{\Sigma_{i=1}^n y_i}e^{-n\theta}$$

The gamma prior has the same form (in terms of θ),

$$Gam(\theta|\alpha,\beta) \propto \theta^{\alpha-1}e^{-\beta\theta} \qquad 5.11$$

Together they result in the gamma posterior (Equation 5.10) having parameters that sum the contributions of prior and likelihood. The preceding example was like the special case of a prior with $\alpha = 1$ and $\beta = 0$, which yielded a posterior dominated by the data. Larger values of α and β would produce priors with more influence.

For conjugate likelihood-prior pairs we can skip the integration of the denominator, because the normalizer is already known. To find the posterior, follow these simple steps:

1. Ignore the coefficients that do not include the parameter of interest, writing the posterior as the product of those coefficients that do contain the parameter. For the Poisson-gamma pair this was

$$p(\theta|\mathbf{y},\alpha,\beta) \propto \theta^{\sum_{i=1}^{n} y_i} e^{-n\theta} \times \theta^{\alpha-1} e^{-\beta\theta}$$

2. Collect coefficients. Again, for the Poisson-gamma,

$$p(\theta|\mathbf{y},\alpha,\beta) \propto \theta^{\sum_{i=1}^{n} y_i + n - 1} e^{-n\theta - \beta\theta}$$

3. Identify the parameters for the posterior based on comparison with the prior. Equation 5.11 shows that the first parameter of the gamma density is equal to the exponent on θ plus 1. For the gamma prior (Equation 5.19) this exponent is $\alpha - 1$, so the first parameter is α. Thus, for the posterior, the first parameter must be $\alpha + \sum_{i=1}^{n} y_i$, or equivalently, $\alpha + n\bar{y}$.

Appendix G includes a table of common conjugate pairs.

I take up Bayesian analysis in more detail in Chapter 7. The limited introduction to Bayesian credible intervals provided here already allows for some comparison with classical methods.

5.3 Likelihood Profile for Multiple Parameters

In Section 5.1.3, the likelihood profile was used to estimate a confidence interval for a parameter value. The profile consisted of a ratio of two likelihoods. One likelihood was evaluated at the MLE. The second was for a hypothesized value of the parameter. The likelihood ratio test was used, because the model having a parameter fixed at the hypothesized value is *nested* within a model where the same parameter is fitted. Nested models are discussed further in Chapter 6. In this section, I discuss how to use the likelihood profile method to obtain confidence intervals for models having multiple parameters. Section 5.4 applies these concepts to regression.

Consider a model with two parameters λ and θ and a set of observations $\mathbf{x}$. The confidence interval for either one of the parameters, say, λ, is obtained as follows. Establish a sequence of values for the target parameter. A single value of this sequence is λ_0. Obtain the MLE for the other parameter (in this case, θ) and calculate the deviance value. This is the deviance for the model with fitted θ and fixed λ_0 compared to that having both parameters fitted (i.e., λ_{ML} and θ_{ML})

$$D(\lambda_0) = -2[\ln L(\mathbf{x};\lambda_0,\theta_{ML|\lambda_0}) - \ln L(\mathbf{x};\lambda_{ML},\theta_{ML})]$$

I use the notation $\theta_{ML|\lambda_0}$ here to indicate the MLE of θ for a fixed value of λ_0. The likelihood ratio test provides the probability that the true value of parameter λ might differ from the MLE by as much as λ_0.

A likelihood profile examines a range of values of λ_0 by testing the hypothesis that the true parameter value might be this large or small. For each hypothesized value of λ_0 determine a P value using the χ^2 distribution with one degree of freedom. Further from λ_{MLE} that probability eventually falls below 0.025. The values at which this occurs on both sides of the MLE define the 95 percent confidence interval.

The likelihood ratio used for each hypothesized value of λ_0 is obtained by fitting all other parameters. It is not correct simply to plug in the values for a parameter when all others are at their MLEs, because the best fit for one parameter depends on the estimate for the other parameter(s).

Example 5.10. Likelihood profiles for the two Weibull parameters discussed in Chapter 3 of the lab manual are constructed in Figure 5.9. To determine a confidence interval for one parameter, calculate P values for the deviance obtained when the second parameter obtains its conditional maximum (i.e., conditional on a fixed value for the other). These profiles are superimposed on the $\ln L$ surface in Figure 5.9a. For parameter λ, the deviance is

$$D(\lambda_0) = -2[\ln L(\mathbf{a};\lambda_0,c_{ML|\lambda_0}) - \ln L(\mathbf{a};\lambda_{ML},c_{ML})]$$

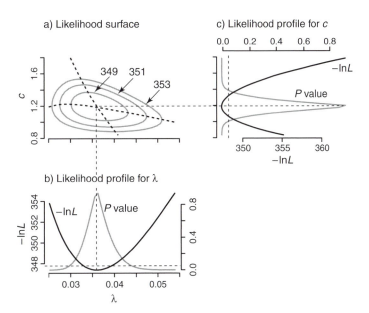

FIGURE 5.9. (a) Likelihood surface, showing the profile transects for parameters λ and c (curved dashed lines), from the example of Chapter 3 in the lab manual. (b) and (c) are the profiles for individual parameters, taken from profile transects in a. In (b) and (c) are also shown the P values associated with the likelihood profiles. The confidence intervals are defined by the intersections of the P-value curves and the dashed lines drawn at $P = 0.025$ in (b) and (c).

This curve is represented as a dashed curve in Figure 5.9a and as the solid, concave curve in Figure 5.9b. Figure 5.9c shows these relationships for parameter c.

5.4 Confidence Intervals for Several Parameters: Linear Regression

Finding maximum likelihood estimates and confidence intervals for models with multiple parameters is a direct extension of those used for one-parameter models. Linear regression is used as an example here because most ecologists are familiar with it, and because some of the intermediate products of this analysis are used in more complex models. This is an extension of ML methods used to find the mean and variance of a normal distribution in Chapter 4.

5.4.1 The Likelihood Function

The least-squares estimates for a linear model are the MLEs that are obtained under the assumption that error is Gaussian. This section demonstrates that fact using methods from Chapter 4. Confidence intervals for parameters follow. Consider the linear model

$$y_i = b_0 + b_1 x_i + \varepsilon_i$$

$$\varepsilon_i \sim N(0, \sigma^2)$$

where (x_i, y_i) are the values of x and y for the ith observation and b_0 and b_1 are fitted parameters. The first two terms on the right-hand side constitute the conditional expectation of y_i given x_i

$$E[y_i|x_i] = \mu_i = b_0 + b_1 x_i$$

The regression equation can also be written this way:

$$y_i = \mu_i + \varepsilon_i$$

The errors ε_i are drawn independently from the same normal distribution. It can be useful to think of the regression model in terms of the errors, ε_i,

$$\varepsilon_i = y_i - \mu_i$$

The normal distribution of errors is

$$\varepsilon_i \overset{i.i.d.}{\sim} N(0, \sigma^2) = \frac{1}{\sigma\sqrt{2\pi}} \exp\left[-\frac{\varepsilon_i^2}{2\sigma^2}\right]$$

$$= \frac{1}{\sigma\sqrt{2\pi}} \exp\left[-\frac{(y_i - \mu_i)^2}{2\sigma^2}\right]$$

This says that the ε_i are independent and identically distributed (i.i.d.) with mean zero and variance σ^2. The expected value of the regression changes with x as

$$\mu_i = b_0 + b_1 x_i$$

and the error is the difference between the observation and the mean value,

$$\varepsilon_i = y_i - \mu_i = y_i - b_0 - b_1 x_i$$

The distribution of errors summarizes the scatter of observations y_i about the regression line. The variance on this scatter is assumed to be the constant σ^2, but the mean changes according to the regression. As before, the likelihood of observing a set of y_i values is the product of observing each value,

$$L(\mathbf{y}; b_0, b_1, \sigma^2) = \prod_{i=1}^{n} N(y_i | \mu_i, \sigma^2) = \prod_{i=1}^{n} N(y_i | b_0 + b_1 x_i, \sigma^2)$$

$$= \prod_{i=1}^{n} \frac{1}{\sigma \sqrt{2\pi}} \exp\left[-\frac{(y_i - b_0 - b_1 x_i)^2}{2\sigma^2} \right]$$

The MLEs are the parameter values that maximize this likelihood. The log likelihood is

$$\ln L = \sum_{i=1}^{n} \left[-\ln \sigma - \frac{1}{2} \ln(2\pi) - \frac{(y_i - b_0 - b_1 x_i)^2}{2\sigma^2} \right]$$

$$-n \ln \sigma - \frac{n}{2} \ln(2\pi) - \frac{1}{2\sigma^2} \sum_{i=1}^{n} (y_i - b_0 - b_1 x_i)^2$$

It is now clear that the regression parameter values that maximize the likelihood of the data are least squares, that is, they minimize the squared deviations about the regression line. This likelihood surface is shown in Figure 5.10c.

5.4.2 Point Estimates

MLEs for b_0, b_1, and σ are obtained by finding the derivatives of $-\ln L$ for each parameter, setting them equal to zero, and solving. Start with the intercept b_0.

$$\frac{\partial \ln L}{\partial b_0} = -\frac{1}{\sigma^2} \sum_{i=1}^{n} (y_i - b_0 - b_1 x_i)$$

$$= -\frac{1}{\sigma^2} \left[\sum_{i=1}^{n} y_i - n b_0 - b_1 \sum_{i=1}^{n} x_i \right]$$

Set this equal to zero, and (to simplify notation) divide through by n

$$0 = \bar{y} - b_0 - b_1 \bar{x}$$

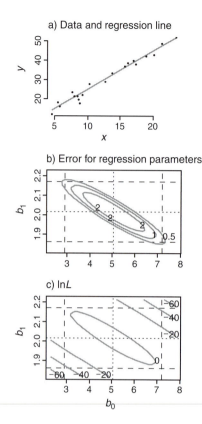

a) Data and regression line

b) Error for regression parameters

c) lnL

FIGURE 5.10. A regression line (a) and parameter estimates. The parameter error density of the intercept and slope is represented as a contour plot in (b). This is the normalized likelihood and is equivalent to a Bayesian posterior with flat priors. Also shown is the log likelihood surface (c). Dashed lines indicate parameter estimates and 95 percent confidence intervals.

or

$$b_0 = \overline{y} - b_1 \cdot \overline{x} \qquad\qquad 5.12$$

Thus, the MLE of the intercept b_0 is equal to the mean of the y's minus the estimate of the slope b_1 times the mean of the x's.

Now estimate the slope.

$$\frac{\partial \ln L}{\partial b_1} = -\frac{1}{\sigma^2} \sum_{i=1}^{n} x_i(y_i - b_0 - b_1 x_i)$$

$$= -\frac{1}{\sigma^2}\left[\sum_{i=1}^{n} x_i y_i - b_0 \sum_{i=1}^{n} x_i - b_1 \sum_{i=1}^{n} x_i^2 \right]$$

Setting this equal to zero and dividing through by n gives

$$0 = \overline{xy} - b_0 \cdot \overline{x} - b_1 \cdot \overline{x^2}$$

or

$$b_1 = \frac{\overline{xy} - b_0 \overline{x}}{\overline{x^2}} \qquad\qquad 5.13$$

We now have both MLEs for regression parameters, each expressed in terms of the other. Substitute the equation for b_0 (5.12) into the equation for b_1 (5.13),

$$b_1 = \frac{\overline{xy} - \overline{x} \cdot \overline{y} + b_1 \overline{x}^2}{\overline{x^2}}$$

Now, collect the b_1's. From Appendix D, recognize the relationship to covariance in numerator and denominator,

$$b_1 = \frac{\overline{xy} - \overline{x} \cdot \overline{y}}{\overline{x^2} - \overline{x}^2} = \frac{\text{cov}[x, y]}{\text{var}[x]} \tag{5.14}$$

Now substitute for b_1 in the equation for b_0 to obtain

$$b_0 = \frac{\overline{x^2} \cdot \overline{y} - \overline{x} \cdot \overline{xy}}{\text{var}[x]} \tag{5.15}$$

Finally, the parameter σ describes the spread of this distribution about the regression line. Again, take the derivative

$$\frac{\partial \ln L}{\partial \sigma} = \frac{n}{\sigma} - \frac{1}{\sigma^3} \sum_{i=1}^{n} (y_i - \mu_i)^2$$

Setting this equal to zero gives

$$\sigma^2 = \frac{1}{n} \sum_{i=1}^{n} (y_i - \mu_i)^2 \tag{5.16}$$

Note this to be a variance (mean square deviation) about the regression, that is, the variance of the residuals. Thus, we see that the familiar least-squares estimators from linear regression are the MLEs under the assumption of normally distributed error. As for the case of a simple mean, due to bias in the estimate, the restricted maximum likelihood estimate of the variance is preferred,

$$\sigma^2 = \frac{1}{n - 2} \sum_{i=1}^{n} (y_i - \mu_i)^2$$

where $n - 2$ accounts for the two fixed effects b_0 and b_1 (McCulloch and Searle 2001).

5.4.3 Parameter Standard Errors and Confidence Intervals

Standard errors for parameter estimates can be obtained from Fisher Information (Section 5.1.4). Recall that the standard error is taken from the curvature of $\ln L$ at the MLE. The second derivative of the ln likelihood function with respect to the intercept is

$$\frac{\partial^2 \ln L}{\partial b_0^2} = \frac{-n}{\sigma^2}\left(1 + \bar{x}\frac{\partial b_1}{\partial b_0}\right)$$

$$= -\frac{n \, \mathrm{var}[x]}{\sigma^2 \overline{x^2}}$$

Using the same approach for the slope parameter,

$$\frac{\partial^2 \ln L}{\partial b_1^2} = \frac{-n}{\sigma^2}\left(\bar{x}\frac{\partial b_0}{\partial b_1} - \overline{x^2}\right)$$

$$= -\frac{n \, \mathrm{var}[x]}{\sigma^2}$$

Thus, the standard errors are

$$se_{b_0} = \sigma\sqrt{\frac{\overline{x^2}}{n \, \mathrm{var}[x]}} \qquad 5.17$$

$$se_{b_1} = \frac{\sigma}{\sqrt{n \, \mathrm{var}[x]}} \qquad 5.18$$

The standard errors for the parameter estimates increase with the residual variance (σ^2 is variation not explained by the regression), and they decrease with sample size and with the spread in the independent (x) values. The derivations above can be treated as a system of equations. Box 5.2 extends this approach to linear models with many parameters using matrix algebra.

Box 5.2 Linear regression in matrix notation

Matrix notation is used for the regression model when there are multiple covariates. The normal distribution can be written as

$$\mathbf{y}|\mathbf{b},\sigma^2 \sim N(\mathbf{Xb}, \mathbf{V})$$

where the ($n \times 1$) response vector is $\mathbf{y} = [y_1 \ldots y_n]^T$, the ($p \times 1$) parameter vector is $\mathbf{b} = [b_0 \quad b_1 \ldots b_{p-1}]^T$, and the ($n \times p$) design matrix is

$$\mathbf{X} = \begin{bmatrix} 1 & x_{11} & \cdots & x_{1,p-1} \\ 1 & x_{21} & \cdots & x_{2,p-1} \\ \vdots & \vdots & \ddots & \vdots \\ 1 & x_{n1} & \cdots & x_{n,p-1} \end{bmatrix}$$

In other words, each row of $\mathbf{X}$ represents the values of covariates or treatments for the corresponding observation in vector $\mathbf{y}$. If these are treatments, columns in $\mathbf{X}$ would contain zeros (control treatment) or ones (other treatments). For i.i.d. errors and constant variance, the $(n \times n)$ covariance matrix is

$$\sum = \sigma^2 \mathbf{I}_n = \begin{bmatrix} \sigma^2 & 0 & \cdots & 0 \\ 0 & \sigma^2 & \cdots & 0 \\ 0 & 0 & \ddots & \vdots \\ 0 & 0 & \cdots & \sigma^2 \end{bmatrix}$$

where $\mathbf{I}_n$ is the identity matrix of dimension n. Taken together, we can think of the model like this:

$$\underset{n \times 1}{\mathbf{y}} = \underset{n \times p}{\mathbf{X}} \; \underset{p \times 1}{\mathbf{b}} + \underset{n \times 1}{\varepsilon}$$

If necessary, review Appendix C. The covariance matrix is the expected value of the squared errors,

$$\sum = E[\varepsilon \varepsilon^T] = \sigma^2 \mathbf{I}_n$$

The off-diagonal elements are zeros, because I am assuming that the errors do not depend on one another. The likelihood function looks like this:

$$L(\mathbf{y};\mathbf{b},\sigma^2) \propto \sigma^{-n} \exp\left[-\frac{(\mathbf{y} - \mathbf{Xb})^T (\mathbf{y} - \mathbf{Xb})}{2\sigma^2} \right]$$

It may help to visualize this model if you begin by thinking of the design matrix one row at a time. The ith row of the design matrix is the vector $\mathbf{x}_i = [1 \; x_{i1} \ldots x_{i,p-1}]^T$. In terms of $\mathbf{x}$ or a single observation i the likelihood function; looks more like the one written in Section 5.4.1,

$$L(y;\mathbf{b},\sigma^2) \propto \sigma^{-n} \exp\left[-\frac{\sum_{i=1}^{n}(y_i - \mathbf{x}_i^T\mathbf{b})^2}{2\sigma^2} \right]$$

The derivatives with respect to parameters are

$$\frac{\partial ln\, L}{\partial \mathbf{b}} = \frac{1}{\sigma^2}\sum_{i=1}^{n}\mathbf{x}_i(y_i - \mathbf{x}_i^T\mathbf{b})$$

Set this equal to zero and use rules for matrix algebra in Appendix C to obtain

$$\hat{\mathbf{b}} = \left(\sum_{i=1}^{n} \mathbf{x}_i\mathbf{x}_i^T \right)^{-1} \sum_{i=1}^{n} \mathbf{x}_i y_i$$

(Continued on following page)

In terms of the full design matrix and vector of observations this is equivalent to

$$\hat{\mathbf{b}} = (\mathbf{X}^T\mathbf{X})^{-1}\mathbf{X}^T\mathbf{y}$$

The first part of this is used to obtain the parameter covariance matrix

$$\mathbf{V_b} = (\mathbf{X}^T\mathbf{X})^{-1}\sigma^2$$

where the residual variance is estimated as

$$\sigma^2 = \frac{(\mathbf{y} - \mathbf{X}\hat{\mathbf{b}})^T(\mathbf{y} - \mathbf{X}\hat{\mathbf{b}})}{n}$$

Again, the REML estimate is preferred,

$$\sigma^2 = \frac{(\mathbf{y} - \mathbf{X}\hat{\mathbf{b}})^T(\mathbf{y} - \mathbf{X}\hat{\mathbf{b}})}{n - k - 1}$$

where $k+1$ is the rank of the matrix $\mathbf{X}$. For a matrix of full rank, the number of columns is equal to the rank. This is equivalent to the number of regression parameters to be estimated, p. A design matrix not of full rank would have redundancy in the columns. The denominator for the estimate of the variance is given by $n - \text{rank}[\mathbf{X}]$.

Example 5.11. Here is the matrix algebra for the slope-intercept (two-parameter) model solved previously without the matrixes (Section 5.4.1). Refer to Appendix C for matrix algebra. The cross product of the design matrix for two parameters gives

$$\mathbf{X}^T\mathbf{X} = \begin{bmatrix} n & \sum_{i=1}^{n}x_i \\ \sum_{i=1}^{n}x_i & \sum_{i=1}^{n}x_i^2 \end{bmatrix}$$

Upon inversion

$$(\mathbf{X}^T\mathbf{X})^{-1} = \frac{1}{n\sum_{i=1}^{n}x_i^2 - \left(\sum_{i=1}^{n}x_i\right)^2} \begin{bmatrix} \sum_{i=1}^{n}x_i^2 & -\sum_{i=1}^{n}x_i \\ -\sum_{i=1}^{n}x_i & n \end{bmatrix}$$

$$= \frac{n}{n^2\overline{x^2} - n^2\bar{x}^2} \begin{bmatrix} \overline{x^2} & -\bar{x} \\ -\bar{x} & 1 \end{bmatrix} = \frac{1}{n \cdot \text{var}[x]} \begin{bmatrix} \overline{x^2} & -\bar{x} \\ -\bar{x} & 1 \end{bmatrix}$$

So the parameter covariance matrix is

$$V_b = \frac{\sigma^2}{n \, \text{var}[x]} \begin{bmatrix} \overline{x^2} & -\overline{x} \\ -\overline{x} & 1 \end{bmatrix}$$

With two more multiplications we have the parameter estimates

$$(X^T X)^{-1} X^T y = \frac{1}{n \cdot \text{var}[x]} \begin{bmatrix} \overline{x^2} \sum_{i=1}^{n} y_i + \overline{x} \sum_{i=1}^{n} x_i y_i \\ \sum_{i=1}^{n} x_i y_i - \overline{x} \sum_{i=1}^{n} y_i \end{bmatrix} = \frac{1}{\text{var}[x]} \begin{bmatrix} \overline{x^2} \cdot \overline{y} + \overline{x} \cdot \overline{xy} \\ \overline{xy} - \overline{x} \cdot \overline{y} \end{bmatrix}$$

or

$$\hat{b} = \begin{bmatrix} \dfrac{\overline{x^2} \cdot \overline{y} + \overline{x} \cdot \overline{xy}}{\text{var}[x]} \\ \dfrac{\overline{xy} - \overline{x} \cdot \overline{y}}{\text{var}[x]} \end{bmatrix}$$

Note that these results agree with those of Section 5.4.1. Matrix algebra is unavoidable for large numbers of parameters.

The (marginal) error distribution of parameter vector b is multivariate Student's t. However, if we condition on a fixed value of σ^2, the distribution of estimates is normal,

$$b | \sigma^2, y \sim N(b, V_b)$$

To see this, return to the matrix form for the likelihood function,

$$L(y; b, \sigma^2) \propto \sigma^{-n} \exp\left[-\frac{(y - Xb)^T (y - Xb)}{2\sigma^2} \right]$$

I will substitute for the MLE's of vector b and complete the square. Considering just the exponent, write the following:

$$y - Xb = y - X\hat{b} + X\hat{b} - Xb$$
$$= y - X\hat{b} + X(\hat{b} - b)$$

Upon substitution into the likelihood exponent,

$$(y - X\hat{b} + X(b - b))^T (y - X\hat{b} + X(b - b))$$

Using relationships from Appendix C, some algebra yields

$$\left[(y - X\hat{b})^T + (X(b - b))^T \right]\left[(y - X\hat{b}) + (X(b - b)) \right]$$
$$= (y - X\hat{b})^T (y - X\hat{b}) + (X(b - b))^T (y - X\hat{b})$$
$$+ (y - X\hat{b})^T X(b - b) + (X(b - b))^T X(b - b)$$

The exponent now contains four terms. The second and third terms equal zero, because $\hat{\mathbf{b}} = (\mathbf{X}^T\mathbf{X})^{-1}\mathbf{X}^T\mathbf{y}$, leaving

$$(\mathbf{y} - \mathbf{X}\hat{\mathbf{b}})^T(\mathbf{y} - \mathbf{X}\hat{\mathbf{b}}) + (\mathbf{X}(\mathbf{b} - \mathbf{b}))^T\mathbf{X}(\mathbf{b} - \mathbf{b})$$
$$= (\mathbf{y} - \mathbf{X}\hat{\mathbf{b}})^T(\mathbf{y} - \mathbf{X}\hat{\mathbf{b}}) + (\mathbf{b} - \mathbf{b})^T \mathbf{X}^T\mathbf{X}(\mathbf{b} - \mathbf{b})$$

Now taken in terms of the parameter vector $\mathbf{b}$, the first term is a constant, and the second term is the exponent for the normal distribution conditional on known σ^2.

$$\propto \exp\left[\frac{(\mathbf{b} - \mathbf{b})^T\mathbf{X}^T\mathbf{X}(\mathbf{b} - \mathbf{b})}{2\sigma^2}\right]$$

$$= N\left(\mathbf{b}|\hat{\mathbf{b}}, (\mathbf{X}^T\mathbf{X})^{-1}\sigma^2\right)$$

The error distribution of the regression parameter estimates (Box 5.2), conditioned on σ^2, is the bivariate normal distribution

$$\mathbf{b}|\mathbf{X},\mathbf{y},\sigma^2 \sim N_2(\mathbf{b},\mathbf{V_b}),$$

where $\mathbf{V_b}$ is given in Box 5.2 (Figure 5.9b). The parameter estimate for σ^2 has error of its own. If we were to integrate over error in σ^2, the distribution of parameter estimates is Student's t with n-p degrees of freedom, where p is the number of regression parameters. This Student's t distribution is a mixture and will be discussed further in later chapters. The confidence intervals for regression parameters can be approximated from standard errors (Sections 5.1.2, 5.1.4).

As discussed in Section 3.8, the design matrix $\mathbf{X}$ contains columns that take on values of a variable, treatment levels, or zeros and ones, for observations on treatment and control individuals. For this reason, analysis of variance (ANOVA) is a linear model.

5.5 Which Confidence Envelope to Use

Throughout this book, I point out some of the considerations that determine the type of analyses that are appropriate for a given setting. Because confidence envelopes are such an important product of an analysis, I use this section to discuss further how techniques presented in this chapter compare to one another.

Confidence intervals based on Fisher Information are commonly used, because their application is broad and calculation can be easy (provided that the likelihood function is differentiable). We saw that Fisher Information provides an asymptotic approximation to a confidence interval, being appropriate when sample size is not too small. It has the advantage of providing a parametric representation of the standard error. A confidence interval is obtained from the normal distribution having this standard error as the standard deviation of the parameter. This parametric representation also allows us to write functions that depend on those parameter estimates using methods of Appendix D.

Greenwood's formula for a survivor function (Example 5.12) is an example. Lande (1987) used this approach to calculate growth rate uncertainty based on parameter estimates reported in the literature. Pacala et al. (1996) used it to propagate error in a simulation model.

Given that it is based on an approximation of asymptotic normality, it is important to ask whether the confidence intervals based on Fisher Information are good enough. Considerations such as sample size and the shape of the sampling distribution emerged from examples in this chapter. The normal approximations in Figure 5.5 were obtained by substituting the MLE and the squared standard error for the mean and variance of a normal distribution, respectively (Example 5.3). Of course, the parameter ρ (Figure 5.4) cannot be normally distributed, because a normal distribution assigns nonzero probability to negative values. Negative values do not make sense for a rate parameter. But, with large n, the normal approximation may be close enough. Conversely, with small sample size, this approximation will be poor. Similar considerations applied to the binomial example in Figure 5.7. In summary, standard errors based on Fisher Information are exact when the likelihood is normal, and they can be used for error propagation in models having multiple parameters that are approximately normally distributed. If not normally distributed, this method will provide a better estimate of the standard error than for confidence intervals, because confidence intervals are usually taken at more than one standard error from the mean parameter estimate.

The likelihood profile usually requires more effort to implement than Fisher Information. It does not depend on the assumption of normally distributed error. It is often the method of choice for confidence intervals when there are few parameters. With more parameters, a bootstrap is often preferable.

The nonparametric bootstrap is easy to use and is often the only practical solution in situations where models are too complex for other methods. For example, if the likelihood is not differentiable (e.g., the gamma function of Example 3.3), we cannot obtain the Fisher Information. Bootstrapped confidence intervals are accurate and can be made more so using some flourishes described in Efron and Tibshirani (1993). A parametric bootstrap can be used (Section 5.6) when raw data are unavailable and for specific situations where bootstrapped data sets cannot be generated nonparametrically (Section 9.5.3).

Bayesian methods allow for external information, and they produce an interval that agrees with how we tend to think about parameter distributions. For noninformative priors, a credible interval may not look much different from a classical confidence interval (Clark and Lavine 2001). Indeed, models are sometimes evaluated using the concept of frequentist coverage (e.g., Ghosh and Mukerjee 1992). Because it allows us to integrate external information, we will usually turn to Bayes for complicated models. Still, it is important to stress that for simple models, frequentist and Bayes confidence intervals produce similar confidence envelopes.

As a simple example of how similar confidence intervals can be, consider the example of a one-ha study region that contains 40,000 individual plants for an average of 4 per m^2. Samples consist of counts of the number of individuals encountered in 1 m^2 samples. The vector of counts is $\mathbf{y}$. A Poisson likelihood is the basis for estimating the density of individuals θ. Figure 5.11a shows histograms of counts $\mathbf{y}$ for samples of size $n = 10$ and $n = 100$. Now suppose

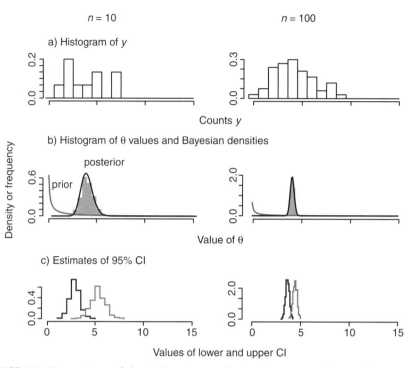

FIGURE 5.11. Comparison of classical estimates obtained by repeatedly sampling a population (histograms in *b*) with those obtained using Bayes (curves in *b*). Details are discussed in Section 5.5.

that this process is repeated 1000 times, each time estimating θ (Figure 5.11b) and a classical confidence interval using Fisher Information (Figure 5.11c). As expected, the distribution of estimates and the confidence intervals thereof converge on the true value of 4 per m². The distribution of estimates in Figure 5.11b comes from the process we attempt to simulate with the nonparametric bootstrap, the difference being that the bootstrap resamples the sample, whereas here I repeatedly sampled the (fake) population itself.

Compared with the classical estimates are the Bayesian posteriors obtained for a noninformative prior $\theta \sim Gam(0.1, 0.1)$, which are overlaid on the distributions of estimates in Figure 5.11b. This example reemphasizes that, for uncomplicated models (including lack of prior knowledge), classical and Bayesian confidence intervals yield about the same answers, and they are subject to the same asymptotic relationships with sample size. I return to this example for predictive intervals in the following section.

To summarize, for large sample sizes, confidence envelopes obtained by different methods usually yield similar results. The standard error of parameter ρ for the sample of ten ages of death was 0.0277 using Fisher information and 0.0214 using the nonparametric bootstrap. The 95 percent confidence interval from the likelihood profile was (0.0391, 0.165), and that for the bootstrap was (0.0613, 0.143). These confidence intervals are compared in Figure 5.8b. Depending on the application, these differences may be important. And the fact that Bayes allows for the use of an informative prior adds a consideration that goes beyond simple goodness of fit. Yet for the case where one has no external knowledge, there may be little motivation for the extra effort needed for Bayes. In later chapters I point out that a principal motivation for Bayes is flexibility to take up complex problems.

5.6 Predictive Intervals

With some basic concepts about confidence interval behind us, we can begin to consider applications and extensions to prediction. Whereas confidence or credible intervals apply to estimates, as part of the model fitting procedure, predictive intervals apply to quantities that have not been estimated but can be predicted from the fitted model. Examples include predicting data that have not yet been collected, forecasts of events that have yet to occur, and relationships that can be calculated from models that have been fitted to data. These extensions involve models and propagation of uncertainty.

5.6.1. Bayesian Predictive Intervals

I find the Bayesian predictive interval to be most intuitive, so I start there. Suppose we wish to predict new values of a variable y' based on observations y and a model M involving parameters θ. The fitted model includes a likelihood $p(y|\theta,M)$, and it yields a posterior density of parameters $p(\theta|y,M)$. I bother to condition explicitly on M to emphasize that a different model would result in different predictions. The prediction involves integrating over the uncertainty represented by the posterior density,

$$p(y'|y,M) \;=\; \int p(y'|\theta,M)p(\theta|y,M)d\theta \qquad\qquad 5.19$$

On the left-hand side is the conditional distribution for the prediction y', which depends on the model and on the observed data. Percentiles from this *predictive density* are termed *predictive intervals*. To reduce notation, I hereafter omit M, but it is important to remember that any prediction is conditioned on the model used to fit the data.

For the example in Figure 5.11, the relevant distributions for Equation 5.19 (Poisson likelihood and gamma posterior) are

$$NB(y'|\alpha+\textstyle\sum_i y_i,\beta+n) \;=\; \int_0^\infty Pois(y'|\theta)Gam(\theta|\alpha+\textstyle\sum_i y_i,\beta+n)d\theta$$

The integration on the right-hand side that leads to this negative binomial distribution is shown in Appendix F. The parameter θ does not appear on the left-hand side, having been integrated away. The effects of the data y enter through the posterior gamma density. The posterior is not the predictive distribution. It expresses uncertainty in a parameter. The spread of the predictive distribution includes not only the effect of Poisson sampling y', but also the degree of uncertainty as to the parameter estimate of this Poisson sampling distribution. Thus, the predictive distribution has more spread than does the likelihood conditioned on a known value of θ. As the sample size increases (Figure 5.12a), the uncertainty in θ decreases, and thus, the predictive distribution converges on the Poisson sampling distribution (Figure 5.12b).

The same approach applies when there are additional variables involved. Consider the model where y depends on x, and we wish to predict y' corresponding to some new values of x'. We apply the same framework,

$$p(y'|x,x',y) = \int p(y'|x',\theta)p(\theta|x,y)d\theta \qquad 5.20$$

The prediction grid for a spatio-temporal process in Section 10.11 is an example where, in the present context, x represents the sample locations, which might be irregularly spaced, and x' represents the new locations, which are arranged on a grid (Chapter 10).

Although we can solve for the predictive distribution in the case of the conjugate Poisson-gamma, the integral needed for a predictive distribution will typically not be available. There will often be many parameters over which we must integrate. Thus, predictive intervals are usually constructed using numerical methods that will be introduced in Chapter 7. A more comprehensive overview of prediction is provided in Chapter 11.

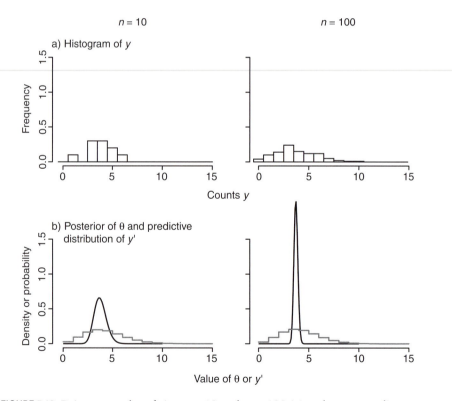

FIGURE 5.12. Poisson samples of size $n = 10$ and $n = 100$ (a) and corresponding posterior densities for θ (continuous curves in b) and predictive distributions of y' (step curves in b). The prior (Figure 5.10) and posterior (b) are gamma, and the predictive distribution is negative binomial (discrete distribution in b).

5.6.2 Classical Error Propagation

Predictive intervals can be constructed within a classical setting. Recall that the growth rate λ is not a fitted parameter, but rather a function of other parameters. In the case of spotted owls, λ is not fitted directly to data as it might be, say, if it were determined from time series data (Chapter 9). Rather, demographic parameters are estimated from data, and the growth rate is calculated using a function based on Lotka's equation or on a matrix model reviewed in Chapter 2. The normal densities in Figure 5.1 are constructed from calculations of λ using estimates of parameters, such as fecundity, age to first reproduction, and survivorship (Clark 2003). These densities are best described as predictive, because they are predicted on the basis of other estimates, as opposed to being directly fitted to data.

The spread of the predictive distribution is determined by an error propagation technique that translates parameter error to error in λ. This technique is based on the method for a "variance of a function of several variables" from Appendix D.5.3. The method provides a variance on λ. It is not based on an assumption that the population is, in fact, variable (Clark 2003). As with standard errors for parameter estimates, it describes a level of uncertainty as to a true fixed value. Its spread derives from the limited sample sizes used to estimate the parameters that went into this prediction. Generally, if a parameter has a large standard error, and if it makes a strong contribution to λ, then λ will have a wide predictive interval. We consider models that do involve direct estimation of growth rates from time series data in Chapter 9. For now, keep in mind that probability statements about λ involve uncertainty in the parameters used to calculate it.

In Appendix D.5.3, the variance of a function of random variables is approximated using a Taylor series. Because parameter estimates are asymptotically normal, this approximation seems appropriate, provided that sample sizes are not too small. Using λ as an example, Equation D.45 looks like this,

$$\text{var}\,[\lambda] \approx \sum_{i=1} \left(\frac{\partial \lambda}{\partial \theta_i} \right)^2 \text{var}\,[\theta_i] \;+\; \sum_{i,j|i \neq j} \left(\frac{\partial \lambda}{\partial \theta_i} \right)\left(\frac{\partial \lambda}{\partial \theta_j} \right)\text{cov}\,[\theta_i \theta_j] \qquad 5.21$$

where the θ_i's are the parameters estimated from data. These include all elements of a matrix model (Chapter 2) or the life history schedule in Lotka's equation (Box 5.3). The parameter covariances may not always be available, because, unless parameters are estimated simultaneously (e.g., Box 5.2), parameter covariances are unknown. For demographic data covariances arise if maturation age, reproductive effort, longevity, and growth are correlated (e.g., Doak et al. 1994). These covariances are unknown, unless parameters are estimated on the same individuals (Chapter 9).

In Figure 5.13 I compare some reasonable error distributions for demographic models (dashed lines) with the normal approximations (solid lines) implicit to the error propagation method Lande (1987) used to place confidence intervals on λ (Box 5.3). The error distributions for the two survival probabilities used in Figure 5.13 are

$$s \sim Beta\,(\alpha, \beta),$$

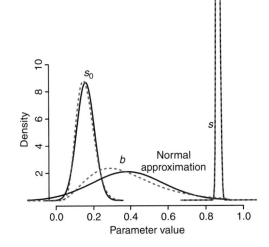

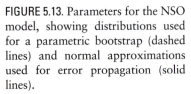

FIGURE 5.13. Parameters for the NSO model, showing distributions used for a parametric bootstrap (dashed lines) and normal approximations used for error propagation (solid lines).

where parameter values are selected by moment matching to produce beta distributions having the parameter means and parameter variances estimated from data (Section 3.11). These are

$$\alpha = \hat{s}\left(\hat{s}\frac{(1 - \hat{s})}{se_s^2} - 1\right)$$

$$\beta = \frac{\alpha(1 - \hat{s})}{\hat{s}}$$

The estimates and variances for s_0 and s are discussed in Box 5.3.

For b I used the sampling distribution

$$b \sim Gam(\alpha, \beta)$$

where moment matching gives the parameter values

$$\alpha = \frac{\hat{b}^2}{se_b^2}$$

$$\beta = \frac{\hat{b}}{se_b^2}$$

The normal approximations based on the same estimates and values are adequate for the two survival parameters, because sample sizes are high (Figure 5.13). The normal approximation is less accurate for b, implying appreciable probability of negative fecundity.

If the asymptotic normal approximation is deemed inadequate, a parametric bootstrap could be used to generate a predictive distribution for λ by simply drawing parameter estimates from these distributions and recalculating λ (Clark 2003). Figure 5.14a shows a histogram of 2000 such estimates with the central 95 percent quantile.

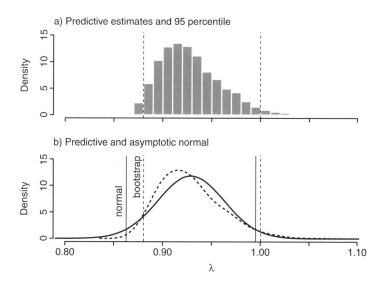

FIGURE 5.14. Comparison of predictive distributions for λ using a parametric bootstrap (*a* and dashed line in *b*) with the normal approximation. From Clark (2003).

Box 5.3 Confidence intervals for northern spotted owl growth rate

The confidence interval for growth rate depends on the confidence intervals of the parameters used to calculate it. In this box, I describe an asymptotic estimate of error propagation to translate the standard error of parameter estimates to a standard error of the growth rate estimate for the NSO example. I use methods from Appendix D to calculate the growth rate variance. It is based on Lande (1987).

Because parameter covariances are unknown, Lande (1987) simplified Equation 5.21 to $var\,[\lambda] = \sum_{i=1}(\frac{\partial \lambda}{\partial \theta_i})^2 \times var\,[\theta_i]$ + contribution of covariances (assumed negligible) where θ_i is the ith parameter in the parameter vector $\theta = [\alpha, b, s_0, s]$, and each sensitivity (derivative) in the series scales the sampling error for that parameter $var\,[\theta_i]$. The sample distribution is binomial for the survival probabilities. Recall Fisher Information to obtain the squared standard error,

$$var\,[s_i] = \frac{s_i(1 - s_i)}{n_i}$$

where n_i is the number of individuals used to estimate s_i. This model has s_0, (juvenile survival) and s (adult survival).

The sampling error for b (birth rate) and α (maturation age) are variances divided by the number of individuals used to obtain those estimates. Recall that these are the standard errors on the mean, assuming the distribution of estimates to be normal. The predicted growth rate and generation time based on the point estimates for each of the four parameters are $\lambda = 0.929$ and $T = 16.2$. These predicted values can be obtained using Lotka's equation or matrix methods (Appendix E). Based on Lotka's equation, the derivatives needed to estimate the growth rate standard error are

$$\frac{d\lambda}{ds_0} = \frac{\lambda}{s_0 T} = 0.360$$

$$\frac{d\lambda}{ds} = \frac{\lambda(bs_0 + 1)(T - \alpha)}{sT} = 0.942$$

$$\frac{d\lambda}{db} = \frac{\lambda}{bT} = 0.150$$

$$\frac{d\lambda}{d\alpha} = \frac{\lambda \ln(s/\lambda)}{T} = 0.00389$$

Because the variance on α is small, I omit this term. The calculation goes like this:

$$var\,[\lambda] = 0.360^2 \times 0.046^2 + 0.942^2 \times 0.008^2 + 0.150^2 \times 0.190^2$$
$$= 0.000274 + 0.0000568 + 0.000812$$
$$= 0.00114$$
$$se_\lambda = 0.0338$$

Sample sizes and associated contributions to variance in λ are shown in Table 5.2. The standard error on λ is the square root of $0.00114 = 0.0338$. Thus a value of $\lambda = 1$ is within two standard errors of the estimate.

Propagation of these error estimates in parameters to error in the estimate of λ can also be accomplished with a parametric bootstrap (Table 5.2), as shown in Figure 5.13.

TABLE 5.2
Elements of Error Propagation for the NSO Example

Parameter	Value	n	Standard Error	Distribution Used in Bootstrap	Contribution to Variance in λ
Juvenile survival: s_0	0.159	63	0.046	$Beta(9.89, 52.3)$	0.000274
Adult: s	0.868	114	0.008	$Beta(1553, 236)$	0.0000568
Fecundity: b	0.382	197	0.190	$Gam(4.04, 10.6)$	0.000812
Growth rate: λ	0.929		0.0338		

*In Mckelvey et al. (1993).

The normal approximation to λ using Equation 5.21 (solid line in Figure 5.14b) shows an upper 95 percent confidence limit slightly less than 1.0, whereas a parametric bootstrap using the more realistic sampling distributions barely includes $\lambda = 1$. Given the propensity to view 95 percent confidence as the threshold between significance and nonsignificance, some audiences might reach different conclusions from these methods. Note that the lower 95 percent confidence limits differ, by larger amounts.

The following example uses Equation 5.21 to propagate error in estimates of age-specific mortality rates to the survivor function. Because the survivor function is discrete, there is an estimate for each age. This method incorporates the error of all estimates into the estimate of survival.

Example 5.12. Confidence intervals for the product limit survivor function Figure 3.6 shows confidence intervals for the survivor function. Here I provide Greenwood's formula, an application of Fisher Information and Equation 5.21.

The likelihood for the age-specific mortality risk ρ is

$$L(d_i; \rho_i) = (\rho_i)^{d_i}(1 - \rho_i)^{n_i - d_i}$$

(Section 3.8.2). Setting the first derivative equal to zero yields MLE

$$\hat{\rho}_i = \frac{d_i}{n_i}$$

This is the MLE for the ith class. For Fisher Information, we also need the second derivative,

$$\frac{\partial^2 \ln L}{\partial \rho_i^2} = -\frac{d_i}{\rho_i^2} - \frac{n_i - d_i}{(1 - \rho_i)^2}$$

Substitute for the MLE, using $d_i = \hat{\rho}_i n_i$ to obtain the negative inverse of Fisher Information,

$$-\frac{\partial^2 \ln L}{\partial \rho^2}\bigg|_{\hat{p}} = \frac{-n_i}{\hat{\rho}(1 - \hat{\rho})}$$

Now the variance on survival can be viewed as the variance of a function. This function is the relationship between survival and age-specific mortality rate,

$$l_a = \prod_{a_i < a}(1 - \rho_i)$$

or, on a log scale,

$$\ln l_a = \sum_{a_i < a}\ln(1 - \hat{\rho}_i)$$

Because the estimates for each interval are independent, the variance of the sum is equal to the sum of the variances,

$$\text{var}[\ln l_a] = \sum_{a_i < a}\text{var}[\ln(1 - \hat{\rho}_i)]$$

This is Equation 5.21, where covariances are zero. Now we need the variances on the right-hand side. From Appendix D.5.2 we have

$$\text{var}[\ln(1 - \hat{\rho}_i)] = \left(\frac{\partial}{\partial\hat{\rho}_i}\ln(1 - \hat{\rho}_i)\right)^2 \text{var}[\hat{\rho}_i]$$

By substitution, the variance on log survival is

$$\text{var}[\ln l_a] = \sum_{a_i < a}\left(\frac{1}{1 - \hat{\rho}_i}\right)^2 \text{var}[\hat{\rho}_i]$$

In terms of the d_i's, this simplifies to

$$= \sum_{a_i < a}\frac{d_i}{n_i(n_i - d_i)}$$

We now have the variance in terms of $\ln l_a$. In terms of l_a we again use the variance of a function

$$\text{var}[l_a] = \left(\frac{\partial}{\partial\ln l_a}l_a\right)^2 \text{var}[\ln l_a]$$

$$= (\hat{l}_a)^2\sum_{a_i < a}\frac{d_i}{n_i(n_i - d_i)}$$

In summary, we obtained variances on the age-specific rates based on Fisher Information, and we used variances of functions to obtain variance on survival. The confidence intervals in Figure 3.6 are obtained as $\hat{l}_a \pm 1.96se_{l_a}$, where the standard error is taken to be the square root of the variance. Note that this is an asymptotic estimate and might be inaccurate in the tail of the survivor function. In most studies, sample size is small for old individuals.

5.7 Uncertainty and Variability

Throughout this book I draw a distinction between estimation error versus variability, which often tends to be overlooked because it can be difficult to estimate. Consider adult survival s. Most models used to estimate survival are based on the binomial model (Chapter 3). A fixed risk $\rho = 1 - s$ is assumed to apply to all individuals of the adult stage in the entire population. Mortality itself is taken to be stochastic, but all individuals are assumed to be at identical risk ρ. Based on the assumption of the binomial process, the variance in the number that die is $var[y] = n\rho(1 - \rho)$.

The binomial model can underestimate the variance in numbers that die, because it entails the assumption that risk ρ is fixed. The model accommodates the variability that results *from* the binomial process, but not the variability *in* the binomial process. In fact, individuals can be exposed to many risks. They differ in age, size, sex, and behavior (Franklin et al. 2000), and environmental risks vary in space and time. Factors that can be measured can be treated as covariates. Alternatively, if there is variability in risk ρ that cannot be tied to covariates, they can be treated as random effects. Unless these sources of variability enter the model in some form, the estimation and prediction error both tend to zero with large sample size n. In other words, the estimate of uncertainty is based on the assumption that risk is constant and the same for all observations. This assumption is different from that of a variable risk.

The distinction between variability and uncertainty arises in many contexts. For example, the negative binomial predictive distribution of Section 5.6.1 arises from mixing the Poisson sampling distribution with a gamma density representing parameter uncertainty, as described by the posterior estimate of θ. According to this model, we believe that there is a single value of θ, and the posterior represents how well we know that fixed value. The negative binomial predictive distribution is broader than the Poisson sampling distribution, because it allows for uncertainty as to the true fixed value of θ. As sample size increases, the posterior for θ will become more peaked, and the predictive distribution will tend to a Poisson. The asymptotic behavior of this posterior density represents how uncertainty tends to decline as sample size increases.

Contrast this interpretation with that of Section D.4.3, which involves the same distributions, but considers the Poisson parameter to be variable—it fluctuates due to factors that are unknown. If θ varies from one individual to the next, larger sample size does not necessarily result in much reduction in the width of the predictive distribution. In Chapters 7 and 8, variability and uncertainty are considered through application of hierarchical structures.

The variability introduced by individuals in two states, immature and mature, is a more widespread concern with ecological data than is realized. In Example 5.6, LaDeau and Clark (2001) allowed that there is a different submodel for zero counts. In fact, many ecological data sets have fundamentally different processes involved in the observation of a zero. For example, abundance, amount, or concentration is often modeled with a continuous likelihood, say, normal or lognormal. These continuous likelihoods have no place for a discrete zero class. In the case of a Gaussian likelihood, there is also the assumption of negative values. Such problems can often be handled hierarchically, with a model for "zero or not" beneath a likelihood that applies specifically to the "not zero" case. Example 5.6 involves discrete classes for the "not zero" case. For the case of continuous counts, Clark et al.'s (2004) model effectively does this:

$$p(y_i|x_i,\mu_i,\sigma^2,Q_i) = \begin{cases} 1 - \theta(x_i) & Q_i = 0 \\ \theta(x_i)LN(y_i|\mu_i, \sigma^2) & Q_i > 1 \end{cases}$$

with $y_i = 0$ seeds for trees that are immature ($Q_i = 0$), and the lognormal likelihood applying to mature individuals ($Q_i = 1$), in which case the μ_i are modeled with covariates. Hierarchical specifications are discussed in Chapter 9.

5.8 When Is It Bayesian?

Without getting too deeply into philosophy, I simply mention here that the distinction between classical and Bayes is not always clear-cut. There can be a murky middle ground that is not frequentist, but does not seem quite Bayesian. In this chapter I described some fundamental differences between a frequentist confidence interval, based on the idea of coverage, and the Bayesian expression of posterior uncertainty. I showed that integrating the likelihood directly (without prior) may not precisely yield a frequentist confidence interval (Box 5.1). Still, the middle ground can be appealing to some who want to avoid specifying a prior. Some might view it as a practical way to estimate a classical confidence interval. Others believe that they avoid philosophical pitfalls by not specifying a prior. Conceptually, integrating a likelihood has more affinity to Bayes, where integrating the likelihood is logical and implies a flat prior, than it does to a classical treatment. Moreover, it often makes use of the Bayesian machinery (Chapter 7). Some statisticians may tolerate either view (Tanner 1996). Integrated likelihoods have apparently been spotted in the physics literature for some time (Cousins 1995). The same can be said for ecology (Pacala et al. 1996; de Valpine 2003). There is a practical consideration that different methods often yield similar results. The motivation for going Bayesian need not be based solely on philosophy. This topic comes up again in Chapter 7.

6 Model Assessment and Selection

6.1 Using Statistics to Evaluate Models

What makes a model useful? The answer depends on the intended application of the model and on how well the model is supported by data. Although we usually speak of model selection as something that is arbitrated by data alone, in fact many important decisions are made before we get around to fitting data. The models under consideration before a study is conducted are chosen to capture the process of interest, with attention to the variables that can be measured and allowance for unknowns. The design of an experiment may be planned with an already small number of models under consideration, before any data are collected. A given design may only permit selection between two models, one of which is a reduced form of the other. In short, the process entails subjectivity, involving the current state of knowledge and the approach believed most likely to advance it. Most of the model selection process does not actually involve the data that eventually enter as part of a formal model selection. Here I consider statistical evaluation of models, but I continue to emphasize that formal statistics is but one of many considerations.

Statistical evaluation of models involves model assessment and model selection. *Model assessment* is used to determine if a model is adequate for the intended purpose. If the data provide little support for the model, there may be little point in using it. *Model selection* asks whether one model is more useful than another, based on a data set. Formal model selection requires a method for comparing models that involve different hypotheses, embodied in a likelihood function. Model simplification represents one aspect of model selection and consists of finding a minimal model that explains the data. The more parameters, the better the fit. So when do we stop fitting? Burnham and Anderson (1998) provide a recent treatment of classical methods that emphasizes information theory and provides many environmental examples. Bayesian methods are more heterogeneous, in part, because Bayesian model structures are diverse.

This chapter considers classical and Bayesian approaches for assessing and selecting among simple models. Because hypothesis testing is central, I begin with some discussion of its place in model evaluation. I summarize the standard likelihood ratio test and penalized likelihood approaches that can be used for non-nested models. I then move to a more in-depth examination of Bayesian methods. These are extended to high-dimensional models in later chapters.

6.2 The Role of Hypothesis Tests

Statistical hypothesis tests have long played a central role in model selection. In Chapter 5, I said that confidence envelopes are not very controversial. Often we can arrive at similar confidence intervals starting from different assumptions. Hypothesis testing is another matter. Much has been said and written about hypotheses and their role in science. Definitions of hypothesis testing are carefully worded, partly due to growing awareness of limitations and potential for confusion. The term applies to a statement about a population based on assumptions (model and parameters), on samples, or both.

A classical hypothesis test involves a null hypothesis and an alternative. The hypotheses are framed as models. Data are used to decide between them. The weight of the evidence may be quantified as a P value (Chapter 1). Let y represent data and θ be one or more (a vector of) parameters about which we wish to draw inference. Data y are taken to be random, and parameters θ are fixed. If model assessment is focused on a value of θ, the null might be $H_0 = \theta_0$, where θ_0 represents specific parameter value(s). The best (e.g., ML) estimate of θ will undoubtedly differ from θ_0. The hypothesis test is intended to help us decide if the fitted value differs importantly from θ_0.

In Chapter 5, the likelihood ratio test was the basis for a classical confidence interval. The maximum likelihood estimate identifies the value of θ that is most likely, but the likelihood is not (in terms of θ) a probability. Because there may be infinitely many potential hypotheses ($\theta = \theta_1, \theta_2, \ldots$), there can be no probability for any one hypothesis that extends beyond the context of that selected as the competing one. Probability statements refer to two competing hypotheses, example a null (say, $\theta = \theta_0$) and an alternative, example $\theta \neq \theta_0$ or $\theta > \theta_0$. As discussed for confidence intervals in Chapter 5, the likelihood ratio test, based on the likelihood ratio

$$LR = \frac{p(data|\theta_0)}{p(data|\theta_{ML})}$$

can be used to obtain a P value. This ratio has the likelihood of the reduced model, with the parameter value fixed at θ_0 in the numerator. The following two sections show how to use the likelihood ratio test to select among nested models of two different types.

6.3 Nested Models

In the classical setting, we compare two models. We fit the models and use the likelihood ratio to determine which fit is best. Often one model is nested within (a reduced form of) a second model. A model is said to be *nested* within another if the more complex model collapses to the simpler model when one or more of the parameters is fixed. I refer to the simple model (i.e., with fewer fitted parameters) as the reduced form. It will often serve as a null hypothesis as basis for deciding whether added complexity is warranted. For instance, the Weibull density

$$f(a) = c\lambda a^{c-1} \exp[-\lambda a^c]$$ 6.1

collapses to the exponential

$$f(a) = \lambda e^{-\lambda a}$$ 6.2

when the Weibull parameter $c = 1$. The exponential is nested within the Weibull. If fitting is done by maximum likelihood, fitting the exponential model would result in the value of λ that maximizes the likelihood of the data. Fitting the Weibull would involve maximizing the likelihood simultaneously over two parameters, λ and c. Together these two models provide a means for testing whether the probability of an event changes over time, as opposed to being constant. If constant, then $c = 1$. If the probability of the event increases with time, then $c > 1$, and vice versa. If the Weibull does not provide a significant improvement over the simpler exponential model, then we cannot conclude that the probability changes with time.

A second example is the model

$$y_i = \beta_0 + \beta_1 x + \varepsilon_i$$

$$\varepsilon_i \sim N(0, \sigma_2)$$

(linear in x) that is nested within the model

$$y_i = \beta_0 + \beta_1 x + \beta_2 x^2 + \varepsilon_i$$

(quadratic in x). Here, model selection involves determining whether β_2 differs from zero (i.e., the nested model). Finding that β_2 differs from zero supports the hypothesis that y depends on x^2. In summary, models are nested if a simple model arises when a fitted parameter in a more complex model is fixed.

6.3.1 Likelihood Ratio Test: When to Stop Fitting

For nested models, the model with more fitted parameters has the largest likelihood value L. But how much greater must it be in order to conclude that the added complexity improves the fit? The likelihood ratio test, used to construct confidence intervals in Chapter 5, is also used to compare competing models.

Consider two models, where the second model is nested within the first. Both models contain parameters $\mathbf{v} = [v_1, \ldots, v_p]^T$. The first model contains an additional q parameters $\theta = [\theta_1, \ldots, \theta_q]^T$. For the second model, only the parameters in $\mathbf{v}$ are fitted. In other words, for the second model θ are fixed at the values θ_0. In the exponential/Weibull example, both models contain $v = [\lambda]$. The Weibull has the additional fitted parameter $\theta = [c]$. For the Weibull, this additional parameter is fixed at the value $\theta_0 = 1$ (the value at which parameter c is fixed in equation 6.1 to obtain the second exponential model 6.2). Similarly, in the regression model $v = [\beta_0, \beta_1]$, $\theta = [\beta_2]$, and $\theta_0 = 0$. The deviance used for the hypothesis test is

$$D(\theta) = -2\ln LR$$

$$= -2[\ln L(\mathbf{a}; \mathbf{v}, \theta_0) - \ln L\ (\mathbf{a}; \mathbf{v}, \theta)] \quad 6.3$$

The notation here describes two likelihoods for observations $\mathbf{a}$. The first term is the likelihood taken for the second (reduced) model, which contains fitted parameters $\mathbf{v}$. The additional fixed parameters θ_0 are constants (they are not estimated). The second likelihood is taken over all $p + q$ parameters in $\mathbf{v}$ and θ (all are fitted). Then $D(\theta)$ is distributed as approximately χ^2 with degrees of freedom equal to the difference in the number of parameters (in this case, q).

For example, in a survivorship experiment plant death is observed to occur on the following days $\mathbf{a} = [5, 7, 10, 10, 12, 15]$. To test the hypothesis that mortality risk is constant, determine whether a changing risk (Weibull) finds more support in the data than does a constant risk (exponential). Recall that the exponential assumes risk is constant. Obtain maximum likelihood estimates for each model and conduct a likelihood ratio test (Table 6.1).

The λ estimates differ for the two models; parameter estimates are always model-dependent. The $\ln L$ value is higher ($-\ln L$ is lower) for the model with more parameters, because the extra parameter allows a better fit. But is it sufficiently better to warrant the additional complexity? The low P value is interpreted as support for the interpretation that the added parameter c provides a better description of the data than does the constant probability of the exponential model.

This approach provides guidance about when to stop fitting. An *overfitted* model is one that describes the data well simply because it has many parameters. (An overfitted model would not be expected to predict well a different data set generated by an identical process.) The goal is to identify a model that

TABLE 6.1.
Elements of a Likelihood Ratio Test for the Weibull Parameter c

	Exponential Model	Weibull Model
λ_{MLE}	0.102	0.0911
c_{MLE}	(fixed at $c = 1$)	3.407
$-\ln L$	19.71	15.47
$D(c)$		8.49
$\Pr\{D\}$		0.0036

performs well because it does a good job of representing the underlying hypothesis and would do so for related data sets that might be collected in the future. The likelihood ratio test is used for this purpose.

Example 6.1. To determine whether fire risk increases with time since the last fire (as, say, if fuels accumulate over time), fire ecologists fit densities to fire interval data. A null hypothesis of no change in risk with time since the last fire is described by an exponential density (equation 6.2), which assumes that the probability of an event is constant. A flexible alternative is the Weibull density (equation 6.3). Figure 6.1a shows the best-fitting exponential and Weibull densities to intervals between fires determined from fire scars on pine trees from northwestern Minnesota (Clark 1990). The Weibull follows the data better, because it has one more parameter. Does the extra parameter significantly improve the fit?

The $n = 82$ intervals used in this analysis are shown in Table 6.2. These data, together with best-fitting models and associated age-specific risk, are plotted in Figure 6.1. Except near zero, differences are slight. The $-\ln L$ values for exponential and Weibull densities are 349.4 and 347.4, respectively, for a deviance of 4.04. The χ^2 probability with one degree of freedom is $P = 0.046$. Thus, the LRT says that data are better described by the Weibull. Table 6.3 shows parameter estimates, standard errors, and 95 percent confidence intervals obtained by a nonparametric bootstrap described in Section 5.1.

The 2000 bootstrap estimates are compared with the likelihood surface in Figure 6.2a. The smoothed parameter densities are also shown with their 95 percent quantiles in Figure 6.2b and 6.2c. Note too that the 95 percent confidence interval for parameter c does not include one and, thus, is consistent with the LRT for this nested model.

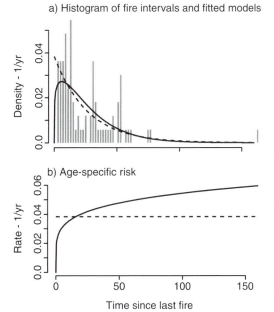

a) Histogram of fire intervals and fitted models

b) Age-specific risk

Time since last fire

FIGURE 6.1. Fitted exponential (dashed lines) and Weibull (solid lines) models fitted to fire interval data (*a*) and implied age-specific rates (*b*). The age-specific rate in (*b*) has the same definition as the age-specific mortality risk h(*a*) used for survival analysis (Chapters 2, 4). It is the probability of an event, given that the event has not yet occurred.

TABLE 6.2.
Number of Fires Observed for Specific Intervals in Years

Interval	# Fires	Interval	# Fires
2		1	
4		5	
5		3	
6		3	
7		2	
8		4	
9		7	
10		1	
11		2	
12		2	
13		5	
14		4	
15		1	
16		2	
17		1	
19		1	
20		1	
21		1	
24		1	
25		2	
30		2	
31		6	
33		2	
34		1	
36		1	
37		1	
39		1	
41		1	
44		1	
45		1	
47		1	
48		1	

Interval	# Fires	Interval	# Fires
51		1	
52		2	
53		5	
57		1	
60		1	
62		1	
76		1	
77		1	
164		1	

TABLE 6.3.
Bootstrap Estimates for Exponential and Weibull Parameters

	Model		Standard Error	Confidence Interval (Weibull)	
	Exponential	Weibull	(Weibull)	0.025	0.975
λ	0.0385	0.0360	0.00353	0.0300	0.0437
c	—	1.19	0.113	1.03	1.46

6.3.2 Do Populations Differ?

Model selection is also used to determine when groups differ, due to, say, being subjected to different treatments. This is an application of nested models, because a hypothesis of "one size fits all" (a simple model) is compared with one of separate fits for each group (a complex model). If the latter represents a significant improvement over "one size fits all," then we conclude that groups differ. The simple model is nested within the complex one, because it represents the special case where parameters for all groups assume the same value. Once again, the likelihood ratio test is the basis for evaluation.

Recall that the likelihood is based on the assumption that samples are independent. Because they are independent, we simply multiply them together—the likelihood of the data set is the product of the likelihoods for each observation. The likelihood of data set $\mathbf{x}_k$ containing observations $[x_{1,k}, \ldots, x_{n,k}]$ is

$$L(\mathbf{x}_k; \theta) = \prod_{i=1}^{n_\kappa} L(x_{i,k}, \theta)$$

We can extend this logic to data sets categorized in q groups (e.g., treatments). If groups are independent, then *the likelihood of several groups is the product of the likelihoods for each group.*

$$L(\mathbf{x}_1, \ldots, \mathbf{x}_q; \theta) = \prod_{k=1}^{q \text{ data sets}} L(\mathbf{x}_k; \theta) \qquad 6.4$$

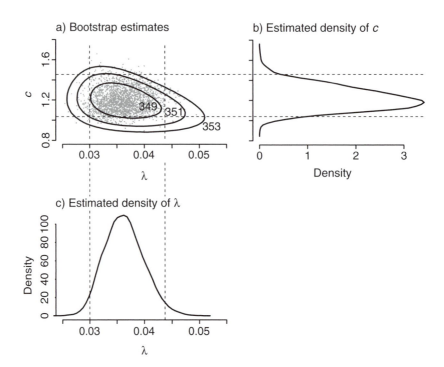

FIGURE 6.2. Bootstrap estimates for the Weibull distribution of fire intervals. The pairs of bootstrap estimates shown with the Deviance surface are shown in *a*. Estimates of marginalized densities are shown in *b* and *c*.

In order to test whether populations differ I construct nested models in three steps:

Step 1. "One size fits all"—This first likelihood assumes that all observations are drawn from the same population. Fit parameters θ to this full data set and calculate the likelihood using these fitted values:

$$L(\mathbf{x}_1, \ldots, \mathbf{x}_q; \theta) = \prod_{k=1}^{q} \prod_{i=1}^{n_1} L(x_{i,k}; \theta) \qquad 6.5$$

Note that θ on the right-hand side has no subscript, because I assume that there is one set of parameters that applies to all data sets. n_k is the number of observations in the kth group or data set. I obtain the MLE's for parameters θ as though this were just one large population.

Step 2. Populations are different—The second likelihood assumes that each group has a separate parameter set θ_k. I now fit separate parameters to each group. For the kth group, this likelihood is

$$L(\mathbf{x}_k; \theta_k) = \prod_{i=1}^{n_1} L(x_{i,k}; \theta_k) \qquad 6.6$$

I find MLE's for k different parameter sets, one for each group k. Under the assumption of a separate parameter set for each group, the full likelihood taken over all groups is

$$L(\mathbf{x}_1, \ldots, \mathbf{x}_q; \theta_1 \cdots \theta_q) = \prod_{k=1}^{q} \prod_{i=1}^{n_k} L(\mathbf{x}_{i,k}, \theta_k) \qquad 6.7$$

TABLE 6.4.
A Likelihood Ratio Test

	Data Set 1	Data Set 2	Combined
ρ_{MLE}	0.102	0.217	0.134
$-\ln L$	19.71	12.63	33.10
$-\ln L_1 - \ln L_2$	19.71 + 12.63 = 32.34		
$D(\rho)$	2[33.10 − 32.34] = 1.514		
P	0.2185		

I do not specify how the q groups differ, I am simply asking if they do differ.
Step 3. The test—I can now use a likelihood ratio test to determine whether the hypothesis that each group has its own parameter set is supported by the data.

$$D(\theta) = -2 \left[\ln L(\mathbf{x}; \theta) - \ln L(\mathbf{x}, \theta_1, \ldots, \theta_q) \right] \qquad 6.8$$

The likelihood ratio test has $q - 1$ degrees of freedom, that is, the difference in number of parameters in the two models.

Example 6.2. Consider the example of mortality, where a first group of observations had deaths at days $\mathbf{a}_1 = (5, 7, 10, 10, 12, 15)$. The days until death for a second group of plants is $\mathbf{a}_2 = (3, 3, 4, 5, 8)$. Does age of death differ for populations 1 and 2? Individually, the data sets each provide estimates for the parameter ρ and associated likelihoods (Table 6.4).

Note that $-\ln L$ is lower ($\ln L$ is higher) when each population has its own parameter estimate ($-\ln L = 32.34$) than when we assume a single estimate for the combined population ($-\ln L = 33.10$). This will always be true. But is the fit improved sufficiently to warrant this extra parameter? The P value is interpreted as saying that the data do not provide support for the conclusion that two populations differ. On the basis of these data, we might conclude that one parameter estimate suffices.

6.4 Additional Considerations for Classical Model Selection

6.4.1 Parameter Identifiability

For a variety of reasons, data may not provide information on all parameters contained in a model. This is true regardless of what approach is taken (e.g., frequentist or Bayes). This occurs when the likelihood is conditionally independent of one or more parameters. Consider the likelihood for data set $\mathbf{a}$ with two parameters. For Figure 6.2 this could be $L(\mathbf{a}; \lambda, c)$. If, given λ, the likelihood is independent of c, then the data provide no information on parameter c. In other words, $L(\mathbf{a}; \lambda, c) = L(\mathbf{a}; \lambda)$. In terms of Figure 6.2, the likelihood surface would be flat in the dimension c. This is not the case here. If it were, we would say that the model is not "identifiable." For example, we cannot identify the parameters in $Y_i \sim N(\mu, \sigma_1^2 + \sigma_2^2)$. This problem cannot be fixed by collecting more data. It can often be fixed by changing the model.

Strong relationships among parameters, such as those evident in Figures 5.9 and 10.11, can result because parameters simply trade off. If so, the model does not well resolve individual parameters, suggesting that there may be a better model for the data.

6.4.2 Penalized Likelihoods and Non-Nested models

The likelihood ratio is widely used. Among its drawbacks is the fact that it is inconsistent. A consistent test would say that, as the sample size tends to infinity, the test statistic should point to the correct model in all cases. The likelihood ratio test is biased toward the more complex model—even as sample size becomes infinitely large, there remains a chance of selecting the full model when it is wrong (Gelfand and Ghosh 1998). Likelihood ratio tests also cannot be applied to non-nested models.

Penalized likelihoods attempt to correct for this bias with a penalty that increases with dimension p (number of parameters) of the model and with sample size. A number of statistics penalize a large model with a term that is increasing in sample size n, in the number of parameters in the model p, or both. Such indexes are also used when models are not nested and, thus, a likelihood ratio test is not an option. The most popular of such indexes is An Information Criterion (AIC), which imposes a penalty for model size,

$$\mathrm{AIC} = -2\ln L + 2p$$

(Akaike 1973, 1974). For model selection, the preferred model is taken to be that having the lowest AIC value. Although the model with more fitted parameters should have an advantage (lower $-\ln L$), the AIC penalizes the more complex model in step with the number of parameters. AIC does not eliminate the inconsistency associated with sample size. Examples of indexes with sample size in the penalty term include Bayes Information Criterion (BIC)

$$BIC = -2\ln L + p\ln n$$

(Schwartz 1978). As with AIC, the model with the lowest value of BIC is taken as that most supported by the data.

6.4.3 Model Uncertainty

Once a model is selected, the uncertainties in estimates and predictions are typically reported as though the model were correct. Model selection methods do not tell us that a model is correct. They tell us that there is more support from this data set for this model than there is for another model. There is still uncertainty associated with the model itself. We might be more confident in one model than another. We cannot say how uncertain any of them are, because the notion of a correct model rarely has meaning.

It is most useful to think of parameter and prediction uncertainty as being conditional on the model. In other words, a confidence or prediction interval reports only part of the uncertainty. The model uncertainty remains. Some ways of addressing model uncertainty are mentioned in Chapter 8.

6.4.4 Some Caveats on Hypothesis Testing

Because model selection is closely linked to hypothesis testing, this is a good time to consider some general implications. We have seen that the classical approach is simple only in the case where there are two hypotheses, and the one is nested within the other. If we entertain a range of hypotheses, and they cannot be represented as simplified (nested) versions of one another, the approach is somewhat different.

The model selection approaches summarized here boil an experiment down to a single number. This number is based on a model fit. One number is rarely enough to capture all of the factors that should influence model choice.

The indexes require some interpretation. I considered the most popular classical example, AIC, which penalizes the large model, but not enough. AIC and BIC require that we can define model size (p_m). BIC relies on the asymptotic assumption that the dimension of the model stays fixed as the sample size becomes infinitely large. For hierarchical models, there is no clear definition of model dimension; model dimension may grow with sample size, each new sample bringing with it additional random effects, latent variables, or both. All of these considerations are revisited in Chapter 8, where the introduction of computation methods opens some new possibilities for dealing with complex relationships.

The classical construct whereby one can only reject or fail to reject the null may seem inefficient. One cannot accept a null hypothesis—a large P value does not mean that the null hypothesis is correct. The P value represents the frequency with which we expect to obtain data at least as extreme as those observed. It is not an error probability (Chapter 1). There might be more than one possibility, suggesting multiple hypotheses (Hilborn and Mangel 1997). And multiple hypotheses can only be compared pairwise. The goal of a study may be quantification—how fast, how much, and so on. This goal might suggest a continuous spectrum of hypotheses.

Related is the problem that the P value violates the likelihood principle, which says that all of the information from the experiment is contained in the likelihood function. This point is often illustrated with an example that goes back at least to Lindley and Phillips (1976). Suppose I flip a coin n times and observe y tails. I wish to calculate a P value for the hypothesis that the probability is equal to θ. If the experiment is "flip the coin n times," the sampling distribution is

$$y \sim Bin(n, \theta) \propto \theta^y (1-\theta)^{n-y}$$

If the design is "flip the coin until y tails occur," the sampling distribution is

$$n \sim NB(y, \theta) \propto \theta^y (1-\theta)^{n-y}$$

With respect to θ, these two likelihoods say the same thing. By the likelihood principle, they contain the same information about θ. However, the frequentist interpretation of the different stopping rules embodied in these two designs makes them different. The P value comes from the summation of everything at least as large as the observation y. For the first experiment, this P value comes from the binomial

$$\sum_{k=y}^{n} \binom{n}{k} \theta^k (1 - \theta)^{n-k}$$

For the second design, the P value comes from the negative binomial

$$\sum_{k=y}^{\infty} \binom{n+k-1}{k} \theta^k (1 - \theta)^n$$

These two P values are different. This example illustrates two undesirable aspects of the P value: (1) the P value is based on values of y greater than observed, and (2) the design of the experiment (in this case, the stopping rule) determines the P value, in violation of the likelihood principle.

The foregoing point contributes to confusion over the concept of significance. Consider that, for a large enough sample size, we are certain to reject the null hypothesis. It is highly unlikely that the null hypothesis is ever exactly true. With large sample size we can reject it even if it is essentially correct.

On the other hand, $\alpha = 0.05$ has been sanctified as the critical P value. As with all things religious, its appeal is invulnerable to rational criticism. Obviously, this value of $\frac{1}{20}$ is aimed at protecting the null. This seeming appeal to parsimony is hardly desirable in many applications. Is there something laudable about rejecting global warming until we are 95 percent sure it has already occurred? How about inaction on threatened species until we are 95 percent sure that $\lambda < 1$? Decision analysis, which combines uncertainty with a loss or utility of different decisions, is one alternative, but it can be hard to apply in many contexts (Chapter 11).

Taken together, P values are not to be taken too seriously and especially not as error probabilities. With caveats in mind, P values can represent a rough index of how well a model fits data.

6.5 Bayesian Model Assessment

Classical approaches to model assessment focus on the likelihood. In a Bayesian framework, the prior is also part of the model. However, the effect of varying the prior is not part of model selection. As with classical approaches, model selection focuses on the likelihood, which applies to the data set at hand. As mentioned above, the prior may be varied to determine sensitivity of the posterior to prior assumptions and, thus, establish model robustness (Berger 1985).

6.5.1 Bayes Factor

As with classical model assessment, hypotheses are part of Bayesian model assessment. A Bayesian analysis combines data with prior knowledge and calculates a posterior probability that a hypothesis is true. If models are nested, then posterior inference would seem to be the logical basis for model selection. The posterior distribution of the parameter of interest indicates whether it differs from a fixed value of a reduced model (e.g., does a regression parameter differ from zero?). In fact, posterior inference is not used for this purpose, because the posterior is conditioned on the model being correct (Box 1980). Moreover, when considering non-nested models, there will be no way to carry out such a comparison. Instead, we would like to know the probability of the model itself, a quantity we can determine in light of data and a prior belief in the model. Of course, this prior belief can be that different models are equally probable.

The Bayes factor is a common basis for model comparison, yielding a posterior odds ratio for competing models. Consider two models, M_1 and M_2. The posterior probabilities for these models are $\Pr\{M_1|\mathbf{y}\}$ and $\Pr\{M_2|\mathbf{y}\} = 1 - \Pr\{M_1|\mathbf{y}\}$. These probabilities are typically expressed as an odds ratio, called the *Bayes factor*,

$$\frac{\text{posterior odds}}{\text{prior odds}} = \frac{\Pr\{M_1|\mathbf{y}\}/\Pr\{M_2|\mathbf{y}\}}{\Pr\{M_1\}/\Pr\{M_2\}}$$

Note that the probabilities in the numerator are conditioned on the data $\mathbf{y}$—they are posterior probabilities.

How do we obtain the posterior probabilities for the models? Write Bayes' theorem in terms of model M_m,

$$p(M_m|\mathbf{y}) = \frac{p(\mathbf{y}|M_m)p(M_m)}{p(\mathbf{y})}$$

Combining these two equations, the priors and $p(\mathbf{y})$ cancel (if priors place equal probability on models) leaving

$$BF = \frac{p(M_1|\mathbf{y})/p(M_2|\mathbf{y})}{p(M_1)/p(M_2)} = \frac{p(\mathbf{y}|M_1)}{p(\mathbf{y}|M_2)}$$

This is the ratio of the probability of the data taken for the two models, represented by the integral

$$p(\mathbf{y}|M_m) = \int \overbrace{p(y|\theta_m, M_m)}^{likelihood} \overbrace{p_m(\theta_m)}^{prior} d\theta_m \qquad 6.9$$

where θ_m is the set of parameters associated with model M_m. This is a marginal distribution of the data for the model m. It consists of the likelihood times prior marginalized over parameters. The prior $p(M_m)$ represents the prior probability that the model is correct (more on this later). If the models differ only in terms of parameter values (they have the same functional form with prior parameter densities concentrated at θ_1 and θ_2), the Bayes factor is equivalent to the likelihood ratio,

$$BF = \frac{p(\mathbf{y}|\theta_1)}{p(\mathbf{y}|\theta_2)} \qquad 6.10$$

For BF values > 1, the data support model 1, and vice versa. The larger the value of the BF, the stronger the support for model 1. As a rough rule of thumb, values of 1 to 3 are taken as weak support and values > 10 of strong support for model 1 (e.g., Spiegelhalter et al. 2004).

There are some limitations to use of the Bayes factor. First, if the prior is improper (Chapter 4), so too is $p(y/m)$. Recall that a proper posterior density can often come from an improper prior density. This is not the case for $p(y/m)$. Second, the Bayes factor is inconsistent. If the models are nested, there is the opposite problem of the likelihood ratio: as n tends to infinity, the reduced model will always be selected. Predictive distributions provide an alternative approach.

6.5.2 Approaches Based on Predictive Distributions

Bayesian model choice can be based on predictive distributions: the probability of the data given the model. These are comparable across models, regardless of model form. Rather than the prior, we integrate over likelihood and posterior (Section 5.6.1). Consider the predictive density for a data set $\mathbf{y}_{(2)}$ on the basis of a model fitted to data set $\mathbf{y}_{(1)}$ for the mth model,

$$p(\mathbf{y}_{(2)}|\mathbf{y}_{(1)},M_j) = \int \overbrace{p(\mathbf{y}_{(2)}|\theta_m, M_m)}^{A}\overbrace{p(\theta_m|\mathbf{y}_{(1)})}^{B}d\theta_m \qquad 6.11$$

The integrand includes the posterior density for parameters fitted to data set $\mathbf{y}_{(1)}$ (density B) times the likelihood of the data set $\mathbf{y}_{(2)}$ in view of the fit to $\mathbf{y}_{(1)}$ (density A). Note that this density is proper, because the posterior B is proper.

Model selection can be based on different choices for the two data sets (Gelfand and Dey 1994). The Bayes factor results when we do not condition on data. For example, if $\mathbf{y}_{(1)}$ is the null set, then equation 6.10 involves the prior for θ_m, rather than the posterior. The ratio of such densities for two models is equivalent to equation 6.10, the likelihood ratio.

If $\mathbf{y}_{(1)}$ and $\mathbf{y}_{(2)}$ are the same data set, we simply have the posterior prediction for the data used to fit the model. For two models this yields the posterior Bayes factor of Aitkin (1991),

$$POBF = \frac{p(\mathbf{y}|\mathbf{y},\theta_1)}{p(\mathbf{y}|\mathbf{y},\theta_2)}$$

Finally, we could hold in reserve a portion of the data $\mathbf{y}_{(2)}$ and determine how well it is predicted by the fit to $\mathbf{y}_{(1)}$. A special case of this approach is termed *cross validation*, where the model is fitted to all but one observation y_k, and the fitted model is used to predict that observation. Let $\mathbf{y}_{(-k)}$ represent the data set containing all observations except y_k, and $\mathbf{y}_{(-)}$ represent the set of n such data sets. Now determine how well the model predicts all observations k based on each of the $\mathbf{y}_{(-k)}$ data sets as the product

$$p(\mathbf{y}|\mathbf{y}_{(-)},M_m) = \prod_{k=1}^{n} p(y_k|\mathbf{y}_{(-k)},M_m)$$

For comparison of two models, determine the pseudo-Bayes factor

$$PSBF = \frac{p(\mathbf{y}|\mathbf{y}_{(-)},M_1)}{p(\mathbf{y}|\mathbf{y}_{(-)}),M_2)}$$

Gelfand and Dey (1994) provide asymptotic approximations to the likelihood ratio,

$$\ln POBF \approx \ln LR + \frac{(p_2 - p_1)}{2}\ln 2$$

and

$$\ln PSBF \approx \ln LR + \frac{p_2 - p_1}{2}$$

where p_1 and p_2 are the number of parameters in the two models. These are penalized likelihoods (compare with AIC and BIC in Section 6.4). I summarize a related index for model selection, predictive loss, following the discussion of hierarchical models in Chapter 8.

The approach is related to prediction of new data sets that involve predictor variables $\mathbf{x}$, as discussed in Chapter 5. Consider a model where a response y_i corresponds to an known explanatory variable x_i. For observations involving $(\mathbf{x}, \mathbf{y})$, the predictive distribution is

$$p(\mathbf{y}_{(2)}|\mathbf{x}_{(2)},\mathbf{x}_{(1)},\mathbf{y}_{(1)}) = \int \overbrace{p(\mathbf{y}_{(2)}|\mathbf{x}_{(2)},\theta)}^{model} \overbrace{p(\theta|\mathbf{x}_{(1)},\mathbf{y}_{(1)})}^{posterior}d\theta \qquad 6.12$$

The model is fitted using the data $(\mathbf{x}_{(1)}, \mathbf{y}_{(1)})$ to obtain the posterior density of parameters $p(\theta|\mathbf{x}_{(1)},\mathbf{y}_{(1)})$. These estimates are used with a new set of covariates $\mathbf{x}_{(2)}$ to predict $\mathbf{y}_{(2)}$. I return to prediction in Chapter 11. In the following section prediction arises in the context of model averaging.

6.5.3 Combining (Averaging) Models

Bayesian model averaging (BMA) provides a way of incorporating model uncertainty into predictions and decisions. Combining models can be viewed as an alternative to model selection, whereby a range of models, each model weighted by its uncertainty, jointly contributes to an estimate or prediction. BMA is used primarily in the context of prediction.

The posterior prediction of $\mathbf{y}_{(2)}$ that comes from combining K models can be written as

$$p(\mathbf{y}_{(2)}|\mathbf{y}_{(1)}) = \sum_{m=1}^{K} p(\mathbf{y}_{(2)}|\mathbf{y}_{(1)},M_m)\mathrm{Pr}(M_m|\mathbf{y}_{(1)})$$

This is a weighted average of posterior predictions, each weighted by the posterior probability of the model. The posterior probability of each of K models is obtained as

$$\mathrm{Pr}(M_m|\mathbf{y}_{(1)}) = \frac{p(\mathbf{y}_{(1)}|M_m)\mathrm{Pr}(M_m)}{\sum_{k=1}^{K} \mathrm{Pr}(\mathbf{y}_{(1)}|M_k)\mathrm{Pr}(M_k)}$$

Hoeting et al. (1999) provide a comprehensive review of model averaging.

A point in favor of model averaging is performance. Model averaging can have better predictive capability than any single model. This can be true in terms of average predictions and in terms of prediction uncertainty (Hoeting et al. 1999). This can even be true for simulation models applied in a noninferential mode (e.g., Hanson et al. 2004). Rather than defend a specific model, one might defend the set of models included in the exercise. This might be especially advantageous if interested parties cannot agree on the model.

There are a number of practical considerations involved with implementation. When should we select models, and when should we combine? Yuan and Yang (2004) recommend model averaging when model selection is unstable. Model selection stability is gauged by the sensitivity of the outcome to minor changes in the data. If instability is low, go with model selection. Model averaging reduces this instability. How do we know the level of stability? They propose an instability index that is obtained by adding small perturbations to the data and assessing how estimates depart from those obtained using the actual data. The increase in effect size with increase in perturbation size is the index used to assess instability.

There are concerns with model averaging. The probabilities are conditional on which models are included in the calculation. How many models should be considered, and how many should be retained for averaging (Hoeting et al. 1999)? Recall that the prior for the model used to calculate the Bayes factor is conditioned on the notion that one of the models is indeed correct. How do we weight the models, when none of the models are correct? Moreover, model averaging has application to estimation problems only when parameters retain their interpretation across models. This is the case in some settings, possibly in some regression problems and the demographic parameters in capture-recapture models of Section 9.19. In many cases, parameter meaning changes from model to model. For example, how does one interpret the model-averaged estimate of the intercept β_0 when averaging over models with and without a slope β_1?

Implementation involves serious computation issues. Bayesian models can be large, involving, for instance, atmospheric circulation, land use, and remote sensing (Wikle et al. 2001; Fuentes and Raftery 2005; Agarwal et al. 2005), demography of a forest of trees over multiple years (Clark et al. 2004), and public health statistics at nested spatial scales (Carlin and Louis 2000) (Chapter 10). The number of candidate structures is enormous, and the implementation of any one model can be challenging. Implementation of a comprehensive set of models is not feasible in such instances. Many of the examples of model averaging come from generalized structures. In their overview, Hoeting et al. (1999) include generalized linear models, survival analysis, and graphical models, such as those that would be used for gene networks. Yuan and Yang's (2004) analysis focuses on regression models. It may be possible to narrow the field of possible models in some logical way or to work with a sample of models. The latter can be accomplished with reversible jump Markov Chain Monte Carlo Computation (MCMC) (Green 1995). I refer to this method again when I discuss MCMC algorithms in Chapter 8.

6.6 Additional Thoughts on Bayesian Model Assessment

The approach to model selection depends on the general modeling framework (e.g., classical, Bayes) and on a number of conceptual and practical considerations. For simple models, there may be less effort involved with a classical approach. The assessment of large Bayesian models remains challenging, but such models typically would not even be attempted in a classical setting. I discuss model selection further with computational methods in Chapter 7. For now, here are several reasons to consider a Bayesian approach:

- Bayes provides a structure to incorporate formally prior information. For most types of analyses, priors can also be made noninformative, allowing focus on information from a specific data set. For large models, parameters will differ in terms of the amount of prior knowledge.
- Bayesian inference is conditional on the actual data, and not on values more extreme than observed. The frequentist P value is calculated based on values more extreme than the data.
- In general, the flexibility of a Bayesian framework liberates the investigator from the rigid constraints of classical design.
- The posterior represents the way we think about parameters. There is a consistent use of probability rules from prior to posterior and from posterior to various summaries, such as marginal distributions, moments, predictions, and decisions.
- The posterior is the goal, and it is obtained directly. For example, we can accept a null hypothesis, rather than simply fail to reject. With Bayes, we have a posterior probability.
- As problems become complex, an inevitable ad hockery creeps into classical approaches. There will be no clear approach to a problem

with the result that methods proliferate, none of which are particularly general. There is a need to appeal to asymptotics (assumptions that apply in the limit as n goes to infinity), and important assumptions may go unrecognized. By contrast, Bayes offers a consistent approach that applies generally.

Some practical advantages of Bayes become obvious where we address complex problems, which require some computational tools. The following chapter provides an introduction.

Part III Larger Models

As the complexity of problems increases, hierarchical structures become increasingly advantageous and computation emerges as a central consideration. The two chapters in this section focus on both aspects. Basic computational methods are the topic of Chapter 7, providing an introduction on how large models are analyzed. With these basic techniques in mind, I return to hierarchical models in Chapter 8, emphasizing how modeling and computation fit together.

7 Computational Bayes: Introduction to Tools Simulation

The genie cannot be put back into the bottle. The Bayesian machine, together with MCMC, is arguably the most powerful mechanism ever created for processing data and knowledge.

—Berger (2000, p. 1273)

Gibbs sampling can be dangerous.

—BUGs project documentation

I$_\text{T IS HARD TO EXAGGERATE}$ the rapid expansion and pervasive impact of modern Bayesian analysis. This explosion of activity can be attributed to the computational tools that allow for formal analysis of complex problems, where much is uncertain (Gelfand and Smith 1990; Berger 2000). Yet, even with increasing availability of software, winBUGS being most prominent, some sophistication with distribution theory and computation is needed to avoid disaster. Whether constructing algorithms or implementing available software, the practitioner needs the some basic understanding of the material covered in this chapter.

7.1 Simulation to Obtain the Posterior

The posterior distribution is the most important product of a Bayesian analysis. From Equation 4.1 Bayes' theorem for the posterior density of θ is written as

$$p(\theta|y) = \frac{p(y|\theta)p(\theta)}{p(y)} = \frac{p(y|\theta)p(\theta)}{\int_{-\infty}^{\infty} p(y|\theta)p(\theta)d\theta}$$

This posterior consists of the product of likelihood and prior distribution divided by the normalization constant. The normalization constant requires integration. Where likelihood and prior are conjugate, this integral is available (Section 5.2). For most complex models, this integral is not available. Thus, distributions are typically simulated. In effect, simulation normalizes the posterior density.

As previously mentioned, some densities cannot be normalized, because the integral is not finite. We typically think of a density as having one or more modes (usually one) and tails at the extremes. These tails tend to zero such that there is a finite area under the curve. For improper distributions, the tails do not converge to zero fast enough, in which case the integral (area under the curve) can be infinite. When I refer to densities that can be normalized I mean that they have a finite area under the curve. When I refer to improper prior densities, I mean a density that does not have a finite integral.

Simulating a distribution entails repeatedly drawing samples at random, in an iterative fashion, and assembling them into a histogram or a smoothed density. For a univariate distribution, a sample of a single variate, $\theta^{(g)}$, is taken at each iteration g. This notation refers to the sample value from the distribution of θ taken at the gth step or iteration of the simulation. Simulation involves drawing $g = 1, \ldots, G$ samples of the parameter θ. Where the posterior involves just this one parameter, then the set of sample values is $\{\theta^{(1)}, \ldots, \theta^{(g)}\}$. For a multivariate distribution of p parameters, a single sample (i.e., one iteration) consists of a vector of length p, represented as $\{\theta_1^{(g)}, \ldots, \theta_p^{(g)}\}$. A simulation would produce a G by p matrix of samples, with each iteration represented by one row of parameter estimates. Some distributions can be directly sampled, often using algorithms available from standard software. Many are not readily available.

In this chapter I consider several methods for simulating posterior distributions, and I illustrate how results can be used for inference and prediction. Whereas previous chapters emphasized modeling, this one is mostly devoted to computation. These technical concerns are taken up here, because enough modeling is behind us to motivate more sophisticated computation methods. These tools allow implementation of more complex models that follow in subsequent chapters. This chapter moves from methods for sampling from univariate distributions, to algorithms that are appropriate for more complex models that can involve a large number of parameters.

7.2 Some Basic Simulation Techniques

Random draws from probability distributions are needed to simulate Bayesian posteriors. Random number generators are now available for most standard distributions (e.g., many of those included in Appendix F). Many other distributions are not standard and, thus, cannot be directly sampled. A number of techniques have been developed for stochastic simulation, a few of which are widely used. Here I introduce several techniques that have broad application. Methods covered in this section demonstrate three approaches to sampling from a nonstandard distribution. The first class of methods is termed *indirect*, because we first sample from a distribution different from the target distribution. This sample is viewed as a proposal that can be accepted or not based on the relationship between the distribution used to generate the proposal and the target distribution. One such method is *rejection sampling* (Section 7.2.1). Later in the chapter I consider another indirect method, *Metropolis* sampling, and a variant on that theme, *Metropolis-Hastings*.

A second method (Section 7.2.2) involves use of the discretized inverse cumulative distribution function to transform a sample drawn from a uniform distribution into one from the target distribution. This method is direct, in the sense that each random draw is a sample from the (discretized) target distribution.

A third approach involves a multivariate distribution where univariate samples can be sequentially drawn. I illustrate the approach with the multinomial and Dirichlet in Box 7.1.

Box 7.1 Simulating the multinomial and Dirichlet

Software is typically available for simulating the binomial,

$$x_i \sim Bin(n, \theta)$$

Random variate generators for the multinomial are readily obtained from the fact that the distribution of a single x_i value is binomial. A multivariate random variate for the multinomial with k parameters (e.g., "bins")

$$x_i \sim Multinom(n; \theta_1, \ldots, \theta_k)$$

can be simulated with a sequence of binomial draws:
Draw a binomial random variate from the first marginal,

$$x_1 \sim Bin(n, \theta_1)$$

Draw x_2 from the new binomial that arises,

$$x_2 \sim Bin\left(n - x_1, \frac{\theta_2}{\theta_2 + \theta_3 + \ldots + \theta_k}\right)$$

Draw subsequent variates $x_j, j = 2, \ldots, k-1$ from

$$x_j \sim Bin\left(n - \sum_{i=1}^{j-1} x_i, \frac{\theta_j}{\sum_{i=j}^{k} \theta_i}\right)$$

The final variate is

$$x_k = n - \sum_{i=1}^{k-1} x_i$$

(Continued on following page)

If the sample size parameter falls to zero, then subsequent variates are zero.

The Dirichlet distribution is the conjugate prior for the multinomial parameters. In a Gibbs sampler, you will draw samples from the conditional posterior. This algorithm is taken directly from Gelman et al. (1995).

Consider a length-p parameter vector that will be drawn from a Dirichlet,

$$\theta = [\theta_1, \ldots \theta_p] \sim Dir(a_1, \ldots a_p)$$

First draw p independent values from the exponential density, $x_k \sim Gorm(\alpha_k, 1)$ (a_k), $k = 1, \ldots p$. The kth element of the random vector θ is

$$\theta_k = \frac{x_k}{\sum_{j=1}^{p} x_j}$$

The methods summarized here are not exhaustive. Additional references include Johnson (1987), Ripley (1987), and Tanner (1996) and a number of recent books specifically on Bayesian modelling and computation (Gilks et al. 1996; Gelman et al. 1995).

7.2.1 Rejection Sampling

Rejection sampling is a simple method for random variate generation that is appropriate in many applications. Suppose we wish to simulate a nonstandard distribution $p(x)$. Rejection sampling is based on random draws from a standard distribution, call it $f(x)$, which can then be compared with the desired distribution $p(x)$ as the basis for accepting the random variate. Do this repeatedly to build up a frequency distribution of accepted draws, which approximates the target distribution.

This method is not restricted to distributions. It can be applied to a function $P(x)$ that need not be a density function, but it must be possible to normalize it. It might be a non-normalized posterior distribution. Here are the constraints:

1. $P(x)$ must have a finite integral over the range of possible values of x. In other words, there is a target density $p(x)$ we wish to simulate that is proportional to $P(x)$,

$$p(x) = \frac{P(x)}{\int P(x)dx}$$

2. $P(x)$ must have nonzero values over the range of x for which the standard density $f(x)$ is defined.
3. It must be possible to select a constant C such that $Cf(x) \geq p(x)$ for all x.

Here is the procedure:

1. Draw a random variate x_i from $f(x)$.
2. Calculate the ratio $a = p(x_i)/Cf(x_i)$. Because of condition (3) above, this ratio will fall between zero and one.
3. With probability a accept this value of x_i as a random draw from $p(x)$. With probability $1 - a$ reject this value of x_i. To do this, draw a random variate from the uniform density $Unif(z|0,1)$. If $z < a$, accept x_i.

This is easiest to see with an example.

Example 7.1. Turning angles in aphids

Hudgens (2000) required turn angles for aphid movement that could be used to simulate their dispersion. This is a common problem encountered by population biologists attempting to simulate movement of individuals and spread of a population (Cain 1990). Ecologists often seek to test for and quantify directional bias in movement (Kareiva and Shigesada 1983; Turchin et al. 1991; Cain et al. 1995). Here I simply use directional data as demonstration of rejection sampling. The data (Figure 7.1a) show a propensity toward low turn angles (a slight tendency to continue in the same direction), but all directions are possible. The data were recorded in bins x_i. I fit a cosine function to the data, with the density of turn angles x being

$$p(x) = \frac{\beta_0 + \beta_1 \cos(\beta_2 + x)}{2\pi\beta_0}$$

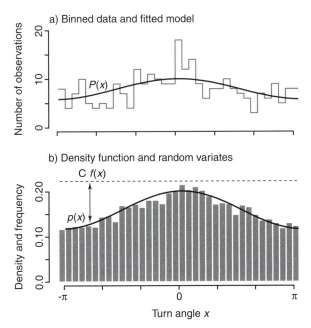

a) Binned data and fitted model

b) Density function and random variates

FIGURE 7.1. Turn angles simulated by rejection sampling. (*a*) Binned data and the model fitted to those data. (*b*) The target density $p(x)$ is less than the envelope for all x, which is proportional to (C times) the uniform sampling distribution $f(x)$.

This happens to be a normalized density (the denominator is a normalization constant), but it need not be. The sampling distribution used to fit this function to the number of turns y_i observed in bin x_i is Poisson

$$y_i \sim Pois(np(x_i)dx)$$

where the Poisson parameter refers to bin x_i, all bins having width dx, and n is sample size. We wish to draw deviates from this distribution, which is not available from standard software. To generate random deviates, $f(x)$ is chosen to be uniform,

$$f(x) = Unif(x|-\pi, \pi)$$

Most software packages do provide uniform random variates. For each uniform deviate drawn, the acceptance probability is calculated. Accepted deviates were used to generate Figure 7.1b.

Rejection sampling can be simple and efficient. The main difficulty with rejection sampling involves fat-tailed distributions. To be efficient, the standard distribution $f(x)$ should have a shape that is not too different from the target distribution. To see this, note that the difference between $Cf(x)$ and $p(x)$ in the lower panel of Figure 7.1 is not large. The relative heights of these two quantities determine the fraction of draws that will be accepted. If the difference is large, most draws are rejected. For this reason, it is efficient to select a sampling distribution that has a shape similar to that of the target distribution. For some target distributions it can be difficult to find a standard distribution from which to sample efficiently—to assure coverage of the tails may require that $Cf(x) >> p(x)$ for values of x near the mode of the distribution. It is also possible to miss violations of this requirement in a complex model.

Adaptive rejection sampling provides an especially efficient envelope in cases where the conditional density is log concave (the second derivative is negative). This condition is often satisfied. A straightforward explanation is given by Gilks et al. (1996).

7.2.2 Inverse Distribution Sampling

Rejection sampling provides a vector of random variates that can be used to assemble a distribution. Although Example 7.1 applied this method to a discrete distribution, it readily applies to a continuous one—I used a discrete distribution of turn angles only because the data were collected that way.

If we are willing to approximate a continuous posterior with a discretized one, and to apply a uniform prior density on an interval $[a, b]$, then it can be sampled directly. By discretized, I mean a continuous density that is approximated by discrete bins (Example 7.1). Consider a sample from the density for a parameter ρ. First, note that a uniform random variate on the interval $[0,1]$

can be directly transformed into a sample from a target distribution $p(\rho)$ using the inverse distribution function $P^{-1}(\rho)$,

$$z_i \sim Unif(0,1)$$

$$\rho_i = P^{-1}(z_i)$$

To see this inverse function, turn your head sideways and look at the upper panel of Figure 7.2. Instead of moving your eyes from the horizontal axis up to the curve, look instead at the vertical axis and follow it sideways to the curve. It has the name *inverse* because we are asking what value of ρ corresponds to $P(\rho)$. This is the inverse of what we usually do. Figure 7.2 illustrates ten values of z_i uniformly spaced along [0,1] (vertical axis) transformed into ten values of ρ (horizontal axis). A large number of samples will approximate the lower density, because most of the random variates on [0,1] are translated by the steep part of the distribution function.

This method for generation of random variates is restricted to target distributions for which the inverse distribution function is available. The inverse distribution function is usually not available, because most distribution functions cannot be inverted. However, discretization allows for an approximation. Here I describe this method in the context of a Bayesian posterior. The approach I describe exploits the same basic relationships as Tanner's (1996) "griddy Gibbs sampler."

Recall that the Bayesian posterior is proportional to the product of likelihood and prior. For the uniform prior that is bounded between values of a and b this is

$$p(\rho|\mathbf{y},a,b) \propto p(\mathbf{y}|\rho)Unif(\rho|a,b)$$

Because the prior is uniform, the posterior is simply proportional to the likelihood within the interval $[a, b]$ and zero elsewhere. Here is a basic implementation

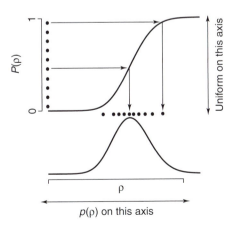

FIGURE 7.2. To sample from the inverse distribution function, generate a uniform random variate on [0,1] (vertical axis labeled $P(\rho)$), and then find the value of ρ to which this random variate applies (horizontal axis). A large number of such samples would approximate the smooth density shown at the bottom. To see this, note how ten uniformly spaced values on the vertical axis translate to the horizontal axis (black dots).

followed by a nuance that should be included when sample sizes are large and the likelihoods are too small for computation.

Step 1. Define a vector of ρ values ranging from a to b. There are k values in the sequence from $\rho_1 = a$ to $\rho_k = b$.

Step 2. Evaluate the likelihood at each of these k values, $p(y|a), \ldots, p(y|b)$.

Step 3. Determine the cumulative sum for these values. The jth value of this sequence is

$$P_j = \sum_{i=1}^{j} p(y|\rho_i) = p(y|a) + p(y|\rho_2) + \cdots + p(y|\rho_j)$$

This sequence of discrete values is approximately proportional to the cumulative distribution function for the likelihood.

Step 4. Draw a uniform random variate from

$$Z \sim Unif(P_1, P_k)$$

Step 5. The value of the random variate is set equal to the value of ρ_j to which Z corresponds.

The additional nuance to include involves rescaling. In Step 2, substitute the log likelihood, computing the sequence $\ln p(y|a), \ldots, \ln p(y|b)$, and then rescale using, say, the maximum value in the sequence (the constant is used only to rescale to a comfortable range of values). I use the maximum value, which leaves a relationship like the normed likelihood of Chapter 5 (Equation 5.11). Substract the maximum value and then exponentiate to obtain

$$R(\rho;y) \equiv \exp[\ln p(y|\rho_j) - \ln p(y|\rho_{ML})]$$

This normalization means that there is no need to exponentiate very large negative numbers. The resulting range of values is [0, 1]. Then Step 3 makes use of the normed rather than actual likelihood. Here is an example.

Example 7.2. A posterior density for the exponential
To demonstrate the method, I draw a sample of $n = 100$ random variates from

$$y_i \sim Exp(\rho),$$

where $\rho = 0.1$. This example is used because it is transparent—the inverse distribution function is available for the exponential and is drawn for comparison. I assume a uniform prior parameter density on [0.05, 0.15], knowing that the likelihood will be concentrated between these values. How do I know this? Clearly the maximum likelihood value will be close to 0.1. From Chapter 6, I know the standard error to be

$$se_\rho = \frac{\rho_{ML}}{\sqrt{n}} = 0.01$$

The interval I choose for my uniform prior will safely cover the distribution, spanning roughly 10 standard errors. Here are the steps outlined above:

Step 1. I choose an arbitrary discretization of thirty increments between $a = 0.05$ and $b = 0.15$. The width of these bins can be identified in the histograms of Figure 7.3c and 7.3d.

Step 2. Determine the log likelihood at each of these values of ρ_j and the normed likelihood (Figure 7.3a).

Step 3. The cumulative sum of values P_j is shown in Figure 7.3b. The range of values on the vertical axis is not important, because I have not bothered to normalize such that $P(b) = 1$, as only relative values matter. The important thing is that it is proportional to the cumulative distribution function for the likelihood.

Step 4. Draw a sample of size m (in this case, $m = 100$) uniform random variates on the range of P_j. These values are shown as horizontal lines in Figure 7.3b. The density of lines along this vertical axis is uniform.

Step 5. The density of ρ is approximated by the frequency of ρ_j values corresponding to these uniform random variates. I have shown them as vertical

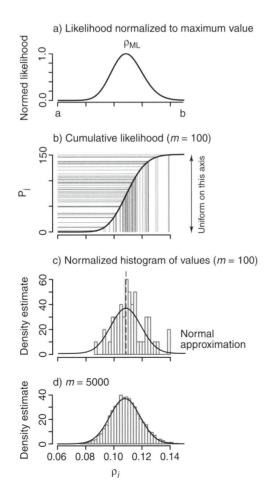

FIGURE 7.3. Sampling the exponential parameter from Example 7.3. The dashed vertical line is the MLE for this sample.

lines in Figure 7.3b. Note how the cumulative distribution function concentrates values where the slope is large. The histogram is shown in Figure 7.3c.

For comparison with previous methods, Figure 7.3c and 7.3d includes the normal approximation from Chapter 5, based on Fisher Information. For large sample size (here $n = 100$) and large m, they will be similar (Figure 7.3d).

I deliberately committed a crime in Example 7.2 to highlight an issue that needs recognition. The data entered the analysis twice in this example, through the likelihood (of course!), but also (and more subtly) for purposes of defining the limits on the uniform prior distribution. Strictly speaking, this is not permitted. One might argue that, from a practical standpoint, the limits on the uniform distribution have no impact (in this example), so this subtle violation does not affect inference. In other words, I could have made this uniform distribution arbitrarily wide and, provided bin widths were the same, the posterior would be essentially the same. While this may often be true, for more complex models, posterior distributions may be very sensitive to limits of this uniform prior.

Discretization circumvents the need for a solution to the inverse distribution function. For computation, one simply needs an algorithm that finds the value ρ_j for a given P_j. For example, in the language R (see lab manual) define the vectors rj (sequence of values of ρ_j from a to b) and Pj (sequence of P_j values from P_1 to P_k). The inversion can be done with the code

```
z <- runif(1,Pj[1],Pj[k])

rj[findInterval(z,Pj)]
```

This algorithm slots the uniform random variate z into the appropriate position in Pj and then locates the value in the rj sequence associated with that interval Pj.

Inverse distribution sampling is direct and easy to program. There is no need to propose and then accept with some probability. It does involve a uniform prior and discretization. There are many instances where discretization is not a drawback.

7.2.3 Other Methods for Nonstandard Distributions

The foregoing examples should provide a feel for stochastic simulation. There are a number of other methods for sampling from nonstandard distributions. The multinomial in Box 7.1 represents still another approach involving sequential steps for each of p elements of a p-dimensional random variate. Additional techniques include *importance sampling* (e.g., West and Harrison 1997) and the *slice algorithm* (Carlin and Louis 2000, p. 168). The latter involves sampling first from a uniform distribution for an auxiliary variable and then retaining only those draws that fall within the envelope defined by a known function of the second variable. Similar options are available when the nonnormalized density is expressed as a product, such as the Gaussian likelihood for growth bounded by a uniform probability that growth in a given year cannot result in a tree size outside the currently estimated sizes for last year and this year (Section 9.11). In practice, most Bayesian posteriors are multivariate, and they cannot be efficiently sampled using a single step. I now consider methods for simulating multivariate distributions within a more general framework. The methods considered thus far will reappear as steps within this framework.

7.3 Markov Chain Monte Carlo Simulation

Markov chain Monte Carlo (MCMC) simulation is used to estimate multivariate distributions. As suggested by the name, this method produces a Markov chain of estimates that represents a random walk through the target distribution. The chain is initialized with values for each of the p parameters represented in the prior and posterior densities. Samples can then be drawn sequentially from conditional densities, with sample values at each step being conditional on the current values of all parameters. After some burn-in period, during which the random walk converges to the target distribution, the chains of estimates are used to assemble a posterior distribution.

As an example, Figure 7.4a shows a contour plot of a bivariate normal density for parameters in the vector $\theta = [\theta_1, \theta_2]^T$. The covariance matrix for this bivariate normal can be written as

$$\Sigma = \begin{bmatrix} \sigma_1^2 & \rho\sigma_1\sigma_2 \\ \rho\sigma_1\sigma_2 & \sigma_2^2 \end{bmatrix}$$

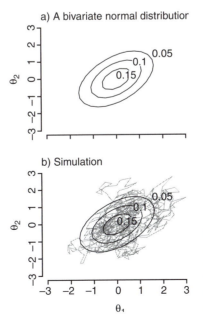

FIGURE 7.4. Simulation of a bivariate normal distribution using MCMC. In b is shown a random walk, where successive steps are taken using a Metropolis algorithm. This is a Markov chain.

with variances on the diagonal and covariances elsewhere. The correlation is ρ. The vector of means is $\mu = [\mu_1\ \mu_2]^T$. The volume under this density equals unity, that is,

$$1 = \int\limits_{-\infty}^{\infty} \int\limits_{-\infty}^{\infty} N_2(\theta|\mu,\Sigma)d\theta_1 d\theta_2$$

For the example in Figure 7.4, the vector of mean parameter values is $\mu = [0\ 0]^T$, the variances are equal to one, and the correlation is 0.5. As demonstration, I simulate this standard density,

$$p(\theta) = N_2\left(\begin{bmatrix} \theta_1 \\ \theta_2 \end{bmatrix} \begin{bmatrix} 0 \\ 0 \end{bmatrix}, \begin{bmatrix} 1 & 0.5 \\ 0.5 & 1 \end{bmatrix}\right) \qquad 7.1$$

This example of Section 7.3.1 follows Gelman et al. (1995).

First, it should be clear that many software packages include samplers for the multivariate normal, in which case there is no need to write an algorithm to simulate it. To approximate the distribution, one could simply draw independent samples directly from the bivariate normal, each draw producing a value for both parameters.

MCMC can be used when a sampler is unavailable. Instead of drawing independent samples directly from the target distribution 7.1, draw samples from a conditional distribution having a mean value centered on the most recent sample drawn. The successive draws represent a random walk (Figure 7.4b). Obviously, this process is less efficient than drawing independent samples directly from the target distribution. Again, MCMC is used when independent sampling directly from the target distribution is not an option.

Here I discuss the conditional distribution and how it is used to approximate the target distribution, that is, the bivariate normal of Figure 7.4. I describe a random walk through the space defined by the parameter vector θ, where the probability of taking a step to a new location in the parameter space of Figure 7.4 is proportional to the density at that new location (shown as contours in Figure 7.4a) relative to that for the current location. This random walk might look something like Figure 7.4b. Because the direction of each step of the random walk reflects the underlying density, the sampler spends most of the time near the mode of the distribution. Excursions deep in the frontier are possible, but eventually find their way back to the center of the distribution. Indeed, the amount of time spent in all sectors of the parameter space can be used to draw a contour map for Figure 7.4b that looks much like 7.4a. The resulting Markov chain approximates the target distribution, which could be a joint posterior. The sampling that goes on within the MCMC algorithm can be done in many ways and can involve some of the techniques already discussed (rejection sampling, inverse distribution sampling) and some others, including Metropolis sampling. In the following section I introduce the Metropolis algorithm for the bivariate normal example, and then discuss it in the context of Gibbs sampling.

7.3.1 The Metropolis Algorithm

Consider again a target density $p(\theta)$ that that can be determined up to a constant, in the sense that we lack the integration constant that makes it a true density. This constant is the normalizer that requires integration in the denominator of Bayes' theorem (Equation 4.1). θ can be a single parameter or a vector of p parameters $[\theta_1, \ldots, \theta_p]$. For the example in Figure 7.4 there are $p = 2$ parameters. Although values can be drawn directly from this standard distribution, I use it as an example for the Metropolis algorithm. The Metropolis algorithm is one method for constructing a random walk having a stationary distribution described by the target distribution (it is an MCMC technique). The algorithm consists of iterative steps. At each step a proposed value is drawn that is conditional on the current location in parameter space—it is a Markov process. Let g be the index of the process. In other words, the parameter value for the gth step of the random walk in Figure 7.4b is indexed as $\theta^{(g)}$. Following Gelman et al. (1995), the algorithm goes like this.

1. Select a symmetric "jump" density or transition kernel from which to draw random candidate samples of θ^* conditioned on the current values $\theta^{(g)}$. I refer to these values as candidates, because, as with rejection sampling, I will not accept all such samples. This jump density is represented as $J(\theta^*|\theta^{(g)})$. For example,

$$J(\theta^*|\theta^{(g)}) = N_p(\theta^*|\theta^{(g)}, \mathbf{V})$$

says that the candidate value is drawn from a multivariate normal density of dimension p with mean values given by the current parameter vector $\theta^{(g)}$ and a covariance matrix $\mathbf{V}$. The covariance matrix determines the size and direction of steps. $\mathbf{V}$ would be set to ensure that the steps are not too small and not too large—more on this in a moment.

2. Now determine the density at the candidate value relative to that of the current value,

$$a = \frac{p(\theta^*)}{p(\theta^{(g)})} \qquad\qquad 7.2$$

In practice the two densities on the right-hand side of Equation 7.1 are not normalized. If they were, we wouldn't bother with this simulation. This example is only for purposes of illustration.

3. If $a > 1$ (i.e., the probability of the candidate value exceeds that of the current value), accept the candidate value. This would be like taking a step toward the center of the distribution in Figure 7.4. If $a < 1$ accept the candidate value as the new value $\theta^{(g)}$ with probability a. Otherwise, set $\theta^{(g+1)} = \theta^{(g)}$ (retain the current value).

Here is the specific example:

Example 7.3. Simulating a bivariate normal distribution using a Metropolis algorithm

The bivariate normal distribution

$$N_2\left(\begin{bmatrix} \theta_1 \\ \theta_2 \end{bmatrix} \begin{bmatrix} 0 \\ 0 \end{bmatrix}, \begin{bmatrix} 1 & 0.5 \\ 0.5 & 1 \end{bmatrix}\right)$$

can be simulated with a Metropolis algorithm. Set an initial value of, say, $\theta^{(0)} = [-1,1]$.

1. Steps should be neither too large nor too small. This scale is set by the covariance matrix of jump density J. Draw a candidate pair of values from J, which I have chosen to also be bivariate normal,

$$J\left(\begin{bmatrix} \theta_1^* \\ \theta_2^* \end{bmatrix} \begin{bmatrix} \theta_1^{(0)} \\ \theta_2^{(0)} \end{bmatrix}\right) = N_2\left(\begin{bmatrix} \theta_1^* \\ \theta_2^* \end{bmatrix} \begin{bmatrix} -1 \\ 1 \end{bmatrix}, \begin{bmatrix} 0.09 & 0.045 \\ 0.045 & 0.09 \end{bmatrix}\right)$$

The mean of the jump distribution is the current value $[-1,1]$. The covariance matrix of J was selected to produce random samples that are smaller than the scale of the target distribution. The implied correlation of proposals, $\rho = 0.5$, means that I will tend to draw values that follow the correlation of the underlying density I wish to simulate. A random draw from this density is

$$\begin{bmatrix} \theta_1^* \\ \theta_2^* \end{bmatrix} = \begin{bmatrix} -0.9321 \\ 0.8906 \end{bmatrix}$$

Note that this proposed new value is closer to the mode of the distribution $[0, 0]$ than is the current value $[-1,1]$. Thus, I accept it. Verify this in Step 2.

2. The relative density is the ratio of candidate versus current parameters,

$$a = \frac{p(\theta^*)}{p(\theta^0)} = \frac{N_2\left(\begin{bmatrix} -0.9321 \\ 0.8906 \end{bmatrix} \begin{bmatrix} 0 \\ 0 \end{bmatrix}, \begin{bmatrix} 1 & 0.5 \\ 0.5 & 1 \end{bmatrix}\right)}{N_2\left(\begin{bmatrix} -1 \\ 1 \end{bmatrix} \begin{bmatrix} 0 \\ 0 \end{bmatrix}, \begin{bmatrix} 1 & 0.5 \\ 0.5 & 1 \end{bmatrix}\right)} = \frac{0.03489}{0.02487} = 1.403$$

Note that the numerator and denominator both contain the target density.

3. The acceptance probability is

$$a = \min(1.403, 1) = 1$$

so I accept this candidate vector. If the acceptance probability were <1, I would draw a random uniform variate on $[0,1]$ and accept the candidate vectors if the acceptance probability was greater than this random draw.

Any time a proposed value falls in a region of higher probability than the current value, the proposal is accepted. If a proposal falls in a lower probability region, it is accepted with a frequency that is proportional to the probability associated with that location relative to the current location. The upshot is a random walk that explores the distribution, spending a fraction of steps in a region proportional to its density.

It should be evident that the efficiency of the algorithm depends on the jump distribution. For the bivariate normal example, the jump distribution (normal with covariance matrix **V**) is efficient, with many candidate values being accepted, yet steps are not excruciatingly small. The Markov chain efficiently explores the parameter space (Figure 7.5a). If steps are too large (large variances in **V**), many candidate values fall in low probability regions of the parameter space, and few are accepted. Figure 7.5c shows a jump distribution having covariances four times larger than those of the target covariance matrix. In contrast to Figure 7.5a, this plot shows only a few steps are accepted, and they are erratic. There are few steps, because most proposals were rejected. Figure 7.5b shows an intermediate proposal density.

It is also good practice to employ more than one chain. The multiple chains come from simulations started from different initial values. Multiple chains can build confidence that convergence has been achieved. Figure 7.5a shows two chains, both of which would eventually map out the same target distribution. If they did not, there may be a convergence problem—more on that later.

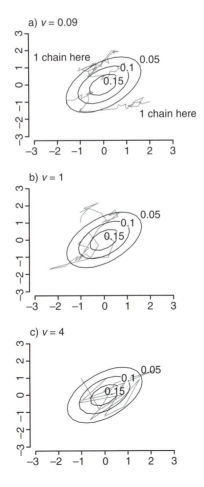

FIGURE 7.5. Metropolis simulation of a bivariate normal using jump distributions with three different covariance matrixes. The first distribution uses a small variance and, thus, requires many steps to explore the distribution. The third distribution has a large variance and, thus, the rejection rate is high. The high rejection rate means that only a few steps have been taken over the same number of iterations of the algorithm. Many steps are evident in *a*.

7.3.2 Metropolis-Hastings

The Metropolis algorithm requires a symmetric jump distribution. In other words,

$$J(\theta_1|\theta_2) = J(\theta_2|\theta_1)$$

For the Gaussian jump distribution used in the last example, the probability of stepping from the current parameter value to the new one is the same as that for stepping in the opposite direction—the Gaussian distribution is symmetric.

There are many cases where a symmetric jump distribution is not appropriate. Consider a target distribution that assumes only positive values (e.g., a gamma or lognormal). If the current parameter value in the simulation is close to zero, the next value should have a greater probability of stepping to the right than to the left; steps should not be negative. One can use an asymmetric jump distribution, provided that the acceptance ratio of Equation 7.2 is corrected for the asymmetry,

$$a = \frac{p(\theta^*)\Big/ J(\theta^*|\theta^{(g)})}{p(\theta^{(g)})\Big/ J(\theta^{(g)}|\theta^*)} \qquad 7.3$$

The correction is achieved by the ratio of jump probabilities between current and proposed values. This Metropolis-Hastings (M-H) algorithm is incorporated in the Gibbs sampler below.

7.3.3 The Gibbs Sampler

Methods described thus far are feasible for low-dimensional problems—problems that involve a small number of parameters. Complex models typically involve more parameters than can be handled with these approaches alone. *Gibbs sampling* makes use of conditional relationships that allow for simulation of high-dimensional problems. The cornerstone of Gibbs sampling involves factoring a high-dimensional posterior density into a collection of low-dimensional (often univariate) densities that can be sampled one or more at a time. The structure makes use of the fact that the collection of full conditional distributions uniquely determines the joint distribution. These relationships are discussed in Appendix D.4. A Gibbs sampler involves alternately sampling from the conditional distributions for each parameter (or block of parameters), updating each value or block of values before proceeding to the next. Perhaps the best way to introduce the method is through example.

In Example 7.4 I consider a posterior of this form:

$$p(b,\alpha,\beta|\mathbf{y}) \propto \overbrace{p(\mathbf{y}|b)}^{A}\overbrace{p(b|\alpha,\beta)}^{B}\overbrace{p(\alpha)}^{C}\overbrace{p(\beta)}^{D} \qquad 7.4$$

The left-hand side is the joint posterior for three parameters. On the right-hand side is the likelihood (labeled A) with parameter b. There follows a density for b (density B), which depends on α and β. Finally, we have densities for α and β (C and D). These densities presumably also involve parameters, but they will be taken as fixed. Thus, I do not write them explicitly. I refer to B as a prior and C and D as hyperpriors. This is an example of a hierarchical model, which I take up in more detail in Chapter 8. For now, I focus on sampling.

To simulate this joint posterior I sample alternately from the conditional posteriors for the three parameters. In contrast to the full model, these posteriors will be low-dimensional, in this case, univariate. One step, or iteration, of the Gibbs sampler involves an update of all three parameter values in the same fashion as described by Figure 7.4b. Because there are three parameters, each of which will be drawn from a different conditional posterior, one Gibbs step involves three draws.

How do we identify the conditional posteriors? Here I use relationships discussed in Appendix D.4. Equation 7.4 is a joint density for three parameters conditioned on $\mathbf{y}$. To simplify notation, ignore $\mathbf{y}$ for the moment. Because I am dealing with proportionalities, I can substitute a conditional density for this joint density. I can do this, because, for parameter b, the following is true:

$$p(b|\alpha,\beta) \propto p(b,\alpha,\beta)$$

In other words, a conditional density is proportional to the joint density. It is not equal to the joint density, just proportional to it. So the joint density on the right-hand side of 7.4 could be replaced with this conditional. This substitution gives

$$p(b|\alpha,\beta,\mathbf{y}) \propto \overbrace{p(\mathbf{y}|b)}^{A} \quad \overbrace{p(b|\alpha,\beta)}^{B} \quad \overbrace{p(\alpha)}^{C} \quad \overbrace{p(\beta)}^{D}$$

The proportionality still holds, but it is more complicated than it needs to be. Any density on the right-hand side that does not contain b is a constant (with respect to b). Densities C and D affect the magnitude of the right-hand side equally for all values of b. So I can omit them, and the proportionality still holds. This gives the simple relationship

$$p(b|\alpha,\beta,\mathbf{y}) \propto \overbrace{p(\mathbf{y}|b)}^{A} \quad \overbrace{p(b|\alpha,\beta)}^{B} \qquad\qquad 7.5$$

In short, the components of each conditional posterior can be identified from the densities in which they appear. We are now ready to discuss sampling.

Suppose I am now at step g of an MCMC simulation, or Gibbs sampler. The gth Gibbs step will update each of the three parameter values. I can sample them in any order. I start with parameter b. I draw a sample from

$$p(b^{(g+1)}|\alpha^{(g)},\beta^{(g)},\mathbf{y}) \propto p(\mathbf{y}|b^{(g)})p(b^{(g)}|\alpha^{(g)},\beta^{(g)})$$

On the right-hand side is the product of two densities (A and B of the joint posterior in Equation 7.1), both of which contain parameter b. The superscript $^{(g)}$ indicates the currently imputed value, and $^{(g+1)}$ indicates the updated value. From this distribution I draw a sample for b. To obtain this draw, I might use one of the techniques discussed above.

I now sample from the conditional posterior for α, which comes from densities B and C of Equation 7.4,

$$p\big(\alpha^{(g+1)}|b^{(g+1)},\beta^{(g)},\mathbf{y}\big) \propto p\big(b^{(g+1)}|\alpha^{(g)},\beta^{(g)}\big)p\big(\alpha^{(g)}\big)$$

Notice on the right-hand side that the density for α uses the updated value for b. I have now updated b and α.

Finally, I update parameter β, which comes from densities B and D of equation 7.4,

$$p\big(\beta^{(g+1)}|b^{(g+1)},\alpha^{(g+1)},\mathbf{y}\big) = p\big(b^{(g+1)}|\alpha^{(g+1)},\beta^{(g)}\big)p\big(\beta^{(g)}\big)$$

For this draw, note that I use updated values for both b and α. I have now completed the gth Gibbs step, with all three parameters updated for step $g + 1$.

The densities that make up the conditional posterior for each parameter can be identified from the graphical models I have been using throughout. Figure 7.6 shows data set $\mathbf{y}$, where each observation y_i is assumed to come from a likelihood with associated parameter b_i. The parameters b_i are, in turn, drawn from a distribution with parameters α and β. The conditional posterior for b_i comes from the arrows that connect it to y_i (density A) and to $\{\alpha, \beta\}$ (density B). The conditional posterior for α comes from the arrows that connect it to all of the $\mathbf{b}$ (density B), and to parameters that define its distribution (density C). And so on.

From initial starting values for the three parameters, produce a chain of values for each parameter, much like the two-parameter example in Figure 7.4b. Following convergence, allow this to run long enough such that a joint distribution can be assembled from these values. Here is an example:

Example 7.4. Variable fecundity
Clark (2003) considered estimation of fecundity for a model that was used for the northern spotted owl, where each owl might have its own fecundity parameter b_i. Let the number of offspring produced by a pair of owls be y_i.

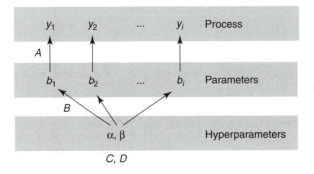

FIGURE 7.6. A model containing random individual effects on birth rates y by including an additional stage. The densities from Equation 7.1 are labeled A, B, C, and D. The conditional densities (A, B) are represented by arrows.

Suppose there are observations on 197 pairs. The likelihood for a data set of owl observations is assumed to be Poisson,

$$p(\mathbf{y}|\mathbf{b}) = \prod_{i=1}^{n} Pois(y_i|b_i)$$

Unlike a standard model that would assume a parameter b that applies identically to all individuals, I allowed that owls might vary. I assumed that there is a population level relationship among owls, but individual owls are not all subject to precisely the same parameter. Because I did not know what that relationship is, I allowed for random effects (Figure 7.6). Variability among individuals is accommodated with a gamma distribution for the Poisson parameters

$$Gam(b_i|\alpha,\beta)$$

This model says that the Poisson parameter for each individual is drawn from a gamma density with parameters that should depend on the entire population. The gamma distribution was used, because it is conjugate with the Poisson likelihood (Section 5.2). The prior on parameter β was taken to be

$$p(\beta) = Gam(\beta|c,d)$$

To illustrate, I begin by assuming a specific value for α that is fixed. I will construct the Gibbs sampler under this simplifying assumption, followed by one that treats α stochastically. The joint posterior for the simplified model is

$$p(\mathbf{b},\beta|\mathbf{y}) \propto \overbrace{\prod_{i=1}^{n} Pois(y_i|b_i)}^{A} \overbrace{\prod_{i=1}^{n} Gam(b_i|\alpha,\beta)}^{B} \overbrace{Gam(\beta|c,d)}^{D}$$

The fitted parameters in this model include the b_i $i = 1, 2, \ldots, n$, (one for each pair) and β. Here is the gth step of the Gibbs sampler:

1. The Gibbs step begins with draws from the conditional posteriors for fecundity parameters. The conditional posterior for the ith fecundity parameter comes from the ith element of density A (the likelihood) and the ith element of density B.

$$p(b_i|y_i,\alpha,\beta) \propto Pois(y_i|b_i)Gam(b_i|\alpha,\beta)$$

To see this, it may help to consult Figure 7.6—the two components on the right-hand side are the only two arrows connected to a given b_i. I have now factored the joint posterior to write a conditional posterior for each of the b_i parameters. The right-hand side is a conjugate Poisson/gamma pair (Appendix G). Thus, the conditional posterior is also gamma,

$$b_i^{(g+1)}|\alpha,\beta^{(g)},y_i \sim Gam(\alpha + y_i,\beta^{(g)} + 1)$$

The parameter α does not have a superscript, because I am (for now) assuming that it is a fixed value. Draw these n parameter values for b_i. These become the updated values for all n parameters. Update these parameters before moving to the next parameter.

2. Parameter β occurs in all n elements of density B and in density D (Figure 7.6). The gamma prior on β allows for direct sampling from the conjugate conditional posterior

$$p(\beta^{(g+1)}|\mathbf{b}^{(g+1)},\alpha) \propto \prod_{i=1}^{n} Gam(b_i^{(g+1)}|\alpha,\beta^{(g)})\,Gam(\beta^{(g)}|c,d)$$

$$= Gam\left(\beta\,\middle|\,c + \alpha n, d + \sum_{i=1}^{n}b_i^{(g+1)}\right)$$

This density thus depends on all parameters $\mathbf{b}$, which depend, in turn, on the observations $\mathbf{y}$ (see Step 1). Again, consult Figure 7.6 to see that β is directly connected to each of the densities included in this conditional posterior.

The process is now repeated. Return to Step 1, and draw a new set of $\mathbf{b}$ using the most recent value of β_b. It is important to remember that each parameter value must be updated before proceeding to the next parameter. The result is a chain of values for each parameter that, together, define the joint posterior and each of the marginals. A chain of β values is shown in Figure 7.7a.

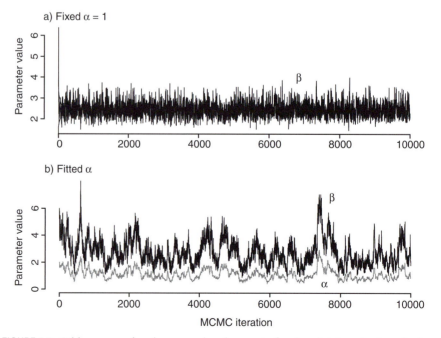

FIGURE 7.7. Gibbs output for the example where α is fixed at 1 (*a*) and where both α and β are fitted (*b*).

In the foregoing example, each Gibbs step involved direct samples from n different b_i conditional densities and one sample from the β conditional density. Direct sampling from each conditional was possible, because each could be expressed in a standard form. To allow for this efficient sampling, I specified prior distributions that are conjugate (Chapter 5). Often it will not be possible to sample directly from a conditional distribution. In such cases, the overall structure of the Gibb sampler is still useful. Each step might involve direct sampling where possible, combined with indirect sampling otherwise. The continuation of Example 7.4 illustrates this hybrid approach.

Example 7.4 (continued). The fecundity Gibbs sampler with an M-H step
Now assume that α is not fixed. For the prior I use the exponential density,

$$p(\alpha) = Exp(\alpha|\rho)$$

I set up the Gibbs sampler as before, defining conditional posteriors for each parameter. Before, I did not sample α, assuming it to be fixed. Now it is not fixed, so the Gibbs sampler requires one more element. I might write the same algorithm as previously, with an additional piece inserted to update α along with the other parameters.

Unlike other parameters, I cannot sample directly from the conditional posterior for α, because it is the product of a gamma and an exponential (densities B and C of Equation 7.4), which are not conjugate. The conditional posterior can be sampled using an M-H step. I use a gamma jump density with a shape parameter of 5 and centered on the current value of α and a rate parameter of $\frac{5}{\alpha_i}$. The jump density is

$$J(\alpha^*|\alpha^{(g)}) = Gam(\alpha^*|5,5/\alpha^{(g)})$$

This density has a mean of $\alpha^{(g)}$ (the current value) and small variance. The two parameters are obtained by moment matching (Appendix F). This produces a candidate value, designated α^*. The acceptance ratio

$$a = \frac{p(\alpha^*|\mathbf{y}) \Big/ J(\alpha^*|\alpha^{(g)})}{p(\alpha^{(g)}|\mathbf{y}) \Big/ J(\alpha^{(g)}|\alpha^*)}$$

determines acceptance of the candidate value with probability $\min(a,1)$, where the conditional posterior for the candidate value is

$$p(\alpha^*|\mathbf{b},\beta) = \prod_{i=1}^{n} Gam(b_i|\alpha^*,\beta)^* Exp(\alpha^*|\rho)$$

I incorporate this third step into the Gibbs sampler.

Gibbs sampler output is shown in Figure 7.7b. The chains for the two parameters α and β show extended autocorrelation. I could modify the algorithm to reduce this auto correlation (Lab Manual). I would run this Gibbs sampler over many iterations to build confidence that it has achieved convergence.

In example 7.4 I introduced an additional level or stage to allow for random effects in fecundity (Figure 7.6). One of the advantages of this structure is the relative ease with which it can accommodate complexity. In principle, many different sources of stochasticity associated with the process, the observations, and the context (e.g., random individual effects) can be incorporated without changing the basic model structure. The next example deals with a different type of complexity associated with observation errors.

Example 7.5. Seedling mortality
Often processes are observed with error. In this example, I demonstrate a Gibbs sampler for estimating mortality of seedlings, when the number of seedlings is estimated with error. This example is modified from Lavine et al. (2002).

Mortality is typically modeled using the binomial likelihood

$$p(\mathbf{k}\,|\,\mathbf{n},\theta) \;=\; \prod_{i=1}^{m} Bin(k_i|n_i,\theta)$$

This model says there are m plots with n_i seedlings per plot. $\mathbf{n}$ and $\mathbf{k}$ are length-m vectors of total and surviving seedlings in each plot. Plot 1 has k_1 survivors of n_1 original seedlings, and so forth. This model assumes that $\mathbf{n}$ and $\mathbf{k}$ are known precisely.

In a study of seedling mortality (Figure 7.8), estimates of the original number of seedlings n_i was imprecisely known (Lavine et al. 2002). Failure to account for observation error results in a biased estimate of the mortality parameter θ, because the observed y_i is less than the true n_i.

Lavine et al. (2002) estimated red maple seedling survival in the southern Appalachians accommodating the error that results from uncertainty in n_i. From this longitudinal study, which spanned multiple years, I extract a single transition for this example. The number of seedlings observed on plot i one year ago is a subset of the true number, introducing a second binomial

$$y_i|n_i \sim Bin(n_i,\phi)$$

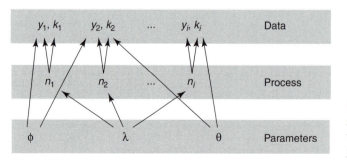

FIGURE 7.8. A seedling survival model where the number of individuals on a sample plot n_i is not precisely known. See Example 7.5.

with some findability (probability of being counted) ϕ. Findability will be large (ϕ approaching one) for species that are conspicuous, for plots that are clearly defined, and so on. But ϕ is not precisely one (Figure 7.9). The likelihood that includes this observation error is

$$p(\mathbf{k},\mathbf{y}|\mathbf{n},\theta,\phi) = \prod_{i-1}^{m} Bin(k_i|n_i,\theta) \times \prod_{i=1}^{m} Bin(y_i|n_i,\phi)$$

Notation is important, because it provides a map for the MCMC algorithm. The likelihood $p(\mathbf{k},\mathbf{y}|\mathbf{n},\theta,\phi)$ implies that both $\mathbf{k}$ and $\mathbf{y}$ might be viewed as data (they are placed on the left side of the bar). The true densities $\mathbf{n}$ are latent variables, in the sense that they must be estimated rather than directly observed. We may only care about θ, but, in order to estimate it, we need a model that includes how findability affects the observations used to estimate θ. The parameter vector includes the true densities and the two binomial parameters. All but θ are "nuisance" parameters.

For binomial parameters I use beta priors, because they are conjugate to the binomial (Appendix G),

$$\theta \sim Beta(\alpha_\theta,\beta_\theta)$$

$$\phi \sim Beta(\alpha_\phi,\beta_\phi)$$

I further assume a Poisson prior for the densities,

$$n_i \sim Pois(\lambda)$$

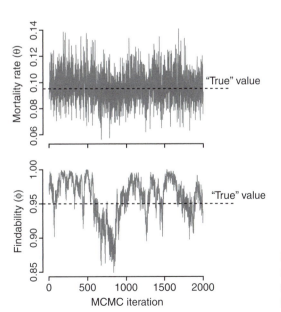

FIGURE 7.9. Gibbs sampler output for the seedling mortality data showing the true parameter values for the simulated data set.

Lavine et al. (2002) considered several assumptions about that process, including spatial autocorrelation and spatial random effects. For this example, I take λ to be fixed. Together with priors for θ and ϕ the joint posterior is

$$p(\theta,\phi,n_i|k_i,y_i,\ldots) \propto \prod_{i=1}^{m} p(k_i,y_i|n_i,\theta,\phi) \prod_{i=1}^{m} p(n_i|\lambda)p(\theta|\alpha_\theta,\beta_\theta)p(\phi|\alpha_\phi,\beta_\phi)$$

The ellipses on the left-hand side represent the prior parameter values, which I omit. Again, the notation is important, so we need to inspect this joint posterior further. We could write five densities on the right-hand side, which, for individual i, are

$$\overbrace{Bin(k_i|n_i,\theta)}^{A}\overbrace{Bin(y_i|n_i,\phi)}^{B}\overbrace{Pois(n_i|\lambda)}^{C}\overbrace{Beta(\theta|\alpha_\theta,\beta_\theta)}^{D}\overbrace{Beta(\phi|\alpha_\phi,\beta_\phi)}^{E}$$

Densities A and B constitute the likelihood. Densities C, D, and E are priors.

For this illustration I use simulated data. This allows for comparison of results with true values. I used values of $m = 20$, $\theta = 0.11$, $\lambda = 30$, $\alpha_\theta = \beta_\theta = 1$, $\alpha_\phi = 9$, and $\beta_\phi = 1$. To limit notational clutter, I suppressed the (g) superscript for "Gibbs step." Here are the components of a Gibbs step:

1. θ—The mortality parameter θ occurs in densities A and D. This beta-binomial combination is conjugate, so I draw directly from the conditional posterior

$$\theta|\mathbf{k},\mathbf{n},\alpha_\theta,\beta_\theta \sim Beta\left(\alpha_\theta + \sum_{i=1}^{m} k_i, \beta_\theta + \sum_{i=1}^{m}(n_i - k_i)\right)$$

2. ϕ—The findability parameter occurs in densities B and E, which together produce the conditional posterior

$$\phi|\mathbf{y},\mathbf{n},\alpha_\phi,\beta_\phi \sim Beta\left(\alpha_\phi + \sum_{i=1}^{m} y_i, \beta_\phi + \sum_{i=1}^{m}(n_i - y_i)\right)$$

3. $\mathbf{n}$—The true abundances appear in densities A, B, and C. They are the sample size parameters in densities A and B, and they are described by a Poisson distribution in C. The conditional posterior is proportional to

$$p(n_i|k_i,y_i,\theta,\phi,\lambda) \propto Bin(k_i|n_i,\theta)Bin(y_i|n_i,\phi)Pois(n_i|\lambda)$$

Unlike the first two conditional posteriors, which assume a standard form (beta), I cannot draw directly from this conditional posterior. Here is a Metropolis step that can be inserted directly into the Gibbs loop. The true density n_i is discrete, so a jump distribution is required that generates discrete values close to the previous ones. A simple possibility is a discrete random walk, which proposes candidate values n_i^* one integer lower or higher than the current value with n_i equal probability, that is,

$$J(n_i + 1|n_i) = J(n_i - 1|n_i) = 0.5$$

At a given Gibbs step I begin by proposing a value n_i^* using the jump distribution. I then calculate the probability of this proposed value relative to the current value, n_i. The conditional posterior is the basis for these acceptance ratios:

$$a = \frac{p(n_i^*|k_i,y_i,\theta,\phi,\lambda)}{p(n_i|k_i,y_i,\theta,\phi,\lambda)} = \frac{Bin(k_i|n_i^*,\theta)Bin(y_i|n_i^*,\phi)Pois(n_i^*|\lambda)}{Bin(k_i|n_i,\theta)Bin(y_i|n_i,\phi)Pois(n_i|\lambda)}$$

The updated value becomes

$$n_i = \begin{cases} n_i^* & \text{with probability } \min(a,1) \\ n_i & \text{otherwise} \end{cases}$$

The Gibbs output for mortality rate and findability is shown in Figure 7.9. Note that the series are autocorrelated and, thus, must be run out for some time to build confidence that convergence has occurred and chains are well mixed. Although the findability parameter tends to straddle the true value, the extended excursion to values values < 0.9 means that a long sequence should be generated. It is also a good idea to initialize with different values. For complex models that contain many parameters there may be no guarantee that convergence will occur. Moreover, trends may be so modest as to make identification of convergence difficult (see below).

The fact that posteriors (Figure 7.10) recover the parameters used to generate the data builds confidence in the model. The findability parameter is more difficult to identify than the survival rate, because it can trade off with n. Note that

$$E[y_i] = \phi n_i$$

Both quantities on the right-hand side are unobserved and, thus, are estimated. ϕ is difficult to identify, because different combinations of ϕ and n_i can result in the same observed value y_i. This does not mean that the two parameters are completely unconstrained. The Gibbs sampler marginalizes over all parts of the model, which, collectively informs these estimates in different ways. Nonetheless, it can be difficult to identify parameters that occur together in this way. Note that the posterior 95 percent credible interval for mortality rate spans an interval of approximately 0.05, whereas the confidence interval for the findability parameter spans an interval > 0.1. Informed priors can be a way of incorporating knowledge of findability based on, say, repeated sampling where true values are known.

7.3.4 Sampling Blocks of Parameters

Efficiency is improved by sampling parameters in blocks. When the conditional posterior is multivariate normal, this comes naturally, because software increasingly includes multivariate normal samplers. But block sampling can also be efficient when parameters are sampled indirectly. High-level languages have many operations that can be efficiently applied to arrays. It is often possible to

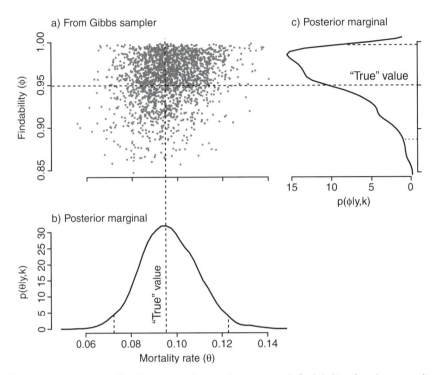

FIGURE 7.10. Posterior distributions of mortality rate and findability for the mortality data. The scatter plot (*a*) shows values generated by MCMC, with estimates of marginal posteriors in *b* and *c*.

propose a large number of parameter values, and evaluate their conditional probabilities with just two function calls. The important point about block sampling is that, for conditionally dependent parameters, if proposed as a block they must be accepted or rejected as a block. Conditional independence means that conditional probabilities do not depend on one another.

As an example, consider the likelihood containing two parameters with independent priors,

$$p(\mathbf{y}|\alpha,\beta)p(\alpha)p(\beta)$$

A Gibbs sampler for the posterior distribution $p(\alpha,\beta|\mathbf{y})$ could involve two steps, one for each parameter. For each step, I could propose a value for one of the parameters, say, α^*, evaluate the conditional posterior for that parameter, $p(\mathbf{y}|\alpha^*, \beta)p(\alpha^*)$, and determine whether to accept it. Upon updating this value, I move to the second parameter. This is the technique discussed above.

Alternatively, I could propose new values for both parameters simultaneously, evaluate $p(\mathbf{y}|\alpha^*,\beta^*)p(\alpha^*)p(\beta^*)$, and accept both or reject both. This latter option is often more efficient, especially when many parameters can be done at once, and the jump distribution can accommodate some of the correlation among parameters. With too many parameters in a block it can become hard to move the sampler, with most proposals being rejected.

7.4 Application: Bayesian Analysis for Regression

In light of the flexibility demonstrated by the last two examples, it may come as a surprise to learn that linear models are just as important in the Bayesian framework as they are in a classical setting. Although modern Bayes eases many of the constraints that have traditionally made for heavy reliance on linear models, they still find wide application within Bayes. For example, hierarchical models (Chapter 8) often have linear equations embedded at lower stages, which allows for more efficient Gibbs sampling than would be the case for nonlinear models (Chapter 8). The efficiency comes from a capacity to sample directly for parameters, latent variables, and variances when error distributions are Gaussian.

The Bayesian treatment of linear regression builds on basics encountered in previous chapters. A standard linear model

$$y_i = b_0 + b_1 x_i + \varepsilon_i$$

has response variable y with both deterministic and stochastic components. The deterministic component comes from the first two terms on the right-hand side of the equation. I start with the assumption that the x_i's are known—they could be administered as treatments. Stochasticity comes from the last term, ε_i. The most common assumption is one of independent errors that are identically, normally distributed,

$$\varepsilon_i \overset{i.i.d.}{\sim} N(0, \sigma^2)$$

Because the mean parameter of the normal distribution is independent of the variance parameter, the $\{y_i\}$ have the same variance as the $\{\varepsilon_i\}$ and a mean given by the linear equation,

$$y_i | x_i, b_0, b_1, \sigma^2 \sim N(b_0 + b_1 x_i, \sigma^2)$$

As before, it is often efficient to write this in matrix notation for a sample of size n

$$\mathbf{y} | \mathbf{X}, \mathbf{b}, \sigma^2 \sim N_n(\mathbf{Xb}, \sigma^2 \mathbf{I}_n)$$

This notation says that the normally distributed errors have a mean value given by the regression and error covariance $\sigma^2 \mathbf{I}_n$. $\mathbf{I}_n$ is the n by n identity matrix. The assumption of independent observations means that off-diagonal elements are zero. There is a single parameter σ^2, which occupies the diagonal of this matrix.

$$\sigma^2 \mathbf{I}_n = \begin{bmatrix} \sigma^2 & 0 & \cdots & 0 & 0 \\ 0 & \sigma^2 & & 0 & 0 \\ \vdots & & \ddots & & \\ 0 & 0 & & \sigma^2 & 0 \\ 0 & 0 & & 0 & \sigma^2 \end{bmatrix}$$

There is a n by p design matrix $\mathbf{X}$ (Section 3.9) and a length-p parameter vector $\mathbf{b}$. In this example, $p = 2$ (there are two parameters). The design matrix consists of a column of ones and a column for the x_i. The likelihood for a data set with intercept and slope terms is

$$p(\mathbf{y}|\mathbf{X},\mathbf{b},\sigma^2) = \prod_{i=1}^{n} N(y_i|b_0 + b_1 x_i,\sigma^2) = N_n(\mathbf{y}|\mathbf{Xb},\sigma^2\mathbf{I}_n)$$

The classical treatment of this model is described in Chapter 4. The Bayesian framework allows flexibility for problems that might have many sources of stochasticity. After introducing the basic model, I consider several examples.

7.4.1 Obtaining the Posterior Distribution

The joint posterior has the form

$$p(\mathbf{b},\sigma^2|\mathbf{X},\mathbf{y}) \propto p(\mathbf{y}|\mathbf{X},\mathbf{b},\sigma^2)p(\mathbf{b})p(\sigma^2)$$

The left-hand side is the joint posterior for regression parameters, and the right-hand side is the likelihood times the prior. The conjugate prior density for regression parameters $\mathbf{b} = [b_0, b_1]^T$ is typically taken to be a normal density

$$\mathbf{b} \sim N_p(\mathbf{b}_0, \mathbf{V}_b)$$

In this case $p = 2$, and this is the length of vector $\mathbf{b}_0$ and the dimensions of matrix $\mathbf{V}_b$. The inverse gamma is conjugate for the variance σ^2,

$$\sigma^2 \sim IG(s_1,s_2)$$

With these priors, the posterior distribution is

$$p(\mathbf{b},\sigma^2|X,\mathbf{y},\dots) \propto \overbrace{N_n(\mathbf{y}|\mathbf{Xb},\sigma^2\mathbf{I}_n)}^{A} \overbrace{N_p(\mathbf{b}|\mathbf{b},\mathbf{V}_b)}^{B} \overbrace{IG(\sigma^2|s_1,s_2)}^{C} \qquad 7.6$$

As before, the ellipses on the left-hand side refer to prior parameter values.

To construct a Gibbs sampler we must identify the conditional posterior distributions that can be used to sample from this joint posterior. There will be a two-dimensional distribution for the regression parameters $\mathbf{b}$ that is conditioned on σ^2 and a univariate distribution for σ^2 that is conditional on the vector $\mathbf{b}$. Here are the conditional distributions.

Regression Parameters—Consider first the regression parameters $\mathbf{b}$, which appear in density A (likelihood) and in density B (prior) from Equation 7.6. The product of these two normal densities is

$$p(\mathbf{b}|\sigma^2,\mathbf{X},\mathbf{y}) \propto \exp\left[- \frac{(\mathbf{y} - \mathbf{Xb})^T(\mathbf{y} - \mathbf{Xb})}{2\sigma^2} \right.$$

$$\left. - \frac{(\mathbf{b} - \mathbf{b}_0)^T\mathbf{V}_b^{-1}(\mathbf{b} - \mathbf{b}_0)}{2} \right] \qquad 7.7$$

I use linear algebra from Box 5.1 to express this exponent in terms of a normal distribution in $\mathbf{b}$. The distribution we desire will have the form

$$p(\mathbf{b}|\sigma^2,\mathbf{X},\mathbf{y}) \propto \exp\left[-\frac{1}{2}(\mathbf{b} - \mathbf{Vv})^T\mathbf{V}^{-1}(\mathbf{b} - \mathbf{Vv}) \right] \qquad 7.8$$

$$= N(\mathbf{b}|\mathbf{Vv},\mathbf{V})$$

so I need to determine $\mathbf{V}$ and $\mathbf{v}$. As in Section 4.2.2, I incomplete the square, now doing so in matrix notation.

Consider just the exponent for the normal distribution in Equation 7.8, ignoring for the moment the coefficient $-\frac{1}{2}$,

$$(\mathbf{b} - \mathbf{Vv})^T \mathbf{V}^{-1} (\mathbf{b} - \mathbf{Vv})$$

with covariance matrix $\mathbf{V}$ and mean $\mathbf{Vv}$. Expand the exponent to obtain

$$\mathbf{b}^T\mathbf{V}^{-1}\mathbf{b} - \mathbf{b}^T\mathbf{V}^{-1}\mathbf{Vv} - (\mathbf{Vv})^T\mathbf{V}^{-1}\mathbf{b} + (\mathbf{Vv})^T\mathbf{V}^{-1}\mathbf{Vv}$$

The second term can be simplified to $\mathbf{b}^T\mathbf{V}^{-1}\mathbf{Vv} = \mathbf{b}^T\mathbf{Iv} = \mathbf{b}^T\mathbf{v}$. The third term can also be simplified:

$$(\mathbf{Vv})^T \mathbf{V}^{-1}\mathbf{b} = \mathbf{v}^T\mathbf{V}^T \mathbf{V}^{-1} \mathbf{b} = \mathbf{v}^T\mathbf{V}\mathbf{V}^{-1}\mathbf{b} = \mathbf{v}^T\mathbf{b} = \mathbf{b}^T \mathbf{v}$$

For this algebra I have made use of the fact that the transpose of a symmetric matrix is equal to itself (e.g., $\mathbf{V}^{-1}$), and the transpose of a scalar is the same scalar (e.g., $\mathbf{b}^T\mathbf{v}$). This gives a form that looks like the univariate case (Equation 4.6)

$$\mathbf{b}^T\mathbf{V}^{-1}\mathbf{b} - 2\mathbf{b}^T\mathbf{v} + \mathbf{v}^T\mathbf{V}^T\mathbf{v}$$

Because I need only identify $\mathbf{V}$ and $\mathbf{v}$, I focus on the first two terms, which contain these two quantities. These terms are

$$\mathbf{b}^T\mathbf{V}^{-1}\mathbf{b} - 2\mathbf{b}^T \mathbf{v} - \cdots \qquad 7.9$$

The inverse of $\mathbf{V}$ is sandwiched within the first term, and $\mathbf{v}$ is premultiplied by $2\mathbf{b}^T$ in the second term. I will now put Equation 7.8 in this form to identify $\mathbf{V}^{-1}$ and $\mathbf{v}$.

Return to Equation 7.7, again considering just the exponent and ignoring the coefficient $-\frac{1}{2}$,

$$\sigma^{-2}(\mathbf{y} - \mathbf{Xb})^T(\mathbf{y} - \mathbf{Xb}) + (\mathbf{b} - \mathbf{b}_0)^T V_b^{-1}(\mathbf{b} - \mathbf{b}_0)$$

Expanding, I have

$$\sigma^{-2}(\mathbf{y}^T\mathbf{y} - \mathbf{y}^T\mathbf{Xb} - \underline{\mathbf{b}^T\mathbf{X}^T\mathbf{y}} + \underline{\mathbf{b}^T\mathbf{X}^T\mathbf{Xb}})$$

$$+ \underline{\mathbf{b}^T V_b^{-1}\mathbf{b}} - \underline{\mathbf{b}^T V_b^{-1}\mathbf{b}_0} - \mathbf{b}_0^T V_b^{-1}\mathbf{b} + \mathbf{b}_0^T V_b^{-1}\mathbf{b}_0$$

I retain only the terms containing $\mathbf{b}^T(\ )\mathbf{b}$ and $\mathbf{b}^T(\ )$, because Equation 7.9 indicates that these terms will contain the quantities $\mathbf{V}$ and $\mathbf{v}$. I have underlined these terms above. Collecting these terms yields

$$\mathbf{b}^T(\sigma^{-2}\mathbf{X}^T\mathbf{X} + V_b^{-1})\mathbf{b} - 2\mathbf{b}^T(\sigma^{-2}\mathbf{X}^T\mathbf{y} + V_b^{-1}\mathbf{b}_0)$$

From Equation 7.9 then

$$\mathbf{v} = \sigma^{-2}\mathbf{X}^T\mathbf{y} + V_b^{-1}\mathbf{b}_0$$

$$\mathbf{V}^{-1} = \sigma^{-2}\mathbf{X}^T\mathbf{X} + V_b^{-1}$$

In other words the mean and variance of this conditional posterior distribution for regression parameters are

$$E[\mathbf{b}|\sigma^2,\mathbf{X},\mathbf{y}] = \mathbf{Vv} = [\sigma^{-2}\mathbf{X}^T\mathbf{X} + V_b^{-1}]^{-1}[\sigma^{-2}\mathbf{X}^T\mathbf{y} + V_b^{-1}\mathbf{b}_0]$$

$$\text{var}[\mathbf{b}|\sigma^2,\mathbf{X},\mathbf{y}] = \mathbf{V} = [\sigma^{-2}\mathbf{X}^T\mathbf{X} + V_b^{-1}]^{-1}$$

Both coefficients have two terms, the first contributed by the likelihood and the second by the prior. If the prior is noninformative, the prior precision will be small and the second terms disappear. To see this, note that as the prior precision becomes small, the posterior mean tends to the MLE's for regression parameters for the classical model (Box 5.1). Thus, the mean of this distribution is a weighted average of contributions from data and prior.

Variance Parameter—The conditional posterior for σ^2 is obtained in the same manner as was done for the variance in Section 4.2.4. That solution is

$$p(\sigma^2|\mathbf{X},\mathbf{y},\mathbf{b}) \propto N(\mathbf{y}|\mathbf{Xb},\sigma^2\mathbf{I})IG(\sigma^2|s_1,s_2)$$

$$= IG(\sigma^2|u_1,u_2)$$

where

$$u_1 = s_1 + n/2 \qquad\qquad\qquad 7.10$$

$$u_2 = s_2 + \frac{1}{2}\sum_{i=1}^{n}(y_i - b_0 - b_1 x_i)^2$$

$$= s_2 + \frac{1}{2}(\mathbf{y} - \mathbf{Xb})^T(\mathbf{y} - \mathbf{Xb})$$

Example 7.6. The posterior conditionals for regression parameters
Here I complete the square for the case where there may be nonzero covariance among observations. Consider the linear model with likelihood

$$N_n(\mathbf{y}|\mathbf{Xb},\Sigma)$$

In this example, there may be correlations between observations—not all off-diagonal elements of the covariance matrix need equal zero, so we have a full covariance matrix Σ. For now, I simply solve for the conditional posterior for regression parameters. As before, take a normal prior on regression parameters

$$N(\mathbf{b}|\mathbf{b}_0, \mathbf{V}_b)$$

The conditional posterior for regression parameters has as the exponent $-\frac{1}{2}$ times

$$(\mathbf{y} - \mathbf{Xb})^T\Sigma^{-1}(\mathbf{y} - \mathbf{Xb}) + (\mathbf{b} - \mathbf{b}_0)^T\mathbf{V}_b^{-1}(\mathbf{b} - \mathbf{b}_0)$$

Now expand and retain terms that contain $\mathbf{b}^T(\ )\mathbf{b}$, and $\mathbf{b}^T(\ )$,

$$- \mathbf{b}^T\mathbf{X}^T\Sigma^{-1}\mathbf{y} + \mathbf{b}^T\mathbf{X}^T\Sigma^{-1}\mathbf{Xb} + \mathbf{b}^T\mathbf{V}_b^{-1}\mathbf{b} - \mathbf{b}^T\mathbf{V}_b^{-1}\mathbf{b}_0$$

Collect terms,

$$[\mathbf{b}^T\mathbf{X}^T\Sigma^{-1}\mathbf{Xb} + \mathbf{b}^T\mathbf{V}_b^{-1}\mathbf{b}] - [\mathbf{b}^T\mathbf{X}^T\Sigma^{-1}\mathbf{y} + \mathbf{b}^T\mathbf{V}_b^{-1}\mathbf{b}_0]$$

and simplify

$$\mathbf{b}^T(\mathbf{X}^T\Sigma^{-1}\mathbf{X} + \mathbf{V}_b^{-1})\mathbf{b} - \mathbf{b}^T(\mathbf{X}^T\Sigma^{-1}\mathbf{y} + \mathbf{V}_b^{-1}\mathbf{b}_0)$$

The two terms give us $\mathbf{V}^{-1}$ and $\mathbf{v}$, respectively

$$\mathbf{V}^{-1} = \mathbf{X}^T\Sigma^{-1}\mathbf{X} + \mathbf{V}_b^{-1}$$

$$\mathbf{v} = \mathbf{X}^T\Sigma^{-1}\mathbf{y} + \mathbf{V}_b^{-1}\mathbf{b}_0$$

Substitute $\sigma^2\mathbf{I}$ for Σ to recover the solution for the special case where observations are independent.

These coefficients provide the conditional posterior of regression parameters, that is, conditional on Σ. A full model still requires a conditional posterior for Σ. A common approach is to use the conjugate inverse Wishart prior, a multivariate version of the inverse gamma distribution. I consider examples in Chapter 9.

The variances associated with observations need not all be identical. The typical covariance matrix for regression is $\Sigma = \sigma^2 \mathbf{I}_n$. Although there are n^2 elements in this matrix, typically only one parameter, σ^2, is estimated. But there could be as many as $n(n + 1)/2$ parameters. (There could not be n^2 parameters, because the matrix is symmetric.) This would be a lot of estimates. Yet it is not unusual to be confronted with the need to consider covariance matrixes with more than one parameter. For instance, if each observation has a unique variance, the covariance matrix has n parameters. This matrix is $Diag[\sigma_1^2, \ldots, \sigma_n^2]$. The $Diag$ notation means that the vector in brackets is placed along the diagonal of a matrix that is otherwise filled with zeros:

$$Diag[\sigma_1^2, \ldots, \sigma_n^2] = [\sigma_1^2, \ldots, \sigma_n^2]I_n = \begin{bmatrix} \sigma_1^2 & 0 & \cdots & 0 \\ 0 & \sigma_2^2 & \cdots & 0 \\ \vdots & \vdots & \ddots & 0 \\ 0 & 0 & 0 & \sigma_n^2 \end{bmatrix}$$

In principle, the covariance matrix could have nonzero elements everywhere but it would be difficult to estimate them. Parametric forms for covariance matrixes are discussed in later chapters. Time series models can have covariance matrixes with parameters for lagged relationships (Chapter 9). Spatial models can have covariances among locations. In all cases, the covariance matrix is symmetric (and positive definite).

In summary, a Gibbs sampler for the basic regression model is simple, because priors and likelihood are conjugate for all parameters. Sample regression parameters form the normal distribution

$$\mathbf{b}|\sigma^2 \ldots \sim N_p(\mathbf{Vv}, \mathbf{V})$$

followed by a sample for the residual variance

$$\sigma^2|\mathbf{b}, \ldots \sim IG(u_1, u_2)$$

Examples come in the form of specific modifications to this general model below.

7.4.2 Errors in Variables

Advantages of a Bayesian approach include a framework that admits many types of stochasticity. A common challenge for ecological models is that many variables may be only partially known. For example, usually treatments are viewed as being precisely known, because the investigator administers them. But, even in this most controlled setting, the treatment may, in fact, be variable and incompletely understood. The investigator may know precisely how much fertilizer is applied, yet field conditions can make nutrient availability highly variable and uncertain. Some is leached away, some is immobilized by microbes, and some may be adsorbed to organics or clays. There may be volatilization losses.

When variables are not controlled or are only partially known, violations of standard assumptions can have a large impact on inference. Consider again the simple regression model of Chapter 4:

$$y_i = b_0 + b_1 x_i + \varepsilon_i$$

$$\varepsilon_i \sim N(0, \sigma^2)$$

x_i is the predictor, and y_i is the response. The model assumes some error or stochasticity is added to the deterministic model. This model makes sense if the y_i's are observed with error (observation error). This model does not have a place for observation errors in x (e.g., the fertilizer application). In a classical framework errors in x are difficult to accommodate. The Bayesian structure allows us to simply add stochasticity to the existing model with insight about x_i coming from the prior. In addition to the equations above, include one more:

$$x_i^{(o)} = x_i + u_i$$

$$u_i \sim N(0, \tau^2)$$

This additional equation says that the observed value of x_i is not the same as the true value. There is observation error in the independent variable, which, in this case, is taken to be normally distributed about the true value of x_i. The full regression model is now

$$p(b_0, b_1, \sigma^2, \tau^2, \mathbf{X} | \mathbf{X}^{(o)}, \mathbf{y}, \dots) \propto N_n(\mathbf{y} | \mathbf{Xb}, \sigma^2 \mathbf{I}_n) N_2(\mathbf{b} | \mathbf{b}_0, \mathbf{V}_0) IG(\sigma^2 | s_1, s_2)$$

$$\times N_n(\mathbf{X}^{(o)} | \mathbf{X}, \tau^2 \mathbf{I}_n) N(\mathbf{X} | \mathbf{X}^{(p)}, \mathbf{H}), IG(\tau^2 | s_3, s_4),$$

where $\mathbf{X}^{(p)}$ is the vector of prior values for $\mathbf{X}$, and $\mathbf{H} = Diag\,[h_1, \dots, h_n]$ has diagonal elements for prior variances. Note that I have simply included two additional distributions, one for the unknown x_i's and another for variance on observations thereof. In other words, each of the x_i's must be estimated along with everything else.

A Gibbs sampler for the regression could be updated by adding two steps, one for each of these new elements. Return to the case where $p = 2$, that is, the vector $\mathbf{b}$ consists of an intercept and a slope. The conditional posterior for a given x_i comes from the product of the factors that contain it,

$$p(x_i | x_i^{(o)}, y_i, b_0, b_1, \sigma^2, \tau^2) \propto N(y_i | b_0 + b_1 x_i, \sigma^2) N(x_i^{(o)} | x_i, \tau^2) N(x_i | x_i^{(p)}, h_i)$$

$$= N(x_i | V_i v_i, V_i)$$

To identify the mean and variance of this normal distribution, write the exponent (neglecting $-\frac{1}{2}$),

$$\frac{(y_i - b_0 - b_1 x_i)^2}{\sigma^2} + \frac{(x_i^{(o)} - x_i)^2}{\tau^2} + \frac{(x_i - x_i^{(p)})^2}{h_i}$$

Expanding and collecting terms in x_i gives

$$\left(\frac{b_1^2}{\sigma^2} + \frac{1}{\tau^2} + \frac{1}{h_i}\right)x_i^2 - 2\left(\frac{(y_i - b_0)b_1}{\sigma^2} + \frac{x_i^{(o)}}{\tau^2} + \frac{x_i^{(p)}}{h_i}\right)x_i + C$$

The coefficients needed for the conditional normal distribution are in brackets above, that is,

$$V_i^{-1} = \frac{b_1^2}{\sigma^2} + \frac{1}{\tau^2} + \frac{1}{h_i}$$

$$v_i = \frac{(y_i - b_0)b_1}{\sigma^2} + \frac{x_i^{(o)}}{\tau^2} + \frac{x_i^{(p)}}{h_i}$$

Note that there is a different mean value for each x_i, each being the weighted average of three terms, the information contributed by the regression, that contributed by the observation, and that contributed by the prior. The observation variance is drawn from

$$p(\tau^2|\mathbf{x}^{(o)},\mathbf{x},\mathbf{y},b_0,b_1,\sigma^2) \propto N_n(\mathbf{x}^{(o)}|\mathbf{x},\tau^2\mathbf{I}_n)IG(\tau^2|s_3,s_4)$$

$$= IG\left(\tau^2\bigg|s_3 + n/2, s_4 + \frac{1}{2}\Sigma(x_i^{(o)} - x_i)^2\right)$$

This approach provides a structure for admitting error on both x and y, as illustrated in the following example.

Example 7.7. Error in variables
Biomass is measured on a balance with an error variance of $\tau^2 = 9.0 + 0.40$. I have intentionally made this balance unreliable to illustrate how the approach can help account for it. Biomass will serve as our predictor variable x for a response variable y. To estimate the linear relationship between the two I allow for the fact that observations of x have uncertainty. I use the model outlined above with the prior parameters on τ^2 of $s_3 = 500$ and $s_4 = 4491$. This gives us a distribution for τ^2 having the mean and variance of 9.0 and 0.16, respectively. The variance in this prior determines how much weight it will have in the model fit. Because the error in the balance is known, I give the prior substantial weight. I further assume no prior information on the x's. Using the Gibbs sampler described above I simulate the posterior, which includes estimates for each of the regression parameters, the two variances, and the unknown x_i's (Figure 7.11). The uncertainty in the predictor variables (e.g., lower panel of Figure 7.11) results in large fluctuations in the chains for regression parameters. Point estimates are recovered for the underlying parameters used to simulate the data (dashed lines in Figure 7.11), but posterior distributions are broad (Figure 7.12). Note too that the model estimates true values for the x_i's that, for the most part, are closer to the correct values than are the observations. This is to be expected when I know that the correct model is, in this case, linear. Given that models themselves will be

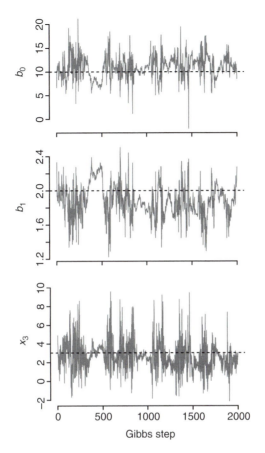

FIGURE 7.11. A portion of Gibbs output for regression parameters and one of the unknown values of x from Example 8.5. Dashed lines are values used for the simulation.

uncertain, one cannot expect that simply allowing for errors in variables will necessarily allow such accurate recovery of the unknown values. A principal advantage here is the formal structure it affords for incorporating the uncertainty in x's.

In the foregoing example, the two sources of uncertainty are difficult to identify (e.g, Bekker 1986). In such cases, the prior provides a means for formally integrating knowledge, in this case, on the error in measurements of **x** and in the x's themselves.

Alternatively, assume the x's are known, but some are missing. Using the same approach, the sample for the kth value of x is

$$x_k | b_1, y_k, \sigma^2, x_k^{(p)}, h_k \sim \mathrm{N}(V_k v_k, V_k)$$

where

$$V_k^{-1} = \frac{b_1^2}{\sigma^2} + \frac{1}{h_k}$$

$$v_k = \frac{b_1(y_k - b_0)}{\sigma^2} + \frac{x_k^{(p)}}{h_k}$$

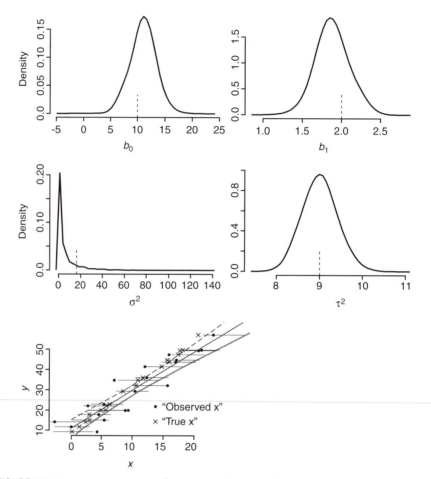

FIGURE 7.12. Posterior estimates of process and observation errors (*above*) with regression model, observations, and true values (*below*). Horizontal lines on the lower panel are 95 percent confidence intervals on estimates of x. They are located at the same height as the observations and true values to allow comparisons.

7.4.3 Combining Data

The prior/likelihood framework provides a natural structure for combining different types of data that may come from diverse sources. Although the term *prior* connotes something known in advance of a particular analysis, it can simply be viewed as information obtained from a source that is different from the likelihood. In fact, one might think of a problem as having multiple likelihoods. Here is an example:

Example 7.8. Data integration
Suppose I wish to draw inference on a velocity u. I have observations of travel times t and travel distances y for objects believed to be traveling at the same,

but unknown, velocity. If the distance traveled is imprecisely known I could write the likelihood as

$$p(\mathbf{y}|\mathbf{t},u,\sigma^2) = \prod_{i=1}^{n_1} N(y_i|ut_i,\sigma^2)$$

I might also have direct observations of velocity $u^{(o)}$, as obtained with a stopwatch as individuals pass. If I assume that individuals travel at the same velocity u, with different estimates resulting from observation error, this second likelihood for observations obtained from the stopwatch is

$$p(\mathbf{u}^{(o)}|u,\tau^2) = \prod_{i=1}^{n_2} N(u_1^{(o)}|u,\tau^2)$$

Here I am assuming that the observations are obtained on different individuals (this affects how I set up the likelihood, but the approach does not depend on this). Because the true velocity is unknown, it has a prior, taken to be $N(u|\mu,\omega)$. A full model might look like this:

$$p(u,\sigma^2,\tau^2|\mathbf{y},\mathbf{t},\mathbf{u}^{(o)}) \propto \prod_{i=1}^{n_1} N(y_i|ut_i,\sigma^2)\prod_{i=1}^{n_2} N(u_i^{(o)}|u,\tau^2)N(u|\mu,\omega)$$
$$IG(\sigma^2|s_1,s_2)IG(\tau^2|s_3,s_4)$$

You might view this model as having two likelihoods and three priors. The conditional posterior for the velocity parameter is

$$p(u|\mathbf{y},\mathbf{t},\mathbf{u}^{(o)},\sigma^2,\tau^2) = N(u|Vv,V)$$

where

$$V^{-1} = \frac{\displaystyle\sum_{i=1}^{n_1} t_i^2}{\sigma^2} + \frac{n_2}{\tau^2} + \frac{1}{\omega}$$

$$v = \frac{\displaystyle\sum_{i=1}^{n_1} y_i t_i}{\sigma^2} + \frac{\displaystyle\sum_{i=1}^{n_2} u_i^{(o)}}{\tau^2} + \frac{\mu}{\omega}$$

This density results from the product of the two likelihoods, both of which contain u, and a prior.
Conditional variances are

$$p(\sigma^2|\mathbf{y},\mathbf{t},u) = IG\left(\sigma^2\bigg|s_1 + n_1/2,s_2 + \frac{1}{2}\sum_{i=1}^{n_1}(y_i - ut_i)^2\right)$$

$$p(\tau^2|\mathbf{u}^{(o)},u) = IG\left(\tau^2\bigg|s_3 + n_2/2,s_4 + \frac{1}{2}\sum_{i=1}^{n_2}(u_i^{(o)} - u)^2\right)$$

These conditionals could be the basis for a Gibbs sampler to estimate velocity that makes use of both types of evidence.

In Chapter 8 I discuss assimilating different types of data. Here I simply mention that some readers may balk at this product of two likelihoods that are clearly not independent. Recall that the assumption of independent sampling allows us simply to multiply the likelihoods for observations. Obviously, observations of travel times and velocities taken on the same individuals must be highly correlated. As I discuss later, the stochastic treatment of u means that we are accommodating the correlations by modeling the joint distribution, which includes stochasticity in the true velocity u. Still, it is important that the two types of observations are conditionally independent. More on this in Chapters 8 and 11.

7.4.4 Linearizing a Nonlinear Model

Because nonlinear models can involve substantial indirect sampling, it can be efficient to consider reparameterizations that minimize the nonlinearities. The example from Chapter 1 involves the familiar Monod or Michaelis-Menton equation as part of the nonlinear model for growth rate y_i,

$$y_i = \mu_i + \varepsilon_i$$

$$\mu_i = g\left(\frac{x_i - x_0}{x_i + \theta}\right)$$

$$\varepsilon_i \overset{i.i.d.}{\sim} N(0,\sigma^2)$$

where μ_i is the mean growth rate expected at a location with resource availability x_i. There are three parameters, including maximum growth rate g, minimum resource requirement x_0, and half-saturation parameter θ. Clark et al. (2003a) rewrite this in terms of linear parameters

$$\mathbf{b} = [\beta_0 \ \beta_1]^T$$

$$\mu_i = \beta_0 + \beta_1 z_i$$

The new variable $z_i = \dfrac{x_i}{\theta + x_i}$ is the basis for a design matrix consisting of a column of ones and a column with each z_i,

$$\mathbf{Z}_\theta_{n\times 2} = \begin{bmatrix} 1 & z_1 \\ \vdots & \vdots \\ 1 & z_n \end{bmatrix}.$$

The new parameters are related to the old ones as

$$g = \beta_0 + \beta_1$$

$$x_0 = -\beta_0\theta(\beta_0 + \beta_1)^{-1}$$

This gives the model

$$p(\beta,\sigma^2,\theta|\mathbf{x},\mathbf{y}, \dots) \propto N_m(\mathbf{y}|\mathbf{Z}_\theta\beta,\sigma^2\mathbf{I}_n)N_2(\beta|\mathbf{b},\mathbf{V}_b) \ IG(\sigma^2|s_1,s_2) \ Unif(\theta|a_\theta,b_\theta)$$

All conditionals can be sampled directly, with the exception of θ. The model is clearly linear in β, which allows direct sampling from a normal distribution. As previously, I can sample from the conditional inverse gamma for σ^2. For θ I use a uniform prior density *Unif* (.001, .3).

Marginal posteriors are shown to the right of a sample MCMC chain in Figure 7.13. The model converges rapidly for all parameters. Because three of four parameters are sampled directly, the algorithm is easy to write and efficient. The tight predictive intervals on growth rate (lower right panel of Figure 7.13) come from the fact that there is a large sample size ($n = 756$). Note that I have plotted σ, rather than σ^2, to have comparable units.

Each of the parameters simulated using MCMC can be transformed back to those of the original model (g and x_0) using the relationships given above. This can be done with the entire chain(s), with further processing being done on the transformed parameters (Clark et al. 2003a).

7.4.5 Flexibility in Bayesian Regression

Linear models and generalized linear models are widely used. Within the Bayesian framework, it is possible to exploit the basic elements of linear equations with Gaussian error as part of a more complex model. For example, the error in variables example was not much different from the combining data example. In

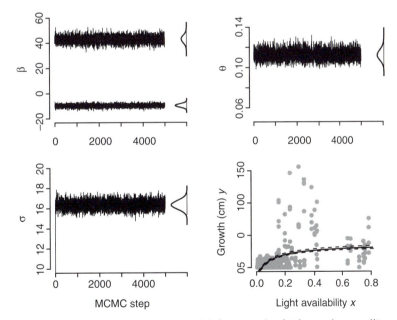

FIGURE 7.13. MCMC output and fitted model for growth of *Ulmus alata* seedlings and light availability (Mohan et al., 2006). Sample chains are shown for regression parameters β, σ^2, and θ. The conditional for θ was drawn using inverse distribution sampling (Section 7.2.2). The predicted model (*lower right*) includes a 95 percent credible interval interval on the mean response μ (dashed lines).

both cases I used the conditional posteriors already derived for the simple regression and simply incorporated relationships for the additional considerations. The conditional distributions were no more complex than those for the basic model, an advantage of working with conditional posteriors. These additional factors can be added as components to a Gibbs sampler previously developed for simpler assumptions. The Bayesian framework allows for multiple sources of stochasticity without resort to massive restructuring. This is true generally, but the advantage is especially apparent for linear models, where stochasticity in y, x, and regression parameters can result in normal conditionals that can be sampled directly.

7.5 Using MCMC

Thus far, I have discussed MCMC with the assumption that the simulation will run for a while, eventually producing a chain of values that approximates the joint posterior. The early iterations that represent preconvergence will be discarded. This early part of the simulation is termed *burn-in* and is typically identified by eye. Following burn-in, the chain of estimates appears stationary. The burn-in portion of the series is viewed as the sequence of estimates leading up to the point where the series becomes stationary. For complex models, values may drift so slowly that quantitative indexes can be helpful in providing an objective (albeit not necessarily correct) assessment of convergence. In short, use of the Markov chain is not as straightforward as, say, the list of estimates that comes from a bootstrap (Chapter 5), because samples are not independent. Here I discuss how long is long enough and what to do with the chain(s).

7.5.1 Has It Converged?

It is important to run an MCMC simulation long enough to obtain a sequence that adequately represents the target distribution. One of the most important aspects of an analysis involving MCMC is determining if and when convergence has been achieved. This can be difficult for models involving many parameters. There are no specific stopping rules for the simulation (e.g., Cappe and Robert 2000). There are some general approaches used to assess convergence, but, for some high-dimensional models, one may never be completely sure if convergence has occurred.

It may be possible to prove that convergence occurs asymptotically. Proofs of convergence must include demonstration that the converged state can be reached by small, incremental steps from appropriate starting points, say, parameter combinations drawn from the prior(s). A reference for Monte Carlo methods is Robert and Casella (1999).

If convergence cannot be proved, it takes a long time, or there is extensive autocorrelation in the series, one resorts to repeated simulations from dispersed initial conditions. Initialization is important. On the one hand, initialization from different parameter combinations builds confidence in convergence. Each such simulation is known as a chain. On the other hand, one cannot expect to obtain useful results from completely arbitrary initial values. For complex models, convergence in practical time may depend on initialization from reasonable parameter values. For a simulation that could eventually converge, if started too far from a combination of realistic values, it may not converge in your lifetime. This can happen, because the target distribution is so flat in the tails that it cannot find its way (in available time) to higher probability ground. There can also be considerable inertia in high-dimensional models, where seriously improbable values for some parameters slow convergence of others. Strong relationships among parameters may make it difficult to identify an efficient jump distribution, example, one that is not constantly proposing values that fall outside some combination of closely correlated parameters. It may also be the case that some initial values are irrelevant. In the accompanying lab manual I discuss a patch model for movement, where not all patches can be reached, a situation that the investigator may not be aware of. There is no point in initializing the MCMC with parameter combinations that are impossible. Finally, some parameters may appear to converge immediately, giving a false impression that the full model has converged. There may be sticky parameters that are much slower to converge than others. These tend to be ones that are sampled indirectly (e.g., M-H) and have high correlations with other parameters. It should be emphasized that apparent convergence may convince you that the algorithm converges to something, but it may not be target distribution. Because the likelihood surface can be complex, apparent convergence should be treated with caution.

The simplest and most basic approach involves initializing the simulation with values that are reasonable, yet different, and observing the behavior of the simulation. Different initial values should lead to the same target. The longer the sequence appears stationary, the more confidence one has that it has converged. Long sequences are important, because a sequence may appear stationary after 1000 iterations, while showing a long-term trend at, say, 10,000 iterations. Large models may require tens or hundreds of thousands of iterations. With some, you may never be sure. The difficulty of assessing convergence is one of the reasons why statisticians say that MCMC can be dangerous (BUGS documentation project).

Several indexes are used to monitor convergence. One such index is Gelman and Rubin's (1992) scale reduction factor. It is based on a comparison of variance within versus among parallel chains. Run the MCMC J times from dispersed initial conditions to produce J different Markov chains, each of length G. Each chain includes a length-G vector for each of p parameters. For each parameter θ, calculate the average variance within each chain,

$$W = \frac{1}{J(G-1)} \sum_{j=1}^{J} \sum_{g=1}^{G} (\theta_{gj} - \bar{\theta}_j)^2$$

where θ_{gj} is the gth estimate of θ in the jth sequence, and $\bar{\theta}_j$ is the mean estimate taken over the jth sequence. Then the variance among chains is

$$B = \frac{1}{(J-1)G}\sum_{j=1}^{J}\left(\sum_{g=1}^{G}\theta_{gj} - \frac{1}{J}\sum_{j=1}^{J}\sum_{g=1}^{G}\theta_{gj}\right)^2$$

These relationships are used to estimate the extent to which the distribution of θ might be improved by continuing the process indefinitely (infinite G) as

$$R = \sqrt{\frac{1}{G}\left(G - 1 + \frac{B}{W}\right)}$$

As the between-sequence variance approaches that within sequences, R approaches one.

Gelman et al. (1995) recommend that convergence may be acceptable for values of $R < 1.2$. R will not be below one. Typically sequences for some parameters converge sooner than others. Convergence should be achieved for all parameters of interest. If there is extended autocorrelation in a series, R may fall in the acceptable range despite the fact that the chain is not actually converged.

7.5.2 Inference Based on MCMC Results

Processing MCMC results involves assembling the marginal distributions and calculation of quantiles and moments. MCMC results readily extend to prediction and decision. After discarding the burn-in steps, the chain for a given parameter can be used to construct a histogram or it can be smoothed to approximate a continuous density. Examples in this chapter include Figures 7.10 and 7.13.

Percentiles—Percentiles are readily extracted from MCMC sequences using standard software. These percentiles are the basis for Bayesian credible intervals (Chapter 5). Of course, the sequence used to determine percentiles must not include the burn-in (preconvergence) portion of the sequence. The autocorrelation in the MCMC chain does not affect percentiles, unless, of course, the chain is too short to result in adequate mixing.

Moments—An MCMC chain can be used to estimate the posterior mean, median, mode (most probable value), and standard error. The mean is obtained in the usual way, as the average value of the converged sequence. The median is determined as the 50th percentile. Determination of the mode might involve approximation.

The standard error involves additional considerations. Recall for the classical bootstrap, I simply determined the standard deviation of the sequence of estimates and used it as an estimate of the standard error (Section 5.3.3). The MCMC series can provide an estimate of the Bayesian standard error only after dealing with the autocorrelation that is typical of most sequences. These trends appear in a number of the MCMC examples presented in this chapter (e.g., Figure 7.7, 7.9, 7.11). Provided the sequence has converged and that it is of sufficient length, this autocorrelation does not impact percentiles.

However, the autocorrelation means that samples from the distribution are not independent and, thus, underestimate the true variance of the distribution.

An efficient way of processing autocorrelated sequences entails calculating the autocorrelation function for the sequence and using it to adjust the sample size and inflate the variance accordingly. Carlin and Louis (2001) provide a transparent description. Alternatively, the chain can be thinned or sampled at intervals sufficiently long to overstep the autocorrelation. This method is obviously inefficient. If thinning is used, it must be sufficiently sparse to overstep any long-period autocorrelation in the series. Despite the inefficiency, if the output is to be saved, the sequence will probably be thinned anyway; whereas one might execute 500,000 iterations or more to ensure adequate mixing for a given model, provided samples are independent, several thousand may be enough to characterize the distribution. Because we often only wish to save a subset of samples, thinning may be done regardless.

Predictive Distributions—Prediction involves inference about data that might be collected some other place at some other time. (I reserve the term *forecast* for predictions that apply to the future.) In Section 5.6 I considered the predictive distribution, which is based on the posterior $p(\theta|y)$, where θ represents the parameters that have been estimated from data y. To predict as-yet unobserved values of y, call them y; I incorporated the uncertainty represented by the posterior as

$$p(y'|y) \;=\; \int \overbrace{p(y'|\theta)}^{A} \;\; \overbrace{p(\theta|y)d\theta}^{B}$$

(Section 5.6). This predictive density of y' has uncertainty determined by the posterior (density B). Density A is the likelihood for new data y'. Typically there will be no solution to this integral, but it can be estimated from MCMC simulation. In most cases, the expectation of y_i is readily determined from the likelihood and can be calculated from the currently imputed parameter values,

$$E[y_i|\mathbf{y}] \;=\; \frac{1}{G}\sum_{g=1}^{G} E[y_i|\theta^{(g)}] \qquad\qquad 7.11$$

Thus, it is possible to predict y_i at each MCMC step and take the mean and quantiles over all G steps. The variance of y_i can also be obtained as described in the following section.

7.6 Computation for Bayesian Model Selection

In Chapter 6 I considered two broad classes of model selection approaches. One is based on penalized likelihoods. I discussed the classical AIC (Section 6.4) and Bayesian alternatives (Section 6.5). Calculation of a penalized likelihood, where the likelihood is penalized by model dimension, requires that we determine model size. There is no simple definition of model dimension for hierarchical models, which were introduced in Section 7.3.3 (e.g., Figure 7.6) and are the subject of Chapter 8. A linear model with Gaussian error might have a random effect β_i associated with each sample unit i. Suppose there are n such random effects. Collectively, these random effects are summarized by a variance, $\beta \sim N_n(0, \tau^2 I_n)$. Do the random effects contribute n parameters (β), one parameter (τ^2), or something in between? There is no obvious answer to this question.

As a second example, suppose each new observation involves a latent variable y_i that must be estimated from an observation z_i. Classical model selection approaches rely on the asymptotic assumption that model dimension p is fixed as sample size n increases (Chapter 6). In this latent variable case, each new observation brings with it an increase in the number of unobservables that must be estimated. Yet, regardless of sample size, the main interest might still focus on a set of parameters θ that describe the process that generates the y's, as opposed to the z's. For a population dynamic model, the y's might be the size of the population, and the z's might be counts. The number of parameters in vector θ does not increase with sample size but the number of z's does increase. Clearly these examples are closely related, in that model dimension lacks simple definition.

Penalized likelihood approaches for hierarchical models attempt to estimate model dimension based on the fit of the model and penalize the likelihood accordingly. An example is the Deviance Information Criterion (DIC) (Carlin and Louis 2000). At present, there is no consensus on model selection for high-dimensional models (Spiegelhalter et al. 2002, with discussion). And many of the proposed alternatives are difficult to compute.

I summarize a practical option that is available from MCMC output. This second approach, predictive loss, is based on prediction and does not require model dimension. Instead, the penalty for overfitting is the predicted variance (Gelfand and Ghosh 1998). This estimate results from minimizing a posterior predictive loss function (residual sum of squares), as motivated by a decision-theoretic framework. The cost of selecting the wrong model is the error sum of squares. For model m, this is

$$G_m = \sum_{i=1}^{n} (E[y_i|\mathbf{y}] - y_i)^2$$

where the expected value comes from Equation 7.8. The penalty term is predictive variance

$$P_m = \sum_{i=1}^{n} \text{var}[y_i|\mathbf{y}]$$

Writing the conditional variance in terms of moments, we have

$$\mathrm{var}[y_i|\mathbf{y}] = E[y_i^2|\mathbf{y}] - (E[y_i|\mathbf{y}])^2$$

From these two moments calculate the expected variance conditioned on **y**. Then select the model with the lowest value of

$$D_m = G_m + P_m$$

The units are y^2, but the absolute values are not given a specific interpretation. For example, both terms increase with sample size. The first term measures goodness of fit. As models become complex, this term will decrease. With increasing model size, the penalty term P_m will decrease at first, but it will eventually increase as overfitting sets in. This index is used like the penalized likelihoods of Chapter 6, but it does not require determination of model dimension. Thus, it has obvious advantages for hierarchical models.

Example 7.9. A linear model
Consider the linear regression

$$y_i \sim N(\mathbf{x}_i\beta,\sigma^2)$$

with prior

$$\beta \sim N_p(0,\mathbf{V}_b)$$

If σ is known and prior variances are large, the predictive loss is obtained from

$$G_m = \sum_{i=1}^{n}(\mathbf{x}_i\beta - y_i)^2 = (\mathbf{X}\beta - \mathbf{y})^T(\mathbf{X}\beta - \mathbf{y}),$$

where the predicted value of y_i is $\mathbf{x}_i\beta$, and

$$P_m \approx (n + p)\sigma^2$$

(Gelfand and Ghosh 1998). The two terms are contributed by the conditional error in the y_i and the error in parameter estimates, respectively. Note that model dimension p appears explicitly here, but it is an approximate result. In Chapter 9, I consider the model of Mohan et al. (2006) for seedling growth under elevated CO_2, where we use predictive loss to assess whether individual differences in growth responses are warranted.

Example 7.10. Discrete states
Agarwal et al. (2005) developed a model for tropical deforestation, where pixels from remotely sensed imagery were classified in one of four land use classes. The hierarchical model related probability of being in a land use class

to covariates involving population density and topography. Let L_{ik} be the probability that pixel i is in land use k, $k = 1, \ldots, 4$. Let $\pi_{ik} \equiv E[L_{ik}|L]$ be the probability of being in state L predicted by the model. Then,

$$G_m = \sum_{i=1}^{n} \sum_{k=1}^{4} (\pi_{ik} - L_{ik})^2$$

where $L_{ik} = 1$ for the one observed state and $L_{ik} = 0$ for the three unobserved states. The penalty is obtained from the multinomial variance,

$$P_m = \sum_{i=1}^{n} \sum_{k=1}^{4} \text{var}[L_{ik}|L]$$

$$= \sum_{i=1}^{n} \sum_{k=1}^{4} \pi_{ik}(1 - \pi_{ik})$$

An example application of predictive loss to discrete states is included for recapture data in Section 9.16.

A transparent approach to predictive loss involves samples of observations based on the MCMC output. If there are latent states, first draw samples of these latent states based on the chains of parameter estimates, including stochasticity that enters through error terms. Then sample new observations $\mathbf{y}$. The penalty term may initially decrease with increasing model complexity, because of the improvement in predictions. But with overfitting, the samples of $\mathbf{y}$ begin to depart from the observations. In other words, the complex model fits well (low G_m), but it can become erratic (large P_m). An example algorithm is provided in Chapter 6 of the lab manual.

For a simple linear regression, there are no latent states, and the approach is transparent. Figure 7.14 shows data simulated from a model with slope and

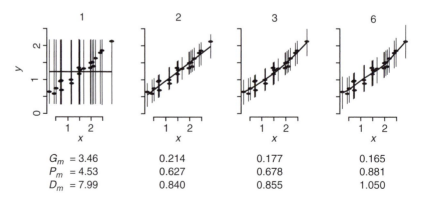

FIGURE 7.14. Predictive intervals for observations y for simulated data as fitted to regression models having dimensions shown above plots. The model used to simulate the data (*second from left*) has dimension $p = 2$.

intercept both equal to 0.5 and $\sigma = 0.1$. To this simulated data set were fitted to polynomial functions of dimension $p = 1$ (intercept only), $p = 2$ (intercept and slope—the correct model), $p = 3$ (including a quadratic term x^2), and $p = 6$ (a polynomial including all terms up to x^5). Estimates of the regression parameters and error variance σ^2 were obtained with a Gibbs sampler. Draws from the Gibbs chains were used to predict new values of $\mathbf{y}$, shown in Figure 7.14 as the predictive mean (regression lines) and 95 percent predictive intervals (vertical lines for each value of $\mathbf{y}$). With increasing model complexity, the predicted values more closely follow the observations, leading to smaller G_m. At the same time, the predictive intervals are reduced from $p = 1$ to $p = 2$ (the correct model), but then increase again beyond $p = 2$. These trends are reflected in P_m. They would be more apparent if the x's were unevenly distributed in Figure 7.14. The predictive loss D_m is minimized at $p = 2$.

7.7 Priors on the Response

Ignorance may not be the only reason why priors are hard to specify. We may have substantial insight about the response y but be unable to specify parameter values that produce that response. The difficulty may stem from the complexity of the relationship between parameters and response, dependencies among parameters, or complex scale relationships. Even if we can produce the correct mean structure for the response, we may be unable to express prior parameter relationships in a way that best describes what is known in advance. In such cases, it is sometimes useful to work directly from priors on the response itself. For example, wavelets represent combinations of parameters that describe scales of variation in a response. We may know what scales should be important without knowing how parameters should combine to produce that response. It is known that mortality rates should not change abruptly from one age bracket to the next, but we may not know much about the specific values of parameters (Girosi and King 2006). For both of these examples, priors on the response were the basis for implicitly specifying priors on parameters. The approach is easiest to grasp with an example.

With many species, mortality rates are difficult to estimate for lack of information on particular strata within a population. In trees, mortality rate increases with declining growth rate and with increasing size. Growth rates can be measured from tree ring widths. Both relationships are difficult to parameterize, because growth rates are difficult to measure once trees are dead, and few trees survive to large size. However, we do know that trees must have at least some diameter increment (it cannot be zero), and the rate at which mortality rate increases with declining growth is related to the frequency with which growth rates are observed in data. We further know that trees do not grow to indefinite size, and upper bounds must be related to the largest individuals observed. The goal is to estimate mortality schedules with growth rate and size in the context of these known patterns in the response. Here is a model:

$$\ln(\theta_i) = \beta_0 + \beta_1 \ln(d_i) + \beta_2 D_i^2$$

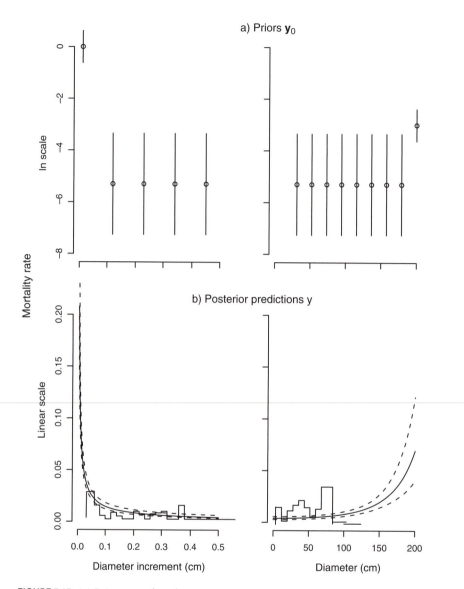

FIGURE 7.15. (*a*) Priors are placed on responses emphasizing the increased mortality risk that attends slow growth and large size, but are broad elsewhere. The vertical lines represent 95 percent of the prior normal density for each value of *y*. Ninety-five percent prediction intervals from the fitted model model corresponding to the two variables in part *a* (solid and dashed lines)with superimposed data showing the fraction of deaths as a histogram.

where θ_i is the mortality rate, D_i is the diameter of tree *i*, and d_i is its growth increment. I use log increments, to increase flexibility. I use the quadratic term for diameter, because I want to model the effect of large diameter only. The prior should express strong insight about θ at small *d* and somewhat weaker insight as D becomes especially large, and be especially weak otherwise. I thus establish sequences of d_j and D_j values, both of length n_0, the prior sample size. For each combination of values I specify a prior mortality rate $\ln \theta$ and prior variance v. These priors could come from different data sets. They are shown as prior means and 95 percent prior belief in Figure 7.15a. Note the broad priors for all y except those at small increment and large diameter. The prior for parameters then becomes $\beta \sim N_3(\mathbf{V}\mathbf{v}, \mathbf{V})$, where $\mathbf{V}^{-1} = \mathbf{X}_0^T \mathbf{V}_0^{-1} \mathbf{X}_0$, $\mathbf{v} = \mathbf{X}_0^T \mathbf{V}_0^{-1} \mathbf{y}_0$, $\mathbf{y}_0$ is the length n_0 vector of priors means, $\mathbf{X}_0$ is the design matrix having columns corresponding to the elements β, and $\mathbf{V}_0 = Diag(v_1, \ldots v_{n_0})$, that is, the diagonal matrix with prior variances.

The estimated values of the parameters appear to be quite different from the prior means (Table 7.1), yet they have been influenced by the full covariance structure of the prior. The posterior predictions of θ show the prior influence at low diameter increment, increasing at the small increments where data give out (trees do not survive at such low growth rates), and at large diameter (few trees survive to large size), but are controlled by the data elsewhere (Figure 7.15b). The data set is large (5342 survivors and ninety deaths), resulting in a narrow prediction interval. However, the priors have a strong impact for low growth increment and on the increase in mortality risk at large size.

Why bother using prior estimates of y for large-diameter trees, beyond where these observations extend? For this application, we wish to predict mortality of trees in simulation models that will produce large trees. The standard approach to such problems would be to lift parameters fitted to a model like this and then add in a new model for situations not covered by it (e.g., the large trees not included in the orginal fit). The problem with this approach is that if we change the assumptions at the prediction stage, the estimates for this fit *at any diameter* are no longer fully relevant. The two sets of assumptions should be compatible with one another. We can accomplish this provided that they are combined at the inference stage. Here inference is done in the context of the full set of assumptions that will be used for prediction. Here we are assuming that the priors on y come from information not contained in this data set. An alternative approach would be to fit multiple data sets simultaneously (Chapter 8).

TABLE 7.1.
Prior Means, Based on the Priors for y, and Posterior Percentiles.

	β_0	β_1	β_2
Prior Mean	-8.49	-1.83	0.000134
50%	-6.09	-0.780	0.0000721
2.5%	-6.52	-0.867	0.0000549
97.5%	-5.68	-0.671	0.0000900

7.8 The Basics Are Now Behind Us

This brief introduction to Bayesian computation illustrates some of the most important techniques for analysis of potentially complex models. Most of the models of later chapters build on the algorithms already covered here. In the remaining chapters of this book I repeatedly return to these techniques with many examples from different types of model structures.

8 A Closer Look at Hierarchical Structures

8.1 Hierarchical Models for Context

Without placing much emphasis on it, I have introduced several levels of structure into Bayesian models. I now consider model structure explicitly, calling attention to how models may be compartmentalized. This compartmentalization is necessary for complex problems that may entail multiple sources of stochasticity that impinge in diverse ways.

In many ecological settings sample units, such as individual organisms, sample plots, lakes, or forest stands, are different, but not independent. The same can be said for the jurisdictional units around which many types of data are organized (county, zip code, state, and so on). In ecological studies of demography, individuals are often classified into different stages, each stage having associated parameter values for growth, survival, and fecundity (Chapter 2). These parameters are treated as independent, which, in turn, affects the classification scheme: classes with few individuals and few transitions from which to estimate parameters must be lumped into larger classes, regardless of ecological considerations. These considerations can apply not only at the stage or group level, but also at the individual level. The mortality risk for one group may be importantly different from others in the same population (e.g., Vaupel et al. 1979; Clark 2003). Yet, the risks have many shared features. Fecundity may depend on resource availability, which varies among populations (e.g., regional context) and within populations (individual differences).

Modeling considerations for mortality risk (Figure 8.1) include whether to assume (1) a single risk parameter θ that applies to all subgroups within a population, (2) fixed differences (independent parameters) for each subgroup θ_j, or (3) a distribution of parameters from which subgroup θ_j is to be drawn. (There is a fourth possibility, differences explained by covariates, which I take up later.) If we assume one size fits all (option A in Figure 8.1), we could miss the important group differences. The mean risk parameter that averages over populations and group might not really apply anywhere. For example, an X percent confidence interval on parameter θ would not represent the true variability

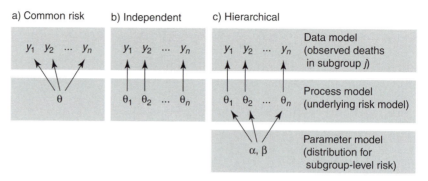

FIGURE 8.1. Three different assumptions about how risk affects the observed mortality rates in a subdivided population.

in risk. If the population varies, it is a confidence interval on a quantity that has no real application (Clark 2003).

At the other extreme, it might be hard to justify treating subgroups as independent effects, and it may be impractical (option B in Figure 8.1). Subgroups might have large differences in sample size (unbalanced data), such that it becomes difficult to classify them in a way that ensures a sufficient sample size for each subgroup. Outliers can have undue influence. An example for capture-recapture data is summarized in Chapter 9. This is a common problem for spatial models, where data come from regions that differ in size and sampling intensity (Chapter 10). It is easy to overfit the data with a large number of independent parameters, many having little support and no predictive potential.

Some of the examples already discussed demonstrate how to add a stage to the model such that effects are drawn from a distribution (option C in Figure 8.1). This assumption admits the relationships that are often realistic for ecological data. The overall structure allows us to borrow strength from the full population for the random effect that applies to a given subgroup. Due to the way in which such models are constructed and analyzed, they are termed *hierarchical*, with the parameter model representing an additional stage. Such models can sometimes be analyzed in a classical setting, for example, using a likelihood that is marginalized over variability associated with random effects, or from a fully Bayesian perspective (Morris 1983; Ver Hoef 1996; Clark 2003). Classical approaches usually involve approximations based on assumptions of asymptotic normality.

A simple simulation can be used to illustrate this point. The correct model for this simulation is a population of $n = 10$ subgroups, from each of which I sample $m_j = 10$ individuals. I observe the number of deaths for each group. I assume that variability among groups enters as $\theta_j \sim Beta(2, 10.4)$ for a mean of 0.16 and variance of 0.01 (these happen to be parameters used for the spotted owl model in Figure 5.12). It is easy to simulate such a data set. Draw n values of θ_j from the beta density followed by n values of $y_j | \theta_j \sim Bin(m_j, \theta_j)$, where m_j is the sample size of the jth group. For this example the simulated data set is $\mathbf{y} = [2, 2, 2, 2, 6, 1, 2, 2, 1, 2]$. If I were attempting a comprehensive evaluation of a new model, I would draw many such data sets. This one is simple, and one sample illustrates how the three assumptions affect inference on risk.

I estimated parameters for this data set under the three assumptions in Figure 8.1. For option A, the posterior density is

$$\theta|\mathbf{y},\alpha,\beta \sim Beta(\alpha + \sum_j y_j, \beta + \sum_j (m_j - y_j))$$

where α and β are prior parameter values. For option B, each group has an estimate, the posterior for which is

$$\theta_j|y_j,\alpha_j,\beta_j \sim Beta(\alpha_j + y_j, \beta_j + m_j - y_j)$$

In both cases, the prior parameter values are treated as fixed constants, so I have not bothered to include them in Figure 8.1. They are not estimated from the data. For option C, estimates borrow strength from the population-level parameters—the parameters α and β are no longer fixed, but are themselves estimated from the data.

Assumption A produces a tight posterior density that misrepresents the variability in the data (dashed line in Figure 8.2a). Options B and C both

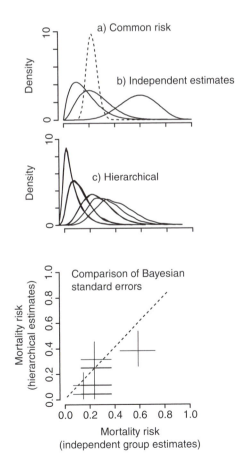

FIGURE 8.2. Posterior estimates of mortality risk under three assumptions of Figure 8.1 (upper two panels) and one standard deviation of the parameter estimate (68 percent confidence interval) for assumptions B and C (lower panel). For option B in panel a, we only see three densities (rather than ten), because there were only three unique outcomes of $y = 1, 2,$ or 6.

admit structure, but in different ways. Because assumption C is known to be the correct one, I attribute the large differences between the two collections of posteriors for θ to the fact that assumption of complete independence can produce misleading results. Under assumption B, estimates may be poor due to the low sample sizes—I might not expect risk to vary among groups as widely as suggested by the estimates in Figure 8.2a, option B. By overfitting the data, I cannot expect to predict as-yet unobserved data sets.

A number of examples already discussed in this text are hierarchical. They share the properties of borrowing strength across groups, while allowing for variability. In the following section I explore models that build on the linear examples of Section 7.4 with focus on hierarchical structures and nonlinear extensions.

8.2 Mixed and Generalized Linear Models

I have discussed how a linear equation can be the starting point for a more sophisticated analysis that allows for errors in variables, combining different data sources, and non-Gaussian errors (Chapter 7). In many environmental applications where traditional assumptions cannot be met, linear equations still play an important role. I return to linear equations here, because they are commonly an important element of hierarchical structures, where there is context based on location and heterogeneity at multiple scales. In fact, hierarchical Bayes is often the easiest and most flexible approach for linear equations once we move beyond the most basic structures.

A full review of linear models is beyond the scope of this book. The literature on classical treatments is vast and accessible. There are many software options available for standard statistical designs. Here I sketch some elements of linear models that will be of general use, emphasizing their role in hierarchical structures.

8.2.1 Generalized Linear Models

Generalized linear models entail a sampling distribution that is non-Gaussian, and the mean of the distribution is a function of a linear equation (as opposed to the linear equation itself). Consider the model of fecundity in Example 7.4. There I assumed random effects in fecundity, with each individual having its own Poisson parameter, each of which is drawn from a standard distribution having parameters that can be estimated. Suppose instead that there exists information on covariates that might explain why fecundity varies among individuals. A standard model might assume that offspring production is still Poisson with an underlying parameter for the mean,

$$y_i \sim Pois(\theta_i)$$

Now use the covariates to describe differences among observations. A common approach would involve a linear model for the log of θ,

$$\ln \theta_i = \mathbf{x}_i\beta = \beta_0 + \beta_1 x_{1i} + \ldots + \beta_{p-1} x_{(p-1)i}$$

where $\beta = [\beta_0, \ldots, \beta_{(p-1)}]^T$ is a vector of parameters and $\mathbf{x}_i = [1, x_{1i}, \ldots x_{(p-1)i}]$ is the vector of covariate values for individual i. It is the ith row of a n by p design matrix $\mathbf{X}$ for a sample of size n. Covariates might be continuous, such as the amount of resource available to the ith individual. The log function of θ may provide a reasonable description of the distributional properties. For example, θ should be positive. This log function of θ is known as the link function. The link function translates a parameter that may have a restricted range to the real line. For a Bernoulli sampling distribution the logit link is commonly used,

$$\text{logit}(\theta_i) \equiv \ln\left(\frac{\theta_i}{1 - \theta_i}\right) = \mathbf{x}_i\beta \qquad 8.1$$

The inverse of this link function is solved in terms of θ_i,

$$\theta_i = \frac{\exp(\mathbf{x}_i\beta)}{1 + \exp(\mathbf{x}_i\beta)} \qquad 8.2$$

A probit function (a normal cumulative distribution function) can also be used for such cases (Example 5.6), as can other cumulative distribution functions—in Section 8.5, I use a gamma cumulative distribution function. These functions translate a parameter with range $(0,\infty)$ (Poisson) or $(0,1)$ (Bernoulli) to $(-\infty,\infty)$. This structure is known as a *generalized linear model* (GLM), because the expected value of θ_i is a function of the linear equation. It is framed by a non-Gaussian sampling distribution.

There is a large literature on generalized linear models from a classical perspective (McCullagh and Nelder 1989). Maximum likelihood parameter estimates can often be obtained numerically. This involves methods mentioned in Chapter 3 and the lab manual. As complexity increases (e.g., with random effects as discussed in the following section), parameter estimation relies on asymptotic approximations, which may not always be reasonable.

With a Bayesian approach, we can work directly with the conditional structure of the model. Although the sampling distribution is nonlinear, we can apply many of the techniques for linear models at the next stage. Here is a simple model for mortality based on a logit link function,

$$p(\beta|\mathbf{x},\mathbf{y}) \propto \prod_{i=1}^{n} Bernoulli(y_i|\theta_i)p(\beta)$$

$$\text{logit}(\theta_i) = \mathbf{x}_i\beta$$

where $p(\beta)$ is a prior density of parameters. The Bernoulli sampling distribution has θ_i as the risk parameter, which depends on the underlying linear equation.

The link-function approach is potentially advantageous in several ways. It can have the advantage over simple data transformations that do not well describe the error structure. For example, ecological count data are frequently modeled as linear regressions on log values. Because counts can be zero, one might add an arbitrary constant (e.g., a one) to each count. This constant changes the estimate of the slope parameter for a covariate. Thus, adding a constant precludes testing of a covariate effect, which is typically the goal of such analyses. Moreover, for count data there is justification for assuming a conditionally Poisson sampling distribution, whereas the simple lognormal likelihood may be difficult to justify for the counts themselves. Finally, the discrete zero class is inherently discontinuous and non-Gaussian.

A link function may also be easier to extend than using, say, a conjugate density at the next stage. Consider, for instance, the beta-binomial pair for mortality or presence-absence. To incorporate covariates or even extra binomial variability, the beta prior becomes clumsy. The alternative logit link can be readily extended to include covariates and additional, non-sampling-related error,

$$\text{logit}(\theta_i) = \mathbf{x}_i \beta + \varepsilon_i$$

$$\varepsilon_i \sim N(0, \sigma^2)$$

With this extension, we now model $\text{logit}(\theta_i)$ at the second stage as Gaussian. This is sometimes termed the *logit-normal*. The extrabinomial variation can accommodate the fact that covariates do not fully account for the mean risk.

There is a growing number of examples in the environmental literature. Because there are often repeated observations on individual sample units over time, which will be correlated, GLM's can be time-dependent and include random effects at the level of these individual sample units. The random effects can involve assumptions about spatial structure. Many examples are presented later in this chapter and especially in Chapters 9 and 10. As a verbal prelude to such models, consider Woodward et al.'s (2003) estimates of turkey harvest in Missouri. As in Example 7.6, where the binomial and Poisson were combined, Woodward et al. (2003) applied a binomial to describe the probability of success of a given hunting trip and a Poisson to describe the number of hunting trips. The sample size parameter of the binomial n_{ij} is the number of hunting trips during a week j in county i. The binomial probability parameter p_{ij} is the success rate. The number of hunting trips is a Poisson variate. To understand how turkey harvest varied with county and week, they applied generalized linear models with a logit link for success and a log link for number of trips. For both linear equations they included an additional term for random error. The random county effects were given a spatial prior specification (Chapter 10). From this analysis they obtained maps of success for the state of Missouri that showed more spatial coherence than was obtained when modeled as a simple frequency estimate. To some extent, the smoothness of the Bayesian estimates results from the spatial prior on county effects. These effects are discussed in Chapter 10. But there was also borrowing of strength, with counties characterized by low sample sizes being influenced by others.

8.2.2 Multinomial Data

Categorical responses can fall in more than two classes. Just as Bernoulli responses can be treated as generalized linear models, so too can multinomial data. Recall that a simple Bernoulli model has a conjugate beta prior (Appendix G). The GLM involves, instead, the logit link for a linear equation. The multinomial extension of the simple Bernoulli often involves a conjugate Dirichlet prior—an example is provided in Section 9.16. With covariates, we have a set of linear equations for each class of the multinomial, again, commonly modeled with a logit link. With the multinomial, probabilities of classes are taken in relation to a baseline category. Gelman et al. (2004) provide a basic overview. Ogle et al. (2006) and Okuyama and Bolker (2005) present ecological examples.

The categories can be nominal (having no particular order) or ordinal. For ordinal classes, the data are viewed as being ordered in terms of adjacency, as in levels of severity (low, intermediate, high) or rating (grades of A, B, C, D, E), but without explicit connotation of distance between them. The categories can represent a latent variable perceived to lie on some continuous scale. Generalized linear models for such data can be constructed in a variety of ways and involve different link functions (Johnson 1996; Lunn et al. 2001; Mwalili et al. 2005). One type is termed a *cumulative logit model*. It is easiest to discuss such models in the context of an example.

Forest ecologists attempt to infer light available to large trees based on simple ordinal rankings assigned by an observer on the ground (e.g., suppressed, intermediate, codominant, dominant). Here I demonstrate how to set up a cumulative logit model for this example assuming the light availability can be observed. This application demonstrates how to estimate the parameters that relate light availability to the status observations. In Section 8.5.2, I extend the model to allow for multiple data types, together with the fact that the true light availability is unknown.

Let λ_i be the light available to tree i, defined on the continuous scale. Ordinal rankings of tree exposure, or status $\lambda_i^{(s)}$, are assigned based on observation from the ground as 1 (suppressed in the understory and not exposed to direct sun), 2 (intermediate, having at least some exposure to direct sunlight), and 3 (codominant and dominant, having a large portion of the canopy exposed to direct sunlight).

There is a logit link to an equation for each of the three status classes. Let $\theta_{i,k}$ be the probability that individual i is assigned status k. As for the Bernoulli model (Equation 8.2), write a logit link for the probability of being assigned a class based on light availability. Because the categories are ordinal and the predictor variable is continuous, the linear equation is written in terms of cumulative probabilities, that is, the probability of being assigned a state less than or equal to k, divided by its complement, the probability of being assigned a state greater than k,

$$\log\left(\frac{\theta_{i,1} + \cdots + \theta_{i,k}}{\theta_{i,k+1} + \cdots + \theta_{i,K}}\right) = \beta_{0,k} + \beta_{1,k}\lambda_i$$

where the total number of classes K is, in this case, 3, and $\beta_{0,k}$ and $\beta_{1,k}$ are intercepts and slopes for the kth class. This looks more like Equation 8.2 upon recognition that

$$\log\left(\frac{\theta_{i,1} + \cdots + \theta_{i,k}}{\theta_{i,k+1} + \cdots + \theta_{i,K}}\right) = \log\left(\frac{\theta_{i,1} + \cdots + \theta_{i,k}}{1 - (\theta_{i,1} + \cdots + \theta_{i,k})}\right)$$

Inverting this link function, the equation for the first class is

$$\theta_{i,1} = \frac{\exp[\beta_{0,1} + \beta_{1,1}\lambda_i]}{1 + \exp[\beta_{0,1} + \beta_{1,1}\lambda_i]}$$

Subsequent categories are

$$\theta_{i,k} = \frac{\exp[\beta_{0,k} + \beta_{1,k}\lambda_i]}{1 + \exp[\beta_{0,k} + \beta_{1,k}\lambda_i]} - \sum_{j=1}^{k-1}\theta_{i,j}$$

The last category is

$$\theta_{i,K} = 1 - \sum_{j=1}^{K-1}\theta_{i,j}$$

Because probabilities sum to one, there are $K - 1$ intercepts $\beta_{0,k}$ and $K - 1$ slopes $\beta_{1,k}$. Because categories are ordered, intercepts are constrained such that $\beta_{0,k} < \beta_{0,k+1}$. Slopes can be likewise constrained. Priors can look like this:

$$\beta_{0,k} \sim \begin{cases} N(b_{0,1},10^4)I(\beta_{0,1} > \beta_{0,2}) & k = 1 \\ N(b_{0,2},10^4)I(\beta_{0,k-1} > \beta_{0,k} > \beta_{0,k+1}) & 1 < k < (K - 1) \\ N(b_{0,K-1},10^4)I(\beta_{0,K-2} > \beta_{0,K-1}) & k = K - 1 \end{cases}$$

$$\beta_{1,k} \sim \begin{cases} N(b_{1,1},10^4)I(\beta_{1,1} < \beta_{1,2}) & k = 1 \\ N(b_{1,k-1},10^4)I(\beta_{1,k-1} < \beta_{1,k} < \beta_{1,k+1}) & 1 < k < (K - 1) \\ N(b_{1,K-1},10^4)I(\beta_{1,K-2} < \beta_{0,K-1}) & k = K - 1 \end{cases}$$

where $I(A)$ is the indicator function, equal to one if A is true and zero otherwise. For this example, $K = 3$, so there are only two intercepts. For regression parameters I use the prior

$$\beta_0 \sim N_2([-.1,0]^T, Diag(10^4,10^4))I(\beta_{0,1} < \beta_{0,2})$$

For the two slope parameters, the prior is

$$\beta_1 \sim N_2([-1,-0.5]^T, Diag(10^4,10^4))I(\beta_{1,1} < \beta_{1,2})$$

The model fitted in Figure 8.3 shows that trees with highest light availability tend to be assigned to class 3, and those with lowest light to class 1. But there is substantial overlap.

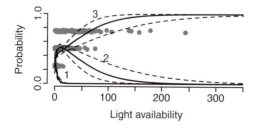

FIGURE 8.3. Canopy status observations 1, 2, and 3 (dots) with the fitted cumulative logistic model (solid lines for each class bounded by dashed 95 percent predictive intervals).

8.2.3 Parametric Survival Models

Survival analysis can involve the elements of a generalized linear model. If survival is observed on the same individuals over multiple censuses, mortality can be modeled using the survivor function. The semi-parametric model, introduced in Section 3.8.3, involves a large number of parameters, one for each time increment. If a parametric form for the survivor function is appropriate, then the model might have few parameters. The *accelerated time model* additionally includes a simple parametric form for covariates (linear). This approach differs from the proportional hazards model of Chapter 4 in that the time until death is written as a linear function with a log link,

$$\ln a_i = \mathbf{x}_i \beta + \varepsilon_i$$

The parameter vector should include an intercept term β_0 (the vector of covariates has a leading 1). The last term is stochastic. The parametric form for this term determines the survival function.

Several parametric models of survival were developed in Chapter 3, including the exponential and Weibull, but other forms can be used. The Weibull survival function

$$l(a_i; \rho, c) = \exp[-\rho a_i^c]$$

can be parameterized to describe increasing risk ($c > 1$ in Figure 2.17c) or decreasing risk ($c < 1$ in Figure 2.17a) with age. The exponential is appropriate when there is no reason to expect that risk will change with age. The Weibull is convenient, because comparisons with the exponential (obtained with $c = 1$) can be used to test whether or not risk changes with age. Other considerations for selecting a parametric model are mostly based on whether it can be parameterized to have an appropriate shape and on convenience. Because survival functions involve integrals of density functions (Equation 2.19), some do not have solutions. Here again the Weibull is convenient, because the survivor function can be solved. The gamma and lognormal distributions have survivor functions that cannot be solved, but they still may be useful. Here I consider how to include covariate effects, focusing on the Weibull.

To obtain a Weibull distribution for survival, assume that the stochastic term of the linear model has the extreme value distribution,

$$p(\varepsilon_i | \sigma) = \frac{1}{\sigma} \exp\left[\frac{\varepsilon_i}{\sigma} - e^{\varepsilon_i / \sigma} \right]$$

To obtain the density of deaths, make a variable change (Appendix D) from ε_i to $\ln a_i$,

$$f(a_i) = \frac{a_i^{\sigma^{-1}-1}}{\sigma} \exp\left[-\frac{\mathbf{x}_i \beta}{\sigma} - \exp\left(\frac{\mathbf{x}_i \beta}{\sigma} \right) a_i^{\sigma^{-1}} \right]$$

with corresponding survivor function

$$l(a_i) = \exp\left[-\exp\left(\frac{\ln a_i - \mathbf{x}_i\beta}{\sigma} \right) \right]$$

$$= \exp\left[-\exp\left(-\frac{\mathbf{x}_i\beta}{\sigma} \right) a_i^{1/\sigma} \right] \qquad 8.3$$

In terms of the Weibull distribution the parameter relationships are

$$\rho_i = \exp\left(-\frac{\mathbf{x}_i\beta}{\sigma} \right)$$

$$c = 1/\sigma$$

If there are no covariate effects, the base survival function has location parameter $\rho = \exp(-\beta_0/\sigma)$. If $\sigma = 1$, then $c = 1$, and the model is exponential.

ML estimation is the standard approach, with numerical methods used to estimate the parameters β, σ (Equation 3.11). This accelerated life model was applied to the data set in Figure 4.6 for covariates ln(light) and ln(height increment), where light is fraction of full sunlight and height increment is in centimeters. I used the Splus function censorReg, which provides asymptotic standard error estimates for β, but not for σ, which has an MLE of 0.347 (see Table 8.1). The deviance for this model is 288. A plot of the survivor function is obtained using Equation 8.1, as shown in Figure 8.4.

There is a vast literature on survival models. Some useful references on classical methods include Cox and Oakes (1984), Kalbfleisch and Prentice (1980), and Klein and Moeschberger (1997). Risk models with individual effects are termed *frailty models* and are often taken up in a Bayesian framework (Spiegelhalter et al. 2003). These models are hierarchical.

8.2.4 Mixed Models

Mixed model is a term used to describe a specific type of hierarchical structure in which some of the parameters in a linear model are treated as realizations of a stochastic process, rather than as constants. Those that are treated as

TABLE 8.1.
Parameter Estimates for Figure 8.4

Estimate	Standard Error	0.025 CL 95% LCL	0.975 CL 95% UCL	z-value	p-value
β_0 1.46	0.124	1.22	1.70	11.8	4.54e−32
β_1 0.395	0.0848	0.228	0.561	4.65	3.29e−06
β_2 0.706	0.0693	0.570	0.842	10.2	2.30e−24

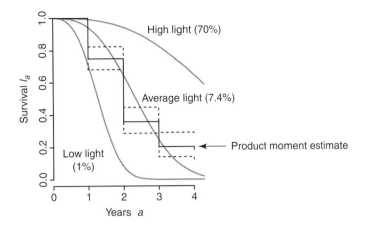

FIGURE 8.4. The accelerated life model, using a Weibull survivor function compared with the product moment estimate from Section 3.8.3. The accelerated life model predictions are shown for three individuals, one growing at the average light availability and two growing near the extreme light availabilities observed during this experiment.

constants are termed *fixed effects*. Those taken to be random are *random effects*. These models are often analyzed with hierarchical Bayes, because the random effects constitute an additional stage.

Consider a study where animals are classified in groups $j = 1, \ldots m$, and these groups are expected to have different mortality risks based on some recognizable attributes. Assume $m = 2$ groups (e.g., male, female). Here is a design matrix with each row corresponding to an observation on one animal,

$$\mathbf{X} = \begin{bmatrix} 1 & 0 \\ 0 & 1 \\ \vdots & \vdots \\ 0 & 1 \end{bmatrix} \begin{matrix} \text{a female} \\ \text{a male} \\ \\ \text{a male} \end{matrix}$$

Columns 1 and 2 are used to identify each observation as male (a 1 in column 1) or female (a 1 in column 2). These zeros and ones are indicators. Then the model for the ith animal might have a second stage that looks like this:

$$\text{logit} \, (\theta_i) = \mathbf{x}_i \beta = \beta_i$$

If female, β_1 applies. If male, β_2 applies. In other words,

$$\mathbf{X}\beta = \begin{bmatrix} 1 & 0 \\ 0 & 1 \\ \vdots & \vdots \\ 0 & 1 \end{bmatrix} \begin{bmatrix} \beta_1 \\ \beta_2 \end{bmatrix} = \begin{bmatrix} \beta_1 \\ \beta_2 \\ \vdots \\ \beta_2 \end{bmatrix}$$

The parameter has a subscript, but it still refers to fixed effects. Gender is a fixed effect, because we suspect that genders may have fixed differences, and we expect that future studies will likewise involve the same identifiable genders. It is not viewed as a random draw from a distribution of different genders, any of which might have appeared in the sample.

Now suppose that animals differ for reasons that cannot be attributed to specific factors or covariates. This was assumption C in Figure 8.1. The m groups might come from different locations. We expect that risk could differ among locations, but we have no information to help explain those differences. We might view them as exchangeable, in the sense that the name or index assigned to each location is not viewed as being important. On the other hand, we have no reason to treat them as independent. In the gender example, males were assumed to have fixed differences from females. We know that differences between the male group and the female group are partly related to the fact that they are of different gender. By contrast, location effects might exist, but it might be reasonable to view them as part of a population of related effects. Future studies will involve different locations. We may not care about the locations specifically used in this study, but we might care about how spatial variability, in a general sense, affects the analysis. Location effects could be represented by a distribution, with the effect for each location drawn at random from that distribution. There are many locations that could have been selected. The ones we did sample were not chosen for their potentially specific (fixed) influences. It is not critical that the sample units have been drawn at random, but their effects are viewed as random. Again, this is an assumption that is useful when the cause of those differences cannot be identified. Both location (e.g., sample plots) and individuals are often treated as random effects.

As a second example, consider a linear model with Gaussian error. A random effects model for a regression parameter can be written with two terms,

$$\beta_i = \alpha + b_i$$

The first term is treated as a constant. In this case, it assumes the same value for all sample units $i = 1, \ldots, n$. This is clear, because it has no subscript. The second term has a subscript, indicating that there is a value for each sample unit. It could be fixed, or it could be random. This second term describes how the ith sample unit departs from the overall effect in the first term. Unless identified as random, it will be treated as a fixed effect.

To specify a random effect, assign it a density. Consider a covariate x_i. Here is a model with a random effect

$$E[y_i|\beta_i] = x_i\beta_i$$

$$\beta_i \sim N(\alpha, V_b)$$

for individual i. We might refer to the mean α of random coefficients β_i. Random effects are exchangeable in that their effects depend on the sample unit to which we are referring only in a probabilistic sense.

I could organize the effects in a different, albeit equivalent, way,

$$E[y_i|\beta_i] = x_i\alpha + x_i\beta_i$$

$$\beta_i \sim N(0, V_b)$$

This representation (i.e., separating the fixed and random components) is often convenient. These two ways of writing the model extend to matrix notation. Suppose I observe animals over time $t = 1, \ldots, T$. I can write a specific type of mixed model, termed a *longitudinal* or *repeated measures* analysis (Chapter 9). If it is convenient to separate the fixed and random effects, write the model for the tth observation on the ith individual as

$$E[y_{it}|\beta_i] = \mathbf{x}_{it}\alpha + \mathbf{w}_{it}\beta_i, \qquad\qquad 8.4$$

where $\mathbf{x}_{it}$ is a p-length vector of covariates for fixed effects, α is the parameter vector for fixed effects, $\mathbf{w}_{it}$ is a q-length vector of covariates for random effects, and β_i is the vector of random coefficients. Typically, random effects are a subset of fixed effects, so $q < p$, and the vector $\mathbf{w}_{it}$ is a subset of effects contained in $\mathbf{x}_{it}$. There could be a fixed effects model ($q = 0$), a random effects model ($q = p$), or a mixed model ($0 < q < p$). For a slope-intercept model, there could be random intercepts, but not random slopes. Then $\mathbf{x}_{it} = [1 \quad x_{it}]$, $\mathbf{w}_{it} = 1$, $\alpha = [\alpha_0 \quad \alpha_1]^T$, and $\beta_i = \beta_{0,i}$. Now every observation depends on the common slope α_1, but each has the random intercept $\alpha_0 + \beta_{0,i}$. Clark et al. (2004) model log fecundity against log tree diameter x_{it} with random slopes, but fixed intercepts, that is, $\mathbf{w}_{it} = x_{it}$ and $\beta_i = \beta_{1,i}$ (Section 9.15.4).

Before proceeding further, we need to be clear on the motivation for this classification of stochasticity. Figure 8.5 illustrates two hypothetical longitudinal studies having the same overall level of stochasticity. A model for this process is

$$y_{it} = \alpha + \beta_i + \varepsilon_{it}$$

$$\beta_i \sim N(0,\tau^2)$$

$$\varepsilon_{it} \sim N(0,\sigma^2)$$

For this example, I use $\alpha = 1$ and a total variance of $\tau^2 + \sigma^2 = 0.1$. The two examples look different, because one assumes that individual differences dominate (Figure 8.5a), and the other assumes that process error dominates (Figure 8.5b). If the model ignores random effects, inference looks like Figure 8.5b, regardless of whether the variance derives from population heterogeneity (described by variance τ^2) or from unexplained fluctuations over years and individuals (variance σ^2). As discussed in Chapter 1, the implications for prediction can be substantial. Include random effects in models when there is reason to believe that sample units may have important differences. They are random, rather than fixed, because they are viewed as being exchangeable.

I discuss longitudinal structures in Section 9.16. Here I take up the basic elements of the approach that could apply whether or not observations accumulate sequentially. Although I represent observations on the same subject with the subscript t, for time, the outline of random effects covered here applies to nonsequential samples. For example, we could think of these as observations on individuals (rather than over time) within groups (rather than individuals).

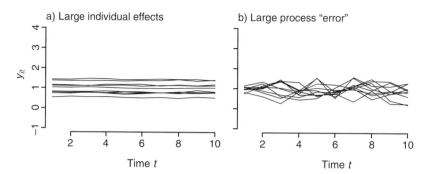

FIGURE 8.5. Two simulated longitudinal data sets with $n = 10$ and the same total variance, but dominated by individual differences (*a*) or process error (*b*). Variance parameters in (*a*) are $\tau^2 = 0.09$ and $\sigma^2 = 0.01$ and in (*b*) are $\tau^2 = 0.01$ and $\sigma^2 = 0.09$.

The model for the $t = 1, \ldots, T$ observations on the ith individual is a length-T vector with conditional expectation

$$E[\mathbf{y}_i|\beta_i] = \mathbf{X}_i\alpha + \mathbf{W}_i\beta_i$$

The design matrixes $\mathbf{X}_i$ and $\mathbf{W}_i$ have dimensions T by p and T by q, respectively. The vector of random effects is $\beta_i^{i.i.d.}\sim N_q(\mathbf{0}, \mathbf{V}_b)$, where $\mathbf{0}$ is a length-q vector of zeros and $\mathbf{V}_b$ is a q by q covariance matrix. I further specify an error covariance matrix for individual i as $\sum_i = \sigma^2\mathbf{I}_T$. Note that, in this case, the same covariance matrix applies to each individual.

The full model for observations can be stacked as a vector of length nT,

$$\mathbf{y} = \begin{bmatrix} \mathbf{y}_1 \\ \vdots \\ \mathbf{y}_n \end{bmatrix}$$

The random effects are contained in a length-nq vector

$$\beta = \begin{bmatrix} \beta_1 \\ \vdots \\ \beta_n \end{bmatrix} = [\beta_{0,1}, \ldots \beta_{(q-1),1}, \beta_{0,2}, \ldots]^T$$

This is a column vector with the q by 1 random effects vector for each individual stacked one upon another. The model for all T observations on n individuals is

$$\mathbf{y}|\beta \sim N_{nT}(\mathbf{X}\alpha + \mathbf{W}\beta, \sigma^2\mathbf{I}_{nT}) \qquad 8.5$$

$$\beta \sim N_{nq}(\mathbf{0}, \mathbf{V}_B)$$

where

$$\underset{(nT \times p)}{\mathbf{X}} = \begin{bmatrix} \mathbf{X}_1 \\ \vdots \\ \mathbf{X}_n \end{bmatrix}, \quad \underset{(nT \times nq)}{\mathbf{W}} = diag(\mathbf{W}_1, \dots, \mathbf{W}_n), \text{ and}$$

$$\underset{(nq \times nq)}{\mathbf{V}_B} = diag(\mathbf{V}_b, \dots, \mathbf{V}_b) = \mathbf{I}_n \otimes \mathbf{V}_b$$

The Kronecker product, indicated by $\otimes$, is a block diagonal matrix. This could be thought of as a n by n matrix, where each "element" is a matrix obtained by multiplying the corresponding element of I_n by the matrix $\mathbf{V}_b$ (Appendix C).

The conditioning on β_i in Equation 8.2 is important. Because this effect is random, the prediction of $\mathbf{y}_i$ is improved by knowing the effect that is associated with that sample unit. For example, if sample units are locations, and location influences $\mathbf{y}_i$, then more is known about $\mathbf{y}_i$ once the random effect associated with the ith location is known. This is the conditional expectation of Equations 8.2 and 8.3. If the effect associated with the sample location is unknown, the estimate of the mean is no better than that of the full population, and the variance is large. To find this marginal distribution of $\mathbf{y}_i$, integrate over the random effects. For a normal sampling distribution with common variance σ^2,

$$\mathbf{y}_i | \beta_i^{i.i.d.} \sim N_T(\mathbf{X}_i \alpha + \mathbf{W}_i \beta_i, \sigma^2 \mathbf{I}_T)$$

and a normal sampling distribution for the random effects with covariance matrix $\mathbf{V}_b$, the marginal distribution is

$$\mathbf{y}_i \sim N_T(\mathbf{X}_i \alpha, \sigma^2 \mathbf{I}_T + \mathbf{W}_i \mathbf{V}_b \mathbf{W}_i^T)$$

This distribution has a mean given by the fixed effects and a large variance that combines random effects and sampling error. This variance can be obtained using Equation D.38b

$$Var[\mathbf{y}_i] = E[Var[y_i | \beta_i]] + Var[E[\mathbf{y}_i | \beta_i]]$$

$$= \sigma^2 \mathbf{I} + \mathbf{W}_i \mathbf{V}_b \mathbf{W}_i^T$$

The second term is the variance among expectations of observations for each individual. The random coefficients β no longer appear, because we have integrated over the individual effects to obtain a global view of variability in $\mathbf{y}$.

We can readily solve for the MLE's of regression parameters conditioned on the covariance matrix. But lack of full, closed-form solutions for the typical case where covariances are unknown restricts classical methods. For all but the simplest covariance matrixes, classical approaches rely on assumptions of asymptotic normality (e.g., McCulloch 2003). Alternatively, these conditional distributions are the basic elements of a sampling-based approach for a Bayesian analysis (e.g., Gibbs sampling).

The Bayesian approach is efficient, because the hierarchical model can be organized in terms of stages and analyzed in terms of conditional densities. Because ecological data sets tend to be unbalanced (not all sample units have the same number of observations), a sample-by-sample representation of the likelihood is often useful. The likelihood and random effects can be written as a product over each individual to highlight different numbers of observations on each individual T_i,

$$p(\alpha,\beta,\sigma^2,\mathbf{V}_b|\mathbf{Y},\mathbf{X}, \text{prior values}) \propto \prod_{i=1}^{n} N_{Ti}(\mathbf{y}_i|\mathbf{X}_i\alpha + \mathbf{W}_i\beta_i,\sigma^2\mathbf{I}_{Ti})N_q(\beta_i|0,\mathbf{V}_b)$$

$$\times\ N_p(\alpha|\mathbf{a},\mathbf{V}_a)IG(\sigma^2|s_1,s_2)$$

$$\times\ W(\mathbf{V}_b^{-1}|(r\mathbf{R})^{-1},r)$$

The third-stage inverse Wishart hyperprior is conjugate with the covariance matrix for the normal distribution of random individual effects. This conjugacy is convenient, but it demands a common degrees of freedom parameter r for all variances.

The conditional posterior for fixed effects parameters is

$$p(\alpha|\beta,\sigma^2,\mathbf{V}_b,\mathbf{X},\mathbf{Y},\mathbf{a},\mathbf{V}_\alpha)=N(\mathbf{Vv},\mathbf{V})$$

where $\mathbf{V}^{-1} = \sigma^{-2}\sum_{i=1}^{n}(\mathbf{X}_i^T\mathbf{X}_i) + \mathbf{V}_\alpha^{-1}$ and $\mathbf{v} = \sigma^{-2}\sum_{i=1}^{n}[\mathbf{X}_i^T(\mathbf{y}_i - \mathbf{W}_i\beta_i)]+ \mathbf{V}_\alpha^{-1}\mathbf{a}$. Note that $\sum_{i=1}^{n}\mathbf{X}_i^T\mathbf{X} = \mathbf{X}^T\mathbf{X}$, where $\mathbf{X}$ is the stack of $\mathbf{X}_i$ matrixes, one atop the other. For random effects, the conditional posterior is

$$p(\beta_i|\alpha,\sigma^2,\mathbf{V}_b,\mathbf{X}_i,\mathbf{y}_i) = N(\mathbf{V}_i\mathbf{v}_i,\mathbf{V}_i)$$

where $\mathbf{V}_i^{-1} = \sigma^{-2}\mathbf{W}_i^T\mathbf{W}_i + \mathbf{V}_b^{-1}$ and $\mathbf{v}_i = \sigma^{-2}\mathbf{W}_i^T(\mathbf{y}_i-\mathbf{X}_i\alpha)$. The conditional posterior for σ^2 is

$$p(\sigma^2|\alpha,\beta,s_1,s_2,\mathbf{X},\mathbf{Y}) = IG\left(\sigma^2\bigg|s_1 + \frac{1}{2}\sum_{i=1}^{n}T_i,s_2\right.$$

$$\left.+ \frac{1}{2}\sum_{i=1}^{n}\sum_{t=1}^{T_i}(\mathbf{y}_{it} - \mathbf{x}_{it}\alpha - \mathbf{w}_{it}\beta_i)^2\right)$$

Finally, the Wishart conditional posterior is

$$p(\mathbf{V}_b^{-1}|\beta,r,\mathbf{R}) = W_2\left(\mathbf{V}_b^{-1}\left[\sum_{j=1}^{n}\beta_i\beta_i^T + r\mathbf{R}\right]^{-1},n + r\right)$$

Examples that build from this basic structure are presented in Chapter 9. I provide an example illustrating random effects after introducing nonlinear models in the following section.

8.2.5 Nonlinear Mixed Models

Nonlinear mixed model is a term applied to a model with Gaussian error (perhaps requiring transformation), the expectation for which is nonlinear in parameters. Nonlinear models can have fixed and random effects incorporated into the expectation function. In this case, data might be modeled as

$$\mathbf{y}_i | \beta_i^{i.i.d.} \sim N_m(f(\mathbf{X}_i\alpha + \mathbf{W}_i\beta_i;\theta),\sigma^2\mathbf{I}_m)$$

where $f()$ is a nonlinear function that may include additional parameters θ, and there are m observations per sample unit. Recall from Chapter 2 that "nonlinear" refers to parameters. For example, Barrowman et al. (2003) estimated reproduction for Coho salmon on the west coast of North America. They had a number of observations from each population and treated individual populations as random effects. One of several models they considered involves the Beverton-Holt function, relating numbers of recruits r_{it} in stream i at time t to the numbers of spawners s_{it} that would have produced them,

$$r_{i,t} = \frac{s_{i,t}}{1/\alpha_i + (s_{i,t}/r_{m,i})} e^{\epsilon_{i,t}}$$

$$\varepsilon_{I,t} \sim N(0,\sigma^2)$$

$$\alpha_i, r_{m,i} \sim N_2(\mu,\mathbf{V})$$

with the bivariate normal specification for random effects having a length-2 mean vector μ and covariance matrix $\mathbf{V}$.

With a log transformation, this remains nonlinear. We cannot solve for maximum likelihood estimates, but classical methods are available that use approximations and appeal to asymptotics (e.g., Lindstrom and Bates 1990). In a Bayesian framework, sampling-based approaches allow us to work directly from conditional densities. For this example, such an approach would entail specification for the prior covariance matrix $\mathbf{V}$. MCMC would require indirect sampling of parameters (e.g., Metropolis-Hastings or inverse distribution sampling). Figure 8.6 shows how fitted models differ under the assumption of random individual effects for streams as opposed to independent fits to each stream. Small sample sizes from some streams make fits highly uncertain. The random effects model borrows strength across the full data set, such that the small sample sizes from each stream do not produce radical differences that result from the overfitting of independent parameters.

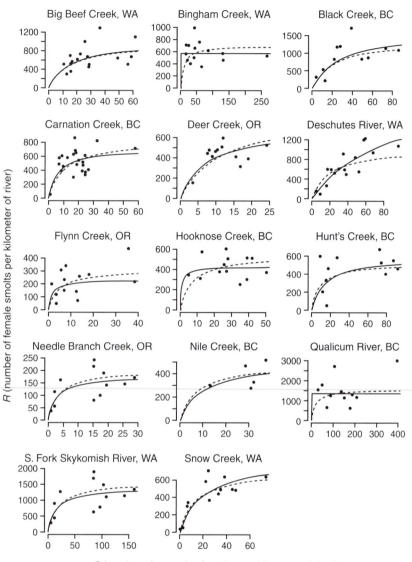

FIGURE 8.6. The Beverton-Holt relationship between coho salmon recruitment and spawning females from streams in the Pacific Northwest. Independent estimates for each stream are shown with solid lines. The random effects model is shown as dashed lines. From Barrowman et al. (2003).

8.3 Application: Growth Responses to CO_2

8.3.1 Model Development and Inference

Because nonlinear models may involve substantial use of indirect sampling, it can be efficient to consider reparameterizations that minimize it. In Section 7.4.4 I showed a linearized version of the familiar Monod or Michaelis-Menton equation. Clark et al. (2003a) included random effects and errors in variables for that model. As in Section 8.2.4, there are cases where linearization is not an option. In this application, I describe a nonlinear hierarchical model for seedling growth that treats one of the predictors (CO_2) as a factor, or treatment, and a second (light availability) as a covariate that is measured with error.

Mohan et al. (2006) tested for the effects of elevated CO_2 on growth rates of seedlings in the understory of loblolly pine stands fumigated with elevated CO_2. Because light levels were expected to influence growth responses to CO_2, light was included in the analysis. Seedlings were grown at ambient (365 ppm) and elevated (565 ppm) CO_2 levels, and seedling diameter and height were measured annually for seven years. Here I focus on height data.

Let y_{ijt} be height growth (cm yr^{-1}) of the ith seedling on the jth plot in year t. There is a mean response μ_{ij}, fixed year effect κ_t, and normally distributed error l_{ijt},

$$y_{ijt} = \mu_{ij} + \kappa_t + \varepsilon_{ijt}$$

$$\varepsilon_{ijt} \sim N(0,\sigma^2)$$

As in Section 7.4.4, the mean response is a saturating function of light availability $l_{j,t}$

$$\mu_{ij} = g_{ij}\left(\frac{l_{j,t} - l_c}{l_{j,t} + \theta}\right)$$

There is an asymptotic growth rate g_{ij}, a minimum light requirement l_c for ambient (l_{350}) and elevated (l_{550}) CO_2 treatments, and a half-saturation constant θ. The likelihood for the ith individual is $N_{T_{ij}}(\mathbf{y}_{ij}|\mu_{ij} + \kappa_{ij},\sigma^2\mathbf{I}_{T_{ij}})$, where $\mathbf{y}_{ij}$ is the vector of $t = 1, \ldots, T_{ij}$ annual growth
rate estimates and κ_{ij} is the length T_{ij} vector of years for which individual ij was included in the study. The total number of years T_{ij} varies among individuals, because many die during the study (survival was also modeled). There is a sum-to-zero constraint on year effects. Because it should have only positive values, the asymptotic rate is lognormal with fixed effect α and variance (on log growth rate) v_g,

$$g_{ij} \sim LN(\alpha,v_g)$$

I have declared this to be a random effect; there is the subscript ij on g, and I place a prior on v_g, making the asymptotic growth rate truly stochastic.

CO_2 effects enter the model only in terms of the low light response. Mohan et al. (2006) did not include a CO_2 effect on, say, g, because there were no data for the high light/high CO_2 combination. Thus, attention focuses on CO_2 effects at low light levels. The random effects in g help account for variability at high light levels that is not accounted for by light.

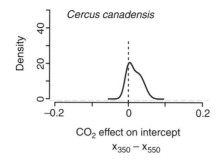

FIGURE 8.7. Difference in intercepts or light compensation point for *Cercus* seedlings grown in ambient and elevated CO_2. The slightly lower intercept at elevated 550 ppm indicates a tendency for positive growth at lower light levels. This difference is obtained for each Gibbs step and plotted as a smooth density.

Light availability is estimated from canopy photos; it varies and is imprecisely known (Clark et al. 2003a). The actual light level is conditionally dependent on repeated canopy photos,

$$l_{j,t} \sim Unif(a_j, b_j)$$

where the uniform interval is centered on the light values typical for understory and for open plots. Remaining priors are uniform on α and θ. The full model is

$$p(\mathbf{g}, \mathbf{x}, l_{350}, l_{550}, \sigma^2, \theta, \alpha, v_m | \mathbf{y}, \mathbf{l}^{(obs)}, \dots) \propto \prod_{j=1}^{m} \prod_{i=1}^{n_j} N_{T_{ij}}(\mathbf{y}_{ij} | \mu_{ij} + \kappa_{ij}, \sigma^2 \mathbf{I}_{T_{ij}})$$

$$\times \prod_{j=1}^{m} \prod_{i=1}^{n_j} LN(g_{ij} | \alpha, v_m) \prod_{j=1}^{m_a} Unif(l_{350} | a_j, b_j) \prod_{j=1}^{m_e} Unif(l_{550} | a_j, b_j) \prod_{t=1}^{T} N(\kappa_t | 0, v_\kappa)$$

$$\times\ Unif(\alpha | a_\alpha, b_\alpha) Unif(\theta | a_\theta, b_\theta) IG(\sigma^2 | a_\sigma, b_\sigma) IG(v_m | a_v, b_v)$$

There are m_a ambient and m_e elevated plots for a total of $m = m_a + m_e$. There are n_j individuals on the jth plot, and T_{ij} observations on the ith individual in the jth plot. Again, the year effect κ has subscript ij; because observations for individuals span different years. κ_{ij} is the vector of years for observations on individual ij. The Gibbs sampler for this model is described by Mohan et al. (2006). With the exception of σ^2, all conditional distributions must be sampled indirectly.

Results of this analysis show a modest effect of elevated CO_2 on growth rate. One way to examine the magnitude of this effect is through the difference in minimum light level at which positive growth is observed (Figure 8.7). Elevated CO_2 reduces this minimum level, causing this difference to be slightly positive. This posterior inference shows the magnitude of the CO_2 effect on one part of the response, but it does not tell us whether the CO_2 effect is significant. We cannot rely on posterior inference, because the posteriors are based on the assumption that the underlying model is correct. Before making a decision on the importance of CO_2, I consider the random effects.

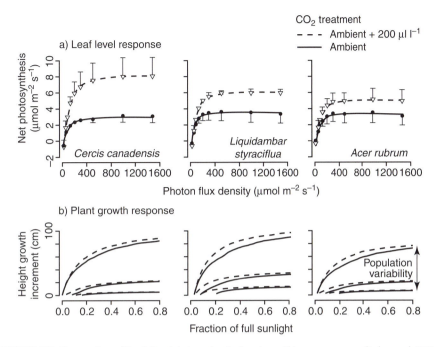

FIGURE 8.8. Examples of leaf level (*a*) and whole plant (*b*) responses to light and CO_2 from forest trees subjected to elevated CO_2 treatment in FACE rings at the Duke Forest. At the leaf level, sensitivity to light is high at low light levels, and response to CO_2 is especially high at high light levels (*a*). At the whole plant level, sensitivity to light spans a broad range of light levels, and the CO_2 effect is swamped by variability within populations. The leaf-level responses come from four "snapshot" estimates, shown as mean and standard deviation for each of eight light levels (DeLucia and Thomas 2000). The plant-level responses integrate effects throughout the year. The fitted model is a longitudinal model for six years of data with fixed and random individual effects and uncertain light levels. Lines represented posterior median and 95 percent credible intervals that include random effects (Mohan et al. 2006).

Predictive distributions for a range of light levels are shown in Figure 8.8b. They are compared with estimates of net photosynthesis from snapshot measurements on leaves (Figure 8.8a). Several aspects of these curves are noteworthy. First, the CO_2 effects appear to be swamped by variability within the population, a point first made in Section 1.2.2. At elevated CO_2, the predictive envelope is shifted slightly higher for all three species (Figure 8.8b). But this effect appears modest in comparison to the population heterogeneity, that is, the random individual effects. This apparently modest CO_2 effect should not be underestimated—it accumulates each year.

Second, this result contrasts with the leaf-level responses, which show large effects at all but the lowest light levels. The contrast between estimates results partly from the scale of the observations and partly from different sources of stochasticity included in the model. The leaf-level observations are not only short-term (minutes as opposed to years for height growth), but they also pertain to a single sample unit, whereas the predictive intervals for height growth integrate over many individuals. Mohan et al. (2006) believed that it was of most interest to gauge the importance of CO_2 effects in the context of the population heterogeneity described by random effects.

8.3.2 Model Assessment

For the Bayesian hierarchical models described here, random effects models may be assessed with the Deviance Information Criterion (DIC) (Spiegelhalter et al. 2002) or predictive loss (Section 7.6). Mohan et al. (2006) used the latter to compare four models. "Only light" includes neither random effects nor CO_2. "CO_2 effect" includes a different intercept for ambient and elevated CO_2. Also tested was a model with random effects and one with both CO_2 and random effects. The lowest values of D in Table 8.2 would indicate that the best models for *Liquidambar* and *Acer* include both random effects and CO_2. Had the random effects for *Cercus* been ignored, the CO_2 effect on growth might have seemed more important than it actually is. With random effects, CO_2 does not seem to improve the model. However, the accumulation of differences in growth over years means that such modest differences can have cumulative effects. In contrast with the leaf-level measurements, CO_2 effects on annual growth rates are small relative to population heterogeneity.

This is a good time to note that in a classical setting, one could select between models using a likelihood ratio test (Section 6.3.2). There is a modification needed for the one-sided test of the hypothesis that the variance among sample units is greater (but not less) than zero. For this model, we are testing whether $v_g > 0$, that is, whether the variance among individuals is positive. It cannot be negative. This one-sided test is accomplished by calculating the P value and then dividing by 2. This is generally the case when testing whether a parameter lies at a boundary of possible values. This approach is problematic when there is a large number of random effects.

As an alternative approach to selecting among models having a conditional linear structure with random effects Chen and Dunson (2003) suggest a

TABLE 8.2.
Predictive Loss for Four Models for the Species Shown in Figure 8.7

Species	Number of Trees (n)	Tree-Years	Predictive Loss D			
			Only Light	Random Effects	CO_2 Effects	CO_2 and Random Effects
Cercus canadensis	256	756	300,195	**33,881**	285,400	34,844
Liquidambar styraciflua	756	793	207,133	28,447	194,670	**27,780**
Acer rubrum	353	1120	175,319	41,588	168,928	**39,104**

decomposition of the random effects covariance matrix $\mathbf{V}_b$ that results in a conditional linear structure and allows one to specify positive prior probability that one or more variances are zero.

8.4 Thinking Conditionally

To appreciate the possibilities of hierarchical approaches it may help to review some of the constraints on traditional modeling options. Experimental design is central to classical methods, in part, because explanatory variables must be precisely known. There are rigid criteria for the data that can be admitted. The form of the likelihood requires independent, identically distributed data. There is typically not much latitude in how the likelihood can be written, making it hard to admit multiple sources of uncertainty and multiple data sources. There may be no obvious way to marginalize overly complex, poorly understood (stochastic) relationships. In this section, I extend some of the previous concepts on modeling hierarchically. We have summarized some of the principal points here in Clark (2005) and Clark and Gelfand (2006a).

8.4.1 Conditional Modeling

Hierarchical structures can free up the investigator to consider how the elements of a complex problem are connected, not just empirically, but also theoretically. There might be several processes that are interrelated in some respects, yet different enough that we would not model them as part of the same analysis within a traditional framework. Hierarchical Bayes provides flexible ways to combine different processes within one analysis. Several types of information (data) may be available that bear on the same process, but in different ways. They may not be independent of each other and thus would violate the assumption of i.i.d. sampling. After all, they emanate from related processes and, sometimes, the same process. Hierarchical Bayes does not require such independence at the sample stage, provided that jointly distributed random variables are modeled as such at a lower stage, by factoring them into combinations of conditionals and marginals. Multiple, interdependent data sources can be used, provided the model is structured appropriately. There are a number of practical advantages. For example, this flexibility can expand the definition of what constitutes data. The output of one model might be treated like data for a second model, in part, because uncertainty and bias for one model can be estimated as part of a second model that uses those predictions. In a hierarchical setting, the connections among model elements are prescribed using combinations of deterministic functions and conditional probabilities. MCMC provides a way to analyze the model.

Now that we have some basic hierarchical tools to exploit, I discuss some of these possibilities in a broader context. In Chapter 5 I introduced the posterior distribution, which for two parameters and data $\mathbf{y}$ might be written as

$$p(\theta_1,\theta_2|\mathbf{y}) \propto p(\mathbf{y}|\theta_1,\theta_2)p(\theta_1)p(\theta_2)$$

To construct a Gibbs sampler, I exploited conditional relationships, alternately sampling from

$$p(\theta_1|\theta_2,\mathbf{y} \propto p(\mathbf{y}|\theta_1,\theta_2)p(\theta_1)$$
$$p(\theta_2|\theta_1,\mathbf{y}) \propto p(\mathbf{y}|\theta_1,\theta_2)p(\theta_2)$$

Such conditional relationships are also used when constructing models.

Consider first the problem of combining data sets, say, $\mathbf{z}_1$ and $\mathbf{z}_2$, that might both provide insight on an underlying process $\mathbf{y}$ that includes parameters (θ_1, θ_2). Fuentes and Raftery (2005) use an example for atmospheric sulfate concentration, where measurements are data $\mathbf{z}_1$ with observation error described by density $p(\mathbf{z}_1|\mathbf{y},\sigma_1^2)$ and $\mathbf{z}_2$ is model output (simulated sulfate concentration) that has a different error and a bias b, with density $p(\mathbf{z}_2|\mathbf{y},b,\sigma_2^2)$. Using conditional relationships, Fuentes and Raftery (2005) allow for these different data models like this:

$$p(\theta_1,\theta_2,\mathbf{y},b,\sigma_1^2,\sigma_2^2|\mathbf{z}_1,\mathbf{z}_2) \propto p(\mathbf{z}_1|\mathbf{y},\sigma_1^2)p(\mathbf{z}_2|\mathbf{y},b,\sigma_2^2) \qquad \text{data models}$$
$$\times \; p(\mathbf{y}|\theta_1,\theta_2) \qquad\qquad\qquad \text{process model}$$
$$\times \; p(\theta_1)p(\theta_2)p(b)p(\sigma_1^2)p(\sigma_2^2) \quad \text{parameter models}$$

The full model indicates that $\mathbf{y}$ will be estimated along with the parameters of the model. $\mathbf{y}$ might be termed a *latent variable*. The data model might be viewed as a joint distribution of the $\mathbf{z}$'s, which can be factored into conditionals,

$$p(\mathbf{z}_1,\mathbf{z}_2|\mathbf{y},b,\sigma_1^2,\sigma_2^2) = p(\mathbf{z}_1|\mathbf{y},\sigma_1^2)p(\mathbf{z}_2|\mathbf{y},b,\sigma_2^2)$$

Someone from a classical background might object to this formulation, saying that data $\mathbf{z}_1$ and $\mathbf{z}_2$ are not independent. How can we just multiply them together? After all, we justified multiplying likelihoods with the argument that they should be independent. $\mathbf{z}_1$ and $\mathbf{z}_2$ cannot possibly be independent—they derive from the same $\mathbf{y}$.

This structure works, because data sets can be *conditionally independent*, and we are modeling their relationship to stochastic $\mathbf{y}$. Once we account for the different ways in which each depends on $\mathbf{y}$, and we allow for stochasticity in $\mathbf{y}$, we can treat them as conditionally independent, with their connection determined at the process level (Figure 8.9). In the Fuentes and Raftery case, conditional independence says that the error in the instrument used to measure atmospheric sulfate concentrations $\mathbf{z}_1$ has not much to do with the error in a model that is used to predict sulfate concentrations $\mathbf{z}_2$. We expect them to be related, but that relationship is taken up by the underlying process $\mathbf{y}$. $\mathbf{y}$ is a node of a graph (Figure 8.9). If we did not recognize that $\mathbf{y}$ is random (because it is unobserved), then the model imposes a deterministic (and perfectly correlated!) relationship between the two z observations for a given y. For $\mathbf{y}$ to be deterministic, it would have to be known. If we could observe $\mathbf{y}$, then we effectively break the graph at

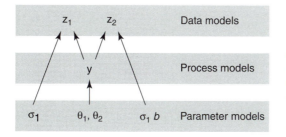

FIGURE 8.9 A model where two data sets provide information on the process **Y** with parameters θ_1, θ_2. One data set has observation error σ_1. The second data set has bias b and observation error σ_2.

that point. In this case, we could treat the **z**'s as independent (given known **y**), but we would not need them—**y** is the variable of interest. If **y** can be seen directly, why rely on indirect **z**? The same arguments apply at other levels. In Figure 8.9, θ_1 and θ_2 are conditionally independent of z_1 and z_2, because we condition on **y**.

Gaggiotti et al. (2004) provide an interesting example where molecular data are combined with demographic information to infer colonization dynamics of gray seal (*Halichoerus grypus*) populations in the Orkney Isles. In this case, there are two likelihoods. There is a likelihood for the genotypes of individuals in new colonies expressed in terms of the probability of assortative mating and the genotypes of potential parents that derive from different source colonies. There is a second likelihood for observed allele frequencies in different source populations. Thus, both likelihoods depend on source population genotypes. The probability of colonization from different source populations depends on environmental variables, principally distance and density. Additional discussion of this example is included in Section 10.10.5.

The full model is complicated, but several things make it approachable. First, the conditional distributions are typically simple and low-dimensional. Constructing this model did not depend on being able to marginalize over complex relationships. Second, complexity is increased by overlaying a simple process model with two new data models. The simple (process) model remains intact. Thus, we can begin with simple relationships and add complexity as needed. There was no need to start over. The temptation to engage in some degree of ad hockery is minimized, because the steps in an MCMC algorithm that will be used for analysis are only as complicated as the conditional relationships. Finally, this structure suggests that we can parameterize identical process models based on different types of data. All of these advantages stem from the simple conditional structure. In the Fuentes and Raftery example, the data sets are at different spatial scales, and space enters (implicitly or explicitly) both data and process models.

The model graphs I have used throughout describe this simple, conditional structure. Models are set up as networks. MCMC techniques provide a simple mechanism for analysis. Algorithms may focus on conditional relationships for each of the unobservables (parameters and latent variables). For example, the conditional posterior for **y** is identified by the arrows connected to it in the graph (Figure 8.9),

$$p(\mathbf{y}|\mathbf{z}_1,\mathbf{z}_2,\theta_1,\theta_2,b,\sigma_1^2,\sigma_2^2) \propto p(\mathbf{z}_1|\mathbf{y},\sigma_1^2)p(\mathbf{z}_2|\mathbf{y},b,\sigma_2^2)p(\mathbf{y}|\theta_1,\theta_2)$$

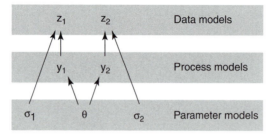

FIGURE 8.10 A model where two processes are informed by a common parameter of interest θ.

If the ultimate goal is to predict new observations $\mathbf{z'}$, this can come directly from the MCMC. Within a classical context, prediction can be difficult, coming from the integration over uncertainty in the parameter estimates,

$$p(\mathbf{z'}|\mathbf{z}_1,\mathbf{z}_2) = \int \int \int \int \int \int p(\mathbf{z'}|\mathbf{z}_1,\mathbf{z}_2,\theta_1,\theta_2,b,\sigma_1^2,\sigma_2^2)$$
$$p(\theta_1,\theta_2,b,\sigma_1^2,\sigma_2^2|\mathbf{z}_1,\mathbf{z}_2)d\theta_1 d\theta_2 db d\sigma_1^2 d\sigma_2^2$$

This integral will not be available. MCMC algorithm does the mixing for us (Chapter 7).

As a second example, consider the case where the quantity of interest, say, parameter θ, may play a role in more than one process, $\mathbf{y}_1$ and $\mathbf{y}_2$, both of which might be partially observed (Figure 8.10). For seed germination, there might be information on seed viability that comes directly from tests of seed characteristics and estimates that derive from field experiments involving seedling emergence (Hille Ris Lambers et al. 2004). A model might look something like this:

$$p(\theta,\mathbf{y}_1,\mathbf{y}_2,\sigma_1^2,\sigma_2^2|\mathbf{z}_1,\mathbf{z}_2) \propto p(\mathbf{z}_1|\mathbf{y}_1,\sigma_1^2)p(\mathbf{z}_2|\mathbf{y}_2,\sigma_2^2) \qquad \text{data models}$$
$$p(\mathbf{y}_1|\theta)p(\mathbf{y}_2|\theta) \qquad \text{process models}$$
$$p(\theta) \qquad \text{parameter model}$$

There is error for both types of observations, $\mathbf{z}_1$ and $\mathbf{z}_2$. Both processes, $\mathbf{y}_1$ and $\mathbf{y}_2$, depend on the underlying parameter of interest (e.g., seed viability). One could argue that this example is essentially like the last, differing only in how we apply the terms *variable* and *parameter*.

The flexibility discussed in reference to data and process extends to parameters. Parameter models allow for context. In a classical setting, parameters are treated as constants. Models often need context that results from relationships among sample units. For sample units that are related to one another, a random effects prior assumes exchangeability. If there are spatial relationships, the prior might specify a structure for how proximity influences sample units. For example, consider observations $\mathbf{z}$ of an obscure process $\mathbf{y}$ described by a model with parameters θ. The full model might look like this:

$$p(\theta, \mathbf{y}, \sigma^2, a | \mathbf{z}) \propto \prod_{i=1}^{n} p(z_i | y_i, \sigma^2) \qquad \text{data model}$$

$$\prod_{i=1}^{n} p(y_i | \theta_i) \qquad \text{process models}$$

$$\prod_{i=1}^{n} p(\theta_i | a) \qquad \text{parameter model}$$

$$p(a) \qquad \text{hyperparameter}$$

I have written out the full products for n-length vectors $\mathbf{z}$, $\mathbf{y}$, and θ to emphasize that individual observations y_i conditionally depend on a single value of θ_i, but the θ_i are interrelated, as specified at the next stage. The hyperparameter a is fitted to the data and determines variability at this lower stage (Clark 2003). As discussed above, random effects are appropriate when we expect variability among sample units, but do not know the cause of that variation. Spatial random effects account for location differences that are not taken up by the process and for proximity to other sample locations. Random effects can be included in classical models, but they are rarely used (e.g., regression) due to the challenges of analysis.

8.4.2 Unrecognized Caveats for Observational Data

It has become common in the environmental sciences to apply the tools originally intended for design-based inference to models fitted with observational data. Because hierarchical modeling can bring in multiple data sets, it is important to emphasize some basic elements of design and data distribution.

With observational data, there often is no randomization across explanatory variables to assure coverage and to break up collinearity. There may be no true replication that could allow for sorting out sources of variation; samples obtained at the same location but at different times are often not viewed as replicates (Section 10.7.1). Covariates are typically not well understood, having been partially observed or derived themselves from models (e.g., GIS layers used as spatial covariates are mostly derived). Limited knowledge makes it difficult to stratify observations to control for group differences. The distribution of data can have powerful impacts on inference that are rarely explored. Different data sets may involve different ranges of covariates and are subjected to different experimental treatments or simply experience a different range of covariates (LaDeau and Clark 2006; Mohan et al. 2005). If any one covariate contains a trend and the response contains a trend, the model will assign the trend to that covariate, regardless of whether or not it has anything to do with the response. Likewise, if more than one covariate shows a trend (collinearity), they will fight for the explanation, regardless of causation.

Information for a response may come from within sample units or among sample units. The distinction is important for understanding where the information derives for inference. For example, the size effect on fecundity of trees (Section 9.14.2) comes in small part from changes in fecundity and size within an individual. Even with more than a decade of data, few trees change by more than a few centimeters. More information on the size effect comes from observations across trees of different size (Figure 9.17). Random effects are needed to take up the large variability among individuals that is not tied to covariates, but it is important to recognize that size differences among individuals can fight with the random effects for the differing fecundities observed for different size trees. In such cases, we could lose all covariate information with the addition of random effects. Priors play a critical role in allowing us to specify the range of variation expected among individuals. When multiple data sets are assimilated, it is useful to first explore model behavior for individual data sets. Data sets may need to be weighted, much like one needs to weight likelihood and prior (Clark et al. 2004). The sensitivity to weightings must be thoroughly examined.

Consider further the case where the response at sites where a covariate is large for reasons not included in the model. Suppose we are interested in temperature effects, but cold sites are also moist. Moisture affects the response, but moisture is not part of the model. If moisture is unknown or unrecognized, the options include fitting a separate model for each site, using only the temperature variation at that site in the model, or model sites together, including a random site effect. The latter would be recommended on sample size considerations, but estimates and uncertainties will be inaccurate. In the case where the random intercept is expected to correlate with the covariate response, Bafumi and Gelman (2006) recommend modeling the effect of the site averages at a lower stage, in the mean structure for the random effect.

The issues related to information within versus between sample units apply to analyses of observational data for inference on climate effects, productivity/diversity relationships, and species area. In each of these cases, analyses appear to give different answers, depending on scale. It is rarely recognized that when individuals/sites differ for reasons other than covariates of interest, the inference can change. For example, the correlation methods involved in climate envelopes and species-area curves usually predict at one scale (change within a site and risk for one species) based on data collected at another scale (differences among sites and changes in number of species) (Ibáñez et al. 2006). Relationships *between* plots, islands, habitats, or regions are often used to make inferences and predictions *within* plots, islands, habitats, or regions. This scale-dependent issue arises when relationships from aggregate data are applied to individuals (Clark 2003), known in other disciplines as the modifiable unit area problem (Openshaw and Taylor 1979), the change of support problem (Cressic 1993; Gelfand et al. 2001), the ecological fallacy (Greenland and Robins 1994), or Simpson's paradox. One of the most famous examples of Simpson's paradox involves analysis of graduate admissions data from Berkeley (Bickel et al. 1975). Based on the aggregate acceptance rates of 44 percent for male applicants and 35 percent for female applicants one could infer a bias *against* admission of females. When taken to the departmental level, the situation was reversed: a bias *toward* admission of females was observed. Given that admissions decisions are made at the departmental level, the conclusion based on aggregated data was wrong. The explanation: females tended to apply to the toughest graduate programs, which had the lowest admissions rates. This fallacy does not entail the complexities that we typically associate with surprise, such as nonlinearities and stochasticity. It results simply from a process at one scale and inference at another.

Does an example like Berkeley admissions mean that one can never translate inference from one scale to another (e.g., campus-wide to department)? Obviously not; averaging over variability is often reasonable. Yet the number of examples in which the practice of inference at the wrong scale results in the wrong conclusion has made the ecological fallacy widely recognized in the social sciences. We could conceive of specific ecological analogies to the example of Berkeley admissions, but the problem extends to a broad range of issues that arise when data, inference, and, by extension, prediction have incompatible reference (Ibáñez et al. 2006).

Decisions concerning what to include and what to leave out of models cannot be left to simple indexes. As pointed out in Chapter 6, model selection requires more than fit to a particular data set. Covariates must be selected to explain systematic patterns, to incorporate theoretical understanding, not idiosyncracies related to the data at hand (Clark and Gelfand 2006a). Limit the number of covariates and random effects, but do not be a slave to criteria like AIC. Plot the random effects to see how they relate to variables in and left out of the model.

The following examples involve different types of information accommodated by models that allow for the combined relationships that affect process and observations. They demonstrate how parameter models allow for context.

8.5 Two Applications to Trees

8.5.1 Fecundity

Fecundity of trees can be estimated from seed trap data, provided we have a transport model to link seeds produced by trees with trap data. I provide details for a more comprehensive treatment of the model from Clark et al. (2004) in Section 10.7. Here I focus on sampling distributions for observations at two different scales. The fecundity process might involve tree diameter and covariates arranged in a design matrix with covariates,

$$\ln(f_i) = \mathbf{x}_i\beta + \varepsilon_i$$

where f_i is seed production by the ith tree. There are multiple observations on each tree, but I temporarily have omitted the time t subscript to minimize clutter. Interest might be in the number of seeds produced by the ith tree, the covariates that control fecundity, represented by parameters β, or both. Unfortunately, $\mathbf{f}$ may not be observable.

If we cannot observe $\mathbf{f}$, how might we learn about it? If trees are grown in the open, it might be possible to count seeds, where the count is taken to be representative of the true number (Koenig et al. 1999). If there is access to the canopy, it may be possible to count reproductive structures, even in closed stands (e.g., LaDeau and Clark 2001). If not, it may be possible to determine whether a tree is reproductive, despite the fact that seeds cannot be counted (Clark et al. 2004). We may also derive information from seeds collected on the ground. In each case there is a different error structure. To illustrate how data can be assimilated from different sources and scales, I consider several data models that can help with inference. We have applied each of these combinations (and more) in our own research.

To draw inference on β and to estimate $\mathbf{f}$, determine how each data type relates to the underlying process. If seeds or reproductive structures can be counted directly, the connection between process and model might involve nothing more than a detection probability,

$$p(\mathbf{z}|v,\mathbf{f}) = \prod_{i=1}^{m} Bin(z_i|f_i,v)$$

where $\mathbf{z}$ are the seed counts for the m trees and v is the probability that a reproductive structure will be counted (Figure 8.11). This model connects unknown $\mathbf{f}$ with observable $\mathbf{z}$. Unless we have external information, we will find it difficult to estimate the fecundities $\mathbf{f}$, because they will tend to trade off with the detection probability v. Useful external information might enter as an informative prior on v. For instance, experience might tell us how successfully an observer identifies reproductive structures.

Alternatively, seed trap data can provide a valuable supplement. To exploit this data set, we need a data model that establishes a connection back to the object of inference, $\mathbf{f}$. This second data set of n seed trap counts is $\mathbf{s}$. The sampling distribution might be Poisson (seeds per unit area)

$$p(\mathbf{s}|\mathbf{f},\mathbf{r}) = \prod_{j=1}^{n} Pois(s_j|A_j g(\mathbf{f},\mathbf{r}_j)) \qquad 8.6$$

where A_j is the area of the jth seed trap and $g(f, \mathbf{r}_j)$ is a function of seed production by m trees within the stand and $\mathbf{r}_j$ is the vector of distances of the

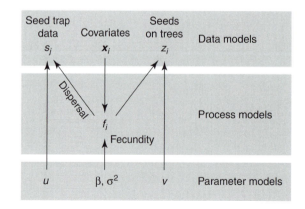

FIGURE 8.11. A model for fecundity that combines two data sources. Parameter values are listed in the text.

m trees from the jth seed trap, $\mathbf{r}_j = [r_{1j}, \ldots, r_m]^T$. This function incorporates dispersal from all m trees to the jth seed trap as

$$g(\mathbf{f},\mathbf{r}_j) = \sum_{i=1}^{m} f_i k(r_{ij};u) \qquad 8.7$$

where $k(r_{ij})$ describes dispersal from the ith tree to the jth seed trap and u is a parameter describing kernel shape (Box 10.1).

By itself, the seed trap data set provides limited insight, because it bears a highly indirect relationship to seed production by individual trees. However, we can combine them. Both data sets are connected to the process of interest, $\mathbf{f}$. The two data sets are certainly not independent. But they are conditionally independent, and $\mathbf{f}$ will be modeled stochastically at the process stage. At this point we have the likelihoods

$$\prod_{i=1}^{m} Bin(z_i|y_i,v) \prod_{j=1}^{n} Pois(s_j|A_j g(\mathbf{f},\mathbf{r}_j))$$

Note that the connection between these data sets comes through the process that generates seed production, that is, $\mathbf{f}$ (Figure 8.11). If we did not model $\mathbf{f}$ (i.e., treated it deterministically) we would have no credible estimates of uncertainty.

There is still additional information we might exploit. The first data set $\mathbf{z}$ allows for error in counts of seeds, cones, or fruits. Because it may be impossible to count seeds on trees, we might need further modification. Clark et al. (2004; in review) extended this general approach to admit observations on tree status. Although we cannot count the seeds on trees in closed stands, we can often identify if they have any seeds. Rather than count the number of seeds per tree, write the probability of observing any seeds conditioned on tree status (all combinations of gender and reproductive state).

The structure for the fecundity section of this model can be written as

$$p(f_{ik,t},Q_{ik,t},\mathbf{s}_{k,t}|\mathbf{q}_{ik,t},\phi_{ik,t},\mathbf{x}_{ik,t}) = p(\mathbf{s}_{k,t}|f_{ik,t},Q_{ik,t})p(f_{ik,t},Q_{ik,t}|q_{ik,t},\phi_{ik,t},\mathbf{x}_{ik,t}) \qquad 8.8$$

where fecundity $f_{ik,t}$ has additional subscripts (seeds produced by tree i in stand k and year t), $Q_{ik,t}$ is reproductive status (zero or one), $q_{ik,t}$ is the observation that seeds are present on a tree (zero or one), $\phi_{ik,t}$ is the probability that the tree is mature, $\mathbf{s}_{k,t}$ is the number of seeds counted at all traps, and $\mathbf{x}_{ik,t}$ is a vector of covariates that affect fecundity (Figure 8.12). The last density on the right-hand side can be factored as

$$p(f_{ik,t},Q_{ik,t}|\mathbf{x}_{ik,t},\phi_{ik,t},q_{ik,t}) = p(f_{ik,t}|\mathbf{x}_{ik,t})p(Q_{ik,t}|q_{ik,t},\phi_{ik,t})$$

The first factor models fecundity as a regression (Section 9.14). If the tree is immature ($Q_{ik,t} = 0$), then no seeds are produced ($f_{ik,t} = 0$). The second factor expresses the probability that a tree is mature based on the observations of status. If seeds are observed on a tree, then the tree is mature,

$$p(Q_{ik,t} = 1|q_{ik,t} = 1) = 1$$

If a tree is not observed in year t and status is unknown (it has never been observed with seeds), it is mature with probability $\phi_{ik,t}$, which increases with tree size, described by estimated parameters. If a tree is observed, but no seeds are seen, maturity depends on the probability of observing seeds when they are present, v,

$$p(Q_{ik,t} = 1 | q_{ik,t} = 0, \phi_{ik,t}) = \frac{\phi_{ik,t}(1 - v)}{1 - v\phi_{ik,t}}$$

Marginally, the probability of being mature also depends on the seed rain data s,—if a large amount of seed is present in traps near an individual, the greater the probability that it is mature.

Efficient sampling from the joint posterior requires blocking of parameters. Note that there are latent variables for every tree year ($f_{ij,t}$ and $Q_{ij,t}$). A naïve approach would involve sequentially updating status and fecundity for each year for each tree and each stand and for each seed trap for each year and each stand. There are too many unknowns to sample each sequentially using a language like R. In these data sets there are up to fifteen years of data for hundreds of seed traps and thousands of trees. We seek blocks of parameters that can be proposed and accepted simultaneously.

For each study year do the following:

1. For those individuals never having been observed bearing reproductive structures and currently imputed to be immature, propose a maturation status with probability 0.5 and a fecundity based on the regression $p(f_{ik,t} | \mathbf{x}_{ik,t})$. If there is a previous observation of reproduction, the status is still reproductive,

$$\Pr\{Q_{ik,t} = 1 | q_{ik,t} = 1\} = 1$$

For these individuals known to be mature ($q_{ik,t} = 1$) and for those currently imputed to be mature ($q_{ik,t} = 0$, $Q_{ik,t} = 1$), propose a new fecundity value centered on the current value. The proposals for all $\mathbf{Q}_{k,t}$ and all $\mathbf{f}_{k,t}$ can each be done with a single call to a function in R. The proposed fecundity for individuals known and proposed to be immature is zero.

2. Determine the acceptance probability for the entire block based on Equation 8.8

$$p(\mathbf{f}_{k,t}, \mathbf{Q}_{k,t} | \mathbf{s}_{k,t}, \mathbf{q}_{k,t}, \phi_{k,t}, \mathbf{X}_{k,t}) = p(\mathbf{s}_{k,t} | \mathbf{f}_{k,t}, \mathbf{Q}_{k,t}) p(\mathbf{f}_{k,t} | \mathbf{X}_{k,t}) p(\mathbf{Q}_{k,t} | \mathbf{q}_{k,t}, \phi_{k,t})$$

inserting current and proposed values for $\mathbf{Q}_{k,t}$ and all $\mathbf{f}_{k,t}$.

The proposals must be accepted or rejected as a block, because they all conditionally depend on the seed rain data through $p(\mathbf{s}_{k,t} | \mathbf{f}_{k,t}, \mathbf{Q}_{k,t})$. In other words, for any given year, the likelihood for seed accumulating in trap j in stand k cannot be taken on a tree-by-tree basis. Both current and proposed statuses and fecundities conditionally determine the likelihood for all seed traps in the stand.

Due to relatively large impacts on the likelihoods $p(\mathbf{s}_{k,t} | \mathbf{f}_{k,t}, \mathbf{Q}_{k,t})$ and $p(\mathbf{Q}_{k,t} | \mathbf{q}_{k,t}, \phi_{k,t})$ of different proposals for statuses $\mathbf{Q}_{k,t}$ (mature or not), it may

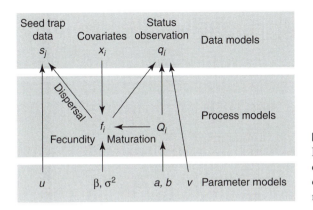

FIGURE 8.12. The model from Figure 8.11 modified to allow for observations on tree status q, as opposed to counts of seeds or reproductive structures.

be difficult to move the Gibbs sampler. If this is the case, a subset of trees can be chosen for updating, holding others at currently imputed statuses and fecundities. Thus, only a loop over years is required. Upon completion of year t, proceed to year $t + 1$. Sampling of other parameters is discussed in Clark et al. (2004). Predictive distributions are shown in Figure 9.17.

8.5.2 Light Availability to Large Trees

As a second illustration of multiple data sets, I extend the canopy status example of Section 8.2.2 to include evidence for light available to large trees from remote sensing and solar geometry models. As mentioned in Section 8.4.1, model output can be treated as data. Model output has bias and uncertainty, which can be explicitly modeled, just as it is for observations.

Light availability to large trees is of interest, because it determines growth rate, but it cannot be directly measured. In this example, there are three data sets that provide differing amounts of information on light availability to each tree in a stand (Figure 8.13). All of these measures are indirect, and they can be combined to inform estimates of canopy coverage collectively (Wolosin et al., in preparation).

Status observations are the easiest data to obtain, but provide the least information (Figure 8.13) and are subject to substantial uncertainty—it is difficult to determine from the ground whether or not many trees are exposed to direct sunlight, and we cannot quantify the level of exposure. From remotely sensed images the exposed canopy area (ECA) $\lambda_i^{(e)}$ can be estimated for those trees having emergent crowns (Wyckoff and Clark 2005; Wolosin et al., in preparation). These data are discontinuous. Trees are either observed on the image, or they occupy a discrete zero class. Trees in the understory are not identified from images, and have little or no direct exposure. Models of solar geometry provide an integrated estimate over day and season of direct and indirect sunlight gathered by a tree, $\lambda_i^{(m)}$ (Govindarajan et al. 2004;

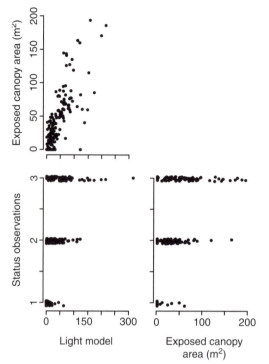

FIGURE 8.13. Pairs plot of three data sets that provide information on light availability for large trees.

Chakraborty et al., in review). Being based on idealized assumptions of allometric relationships between diameter and canopy characteristics, they are subject to substantial uncertainty. In summary, then, the data types are

- status observations $\qquad \lambda_i^{(s)} \in [1,2,3]$
- exposed canopy area in m² $\qquad \lambda_i^{(e)} \in [0, \infty)$
- integrated light availability in arbitrary units $\qquad \lambda_i^{(m)} \in [0, \infty)$

There are different observations available for each tree. Only some of the stands analyzed by Wolosin et al. (in preparation) included remotely sensed data and solar geometry models. Status observations were available for all stands.

There is a data model for each type of observation (Figure 8.14). For status observations Wolosin et al. used the cumulative logit model discussed in Section 8.2.2. For ECA, there is a two-part likelihood, with a probability p_i that tree i is observed in an image, and, if observed, a lognormal for the canopy area,

$$p(\lambda_i^{(e)}) = \begin{Bmatrix} 1 - p_i & \lambda_i^e = 0 \\ p_i N(\ln(\lambda_i^{(e)})|\ln(\lambda_i), \nu_e & \lambda_i^{(e)} > 0 \end{Bmatrix}$$

where λ_i is the true value. The model for detection on the image is

$$\text{logit}(p_i) = c_0 + c_{12}\lambda_i$$

The two components allow that probability of being observed in the image and measurement of ECA increase with the true light availability λ_i. The model based on solar geometry allows for bias in model output, with logarithmic function

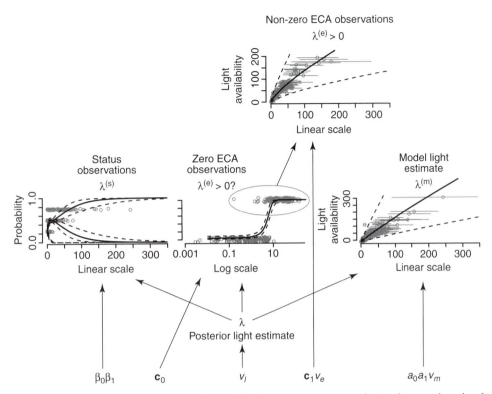

FIGURE 8.14. The graph for light model showing parameter relationships and each of three data types plotted against the posterior estimate of true light availability. In each case the horizontal axis is the posterior estimate of light. The predictive interval for each data type is a solid line bounded by the 95 percent predictive interval (dashed lines). To minimize clutter, one standard error for posterior light estimates is shown as solid horizontal lines in only two of the plots. Note the two-stage model for ECA values, having a Bernoulli probability of being identified in the image ($\lambda^{(e)} > 0$) and, if identified, a lognormal model for canopy area.

$$\lambda_i^{(m)} = a_0\lambda_i^{a_1}$$

and lognormal error

$$p(\lambda_i^{(m)}) = N(\ln(\lambda_i^{(m)})|\ln(a_0) + a_1\ln(\lambda_i), v_m)$$

Finally, because observations are correlated, the true light availability is random. A simple prior specification comes from previous studies that show a relationship with tree diameter (Wyckoff and Clark 2004). A weak prior based on this relationship is

$$p(\lambda_i) = N(\ln(\lambda_i)| - 5.5 + 2.5\ln(d_i), v_i)$$

Wolosin et al. (in preparation) explored different degrees of weight applied to each of the data types. This is readily done through inverse gamma prior distributions for each of the variance parameters v_e and v_m. There were $n_s = 310$

TABLE 8.3.
Parameter Estimates for the Light Model

Parameter	Posterior Median	95 percent Credible Interval	
		0.025	0.975
a_0 – model coefficient	0.989	0.958	1.02
a_1 – model exponent	0.902	0.873	0.931
β_{01} – class 1 intercept	−0.488	−0.699	−.299
β_{02} – class 2 intercept	0.431	0.248	0.627
β_{11} – class 1 slope	−0.296	−0.409	−0.207
β_{12} – class 2 slope	−0.0236	−0.0378	−0.0127
c_0 – detection intercept	−3.24	−4.27	−2.34
c_1 – detection slope	0.602	0.408	0.838
v_c – ECA error	0.205	0.184	0.232
v_m – model error	0.292	0.224	0.383

observations for status, $n_m = 410$ light model calculations, and $n_e = 285$ measurements of ECA. Of the total 435 trees, 218 had 3 types of observations, 134 had 2 types of observations, and 83 had one type of observation. Weights for ECA and the light model were used to roughly balance the likelihood for status observations. To scale light estimates roughly like those for the canopy model and ECA, strong priors were applied on the light model parameters $[a_0 \quad a_1]^T \sim N_2(1_2, n_m^{-1} Diag(1,1))$ and $[c_{11} \quad c_{12}]^T \sim N_2(1_2, n_e^{-1} Diag(1,1))$, where 1_2 is the length-2 vector of ones.

Although all of the estimates were well identified (Table 8.3), interest focuses on the light values themselves, λ. These estimates are shown as posterior means ± 1 standard error in Figure 8.14. The relationships between estimated light availability and each of the three data sets indicate that each has information to offer. In general, standard errors are large for trees having only one or two types of observations.

8.6 Noninformative Priors in Hierarchical Settings

As previously mentioned, improper priors may lead to improper posteriors. Weak priors can be especially undesirable for hierarchical models, even when proper. For example, variance parameters on random effects may be extremely sensitive to the prior, even when weak. A rather flat, but proper prior, with bounds seemingly wide enough to have no impact may, in fact, yield a posterior that depends on what those bounds are. Likewise, the standard inverse gamma with prior parameter values of, say, 0.01, may result in a different posterior from one having prior parameter values of 0.1 or 0.001. Consider the simple hierarchical model $y_i \sim N(x_i, \sigma^2)$ with second stage $x_i \sim N(0, \tau^2)$. The marginal distribution is $y_i \sim N(0, \tau^2 + \sigma^2)$. Clearly, the variance parameters are not identifiable. Maximum likelihood is not a good idea, yielding MLE's $\tau^2_{ML} = \overline{x^2} - \sigma^2$ and $\sigma^2_{ML} = \overline{x^2} - \tau^2$. Estimates can be negative, but variances cannot be negative. The Bayesian approach does not suffer from this problem, and the formal incorporation of informative priors is a big advantage.

With some exceptions, a rule of thumb makes use of insight as to the range of variability to expect. Returning again to a variance on random effects, the range of individual differences may be known from prior observation. If so, prior parameter values for the variance can be specified to describe this range. When in doubt it is important to examine sensitivity to the prior.

The reluctance of some ecologists to embrace informative priors can result in missed opportunity to incorporate information. Knowledge can enter the analysis through data or prior. For example, stable isotopes are used to infer the depth of water taken up by plants based on variation in isotopic ratios of hydrogen/deuterium and oxygen with depth. Because the isotopic ratios within plants result from the integration (mixing) of waters obtained over a range of depths, there is no single answer to the question "From what depth did the water come?" That depends not only on the variation in isotopic values, but also on the variation in root density with depth. Although not all information on relevant variables may be collected in a given study, much is known from a long history of analysis from many investigators. Ogle et al. (2004) combined isotopic data with informed priors on rooting depth to deconvolve the root uptake profile for *Larrea tridentate*, demonstrating how models and data combine to provide detailed insight.

8.7 From Simple Models to Graphs

As models increase in complexity, the advantages of the graphical representation become apparent, providing the road map for computation. Graphical representation of models dates at least to Wright (1921), who used it to conceptualize path analysis. Particular interest in the Bayesian setting for graphical models has been in the context of expert systems or machine learning, where the model can be updated as experience accumulates (e.g., Pearl 2000).

Graphical models consist of variables, relationships, often including both deterministic and stochastic parts (arrows or edges), observations, parameters, and priors. In environmental applications, variables will often be partially known, through observations. A data model is represented by an arrow that points from a variable to an observation. Parameters to be estimated are unknown. There are arrows pointing from parameters to variables and observations—both process models and data models may contain parameters to be estimated. Priors are prescribed. To limit clutter, I have not included priors in the model graphs. If I were to include them, they would be represented as roots of the model, with outgoing arrows pointing to parameters. If a variable is precisely known (it is observed without error), the graph is broken at this point. Many of the arrows have both deterministic and stochastic elements, which together summarize the knowns and the unknowns.

Graphical models are rapidly being adopted in fields that involve analysis of causality when there can be many variables involved. The directed graphs used to represent models in this book are also termed *Bayesian networks*. Applications include diagnosis involving diseases that share overlapping sets of symptoms (Madigan and York 1995), and inference on the structure of networks of interaction based on gene expression data (Hartemink et al. 2002). Shipley (2000) argues for increased application of structural equation models in ecology. These can be viewed as a particular type of state space model, which I consider using graphical models in Chapter 9.

Part IV. More Advanced Methods

Many ecological processes are inherently spatio-temporal. The state variable is viewed as a product of its recent past, and it depends on its surroundings. If we were to extract a population, a community, or an ecosystem from its spatio-temporal context, we would not expect it to behave as it does in nature. Although not all ecological process models are written with space and time as explicit variables, most could be.

Spatio-temporal dependence requires models that allow for distance (or location) and time relationships. These relationships may be part of the process of interest, part of the unknown variability that must be accommodated for purposes of parameterization, or both. In many cases, the spatial and temporal effects are a goal in themselves. Process models may be spatio-temporal. We might fit a model with dispersal, with diffusion or advection, or with relationships that depend explicitly on time. In other cases, space and time may be necessary evils. They may not appear in the deterministic part of a model, yet arise in variance terms, acting more like nuisance parameters—we need to allow for the fact that observations are not independent in time and space. If the process model fully accounts for spatio-temporal effects, then stochastic components might not need to.

Chapters 9 and 10 focus on models and data that are inherently temporal. In Chapter 9, I outline some general features of time series, within the context of population dynamics. Spatial and spatio-temporal processes are taken up in Chapter 10.

9 Time

9.1 Why Is Time Important?

Time is an explicit variable in many ecological models, because the process involves how the state variable(s) change(s) from one time to the next. Correlations in data over time can range from nonexistent, to weak, to severe and episodic. The large and perplexing fluctuations in population abundances have been a source of speculation, experiment, and analysis for many decades (Turchin 2003 provides a recent summary). Complex dynamics result when density dependence is strong (it "overcompensates") and when population structure and life history (Kendall et al. 1999, 2005; Lande et al. 2002) or interactions (e.g., Royama 1992; Turchin et al. 1999) produce lags and, potentially, cycles. Regular fluctuations are a common feature of many population data sets. Robert May (1974) showed that deterministic forces (density dependence) could result in complex dynamics that might appear stochastic (forced by fluctuations external to the system) (Chapter 2).

The possibility of chaos or complex dynamics has fueled a long-standing interest in attempting to identify when fluctuations might result from external influences as opposed to the intrinsic dynamics of the populations themselves. Multiple stable states can arise because there are multiple attractors (Appendix B). Stochasticity can move a state variable among attractors. In some sense, nonlinearities can magnify stochastic fluctuations (Higgens et al. 1997). Local Lyapunov exponents are used to quantify this tendency for perturbations to dissipate or magnify at specific locations in the variable space (Turchin and Ellner 2000). Quasiperiodic variation, with some level of synchronicity among individuals, is well known in the masting of cycles of trees (Koenig et al. 1999; Kelly and Sork 2002; Rees et al. 2002). Detection of delayed density dependence can suggest the importance of trophic interactions (Turchin 1995; Bjørnstad et al. 1995; Stenseth et al. 1996). By manipulating cannibalism rates in flour beetles (*Tribolium* spp.), Costantino et al. (1997) confirmed the transitions from stable dynamics through cycles and chaos predicted from models. Analyses of long-term surveys of cod (*Gadus morhua*) from the Norwegian Skagerrak coast suggest strong asymmetric competition, and cannibalism of young can generate cycles (Bjørnstad et al. 1999). Populations of the same size can respond differently to fluctuating climate, depending on age structure and sex ratio (Coulson et al. 2001). A chaotic system may have more than one attactor (McCauley et al. 1999).

Recent analyses of population data concern the combined effects of external influences that fluctuate with nonlinear relationships, which can stabilize or destabilize (Bjørnstad and Grenfell 2001). These two types of influences interact in ways that are difficult to identify, analyze, and predict (Murdoch 1994). They require models that explicitly accommodate the time series nature of population density data (Dennis et al. 1995; Coulson et al. 2001). The dynamics can be obscure or hidden from view, due to incomplete information (Pascual and Kareiva 1996; Hilborn and Mangel 1997; Bjørnstad et al. 1999; Calder et al. 2003; Clark et al. 2003a). Early efforts to identify simple density dependence explored the hypothesized negative relationship between density and growth rate (Hassell et al. 1976) (Chapter 2). Dennis and Taper (1994) describe classical methods for detecting density dependence when observation error is small enough to be ignored. I take up examples in this chapter.

Many time series models include discrete state variables. At high population densities, models tend to ignore the random variation contributed by birth and death (these processes are represented by averages). At low densities, demographic stochasticity cannot be so readily ignored (May 1974). Grenfell et al. (2002) consider its consequences for spread of measles. Models that allow for extinction often represent population size as a discrete variable and are used to estimate probabilities of extinction. I include several examples of discrete states in models.

As in previous chapters, covariate effects that are known can be modeled as such and necessarily reduce the contribution delegated to stochastic elements. Environmental effects that might be treated deterministically include periodic phenomena such as El Niño (Leirs et al. 1997; Lima et al. 1999; Turchin and Ellner 2000). The interacting effects of these different forces are challenging, particularly in light of the fact that the observation errors can be large (Grenfell et al. 1998; Bjørnstad and Grenfell 2001; Coulson et al. 2001; Calder et al. 2003; Clark and Bjørnstad 2004).

I begin this chapter with some basic concepts and terminology, followed by descriptions of different classes of models, with examples and suggestions for computation.

9.2 Time Series Terminology

The term *time series* has traditionally been applied to one or a few sequences that are long and may have a small number of parameters. Data sets typically do not derive from experiments. Time series models are often used for data description, example, identifying trends, autocorrelation, periodicity, and so forth. Due to the length of such series, analyses may consider relationships that span long lags. I begin this section with a summary of the basic elements of such models.

There is a large literature on models that follow state variables, such as population densities, climate variables, and financial markets, through time. In these models the state variable is continually updated from one time increment to the next. Observation and process errors are handled differently in different types of models. The *Kalman filter* refers to a classical technique widely applied to linear models with Gaussian error. The Bayesian *state-space* framework is more flexible and will be the focus here. Some classical (Section 9.5) and Bayesian (Sections 9.6–9.12) treatments of population time series models are discussed.

Longitudinal or *repeated measures* are terms applied to observational or experimental studies that usually involve many time series of short duration (Section 9.13). Autocorrelation is often a part of such models, but, due to short duration, receives less emphasis than in the traditional time series models. Longitudinal studies with imposed treatments involve *intervention analysis* (Section 9.14).

Capture-recapture studies entail a specific type of sampling, where marked individuals are observed over time. These are widely used in wildlife management. They have also been applied to plant populations (Section 9.15).

When there is substantial structure in time series models that are posed to describe a particular process, the term *structural model* is often applied. Structural models can be large and contain many parameters.

9.3 Descriptive Elements of Time Series Models

In Chapter 2 I discussed simple models of population dynamics and pointed out that stochasticity can have an important role. The simplest form of stochasticity is a random walk. Most examples from the early part of this chapter involve continuous state variables. I extend these results to discrete states later in this chapter (capture-recapture studies and spread of infectious disease) and take up several types of population interactions and population spread in Chapter 10.

9.3.1 The Treatment of Time

Discrete time (difference equation) models are not used exclusively for processes that occur at continuous intervals. In fact, most models can only be analyzed in discrete form (by computers), and data are obtained at discrete intervals. Many processes that are posed first as differential equations, such as general circulation models of the atmosphere, are discretized in space and time (Wikle et al. 2001; Berliner et al. 2000). For processes that are more-or-less continuous, it is important that the underlying process model should accurately represent how the system steps forward. If the process equation is nonlinear, this means that integration errors (typically subsumed as process error) should be small relative to other sources of variability. If time steps are defined by sample intervals, and the process equation is highly nonlinear, important dynamics may be misrepresented. Any time a continuous process is modeled in discrete time, careful thought is needed concerning the scaling of both deterministic and stochasitic elements (Section 9.7.2).

9.3.2 Stationarity

A stationary time series is one for which the joint distribution of states does not depend on when we observe the process. This definition of *strict stationarity* may be difficult to apply, because we typically cannot say whether distributions are precisely unchanging. We can often draw inference on the first few moments, and we focus on just the mean and covariance. Second-order stationarity holds when the mean is constant and covariance among states depends only on the lag j,

$$\text{cov}[x_i, x_{i-j}] = \gamma_j \qquad\qquad 9.1$$

where γ_j is termed the *autocovariance function*. Thus, not only the mean, but also the variance (given by γ_0) is constant. Autocovariance is discussed in more detail in Section 9.3.7.

9.3.3 Random Walk

In this book, the term *random walk* arises in several contexts. Here I apply the term to changes in abundance, where a continuous state variable represents abundance at time t, such as n_t or log abundance x_t. In this context, the process is a random walk among potential values of abundance. For populations that are structured in some way, there can be a random walk among states, where state can refer to stage of the population (e.g., stage transitions) or location (e.g., population spread).

A process model describing purely random fluctuations is a random walk

$$x_t = x_{t-1} + \varepsilon_t$$

where the vector of random variables ε_t are i.i.d. with mean μ and variance σ_ε^2. At any given time the process has a value that is simply the accumulation of these random changes,

$$x_t = \sum_{j=1}^{t} \varepsilon_j$$

If $\mu = 0$, the random walk is unbiased. The mean and variance of the process are

$$E[x_t] = t\mu$$

$$\text{var}[x_t] = t\sigma_\varepsilon^2 \qquad\qquad 9.2$$

This variance can be obtained by induction. If the process begins with a value of x_0, then at time $t = 1$,

$$\mathrm{var}[x_1] = \sigma_\varepsilon^2$$

Because the ε_t are independent, at time $t = 2$,

$$\mathrm{var}[x_2] = \mathrm{var}[x_1] + \sigma_\varepsilon^2 = 2\sigma_\varepsilon^2$$

(for independent random variables, the variance of the sum is the sum of the variances—Appendix D). Clearly, we can continue in this fashion. Because the variance changes over time (it is proportional to it), the process is not stationary. Trends in variance over time have been the basis for inference on population regulation, from both single times series (Murdoch 1994) and multiple time series (Clark and McLachlan 2003).

9.3.4 Filtering

A linear filter transforms the time series $\mathbf{x}$ into a time series $\mathbf{y}$

$$y_t = \sum_{j=-q}^{s} a_j x_{t+j} \qquad\qquad 9.3$$

having weights a_j. If the weights sum to one, and they are symmetric about t (i.e., $s = q$ and $a_j = a_{-j}$), then $\mathbf{y}$ is the moving average of $\mathbf{x}$. This removal of high frequency variation is termed a *low pass filter*. A *high pass filter* removes the low frequency variation and is accomplished by taking the residuals,

$$y_t - \dot{x}_t$$

9.3.5 Differencing

Trends can be removed by differencing. The first difference of a time series is

$$\Delta x_t = x_t - x_{t-1}$$

Example 9.1. The first difference of a random walk is

$$\Delta x_t = x_t - x_{t-1} = \varepsilon_t$$

Unlike the random walk itself, the differenced series is stationary, having variance $\mathrm{var}[\Delta x_t] = \sigma_\varepsilon^2$, obtained as the variance of a difference (Section D.5.3),

$$\mathrm{var}[x_t - x_{t-1}] = \mathrm{var}[x_t] + \mathrm{var}[x_{t-1}] - 2\mathrm{cov}[x_t, x_{t-1}]$$

$$= t\sigma_s^2 + (t - 1)\sigma_s^2 - 2(t - 1)\sigma_s^2 = \sigma_s^2$$

The spatial analog of this relationship motivates the semivariogram (Section 10.7).

The MLE for parameter r in Section 3.6.1 was obtained as the sum of the first differences.

9.3.6 Moving Average Process

The moving average process is a specific type of linear filter, where $s = 0$ in Equation 9.3 and ε_t is a zero-mean random variate,

$$y_t = \sum_{j=0}^{q} a_j \, \varepsilon_{t-j} \qquad 9.4$$

The value of the series at time t is a moving average that depends on the previous q values of the series, as weighted by the coefficients a_j. The order of the process is q. I use the shorthand MA(q). The process has mean and variance

$$E[y_t] = 0$$

$$\mathrm{var}[y_t] = \sigma_\varepsilon^2 \sum_{j=1}^{q} a_j^2 \qquad 9.5$$

Once again, induction can be used to obtain the variance. At time $t = q = 1$,

$$\mathrm{var}[y_1] = a_1^2 \, \sigma_\varepsilon^2$$

This relation says that the variance of a constant times a random variable is the square of the constant times the variance of that random variable. Recall that $\mathrm{var}[ay] = E[ay - E[ay]]^2$ (Appendix D). Expand and collect terms to obtain $\mathrm{var}[ay] = a^2 \mathrm{var}[y]$ (Equation D.38). At time $t = 2$,

$$\mathrm{var}[y_2] = \mathrm{var}[y_1] + a_1^2 \, \sigma_\varepsilon^2 = (a_2^2 + a_1^2)\sigma_\varepsilon^2,$$

again, relying on the assumption of uncorrelated errors. Then, by induction, obtain the desired result.

9.3.7 Autoregressive Process

Autoregressive models can be written in several ways. One way incorporates autoregression in the mean term. This is like a linear regression on the previous p values of the series

$$y_t = \sum_{j=1}^{p} \rho_j \, y_{t-j} + \varepsilon_t \qquad 9.6$$

This is the AR(p) model, with coefficients ρ_j up to lag p. The AR(1) model is the first-order Markov process. To find the mean and variance of the AR(1) model note that

$$y_1 = \rho y_0 + \varepsilon_0$$

At this point the mean and variance are

$$E[y_1] = \rho y_0$$

$$\text{var}[y_1] = \sigma_\varepsilon^2$$

The next time step is

$$y_2 = \rho y_1 + \varepsilon_1 = \rho(\rho y_0 + \varepsilon_0) + \varepsilon_1$$

The mean and variance are now

$$E[y_2] = \rho^2 y_0$$

$$\text{var}[y_2] = \rho^2 \sigma_\varepsilon^2 + \sigma_\varepsilon^2$$

Continuing in this fashion we arrive at

$$E[y_t] = \rho^t y_0$$

$$\text{var}[y_t] = \sigma_\varepsilon^2 \sum_{j=0}^{t} \rho^{2j} \tag{9.7}$$

If $|\rho| < 1$, then the mean converges to zero and the variance converges to

$$\text{var}[y] \rightarrow \frac{\sigma_\varepsilon^2}{1 - \rho^2}$$

The covariance between states of lag j is

$$\gamma_j = \sigma_\varepsilon^2 \frac{\rho^j}{1 - \rho^2} \tag{9.8}$$

A covariance matrix for states y_t, $t = 1, \ldots, T$, is

$$\Sigma = \sigma_\varepsilon^2 \mathbf{R},$$

where

$$
\mathbf{R} = \frac{1}{1-\rho^2}
\begin{bmatrix}
1 & \rho & \rho^2 & \cdots & \rho^{T-1} \\
\rho & 1 & \rho & & \\
\rho^2 & \rho & 1 & & \\
\vdots & & & \ddots & \rho \\
\rho^{T-1} & & & \rho & 1
\end{bmatrix}
\qquad 9.9
$$

This symmetric matrix results from symmetry in the autocovariance function, $\gamma_{-j} = \gamma_j$.

If $\rho = 1$, then the model is not stationary, as we recover the random walk. Thus, there has traditionally been interest in testing the null hypothesis that $\rho = 1$ (e.g., Dickey and Fuller 1979).

Another way of writing a AR(p) model places the autocovariance in the error term. We might have the model

$$
y_t = \mu_t + \varepsilon_t
$$

$$
\varepsilon_t \sim N\left(\sum_{j=1}^{p} \rho_j \varepsilon_{t-j}, \sigma^2 \right),
$$

where $\rho_1, \ldots, \rho_p$ are correlations up to lag p. As before, stationarity of the AR(1) model requires that $|\rho| < 1$. This general form is commonly used to fit temporal and spatio-temporal models, where the mean function μ_t might involve covariates. This model can be written as

$$
\mathbf{y} \sim N(\boldsymbol{\mu}, \boldsymbol{\Sigma}),
$$

where $\mathbf{y}^T = [y_1, \ldots y_T]$, $\boldsymbol{\mu}^T = [\mu_1, \ldots \mu_T]$, and $\boldsymbol{\Sigma} = \sigma^2 \mathbf{R}$ is the covariance matrix, with $\mathbf{R}$ from 9.9. The stationarity assumption is needed to ensure that the covariance matrix $\boldsymbol{\Sigma}$ is positive definite. The mean function can be a regression on covariates. Chib (1993) describes application of Gibbs sampling to this broad class of models.

For higher-order AR processes, the stationarity criterion is more complicated. Stationarity requires that the covariance decreases with elapsed time (an analogous criterion exits for spatial covariance—Chapter 10). Depending on the order of the process, the autoregressive coefficients can assume a restricted set of values such that roots of the characteristic equation $\lambda^p - \rho_1 \lambda^{p-1} - \ldots - \rho_p = 0$ fall between -1, and 1. For the AR(2) process we must have $|\rho| < 2$ and $-1 < \rho_z < 1 - |\rho_1|$. These constraints are used to construct priors for autoregression parameters (e.g., McCulloch and Tsay 1993; Johnson and Hoeting 2003). Huerta and West (1999) provide a flexible approach for prior specification when model order is uncertain.

Example 9.2. Gibbs sampler for the AR(1) model with explanatory variables
The autoregression with covariates can be written as

$$y_t = \mu_t + \varepsilon_t, \quad t = 1, \ldots, T$$

The mean term is a regression on covariates, which can change over time. The vector of means $\mu = [\mu_1, \ldots, \mu_T]^T$ is given by

$$\mu = \mathbf{X}\beta,$$

where $\mathbf{X}$ is a T by k design matrix and β is a length-k vector of parameters. Each row of $\mathbf{X}$ includes a leading 1 followed by values for covariate values for time t. The AR(1) model has error term $\varepsilon_t \sim N(\rho\varepsilon_{t-1}, \sigma^2)$.

The covariance matrix is given by Equation 9.9. With a Gaussian prior on regression coefficients, an IG prior on the variance, and a uniform prior on the autocorrelation parameter, we have the model

$$p(\beta, \sigma^2, \rho | \mathbf{y}, \mathbf{X}, \ldots) \propto N_T(\mathbf{y}|\mathbf{X}\beta, \Sigma) \ N_k(\beta|b, V_b)$$

$$IG(\sigma^2|a_s, b_s) \ Unif(\rho|-1, 1)$$

Here are Gibbs sampler steps:

1. The regression parameters have conditional posterior

$$p(\beta|\mathbf{y}, \mathbf{X}, \sigma^2, \rho, \ldots) \propto N_T(\mathbf{y}|\mathbf{X}\beta, \Sigma) \ N_g(\beta|\mathbf{b}, \mathbf{V}_b)$$

which is distributed as $N(\mathbf{Vv}, \mathbf{V})$, with

$$\mathbf{v} = \mathbf{X}^T \Sigma^{-1} \mathbf{y} + \mathbf{V}_b^{-1}\mathbf{b}$$

$$\mathbf{V}^{-1} = \mathbf{X}^T \Sigma^{-1} \mathbf{y} + \mathbf{V}_b^{-1}$$

This result is obtained by incompleting the square (Section 7.4).

2. The conditional posterior for the variance is

$$p(\sigma^2|\mathbf{y}, \mathbf{X}, \beta, \rho, \ldots) \propto N_T(\mathbf{y}|\mathbf{X}\beta, \Sigma) \ IG(\sigma^2|a_s, b_s)$$

$$= IG\left(\sigma^2 \middle| a_s + \frac{T}{2}, b_s + \frac{1}{2}(\mathbf{y} - \mathbf{X}\beta)^T \mathbf{R}^{-1}(\mathbf{y} - \mathbf{X}\beta)\right)$$

where $\mathbf{R}$ is given by Equation 9.9.

3. The AR(1) parameter ρ is sampled indirectly, such as Metropolis-Hastings or inverse distribution sampling.

A simulated data set for an intercept and a single covariate for parameter values $\beta = [2,3]^T$, $\sigma^2 = 3$, and $\rho = -0.5$ is shown in Figure 9.1. The response y_t depends not only only x_t, but also on past values of y_t, through the autocorrelated error, which, in this case, is negative. MCMC output is shown as a pairs plot of thinned values in Figure 9.2. As expected, we see the negative correlation of regression parameters.

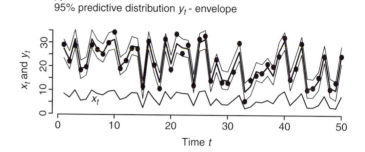

FIGURE 9.1. A simulated AR(1) series with covariate x_t (lower line). The response variable y_t is shown as simulated values (dots) and a 95 percent predictive distribution, represented by the median prediction (thick line) and upper and lower percentiles.

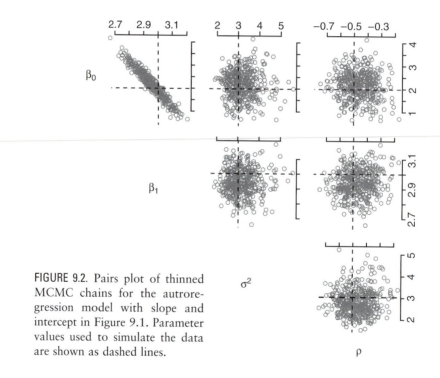

FIGURE 9.2. Pairs plot of thinned MCMC chains for the autroregression model with slope and intercept in Figure 9.1. Parameter values used to simulate the data are shown as dashed lines.

9.3.8 ARMA Models

MA and AR models can be combined:

$$y_t = \sum_{j=1}^{p} \rho_j \, y_{t-j} \; + \; \varepsilon_t \; + \; \sum_{j=0}^{q} a_j \varepsilon_{t-j}$$

This is the ARMA(p,q) model. As with the AR model, we can have a mean function that includes covariates,

$$y_t = \mu_t + \sum_{j=1}^{p} \rho_j y_{t-j} + \varepsilon_t + \sum_{j=0}^{q} a_j \varepsilon_{t-j}$$

Chib and Greenberg (1994) apply a Gibbs sampler to ARMA models.

With a Bayesian approach, we can allow that the parameter values themselves evolve over time. The parameters in the ARMA model thus have an additional subscript for time. Beckage and Platt (2003) used this model to predict area burned in Florida, with a covariate for the Southern Oscillation Index (SOI) and ARMA(2,2) error structure. Each of the five parameters (SOI plus autoregressive-moving average parameters) were assigned Gaussian distributions with variance parameters that determine how rapidly parameters can evolve. Predictions were good, but highly correlated parameter estimates suggested that the model may have been overparameterized (although AIC values pointed to the ARMA(2,2) model). Beckage and Platt (2003) suggested that a longer time series might allow for better estimates.

9.3.9 Autocovariance and Autocorrelation

Correlation in time series affects inference. Models that include autocorrelation and autocovariance have wide application. The autocovariance function arose in Equation 9.1. For a stationary series y, the autocovariance function is the covariance of points separated by lag j,

$$\gamma_j = \text{cov}[y_t, y_{t-j}]$$

The series must be stationary—the autocovariance must not depend on t. The variance of the series is γ_0, and the autocorrelation function is

$$\rho_j = \gamma_j / \gamma_0$$

Like the autocovariance, the autocorrelation is symmetric about zero ($\rho_{-j} = \rho_j$) and lies between -1 and 1. The cross covariance between two time series is

$$\gamma_{xy,j} = \text{cov}[x_t, y_{t-j}] \qquad\qquad 9.10$$

and the cross correlation is

$$\rho_{xy,j} = \frac{\gamma_{xy,j}}{\sqrt{\gamma_{xx,j}\,\gamma_{yy,j}}} \qquad\qquad 9.11$$

Unlike the autocorrelation, the cross correlation is not symmetric about zero. A sample autocovariance is estimated as

$$g_j = \frac{1}{n}\sum_{t=j+1}^{n} (y_t - \bar{y})(y_{t-j} - \bar{y}) \qquad\qquad 9.12$$

with autocorrelation estimate

$$r_j = g_j/g_0.$$ 9.13

The sample cross covariance is

$$g_{xy,j} = \frac{1}{n}\sum_{t=1}^{n-j} (x_t - \bar{x})(y_{t+j} - \bar{y})$$ 9.14

for positive lags ($k \geq 0$). For negative lags the summation proceeds from $1 - k$ through n. The sample cross correlation is

$$r_{xy,j} = \frac{g_{xy,j}}{\sqrt{g_{xx,j}g_{yy,j}}}$$ 9.15

9.4 The Frequency Domain

In view of the fact that we are often interested in scales of variation, it can be useful to examine variance from the perspective of frequency. The autocorrelation function can be expressed as the Fourier transform of the *spectral density* $f(\omega)$

$$\rho_k = \int_{-\pi}^{\pi} e^{i\omega k} f(\omega) d\omega$$ 9.16

The frequency domain can be useful for identifying cycles in population time series (e.g., Bjørnstad et al. 1999) and for simplification of high-dimensional spatial problems (Section 10.9.8). Power spectra for cyclic time series show a peak in variance at the relevant frequencies. Likewise, phase lags between two stationary time series can be explored by taking the Fourier transform of the cross correlation function (Clark et al. 2002). Wavelet analysis is used to assess how the strength of periodic phenomena varies over time (Grenfell et al. 2001). Both Fourier analysis and wavelet analysis summarize information in terms of some functions that may be much more compact than the original data (Section 10.9.4).

9.5 Application: Detecting Density Dependence in Population Time Series

Ecologists have long debated the importance of density dependence in natural populations. The possibility that population densities might stabilize at some carrying capacity implies that population numbers might be regulated by competition for limiting resources (Chapter 2). Alternatively, populations might be more strongly regulated by continuing environmental variability, natural enemies, or episodic disturbances.

Because experiments on natural populations are difficult, the use of time series data for detection of density dependence has a long legacy (Hassell et al.

1976; Berryman and Turchin 2001). The approach often begins with an hypothesis concerning population regulation (e.g., are growth rates controlled by density?) and then applies a statistical model to test the hypothesis. The approach raises the question of what statistical models are appropriate, in terms of their ability to detect density dependence, where it exists, and their ability to rule it out, where it does not. Here I summarize several methods used to identify population regulation in nature.

The potential for population regulation must be assessed in the context of stochasticity, which stands in for the factors that affect density and cannot be explicitly included in the model, because they are unknown or they cannot be measured. This process error represents the extent to which the deterministic model fails to capture change in population density.

Recall the difference equation for logistic growth,

$$n_{t+1} = n_t \exp\left[r\left(1 - \frac{n_t}{K} \right) \right] \qquad 9.17$$

(Chapter 2). The density dependence (DD) in this model is given by the second term, which includes the carrying capacity K. Without this feedback we have the density independent (DI) model

$$n_{t+1} = n_t e^r$$

These nested models suggest a test for DD. By allowing K to tend to ∞ we recover the DI model. For a model that includes process error, a likelihood ratio can be used to test for the effect of K, but there are a few more steps involved with time series data. I follow the approach of Dennis and Taper (1994). This is a classical method that can be useful when observation errors are small.

9.5.1 A Classical Analysis with Process Error

Recast the logistic growth model in a form more amenable to analysis,

$$n_{t+1} = n_t \exp\left(r - \frac{r}{K} n_t \right) = n_t \exp(b_0 + b_1 n_t)$$

The parameters b_0 and b_1 are equated with the model parameters r and $-r/K$, respectively. The nested DI model is recovered if b_1 is zero. The test for DD consists of determining whether b_1 does, in fact, differ from zero.

Before we can proceed to a test we need a distribution of error as basis for the likelihood function. I assume that process error is distributed normally about the log of density. I use this assumption for at least two reasons. First, negative densities are impossible. Second, population growth is multiplicative, not additive, so logs make sense. If log density has a normal distribution of error, then the model is

$$\ln n_{t+1} = \ln n_t + b_0 + b_1 n_t + \varepsilon_t \qquad 9.18$$

$$\varepsilon_t \sim N(0, \sigma^2)$$

In other words, the error term is normal with mean zero. It is equivalent to

$$n_{t+1} = n_t \exp(b_0 + b_1 n_t + \varepsilon_t) \qquad 9.19$$

I hereafter refer to this as Model 2 (two fitted parameters). There are two other models nested within this DD model. The DI model (Model 1) has $b_1 = 0$,

$$n_{t+1} = n_t \exp(b_0 + \varepsilon_t) \qquad 9.20$$

I have substituted b_0 for r in this model of exponential growth to emphasize its relationship to Model 2. A random walk (Model 0) has $b_0 = b_1 = 0$,

$$n_{t+1} = n_t \exp(\varepsilon_t) \qquad 9.21$$

9.5.2 A Likelihood Function

To simplify notation, let x_t represent $\ln n_t$. Then the discrete logistic model (9.20) is

$$x_{t+1} = x_t + b_0 + b_1 n_t + \varepsilon_t$$

$$\varepsilon_t \sim N(0, \sigma^2)$$

To simplify futher, I work with the differenced series, the *change* in density from one time step to the next:

$$y_t \equiv x_t - x_{t-1} = b_0 + b_1 n_{t-1} + \varepsilon_t \qquad 9.22$$

(Section 9.3.5).

For now, I condition on the first element of the series, because I choose not to write a model for how n_1 derives from unobserved n_0. Thus, the first element y_1 is the difference between x_2 and x_1. For notational convenience, I define the length of the series T to be the number of values in vector **y**. For this example, T is one less than the number of observations, because I condition on the first value. The likelihood function is

$$L(\mathbf{y}; a, b, \sigma^2) = \prod_{t=1}^{T} N\left(y_t | b_0 + b_1 n_{t-1}, \sigma^2\right)$$

$$= \sigma^{-T}(2\pi)^{-T/2} \exp\left[-\frac{1}{2\sigma^2} \sum_{t=1}^{T} (y_t - b_0 - b_1 n_{t-1})^2\right]$$

The $-\ln L$ for Model 2 is

$$-\ln L = T \ln \sigma + \frac{T}{2}\ln(2\pi) + \frac{1}{2\sigma^2}\sum_{t=1}^{T}(y_t - b_0 - b_1 n_{t-1})^2 \qquad 9.23$$

9.5.3 ML Estimators

The MLE's are obtained as for the linear model with Gaussian error (Section 5.5). Beginning with the variance, set $\frac{\partial \ln L}{\partial \sigma} = 0$ and obtain the mean squared deviation for Model 2 as

$$\hat{\sigma}_2^2 = \frac{1}{T}\sum_{t=1}^{T}(y_t - \hat{b}_0 - \hat{b}_1 n_{t-1})^2 \qquad 9.24$$

I have inserted the subscript 2 to indicate that this is the variance for Model 2. This is the residual variation around the regression.

For the intercept b_0 determine

$$\frac{\partial \ln L}{\partial b_0} = 0$$

$$0 = \sum_{t=1}^{T}(y_t - b_0 - b_1 n_{t-1}) = \sum_{t=1}^{T} y_t - Tb_0 - b_1 \sum_{t=1}^{T} n_{t-1}$$

Divide through by T (to simplify notation) and rearrange to obtain

$$\hat{b}_0 = \bar{y} - \hat{b}_1 \bar{n} \qquad 9.25$$

For slope b_1 set the derivative to zero and divide by T,

$$0 = \sum_{t=1}^{T} n_{t-1}(y_t - b_0 - b_1 n_{t-1}) = \overline{ny} - \hat{b}_0\bar{n} - b_1\overline{n^2}$$

Now substitute for the ML estimate of the intercept,

$$(\bar{y} - b_1\bar{n})\bar{n} + b_1\overline{n^2} = \overline{ny}$$

Rearrange to solve for b_1,

$$\hat{b}_1 = \frac{\overline{ny} - \bar{n}\cdot\bar{y}}{\overline{n^2} - \bar{n}\cdot\bar{n}} = \frac{\text{cov}(n,y)}{\text{var}(n)}$$

The same approach for Model 1 gives

$$\hat{b}_0 = \bar{y}$$

that is, the mean change in log density. The associated residual variation is

$$\hat{\sigma}_1^2 = \frac{1}{T} \sum_{t=1}^{T} (y_t - \hat{b}_0)^2 \qquad\qquad 9.26$$

These estimates will be familiar from the regression example of Section 5.5.2.

9.5.4 Hypothesis Tests

Dennis and Taper (1994) test for density dependence using three models. The first model relates to the hypothesis H_0 that density is a simple random walk and so has parameters b_0 and b_1 equal to zero. One alternative hypothesis H_1 is that of density-independent growth, with growth rate b_0 and $b_1 = 0$. The second alternative hypothesis H_2 is that density is regulated by growth rate b_1. Likelihoods for the three models used to test for density dependence are

$$\text{Model 0:} \quad L_0 = \sigma_0^{-T}(2\pi)^{-T/2} \exp\left[-\frac{1}{2\sigma_0^2} \sum_{t=1}^{T} y_t^{\,2} \right]$$

$$\text{Model 1:} \quad L_1 = \sigma_1^{-T}(2\pi)^{-T/2} \exp\left[-\frac{1}{2\sigma_1^2} \sum_{t=1}^{T} (y_t - b_0)^2 \right]$$

$$\text{Model 2:} \quad L_2 = \sigma_2^{-T}(2\pi)^{-T/2} \exp\left[-\frac{1}{2\sigma_2^2} \sum_{t=1}^{T} (y_t - b_0 - b_1 n_{t-1})^2 \right]$$

We can test the hypothesis of density-independent population growth against a null model of a random walk by comparing the first two models using a likelihood ratio (LR) test (Section 6.3). The likelihood ratio is simple, because the summations in the exponents are equal to $T\hat{\sigma}_m^2$, for the estimate of the variance parameter of a given model m. Thus,

$$LR_{01} = \frac{L_0}{L_1} = \frac{\hat{\sigma}_0^{-T} \exp\left[-\dfrac{T\hat{\sigma}_0^2}{2\hat{\sigma}_0^2} \right]}{\hat{\sigma}_1^{-T} \exp\left[-\dfrac{T\hat{\sigma}_1^2}{2\hat{\sigma}_1^2} \right]} = \left(\frac{\hat{\sigma}_1}{\hat{\sigma}_0} \right)^T$$

The corresponding deviance is

$$D_{01} = -2 \ln \frac{L_0}{L_1} = -2T(\ln \hat{\sigma}_1 - \ln \hat{\sigma}_0)$$

However, we do not use this as the basis for a LR test for population regulation (i.e., Model 2 versus Model 1), because observations are not independent. Instead, Dennis and Taper (1994) use a parametric bootstrap.

9.5.5 Parametric bootstrap

Dennis and Taper (1994) describe a parametric bootstrap for the discrete logistic time series (Models 1 and 2). For parameter error do the following:

1. Determine ML estimates for parameters in Model 2.
2. Generate a new series of Model 2 data, conditioning on the first observation. Each subsequent value in the simulated time series has an expected value given by the fitted model plus random variation. The random variate is normally distributed with mean zero and variance equal to the MLE of σ^2. In other words, the next value in the simulated time series is

$$x_t = x_{t-1} + \hat{b}_0 + \hat{b}_1 n_{t-1} + \varepsilon_t$$

$$\varepsilon \sim N(0, \hat{\sigma}_2^2)$$

3. Estimate parameters for Model 2 from this new series and save them.
4. Repeat 2000 times.

Confidence intervals for the parameters are quantiles for these bootstrapped estimates.

With this algorithm one can also generate bootstrap estimates of the deviance. Assume that data are generated by a null model (Model 1), fit both models, and determine the distribution of deviance values. The percentile of this bootstrapped distribution represented by the deviance for the original data provides a P value for Model 2 given data generated by Model 1. Use the following steps:

1. Determine ML parameter estimates and the deviance for the comparison of the DI (Model 1) with the DD (Model 2) model as

$$D_{01} = -2T(\ln \sigma_2 - \ln \sigma_1)$$

2. Generate a new series of Model 1 data (using parameters from Model 1) as

$$x_t = x_{t-1} + \hat{b}_0 + \varepsilon_t$$

$$\varepsilon \sim N(0, \hat{\sigma}_2^2)$$

3. Fit both Models 1 and 2 to this series and calculate a deviance (and save it).
4. Repeat Steps 2 through 4 many (e.g., 2000) times.

The probability of a deviance as large as observed from the data is given by the percentile for the bootstrapped sample of deviances.

Example 9.3. Turchin and Taylor (1992) analyzed several times series in an effort to identify lagged effects of density on population growth. Here is an analysis of the simple DD model applied to one of those series, fall webworm, sampled from 1937 to 1958 (Morris 1983) (Figure 9.3a).

The negative slope (parameter b_1) in Figure 9.3b suggests density dependence. The deviance for the comparison of the DD and DI models (vertical line at 12.0) is well above the 95 percent values determined by parametric bootstrap (the histogram with vertical line at 7.86 for 95 percent confidence) and by a naïve application of an LRT (the dashed lines, including the vertical line at 3.84 for 95 percent confidence) (Figure 9.3c).

Here are the parameter estimates obtained from the parametric bootstrap in Figure 9.4:

	DI estimate	DD estimate	se	0.025	0.975
b_0	0.0548	0.812	0.245	0.433	1.38
b_1		−0.160	0.0480	−0.230	−0.0992
error	1.10	0.855	0.214	0.310	1.15

The parameter distributions are shown in Figure 9.4.

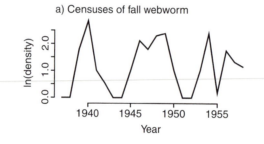

a) Censuses of fall webworm

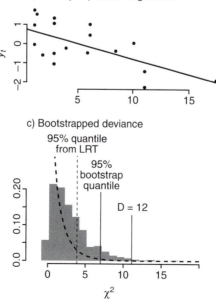

b) Density dependent regression

c) Bootstrapped deviance

FIGURE 9.3. Fall webworm density-dependent model with deviance. (*a*) Time series. (*b*) Growth rate declines with density. (*c*) Comparison of bootstrapped distribution of *D* values (histogram) with the χ^2 distribution for the corresponding degrees of freedom (dashed line) and the value of $D = 12$ for this data set.

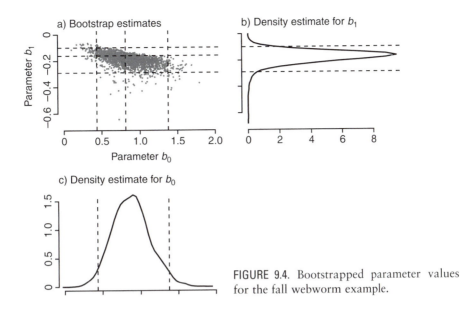

FIGURE 9.4. Bootstrapped parameter values for the fall webworm example.

9.5.6 Other Approaches for Detecting Density Dependence

There is a large literature on the application of classical hypothesis-testing methods to the problem of detecting density dependence in time series data. A series of papers including Turchin and Taylor (1992) model change in log density with linear, nonlinear, lag terms, and combinations thereof (also Turchin 1996; Lindstrom et al. 1999). This has been termed the *response surface methodology* (Box and Draper 1987). It represents an attempt to admit density feedback on growth rate in a flexible fashion. Model selection techniques are used to identify the order of the model (number of lags) and the nature of nonlinearities. One of the techniques is based on cross validation with the r^2 for predicted values used as a goodness of fit index (Turchin 1996).

This classical approach to estimation and hypothesis testing can be useful when observation errors can be ignored. Of those tested, the best model for the fall webworm data is not very convincing. Counts show order of magnitude fluctuations over short time periods (Figure 9.3a) that are not well described by the simple model of density dependence (Figure 9.3b). The classical approach suggests that we can reject the simple exponential model in favor of the density-dependent one, but there is much more going on here. There are additional classical methods for accommodating contingencies. For example, combinations of lagged and nonlinear forms can be included. Staples et al. (2004) present an approach, whereby an observation equation combined with exponential growth is treated as a linear random effects model. In general, as the complexity of the problem increases, it can be difficult to incorporate the additional effects within this classical framework. A Bayesian framework can help in these cases, because additional sources of stochasticity do not involve severe modifications to the model, and sampling-based approaches permit analysis. The simplicity of the Bayesian framework rests in large part with the capacity to focus on conditional relationships.

9.6 Bayesian State Space Models

9.6.1 General Structure

Proper treatment of uncertainty can be especially important for time series data, because process error propagates forward with the process, whereas observation error does not (Carpenter et al. 1994; Pascual and Kareiva 1996; Turchin 1995; Hilborn and Mangel 1997; Stenseth et al. 1999; Bjørnstad and Grenfell 2001; Coulson et al. 2001; Dennis et al. 2001; de Valpine and Hastings 2002; Calder et al. 2003; Clark and Bjørnstad 2004; Staples et al. 2004). As with previous models we acknowledge that superimposed on the process is a data-generating mechanism. Dynamics can be obscured by observation errors, sample information being anecdotal, unreliable, or missing entirely. Methods and sampling effort can change over time.

There have been many efforts to handle the combination of process and observation error in time series. Classical approaches to state space models include the Kalman filter (Harvey 1989), an updating procedure whereby the prior estimate of the state is corrected by how well we can predict the next observation. It is carried out as an iterative procedure. The approach is motivated by the need to know the underlying state x_t, given that y_t is observed, to estimate the transition parameters in the function $f(x_t)$, and to predict y_{t+k} for some time k in the future.[1] Why predict y_{t+k}, rather than just x_{t+k}? Because we may eventually observe y, but we will not observe x. To compare observation with prediction, we need y. Recent applications in ecology provide valuable insight (Ives et al. 2003; Williams et al. 2003). The Kalman filter is not ideal for nonlinear models and when errors are non-Gaussian. In this section I describe how the Bayesian framework used for previous models is readily adapted to time series data.

Among the most general and flexible alternatives is the Bayesian state space model. In principle, any functional form and error distribution is admissible. In practice, there will be many considerations that influence identifiability of parameters and efficiency. Here I summarize basic elements of the approach from Carlin et al. (1992) and Calder et al. (2003), with extensions for limitations of ecological data from Clark and Bjørnstad (2004). I consider models of the general form given in Equation 9.5.

Consider the pair of equations

$$x_t = f_t(x_{t-1}) + \varepsilon_t$$
$$y_t = g_t(x_t) + w_t$$

9.27a, b

---

[1] Unlike the last section, here y refers to an observation of variable x, rather than the difference between two values of x.

The first equation is the *process model, system equation,* or *evolution equation.* The process model includes a deterministic component $f_t(x_{t-1})$, which encapsulates understanding of how density affects growth. We saw that the stochasticity in this equation, the *process error,* affects the process itself, because variability introduced at one time step affects the trajectory of change in the future. Used by itself, the process model assumes that the only type of variability is that which directly affects change in density (Figure 9.5). For a simple random walk, the best estimate of density is whatever density existed at time $t - 1$,

$$f_t(x_{t-1}) = x_{t-1} \qquad\qquad 9.28$$

This equation is obtained by taking the log of Equation 9.22. For exponential growth, log density changes by a constant, and is obtained as the log of Equation 9.21,

$$f_t(x_{t-1}) = x_{t-1} + b \qquad\qquad 9.29$$

We first saw this as Equation 3.9, where I used maximum likelihood to estimate the growth rate parameter based on a likelihood that allowed for process error. For the logistic, the process model of Equation 9.23 is

$$f_t(x_{t-1}) = x_{t-1} + b_0 + b_1 e^{x_{t-1}} \qquad\qquad 9.30$$

$$= x_{t-1} + b_0 - b_1 n_{t-1}$$

To the deterministic component is added a stochastic term ε_t that allows for the fact that the deterministic model will not precisely describe change in density. As before, I assume a simple lognormal error (i.e., normal error on log density x)

$$\varepsilon_t \overset{i.i.d.}{\sim} N(0,\sigma^2)$$

but these methods can be applied under a range of assumptions regarding process error (Carlin et al. 1992; Stroud et al. 2003).

Equation 9.25b is the *observation equation.* In most cases we do not observe n, but rather some sample of it. There is random variability w_t that results from observer error. This error is often modeled as the Gaussian distribution

$$w \overset{i.i.d.}{\sim} N(0,\tau^2)$$

but, again, it does not have to be. In contrast to process error, observation error does not affect the process itself.

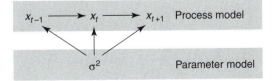

FIGURE 9.5. Population growth based on the random walk model with process stochasticity described by variance σ^2 (Equation 9.25a).

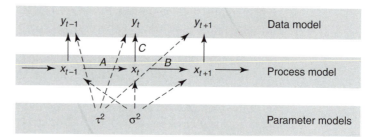

FIGURE 9.6. The state space model of a random walk where observations y are drawn from the underlying process x. There is an additional level above the process stage.

Figure 9.6 illustrates the relationships between the process model and observations. You can think of the observation equation as representing the data model that overlays the underlying, unobserved process. So far, the model assumes an extremely simple assumption that observations have the same mean as the process, with observation errors represented by the variance τ^2. In other words, I assume that there is no inherent bias in observations. There often is systematic bias. If the sample represents, on average, a constant fraction of the population, then I could write

$$y_t = gx_t + w_t,$$

where $g < 1$. Sampling efficiency may depend on density, for example, when population estimates come from hunting (Hudson et al. 1998; Stenseth et al. 1998) or fisheries catch effort (Meyer and Millar 1999). Stenseth et al. (1998) show a relationship that suggests a linear form for $g_t(x_t)$ (Figure 9.7). Note that with the state space formulation I include the trend and allow for the additional variance in terms of w_t. In the initial exposition, I assume that samples entail constant variance (a single value of τ^2 applies to all observations), as might

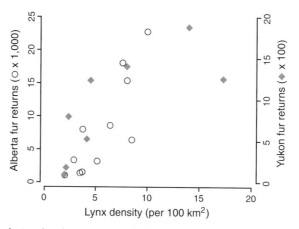

FIGURE 9.7. A relationship between population density and fur returns for lynx in two Canadian provinces. From Stenseth et al. (1998).

occur if methods are consistent throughout the study. An extension of this model that allows not only for non-normal errors, but also for errors that vary over time, includes a scale-mixture structure for errors (Carlin et al. 1992). I return to these considerations later.

In summary, the state space model recognizes that the process itself is not directly observable. Because the x's are hidden from view, we cannot estimate the parameters contained in the process model without estimating the x's themselves. We have as many x's as there are sample dates (or more), in addition to parameters contained in $f(x)$. Obviously, there can be many latent variables, as well as parameters, to estimate. The complexity of the problem begins to highlight the advantages of a Bayesian framework.

In Chapter 8 we approached complex problems by factoring them into low-dimensional ones. A marginal posterior for the parameters and all latent variables in the state space model might be difficult to think about, let alone write down or solve. Sampling-based approaches can operate from low-dimensional (often univariate) conditional relationships. The graph for the state space model (Figure 9.6) identifies the conditional relationships needed to apply these methods—conditional relationships are arrows. Thus, although we have moved to models having important (and often intractable) marginal dependencies, we apply the same tools. Three common ecological models that follow demonstrate the approach.

9.6.2 A Random Walk

Before considering the likelihood explicitly, recall the joint probability, now in the context of a time series. I begin with three states, having joint probability $p(x_1, x_2, x_3)$. Because joint densities are unwieldy, I factor this density to a series of univariate densities,

$$
\begin{aligned}
p(x_1, x_2, x_3) &= p(x_3|x_2, x_1)\ p(x_2, x_1) \\
&= p(x_3|x_2, x_1)\ p(x_2|x_1)\ p(x_1)
\end{aligned}
$$

You can already appreciate that, for a long time series, this factorization comes at the cost of conditioning on many things. For example, I have to specify the relationship between x_1 and x_3. As discussed in Section 7.3.3, this can be simplified. In this case, if x_3 has no direct effect on x_1, there is no need to include this conditional relationship. I am not saying that they are independent—they certainly are dependent, because each depends on x_2. But if they are conditionally independent, this allows us to write

$$
p(x_1, x_2, x_3) = p(x_3|x_2)\ p(x_2|x_1)\ p(x_1)
$$

or, for T total observations $\mathbf{x} = [x_1, \ldots, x_T]$,

$$
p(\mathbf{x}) = p(x_1) \prod_{t=2}^{T} p(x_t|x_{t-1})
$$

This Markov assumption means that each observation is conditioned only on the previous one.

The random walk is a good place to begin understanding how time series data are represented in the state space model. Consider equation 9.26. For simplicity I assume normally distributed errors. As I develop this model it can be helpful to think of each value x_t as acting like a prior for the subsequent value, x_{t+1}.

Likelihood—The likelihood expresses the probability of obtaining the ensemble of true and observed values. It can be written as three parts:

$$p(\mathbf{x},\mathbf{y},x_0|\sigma^2,\tau^2\ \mu_0,\sigma_0^2) \;=\; \overbrace{p_1(x_0|\mu_0,\sigma_0^2)}^{A}\overbrace{\prod_{t=1}^{T}p_1(x_t|x_{t-1},\sigma^2)}^{B}\overbrace{\prod_{t=1}^{T}p_2(y_t|x_t,\tau^2)}^{C}$$

The left-hand side identifies the first density x_0 as a latent variable, which could be viewed as serving as a prior for x_1. The right-hand side includes two types of densities. $p_1(\ )$ describes the process (the x's). I assume standard distributions—normal for the x_t's and inverse gamma prior for variances. Then density A (shown as an arrow in Figure 9.6) says that the first observation is a random variate drawn from the normal distribution

$$x_0 \;\sim\; N(\mu_0,\sigma_0^2)$$

Density B is the joint probability of the true densities. Conditioned on the preceding density, write the distribution for the density at time t as

$$x_t|x_{t-1},\sigma^2 \sim N(f(x_{t-1}),\sigma^2),$$

where the predicted value for this random walk is

$$f(x_{t-1}) \;=\; x_{t-1}$$

Together A and B describe the joint probability of x_0 and $\mathbf{x}$.

Distribution C is the probability of obtaining the observed values $\mathbf{y}$. For normally distributed observation errors,

$$y_t|x_t,\tau^2 \sim N(x_t,\tau^2)$$

The priors for σ^2 and τ^2 are conjugate for this normal likelihood

$$\sigma^2 \sim IG(\alpha_\sigma,\beta_\sigma)$$

$$\tau^2 \sim IG(\alpha_\tau,\beta_\tau)$$

Pulling this together, the joint posterior then is proportional to

$$p(\sigma^2,\tau^2\ \mathbf{x}|\mathbf{y}) \;\propto\; \overbrace{N(x_0|\mu_0,\sigma_0^2)}^{A}\overbrace{\prod_{t=1}^{T} N(x_t|f(x_{t-1}),\sigma^2)}^{B}\;\overbrace{\prod_{t=1}^{T} N(y_t|x_t,\tau^2)}^{C}$$

$$\underbrace{IG(\sigma^2|\alpha_\sigma,\beta_\sigma)}_{D}\underbrace{IG(\tau^2|\alpha_\tau,\beta_\tau)}_{E}$$

9.31

As usual, I have omitted the normalization constant (hence, the proportionality), which involves an integral over all parameters.

In most cases we will be unable to solve the posterior distribution, but we can draw samples from all of the conditional distributions. Thus, we can construct an efficient algorithm to simulate the posterior distribution.

Conditional Posterior for x_t—Consider all but the first and last values of x_t: that is, ignore for now x_0 and x_T. To construct the conditional posterior for x_t, identify all of the densities in Equation 9.29 that contain x_t. These densities will be the ones that describe the arrows that are directly connected to x_t in Figure 9.6. Those arrows connect x_t to x_{t-1}, to x_{t+1}, and to y_t. These arrows are found in B and in C of Equation 9.29. In B, we find the arrow connecting x_t to x_{t-1}. This arrow is represented by the density $N(x_t|x_{t-1},\sigma^2)$. In B, we also find the arrow connecting x_t to x_{t+1}. This arrow is represented by the density $N(x_{t+1}|x_t, \sigma^2)$. Finally, the arrow connecting x_t to y_t, from C is $N(y_t|x_t, \tau^2)$. Then the conditional posterior includes just a few elements of B and C. Distribution B occurs twice, because x_t can occur to the left or right of the bar.

On the left-hand side, the conditioning on all values of x "not equal to x_t" boils down to simply x_{t-1} and x_{t+1}, because there are no arrows linking x_t to any other value of x. This result could seem strange for at least two reasons. First, you might wonder why x_t depends on x_{t+1}. After all, x_{t+1} "has not happened yet." Remember that we are estimating the latent states x_t, which means that the *estimate* of each x_t is informed by the full data set, forward and backward. We will see that the last value in a series can be less confidently estimated than can one that lies in the interior of the series, because the last value is conditionally constrained only by x_{T-1} and y_T.

Second, if all values depend on all others, why do we only include those values adjacent to x_t in the series? Remember, we will marginalize over the entire series (by simulation). The draws of individual values of x_t come from conditional distributions, and that is all we need to specify.

Now solve for this product of three normal densities by incompleting the square. To increase transparency, I start by ignoring the observation error and focus on the connections from x_t to x_{t-1} and x_{t+1}. Then the conditional posterior is proportional to

$$p(x_t|x_{j \neq t},\sigma^2) \propto \exp\left[-\frac{1}{2}\left(\frac{(x_t - x_{t-1})^2}{\sigma^2} + \frac{(x_{t+1} - x_t)^2}{\sigma^2}\right)\right]$$

Expanding this exponent, collecting terms in x_t, and ignoring the $-\frac{1}{2}$, we have

$$x_t^2\left(\frac{2}{\sigma^2}\right) - 2x_t\left(\frac{x_{t-1} + x_{t+1}}{\sigma^2}\right) + \frac{x_{t-1}^2 + x_{t+1}^2}{\sigma^2} \qquad 9.32$$

To find the normal distribution for x_t, we need to incomplete the square in x_t (see Section 4.2) such that this exponent takes the form

$$\frac{1}{V}(x_t - Vv_t)^2 \qquad 9.33$$

In other words, the new normal distribution will have a mean of Vv_t and a variance of V. The expanded exponent is

$$x_t^2\left(\frac{1}{V}\right) - 2x_t\left(\frac{Vv_t}{V}\right) + \frac{(Vv_t)^2}{V} \qquad\qquad 9.34$$

indicating that the variance for this new distribution can be found as the coefficient of x_t^2, and the mean for this new normal distribution can be found in the second term, multiplied by $-2x_t/V$. Comparing this new distribution (9.32) with that for x_t (9.30) shows that

$$\frac{1}{V} = \frac{2}{\sigma^2}$$

that is, half the process error,

$$V = \frac{\sigma^2}{2} \qquad\qquad 9.35$$

The posterior variance is less than the process variance, because it is constrained by two values.

The mean can be found from

$$\frac{Vv_t}{V} = \frac{x_{t-1} + x_{t+1}}{\sigma^2}$$

or

$$Vv_t = V\left(\frac{x_{t-1} + x_{t+1}}{\sigma^2}\right) = \frac{x_{t-1} + x_{t+1}}{2} \qquad\qquad 9.36$$

Thus, the estimate of the mean is simply the linear interpolation of the values that precede and follow it.

Observation Error—This framework readily accommodates the observation error. Returning to the conditional posterior, now write the exponent with the inclusion of the observation error,

$$p(x_t|x_{j\neq t}, y_t, \sigma^2, \tau^2) \propto \exp\left[-\frac{1}{2}\left(\frac{(x_t - x_{t-1})^2}{\sigma^2} + \frac{(x_{t+1} - x_t)^2}{\sigma^2} + \frac{(y_t - x_t)^2}{\tau^2}\right)\right]$$

Expanding the exponent and ignoring the coefficient $-\frac{1}{2}$, we have

$$\frac{2x_t^2 - 2x_t(x_{t-1} + x_{t+1}) + x_{t-1}^2 + x_{t+1}^2}{\sigma^2} + \frac{y_t^2 - 2x_ty_t + x_t^2}{\tau^2}$$

Collecting terms in x_t this becomes

$$x_t^2\left(\frac{2}{\sigma^2} + \frac{1}{\tau^2}\right) - 2x_t\left(\frac{x_{t-1} + x_{t+1}}{\sigma^2} + \frac{y_t}{\tau^2}\right) + C$$

By incompleting the square we find the estimates of mean and variance

$$Vv_t = V\left[\frac{x_{t-1} + x_{t+1}}{\sigma^2} + \frac{y_t}{\tau^2}\right] \qquad 9.37a$$

$$V^{-1} = \frac{2}{\sigma^2} + \frac{1}{\tau^2} \qquad 9.37b$$

Thus, the mean is a weighted average of the observed value y_t and the two values of x that bracket x_t. Their contributions are weighted by their variances. Also note that this distribution differs from that lacking observation error (Equations 9.33, 9.34) in that observations now weigh in, much like the adjacent values of x.

The result for this time series of observations is intuitive: the best estimate of density for a random walk is interpolated from neighboring ones and from the corresponding observation. The conditional variance is half that of the process error (because the estimate is constrained by two neighboring values) added to the observation variance.

Now consider the endpoints x_0 and x_T. The conditional posterior for the initial x_0 contains the distributions A and B, and distribution B only occurs once (Equation 9.29). If there is no observation at $t = 0$ (there is no observation y_0),

$$p(x_0|x_1,\sigma^2,\tau^2) = p_1(x_0|\mu_0,\sigma_0^2)p_1(x_1|x_0,\sigma^2)$$

Using the foregoing methods we have the conditional posterior

$$x_0|\mu_0,x_1,\sigma_0,\sigma \sim N(Vv_0,V)$$

with variance and mean

$$V_0^{-1} = \frac{1}{\sigma_0^2} + \frac{1}{\sigma^2} \qquad 9.38a$$

$$V_0v_0 = V_0\left(\frac{\mu_0}{\sigma_0^2} + \frac{x_1}{\sigma^2}\right) \qquad 9.38b$$

For the last value, having conditional posterior

$$p(x_T|x_{T-1},\sigma^2,\tau^2) = p_1(x_T|x_{T-1},\sigma^2)p_2(y_T|x_T,\tau^2)$$

we obtain

$$V_T^{-1} = \frac{1}{\sigma^2} + \frac{1}{\tau^2}$$

$$V_T v_T = V_T \left(\frac{x_{T-1}}{\sigma^2} + \frac{y_T}{\tau^2} \right)$$

Conditional Posteriors for Variances—Using the inverse gamma priors, the posteriors are

$$\sigma^2|\mathbf{x},x_0 \sim IG\left(\alpha_\sigma + \frac{T}{2}, \beta_\sigma + \frac{1}{2}\sum_{t=1}^{T} (x_t - x_{t-1})^2 \right)$$

$$\tau^2|\mathbf{x},\mathbf{y} \sim IG\left(\alpha_\tau + \frac{T}{2}, \beta_\tau + \frac{1}{2}\sum_{t=1}^{T} (y_t - x_t)^2 \right)$$

These posteriors come from B and D (for σ^2) and C and E (for τ^2) and are the conjugate posteriors (Appendix F).

A Gibbs Sampler—A Gibbs sampler for the state space model is arranged such that the sample for each latent variable is updated before the next sample is drawn. Suppose I have just entered the gth Gibbs step. I might begin by drawing an initial value

$$x_0^{(g)}|\mu_0, x_1^{g-1}, \sigma_0, (\sigma^2)^{(g-1)} \sim N(V_0 v_0, V_0)$$

where the value used for x_1 has yet to be updated. Parameters are from Equation 9.36. The draw for x_1 is now conditioned on this new value for x_0 and the old value for x_2,

$$x_1^{(g)}|x_0^{(g)}, x_2^{(g-1)}, y_1, (\sigma^2)^{(g-1)}, (\tau^2)^{(g-1)} \sim N(V v_1, V)$$

(9.35). Once each of the states has been updated, update the variance parameters.

9.6.3 Exponential Growth

Exponential growth is a simple extension of the random walk. The modifications are minor, the principal one being a conditional posterior for the growth rate parameter. Following the outline for the basic model, I demonstrate how to include unequal sample intervals, changing observation errors, and missing values.

The basic exponential model is given by Equation 9.25 with the process equation from 9.27,

$$x_t = x_{t-1} + b + \varepsilon_t$$

$$y_t = x_t + w_t$$

Retaining previous assumptions for error terms, the conditional posterior for log density is

$$p(x_t|x_{j\neq t}, y_t, b, \sigma^2, \tau^2)$$
$$\propto \exp\left[-\frac{1}{2}\left(\frac{(x_t - x_{t-1} - b)^2}{\sigma^2} + \frac{(x_{t+1} - x_t - b)^2}{\sigma^2} + \frac{y_t - x_t}{\tau^2} \right) \right]$$

Expanding the exponent, and ignoring the factor $-\frac{1}{2}$, we have

$$\frac{x_t^2 - 2x_t(x_{t-1} + b) + (x_{t-1} + b)^2}{\sigma^2} + \frac{(x_{t+1}^2 - b)^2 - 2x_t(x_{t+1} - b) + x_t^2}{\sigma^2}$$
$$+ \frac{y_t^2 - 2x_t y_t + x_t^2}{\tau^2}$$

Now collect terms in x_t,

$$x_t^2\left(\frac{2}{\sigma^2} + \frac{1}{\tau^2} \right) - 2x_t\left(\frac{x_{t-1} + x_{t+1}}{\sigma^2} + \frac{y_t}{\tau^2} \right) + C$$

Some algebra returns us to the same result we obtained for the random walk (equation 9.35): the weighted average of values from each of the arrows on Figure 9.6. This makes sense in view of the fact that the model is linear.

The conditional variance for observation error τ^2 is the same as before. For the process error, it differs only slightly,

$$\sigma^2|\mathbf{x}, x_0, b \sim IG\left(\alpha_\sigma + \frac{T}{2}, \beta_\sigma + \frac{1}{2}\sum_{t=1}^{T}(x_t - x_{t-1} - b)^2 \right)$$

For a normal prior, the conditional posterior for the parameter b is also normal,

$$p(b|\mathbf{x}, \sigma^2) \propto \prod_{t=1}^{T} N(x_t|x_{t-1} + b, \sigma^2)N(b|b_0, v_b)$$
$$= N(b|Vv, V)$$

We require V and v. The exponent of this product of normal distributions (without $-\frac{1}{2}$) is

$$\frac{1}{\sigma^2}\sum_{t=1}^{T}(x_t - x_{t-1} - b)^2 + \frac{(b - b_0)^2}{v_b}$$

Expanding and collecting terms containing b we have

$$b^2\left(\frac{T}{\sigma^2} + \frac{1}{v_b} \right) - 2b\left(\frac{\sum_{t=1}^{T}(x_t - x_{t-1})}{\sigma^2} + \frac{b_0}{v_b} \right) + C$$

yielding mean and variance

$$V\upsilon = V\left(\frac{\sum_{t=1}^{T}(x_t - x_{t-1})}{\sigma^2} + \frac{b_0}{\upsilon_b}\right) \qquad 9.39a$$

$$= V\left(\frac{x_T - x_0}{\sigma^2} + \frac{b_0}{\upsilon_b}\right)$$

$$V^{-1} = \frac{T}{\sigma^2} + \frac{1}{\upsilon_b} \qquad 9.39b$$

Note that, for a noninformative prior (large υ_b), the posterior conditional mean tends to the average log density change over a time increment, which, in turn simplifies to equation 3.11.

9.7 Application: Black Noddy on Heron Island

Ecological data are heterogeneous and often do not permit a direct application of simple models. Fortunately, the basic model is readily extended to allow for contingencies. As an example, I return to Heron Island on the southern end of the Great Barrier Reef, which hosts a nesting population of black noddy (*Anous minutus*). Through the twentieth century, this population increased at the rapid rate of nearly 10 percent per year (i.e., *b* is roughly 0.1) (Chapter 2). The earliest estimate of about fifty breeding pairs comes from the beginning of the twentieth century. Subsequent estimates were based on a variety of methods taken at irregular intervals. They extend through 1992, when Ogden (1993) estimated 63,140 + 7043 birds. Despite the explosive increase in this population and the apparent good fit with the exponential model, there are several reasons to question this simplicity, its implications for demography, and its utility for prediction.

First, the precision of estimates must vary from sample to sample. The estimates span many decades, and many are anecdotal (Barnes and Hill 1989). For the analysis here I had access to standard error estimates of population size for the two most recent surveys, but remaining censuses are difficult to compare, due to survey methodology (Barnes and Hill 1989). Errors are not reported for all censuses, so I estimated some from Ogden's (1993) Figure 3. Thus, my uncertainty exceeds that of the authors who previously applied the exponential model to these data.

The foregoing factors are difficult to accommodate in traditional models, but the state space approach is flexible. Effects can be summarized in terms of those that might be treated deterministically and the unknowns that we should allow for in terms of stochasticity. The different observation errors suggest that we should assume a different value of τ^2 for each census (Figure 9.8). The uneven sample intervals require extension of the basic model, because variability in the process must accumulate and, thus, depend on time elapsed since the last census. In other words, if there is a variance σ^2 associated with one time step of the process, there must be some larger variance when samples are obtained farther apart. I consider two approaches to this problem, one as unequal sample intervals and a second as missing values.

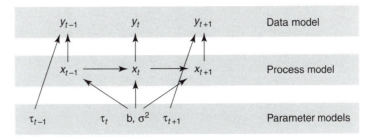

FIGURE 9.8. The state space model used to model exponential growth with different errors for each census.

9.7.1 Unequal Observation Errors

The modifications required to allow for differing observation errors are not complicated. For the conditional posterior on x_t, I simply replace τ^2, with the value for that specific observation, τ_t^2. The conditional posterior for this variance is

$$\tau_t^2|x_t,y_t \sim IG\left(\alpha_t + \frac{1}{2},\beta_t + \frac{1}{2}(y_t - x_t)^2\right)$$

The prior estimates for each time increment can be selected to accommodate what is known about observation errors. In this example, there are samples for which little is known (the early ones) and others for which there are estimates of the standard error. In these cases, the priors can be selected to have a mean value that matches the reported standard error and a variance that might reflect the number of samples that contributed to that standard error estimate (our level of confidence in it). This can be done by moment matching (Section 3.10). For sample dates having no estimated errors, use prior values that allow for the range of error that might be expected. Other parts of the model remain unchanged.

9.7.2 Unequal Sample Intervals

Consider a population model for exponential growth where samples are not obtained at equally spaced intervals. There are several ways to approach this problem. Here I assume that process error is constant and, thus, the variance in the process depends on the sample interval. The growth model is modified for unequal intervals from equation 9.21,

$$n_t = n_{t - \delta_t}(e^{b + \varepsilon_t})^{\delta_t}$$

I assume a process error that applies per unit time, where the census interval δ_t is the time elapsed between the sample time t and the previous one. For log density x_t, we have the following,

$$x_t = x_{t - \delta_t} + \delta_t b + \delta_t \varepsilon_t$$

$$y_t = x_t + w_t$$

Recall that the variance of a constant times a random variable is that constant squared times the variance of the random variable (equation D.38). In this case the likelihood for the tth value of x is

$$p(x_t|x_{t-\delta_t},x_{t+\delta_{t+1}},y_t,b,\sigma^2,\tau^2)$$

$$\propto N(x_t|x_{t-\delta_t} + \delta_t b,\delta_t^2\sigma^2)N(x_{t+\delta_{t+1}}|x_t + \delta_{t+1}b,\delta_{t+1}^2\sigma^2)N(y_t|x_t,\tau^2)$$

Completing the square, obtain posterior mean and variance

$$V_t\nu_t = V_t\left(\frac{x_{t-\delta_t} + \delta_t b}{\delta_t^2\sigma^2} + \frac{x_{t+\delta_{t+1}} - \delta_{t+1}b}{\delta_{t+1}^2\sigma^2} + \frac{y_t}{\tau^2}\right)$$

$$V_t^{-1} = \frac{1}{\delta_t^2\sigma^2} + \frac{1}{\delta_{t+1}^2\sigma^2} + \frac{1}{\tau^2}$$

Recognize this to be identical to the result I obtained before, with the exception that the time intervals and, thus, the variances differ.

The conditional posterior for the growth rate parameter

$$p(b|\mathbf{x},\mathbf{y},\sigma^2) \propto \prod_{t=1}^{T}N(x_t|x_{t-\delta_t} + \delta_t b,\delta_t^2\sigma^2)N(b|b_0,V_0)$$

$$= N(b|V\nu,V)$$

having mean and variance

$$V\nu = V\left(\frac{1}{\sigma^2}\sum_{t=2}^{T}(x_t - x_{t-\delta_t})/\delta_t + \frac{b_0}{V_0}\right)$$

$$V^{-1} = \frac{T}{\sigma^2} + \frac{1}{V_0}$$

The identifiability of parameters in time series models with observation error is the subject of a substantial literature that predates modern Bayes. Parameters are identifiable if different parameter values result in different observations. If different combinations of the same variables produce identical outputs, parameters are not identifiable. Signal processing in engineering applications has motivated substantial effort, with initial attention to linear models (Walter 1982) having expanded to diverse types of nonlinear models. For the state space models discussed here, the Bayesian approach provides opportunity to incorporate prior information on the range of variation expected for process (e.g., life history considerations) and observations (e.g., through repeated observation).

Estimates for growth of this population allow for how the uneven sample intervals affect process error (the exponential model is a crude approximation) and observation errors. The best estimate of growth rate is 8.0 percent per

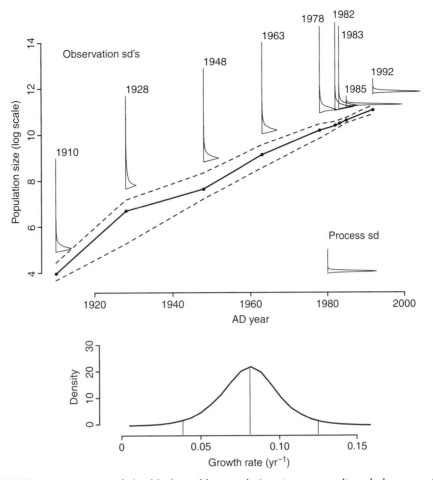

FIGURE 9.9. Estimates of the black noddy population (upper panel) and the rate of increase (posterior density in lower panel). Dashed lines show the 90 percent credible intervals for log population size x_t. *Above* are posterior densities for the observation errors, given as standard deviation τ and plotted sideways to have the same scale as the log densities. *Lower right* is the estimate of process error σ, also plotted to have the same log density scale. From Clark and Bjørnstad (2004).

year (Figure 9.9, lower panel). The dashed lines in Figure 9.9 are 90 percent credible intervals for population size x_t. Of course, these intervals are conditioned on the model: perhaps there is exponential growth, but it is only approximately correct, and the true population size is unknown. If I used a different model I would obtain different estimates of the population size.

The estimate of process error summarizes the extent to which the exponential model fails to describe growth. This mean estimate of process standard deviation (posterior mean = 0.049 on the lower right side of Figure 9.9) is more than half as large as the mean estimate of b (posterior mean = 0.081). So the model leaves much unexplained. Note that both b and σ have the same units: change in log density per unit time.

The posterior densities for observation errors are shown above the corresponding observations (dots) and posterior mean estimates of x_t (lines) in Figure 9.9a. As is the case for all estimates, the high observation error assigned to the 1928 observation is predicated on the assumption that the population is growing exponentially. As a consequence of this assumption, the observation y_{1928} lies near the edge of the credible interval for the 1928 estimate x_{1928}. The model could be importantly wrong if the population were indeed as large as the observation of 1928, in which case the observation error might be small. If we could know with some certainty that the 1928 observation is accurate, that insight could enter in the form of prior parameter values for τ^2_{1928}. This would, in turn, have the effect of increasing the estimate of the process error—this added confidence in the observation for 1928 means that the process model (simple exponential growth) is worse than we thought. If we know why densities are higher in 1928 than predicted by an exponential model, that information might enter as part of the process model. This would reduce the process error, as more is explained by the deterministic process itself. Thus, it is important to allow a structure for the unknowns and to incorporate what is known through priors.

9.7.3 Missing Values

Although I have represented estimates and credible intervals for x_t as lines in Figure 9.9, I have only estimated them at times for which there are observations. Estimates of population size should be more certain in years for which there are observations than for years in which there are no observations. For example, the year 1940 is not within a decade of any observation. In other words, the credible intervals in Figure 9.9 represent minimal uncertainty, in the sense that lines connect intervals for years when observations are available. How well can we estimate population size in, say, 1940? To answer this question, we can treat the years for which information is not available as missing values. Fortunately, the state space model readily accommodates missing values.

The principal modifications needed to accommodate missing values involve allowing for estimates of x_t in years for which there are no observations, with attendant modifications to several other distributions. I now consider the case where the vector **y** is shorter than vector **x**.

In the Gibbs sampler, I distinguish between times for which observations are available from those where observations are not. To sample x_t in years for which observations are available, I draw new values of x_t as previously (Equation 9.35) from the conditional normal with

$$V v_t = V \left[\frac{x_{t-1} + x_{t+1}}{\sigma^2} + \frac{y_t}{\tau^2} \right]$$

$$V^{-1} = \frac{2}{\sigma^2} + \frac{1}{\tau^2}$$

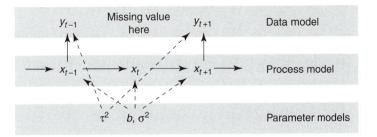

FIGURE 9.10. There is no observation in year t. This is a missing value.

To sample x_t for years in which no observations are available (there is no arrow connecting x_t to y_t (Figure 9.10) the conditional means and variance are

$$V v_t = V\left[\frac{x_{t-1} + x_{t+1}}{\sigma^2}\right] = \frac{x_{t-1} + x_{t+1}}{2}$$

$$V^{-1} = \frac{2}{\sigma^2}$$

So there are two ways of drawing estimates of x_t, depending on whether or not there is an observation available in year t.

Because it conditionally depends only on the x_i's, the sampling for parameter b is unchanged (Equation 9.37). The estimates of τ^2 are based on observations, so the parameters for the conditional posterior inverse gamma contain only the terms for which observations are available.

If there are large numbers of missing values, we can expect the credible intervals to expand as intervals between observations widen. To illustrate this effect, Figure 9.11 shows credible intervals for estimates of x_t in every year from the first to most recent observation. This example makes an important point concerning modeling decisions. The models based on unequal sample intervals and the missing values appear to be the same. Why don't the credible intervals for the x's look identical (Figures 9.9, 9.11)? In fact, these models are different. The one assumes an unknown x occurs every year, with associated uncertainty that absorbs some of the variability in the model. The marginal dependence across the full model means that uncertainty in the x's affects estimates of one another and of other parameters in the model. The alternative model assumes that x occurs only at irregular and infrequent intervals, with uncertainty associated only with those years. These relationships highlight the importance of the fact that models are not correct, and inference on the process depends on the full structure of the model, including details of how data are handled. In this sense, the notion of process error versus observation error is overly simplistic.

9.7.4 Assessment of the Model

Is the simple exponential model sufficient to describe growth of natural populations? This population increase may result not only from offspring produced by the resident population, but also from arrival of birds from nearby

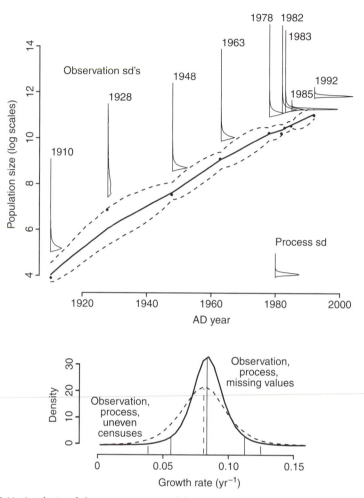

FIGURE 9.11. Analysis of the state space model assuming missing values for each year in which censuses are not available. In *b* is a comparison of posterior estimates for parameter *b* for two models of exponential growth. From Clark and Bjørnstad (2004).

islands. The exponential model leads us to infer that births are adding new individuals 10 percent faster than mortality is removing them. If, instead, population increase is partly supported by immigration, then the model is importantly wrong. Immigration rates may have little to do with size of resident population (Chapter 2), particularly when that population is small. The long-term implications of the exponential and immigration models are different.

Moreover, despite a good fit to the exponential growth model, observational evidence suggests that density dependence is occurring. Densities appear to be limited by nesting sites. At the last sample date included here many large trees supported dozens of nests, and tree mortality even seemed to result from heavy nesting pressure (Ogden 1993). This observational evidence suggests that the density-independent model does not describe the true dynamics of the population. The following section describes how to include density dependence in a state space model for population growth.

9.8 Nonlinear State Space Models

9.8.1 General Structure

To illustrate the state space formulation of nonlinear models, I use the logistic difference equation model considered from a classical standpoint in Section 9.5, but now including observation error. The approach here is not restricted to specific assumptions about how density dependence operates. The nonlinear model does not require any new techniques from the previous state space models.

For normally distributed errors and priors I wrote the joint posterior as

$$p(\mathbf{b},\sigma^2,\tau^2,\mathbf{x}|\mathbf{y}) = \overbrace{N(x_0|\mu_0,\sigma_0^2)}^{A}\overbrace{\prod_{i=1}^{T}N(x_t|f(x_{t-1}),\sigma^2)}^{B}$$

$$\overbrace{\prod_{i=1}^{T}N(y_t|x_t,\tau^2)}^{C}\ \overbrace{IG(\sigma^2|a_0,b_0)}^{D}\overbrace{IG(\tau^2|c_0,d_0)}^{E}$$

If $f(x_t)$ is nonlinear, we cannot solve the conditional posterior for x's. Here I demonstrate this and provide some ways to deal with it. Ignoring x_0 for the moment, the conditional distribution for the x's involves B and C. The conditional posterior for x_t is proportional to

$$p(x_t|x_{j\neq t},b,\sigma^2,\tau^2) \propto \overbrace{N(x_t|f(x_{t-1}),\sigma^2)}^{B_1}\overbrace{N(x_{t+1}|f(x_t),\sigma^2)}^{B_2}\overbrace{N(y_t|x_t,\tau^2)}^{C}$$

The product of B_1 and C has an exponent that is quadratic in x_t. This product is

$$N(x_t|f(x_{t-1}), \sigma^2)N(y_t|x_t,\tau^2) = N(x_t|Vv_t,V)$$

with mean and variance

$$Vv_t = V\left(\frac{f(x_{t-1})}{\sigma^2} + \frac{y_1}{\tau^2}\right) \qquad 9.40a$$

$$V^{-1} = \frac{1}{\sigma^2} + \frac{1}{\tau^2} \qquad 9.40b$$

For the logistic model, this mean is

$$Vv_t = V\left(\frac{x_{t-1} + b_0 + b_1 e^{x_{t-1}}}{\sigma^2} + \frac{y_t}{\tau^2}\right) \qquad 9.41$$

This is not yet the full conditional distribution, because I have not included distribution B_2. The full conditional distribution can now be written as the product of two normal distributions, one of which (B_2) has an exponent that is not quadratic in x_t,

$$p(x_t|x_{j \neq t},\mathbf{b},\sigma^2,\tau^2) = \overbrace{N(x_t|Vv_t,V)}^{B_1 \times C}\overbrace{N(x_{t+1}|f(x_t),\sigma^2)}^{B_2} \qquad 9.42$$

This target distribution requires indirect sampling. This can be accomplished by embedding within the Gibbs sampler either a rejection step or a Metropolis step.

Rejection Step—Because we can sample from the normal distribution that results from the product of distribution B_1 and C, we can use that to generate normally distributed random variates that will be centered near the true conditional posterior. Carlin et al. (1992) suggest rejection sampling (Section 7.2.1). Propose with density $N(x_t|Vv_t,V)$ (Equations 9.38b, 9.39) and accept with probability.

$$a = \exp\left[-\frac{1}{2\sigma^2}(x_{t+1} - f(x_t))^2\right]$$

Some modification is needed for the 0th and T^{th} values. For the 0th value

$$p(x_0|x_1,\mathbf{b},\sigma^2,\tau^2) = N(x_0|\mu_0,\sigma_0^2)N(x_1|f(x_0),\sigma^2)$$

the second density is nonlinear in x_0, so draw from the first, and accept with probability

$$a = \exp\left[-\frac{1}{2\sigma^2}(x_1 - f(x_0))^2\right]$$

For the Tth value,

$$p(x_T|x_{T-1},y_T,\mathbf{b},\sigma^2,\tau^2) = N(x_T|f(x_{T-1}), \sigma^2)N(y_T|x_T,\tau^2)$$

This can be sampled directly from the normal distribution having mean and variance

$$Vv_T = V\left(\frac{f(x_{T-1})}{\sigma^2} + \frac{y_T}{\tau^2}\right)$$

$$V^{-1} = \frac{1}{\sigma^2} + \frac{1}{\tau^2}$$

For the Tth value, no rejection step is needed.

Metropolis Step—The Metropolis step represents an alternative to the rejection step (Geweke and Tanizaki 2001). Here I could use the same distribution we used for the rejection step as the jump distribution (Section 7.3.2). A random draw from this distribution is the candidate value

$$x_t^* \sim N(x_t | V v_t, V)$$

The acceptance probability is

$$a = \min\left(1, \frac{p(x_t^* | x_{j \neq t}, y_t, \mathbf{b}, \sigma^2, \tau^2)}{p(x_t | x_{j \neq t}, y_t, \mathbf{b}, \sigma^2, \tau^2)}\right)$$

where the denominator evaluates the target distribution for the current value of x_t, and the numerator evaluates it for the proposed value.

Variances—The conditionals for variances are generated from

$$\sigma^2 | \mathbf{x}, \mathbf{y}, \mathbf{b}, x_0 \sim IG\left(a_0 + \frac{T}{2}, b_0 + \frac{1}{2}\sum_{t=1}^{T}(x_t - f(x_{t-1}))^2\right)$$

$$\tau^2 | \mathbf{x}, \mathbf{y} \sim IG\left(c_0 + \frac{T}{2}, d_0 + \frac{1}{2}\sum_{t=1}^{T}(y_t - x_t)^2\right)$$

where a_0, b_0, c_0, and d_0 are prior parameter values.

Regression Parameters—There are two regression parameters. Using conjugate normal priors, $N(b_0 | B_0, V_0)$ and $N(b_1 | B_1, V_1)$, I can sample directly for both. For the intercept parameter, the conditional posterior mean and variance

$$Vv = V\left(\frac{B_0}{V_0} + \frac{\sum_{t=1}^{n}(x_t - x_{t-1} - b_1 e^{x_{t-1}})}{\sigma^2}\right)$$

$$V^{-1} = \frac{1}{V_0} + \frac{n}{\sigma^2}$$

For the slope, the mean and variance are

$$Vv = V\left(\frac{B_1}{V_1} + \frac{\sum_{t=1}^{T} e^{x_{t-1}}(x_t - x_{t-1} - b_0)}{\sigma^2}\right)$$

$$V^{-1} = \frac{1}{V_1} + \frac{\sum_{t=1}^{T} e^{2x_{t-1}}}{\sigma^2}$$

The parameters B_0, V_0, B_1, V_1 are prior parameter values. The example of Section 9.8.3 applies this model to a population time series.

9.8.2 Model Selection with Time Series

The predictive loss approach to model selection used by Clark and Bjørnstad (2004) is demonstrated in Chapter 7 of the lab manual. From the posterior samples, predictions of y are generated and used to calculate means and variances. These are then used to calculate the goodness of fit and penalty terms of predictive loss.

9.8.3 Application: Density Dependence and Variability

The logistic model is a favorite of ecologists for many of the reasons that it appealed to Pearl (Section 2.1)—it is simple enough to seem general. I now return to the example from Chapter 2, which poses some typical challenges encountered with actual data sets. Census data for moose (*Alces alces*—note that the authors refer to these animals as elk) come from the Bialowieza primeval forest (BPF), which straddles the border of Poland and Belarus (Jedrzejewska et al. 1997). Estimates attempt to account for the errors that result from inaccuracies, including nonuniform methods that evolved over time and differed between the two jurisdictions. The population was locally eradicated after World War I and began to increase after 1945, when moose recolonized the park.

Unlike the U.S. census data (Figure 2.1), the BPF data look noisy (Figure 9.12a). Additional information indicates that the population has been influenced by many factors that changed over time. These include human harvest (e.g., the decline after 1993), warfare (negative correlation with density), temperature (positive correlation), and wolf density (negative correlation). Many of the wiggles in this population time series result from these influences. If we had precise information on all of the important variables, and we understood how they affect growth of moose populations, a model with many deterministic elements could be used to quantify these relationships.

Jedrzejewska et al. (1997) possessed some of the relevant data, yet they did not fit a complex model containing many deterministic dependencies. Even with many data it may be difficult to construct a model that integrates them in a realistic way. Does a positive correlation with temperature reflect survival during mild winters, or high productivity during the growing season? Does it affect all age classes, or perhaps only juveniles? How does it impact other herbivores with which moose compete, or natural enemies? What is the time scale of the temperature effect on total population size? Might temperatures affect the accuracy of census procedures, much of which is conducted by snowtracking? What estimate of temperature do we use? The annual mean? Minimum winter temperature?

Perhaps because the many influences would be difficult to describe, the correlations mentioned above were not identified through use of a population dynamic model that included them all. They were discovered by an alternative approach, which can be roughly described as an exploration of low-dimensional relationships. Simple plots of how growth rate relates to potentially important variables can lend insight. Such plots can help to validate a role for processes already known (e.g., the fact that wolves eat moose has been important at BPF). They may not provide much guidance for those we do not (e.g., the ways in which annual temperature impacts moose populations).

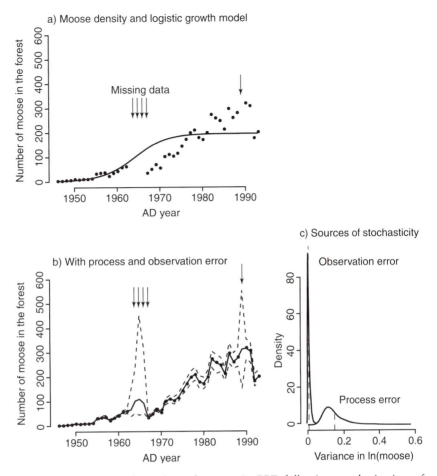

FIGURE 9.12. Estimates of number of moose in BPF following recolonization after the Second World War. (*a*) Estimates together with a noise-free logistic growth model. (*b*) A logistic model fitted with observation and process error. (*c*) Posterior estimates of process and observation error. From Clark and Bjørnstad (2004).

The simple logistic model used here includes in the deterministic process only population size. This relationship does not explain much (Figure 9.13). The authors recognized many of the important influences, but quantification can still be a challenge. Rather than attempt a detailed process model that includes all of these effects, they enter in terms of stochastic relationships. The estimates of the population densities are very close to those reported by the original authors—the solid line in Figure 9.12b represents the best estimates of density, and it is close to all observations. The large process error allows for the fact that population growth is far more irregular than depicted by the smooth curve in Figure 9.12a. The uncertainty in population densities is especially large for years of missing data. Due to the large process error, we have limited ability to predict densities in the years where data are absent. If we ignored this process error, a simple projection of the model would not be close to the data (solid line in Figure 9.12a). It is common to plot lag-one data and the fitted (deterministic) model (Figure 9.13). A cobweb diagram using these parameter values, as discussed in Appendix B, converges to the point of intersection at $K = 193$. This best-fitting model is shown as the solid line in Figure 2.5a. Scatter in data makes this interpretation uncertain; the deterministic model is only part of the story. Lacking data or the knowledge to express in realistic ways factors known to be important, we develop a structure that allows for their effects as stochasticity. The opportunity to admit known ranges of variation, by way of priors, should not be missed. Of course, the process component of this model for moose (Figure 9.12b) is the same one used by Pearl for humans (Figure 2.1). The model misspecification, or process error, allows for the fact that the process model is much too simple to capture all of the influences known to occur here. The change in density from one time step to the next is partly explained by the deterministic density effect. Stochasticity stands in for everything else.

Of course, just because we have left room for process error does not mean that the process equation is now correct. It is certainly not. The error distribution allows for unknowns, but it cannot do so precisely. For example, different error distributions that seem equally plausible will result in different estimates. Priors should be used to help with estimation of errors.

The observation error has a different structure than the process error, because it affects the data in a different way. We do not want to confuse the two, because the change in population density from one time step to the next should not depend on whether a census taker had a bad year. In fact, for five years we have no observations at all (Figure 9.12b). This is not part of the process. In addition to the process parameters b_0 and b_1, we have an additional variance parameter τ^2 describing the factors that cause observations to differ from the true process. This diagram makes it clear that observations have errors that do not affect the process itself. Additional discussions of these types of error are found in Pascual and Kareiva (1996), Hilborn and Mangel (1997), Calder et al. (2003), and Clark and Bjørnstad (2004).

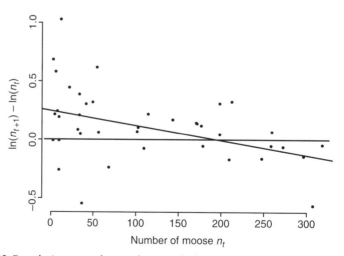

FIGURE 9.13. Population growth rate shows a slight tendency to decline with population size. Change in log density reflects change in per capita growth rate. From Clark and Bjørnstad (2004).

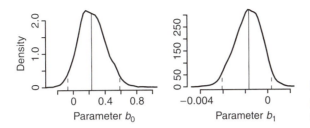

FIGURE 9.14. Regression parameter posteriors for the BPF moose analysis.

Although the model specifies observation and process errors as independent, it is always important to remember that the *estimates* of those errors are not independent. It would be a mistake to interpret the posterior estimates of observation error as being independent of the model used to estimate the process. Perhaps because authors already took pains to correct for observation errors, the remaining observation inaccuracies are estimated to be small. The density of observation error estimates is clustered near zero in Figure 9.13c. By contrast, the many variables known to be important that are omitted from the deterministic part of the model result in large estimates of the process error. Regression parameters do not provide strong indication of density dependence (Figure 9.14). Note that the slope parameter estimate is roughly zero. As with the black noddy data set, models require thoughtful treatment of both deterministic (process) and stochastic components. Stochasticity can help accommodate unknowns, but, like the process itself, model structure is critical.

9.8.4 Application: Stock Assessment

Fisheries stock assessment is conducted, in part, using population models and data on fish biomass f_t, catch c_t, and estimates of observation error (e.g., variance τ^2). Goals include estimates of actual population biomass b_t, population growth r, catch per unit effort q, carrying capacity K, and process error (variance σ^2). Millar and Meyer (2000) use the following process model for change in log biomass:

$$\log(b_t) = \log\left(b_{t-1} + rb_{t-1}\left(1 - \frac{b_{t-1}}{K} \right) - c_{t-1} \right) + \varepsilon_t$$

$$\log(b_1) = \log(K) + \varepsilon_t$$
$$\varepsilon_t \sim N(0,\sigma^2)$$

The observation equation is

$$f_t = qb_t e^{w_t}$$

$$w_t \sim N(0,\tau^2)$$

Note that the true biomass is taken be a latent variable, and catch is assumed to be precisely known.

Millar and Meyer (2000) found that estimates of K and q were highly correlated. This should be apparent if you consider that, if the population is at carrying capacity, then

$$f_t = qKe^{W_t}$$

The authors point out that prior information can be valuable. If external information places bounds on q, K, and observation errors, then more precise estimates of parameters and of the actual stock b_t can be obtained.

9.8.5 Uneven Sampling in Nonlinear Models

The treatment of unequal sample intervals for the black noddy example was easy, because the linear growth model had a solution. Nonlinear models often cannot be solved, particularly when process stochasticity is present. Some progress may be possible using stochastic differential equations. For example, Jones and Ackerson (1990) provide estimators for the continuous time ARMA model with random individual effects.

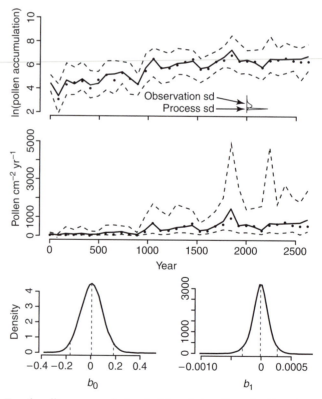

FIGURE 9.15. Fossil pollen accumulation in Nutt Lake, Ontario (dots in upper panel), and posteriors for regression parameters (lower panels). Dashed lines are 95 percent confidence intervals. From Clark and Bjørnstad (2004).

In many cases a discrete approximation may suffice, using process error to absorb the error of discretization (e.g., Wikle 2003a). Clark and Bjørnstad (2004) provide results for the discrete logistic model and applied this model to fossil pollen data, which typically have unevenly spaced samples and large observation errors (Figure 9.15). Application of a nonlinear model must take cognizance of the fact that nonlinearities can have different impacts on model behavior, depending on the time increment. A model that is well behaved for the short sample intervals may not be for the longer intervals.

9.9 Lags

Population growth models are written in terms of simple state variables that can be observed, albeit indirectly, such as density. Structured population models admit more detail (Chapter 2). Yet many of the factors that control population growth cannot be readily measured. Typically, unobservables are accommodated with stochastic elements of models. In some cases, unobservables are accommodated through lag terms (Turchin and Taylor 1992; Lindstrom et al. 1999). The current density can affect population growth, not only immediately, but also in the future (Nisbet and Gurney 1982; Lande et al. 2002). This can occur because current births will not begin to reproduce until they reach maturity. Competitors, resources, and natural enemies respond to current densities and those responses will influence population growth at some point in the future (Turchin 2003; Murdoch et al. 2003). There is a large literature on lagged models. Here I simply point out that such models are readily accommodated within the Bayesian state space framework. This treatment follows Clark and Bjørnstad (2004).

The lagged logistic model can be written as

$$x_t = f(x_{t-p}, \ldots, x_{t-1}) = x_{t-1} + b_0 + \sum_{j=1}^{p} b_j e^{x_{t-i}} + \varepsilon_t$$

$$y_t = x_t + w_t,$$

where p is the order of the model (number of lags). Analysis is similar to the simple logistic, with some modifications. The state x_t is conditionally dependent on $2p + 1$ terms, one for each of the states $(x_{t-p}, \ldots, x_t, \ldots, x_{t+p})$. The conditional posterior for x_t is thus

$$p(x_t | x_{t-p}, \ldots, x_{t+p}, \ldots) \propto \prod_{j=0}^{p} N(x_{t+j} | f(x_{t-p+j}, \ldots, x_{t-1+j})) N(y_t | x_t, \tau^2)$$

Most of these coefficients are nonlinear in x_t and, thus, cannot be directly sampled. Metropolis-Hastings is an option (previous section).

Remaining elements can be directly sampled. For regression parameters, sample from

$$\mathbf{b} | x, y, \sigma^2 \sim N(\mathbf{Vv}, \mathbf{V})$$

with mean and variance

$$\mathbf{Vv} = \mathbf{V}(\sigma^{-2}\mathbf{X}^\mathsf{T}\Delta + \mathbf{V}_b^{-1}\mathbf{b}_0)$$

$$\mathbf{V}^{-1} = \sigma^{-2}\mathbf{X}^\mathsf{T}\mathbf{X} + \mathbf{V}_b^{-1}.$$

There is a response vector

$$\Delta = \begin{bmatrix} x_1 - x_0 \\ \vdots \\ x_T - x_{T-1} \end{bmatrix}$$

a design matrix

$$\mathbf{X} = \begin{bmatrix} 1 & e^{x_{1-2}} & \cdots & e^{x_{1-k}} \\ 1 & e^{x_{2-2}} & \cdots & e^{x_{2-k}} \\ \vdots & \vdots & & \vdots \\ 1 & e^{x_{T-2}} & \cdots & e^{x_{T-k}} \end{bmatrix}$$

and parameter vector $\mathbf{b}^\mathsf{T} = [b_0, \ldots, b_p]$. The prior is $N(\mathbf{b}_0, \mathbf{V}_b)$.

The first p terms can be handled in one of two ways. We may specify priors for unobserved terms, indexed by time intervals $(-p, \ldots, 0)$. There could be a normal prior for each of $(x_{-p}, \ldots, x_0)$. Alternatively, we could simply condition on the first p values of the series and only draw inference on the observations obtained after time $t = p$. The last p values of the series need not condition on unobserved, future values of the series. Variances are sampled as discussed for the simple logistic.

Stenseth et al. (2003) modeled winter and summer densities of voles (*Clethrionomys rufocanus*) on Hokkaido using a nonlinear process model with autoregressive terms embedded within a Poisson sampling distribution. Fitted coefficients in the model describe density dependence separately for winter and summer, when diets shift to match seasonal resources. They had an unusually large data set, consisting of eighty-four time series of spring and winter censuses spanning thirty years. Priors were noninformative. Of particular interest were parameters describing density dependence. Also fitted were error terms. For comparison they fitted a model that ignores observation error and found that it overestimated the level of density dependence. They concluded that longer winters have the effect of shifting stable dynamics to long-period fluctuations.

9.10 Regime Change

Time series models can incorporate extreme nonlinearities through the specification of different regimes (Tong 1978, 1990). These models are characterized by step changes in dynamics, usually by allowing that parameters assume a different set of values depending on two or more regimes, sometimes represented by threshold values. For a threshold model, we might fit one parameter set when log density is less than x^* and a different set otherwise. Then x^* is a threshold that defines two regimes. There can be an autoregressive component that allows lag effects. These models can be termed *threshold autoregressive models*. The term *self-exiting autoregressive* (SETAR) model is sometimes used (Tong 1990).

Threshold models have been applied to several ecological data sets. Stenseth et al. (1998) use a time series approach to model the ten-year cycles in lynx returns from Canada. The model is a second-order nonlinear autoregressive process that includes a threshold, which allows the magnitude of density dependence to change, depending on the density present in the past. The autoregressive model of $y_t = \ln(\text{density}_t)$ is

$$y_t = \begin{cases} \beta_{1,0} + \beta_{1,1}y_{t-1} + \beta_{1,2}y_{t-2} + \varepsilon_{1,t} & y_{t-d} \leq x^* \\ \beta_{2,0} + \beta_{2,1}y_{t-1} + \beta_{2,2}y_{t-2} + \varepsilon_{2,t} & y_{t-d} > x^* \end{cases}$$

where the $\beta_{i,j}$ of are fitted parameters, with i applying to low ($i = 1$) and high ($i = 2$) density regimes, and j corresponding to the intercept ($j = 0$) and lag terms, and errors $\varepsilon_{i,t}$ being normally distributed as $N(0,\sigma_i^2)$. Lag $j = 1$ parameters ($\beta_{i,1}$) are direct density dependence, and lag $j = 2$ parameters ($\beta_{i,2}$) are delayed density dependence. Note that there is a fitted parameter set for both upper and lower regimes.

Classical treatment of these models is difficult, because the likelihood must accommodate the discontinuity. They are typically assumed to include no observation error or deterministic error, a known relationship between observations and the true density. Conditional on the states **x** and on the threshold x^* and order p, maximum likelihood estimates of the parameters **b** are available. Tong (1990) describes parameterization and model selection using AIC to discriminate among the conditional models. A Bayesian approach can simplify analysis. For example, Geweke and Terui (1993) derive the joint posterior for threshold and delay parameters (x^*,p) for the case of normally distributed errors on log density.

Models that contain thresholds can display complex behavior. An application to the Soay sheep populations in the St. Kilda archipelago indicated that population growth was influenced by environmental forces in different ways, depending on the regime (Grenfell et al. 1998). The two isolated populations (on different islands) were correlated to some degree in terms of their responses to environmental fluctuations, but external forces, represented by stochasticity, played an interesting role. When both populations are either below (positive population growth) or above (negative population growth rate) the fitted threshold, the same basic dynamics apply. Stochasticity means that populations can cross the thresholds at different times. Grenfell et al. (2000) point out that, if the threshold were permitted to be random, rather than fixed, this could have a large impact on model dynamics.

Regimes can be represented in other ways. For example, there may be a smooth transition between regimes or there may be unobserved variables that determine regime change. A summary is given by Stock and Watson (1995). Hidden Markov models have hidden states, transitions between them being unknown. These states determine change in the structural model. McCulloch and Tsay (1993) refer to random level shift and random variance shift models where mean and variance, respectively, can undergo stochastic change. They use a Gibbs sampling framework to demonstrate how the parameters affecting shifts can be estimated. Clark and LaDeau (2006) describe a partially hidden Markov model for the case where one of the states is hidden and a second state is partially observed. When an individual is observed to be in state 2, then state is known (of course!). If state is not observed, the individual could be in either of the two states. The transition between states can be estimated with a hierarchical Bayes model.

9.11 Constraints on Time Series Data

A temporal process may be constrained in ways that the data are not. Growth data are a typical example, where individual size is known to increase over time, yet observation errors result in apparent decreases in size. This is a common problem with tree census data (Biging and Wensel 1988; Gregoire et al. 1990), where apparently negative diameter growth increments can result from small differences in how and at what height diameter is measured, changing diameter with bole moisture content, and abrasion of bark. Investigators face the dilemma of either eliminating from data sets any measured diameters that are smaller than the previous one or adding some arbitrary constant to each observation (e.g., Clark and Clark 1999). Either practice results in biased estimates.

Clark et al. (in review) developed a state space approach that accommodates the fact that diameter growth is positive, whereas measurements on both diameter and diameter increment have error. The model for $D_{ij,t}$, the diameter of tree i in stand j at time t, is

$$N(\ln(D_{ij,t+1} - D_{ij,t})\|\ln\mu_{ij,t}, \sigma^2)$$

$$\ln\mu_{ij,t} = \beta_0 + \beta_{ij} + \beta_i$$

$$N(\ln X_{ij,t}^{(o)}\|\ln(D_{ij,t+1} - D_{ij,t}), v)$$

$$N(D_{ij,t}^{(o)}|D_{ij,t}, W)$$

$$\beta_{ij} \sim N(0, \tau^2)$$

The model for log increment $\mu_{ij,t}$ has population mean log growth rate β_0, random individual effect β_{ij}, and year effect β_t. They applied data models for both diameter measurements $D_{ij,t}^{(o)}$ and tree ring measurements $X_{ij,t}^{(o)}$. The constraint for positive growth is on the process, not on the data. Thus, the model fully accommodates negative increments in data while maintaining the assumption that the underlying process has positive increments.

9.12 Additional Sources of Variability

As with previous models we can allow for factors that affect population growth beyond the density of the population itself. In the Bayesian state space framework, this requires no new methods. These factors can be represented as covariates or as random effects. For instance, Engen et al. (1998) consider a model that could be viewed as an extension of the one described in Section 4.6.2. Suppose that generations do not overlap and individuals vary in terms of reproductive output. We could extend the Poisson model to include covariates and extra-Poisson variability,

$$n_t | n_{t-1} \sim \sum_{i=1}^{n_{t-1}} Pois(\lambda_i)$$

$$\ln \lambda_i = \mathbf{x}_i \beta + \varepsilon_{it}$$

with normally distributed ε_{it}. The design vector $\mathbf{x}_i$ can include density and environmental variables. A Gibbs sampler is readily constructed for this model.

Saether et al. (2000) modeled change in songbird populations using a process model

$\ln(n_t - m_t) = \ln n_{t-1} + \mathbf{x}_{t-1}\alpha + \varepsilon_t$ change in density due to nonimmigrants

$m_t \sim Pois(\lambda_t)$ number of immigrants . . .

$\ln\lambda_t = \mathbf{w}_{t-1}\beta + v_t$ . . . depends on covariates

$\varepsilon_t \sim N(0,\sigma^2)$

$V_t - N(0,\tau^2)$

The first equation describes change in the number of birds that come from the residents, previously numbering n_{t-1}. There is a vector of covariates $\mathbf{x}_{t-1} = [1 \; n_{t-1} \; c_{t-1}]^T$, where c_t is mean winter temperature. The three parameters in α are (1) density-independent growth rate, (2) density dependence (negative values), and (3) the climate effect. The number of immigrants m_t is Poisson distributed with the Poisson mean depending on winter temperatures, with $\mathbf{w}_{t-1} = [1 \; c_{t-1}]^T$ and β having an intercept and slope. Variance in the process error term σ^2 was treated as a sum, one term of which was assumed known (based on other data) and ascribed to demographic variability. Noninformative priors were used. There was no model for the data themselves, and authors believed that censuses accounted for greater than 90 percent of the population size.

Parameters were estimated using BUGs. The posterior mean estimate for DI growth rate (β_0) was near zero, and there was evidence for density dependence ($\beta_1 < 0$). They found that both reproduction by residents and immigration rates were higher in years following high mean winter temperatures—posterior means were $\alpha_3 = 0.15$ and $\beta_2 = 0.11$. Based on the model they generated predictive distributions of the population carrying capacity for different winter temperatures and related those results to Intergovernmental Panel on Climate Change (IPCC) climate scenarios.

Within the state space framework, it is also straightforward to admit variability in parameters over time. I discussed this with respect to the process variance in Section 9.7.1. Parameters also change among regimes in hidden Markov models (Section 9.10). Parameters for the process model or system equation can also change with each time step (West and Harrison 1997). Gross et al. (2005) provide an example where demographic rates vary over time.

9.13 Alternatives to the Gibbs Sampler

The Gibbs sampler may not be the method of choice where real-time prediction is needed, especially where arrival of the data stream outpaces convergence. There have been a number of techniques developed to assimilate data and predict change on-line. Sequential importance sampling is a technique for sequentially updating predictions described by Liu and Chen (1998).

9.14 More on Longitudinal Data Structures

Models introduced in Section 8.2.3 involved multiple observations on each of *n* sample units. Such observations can accumulate sequentially. In ecological studies, measurements may be taken repeatedly on the same organisms or at the same locations. Such models are termed *repeated measures* or *longitudinal* (Lindsey 1999; Diggle et al. 1996). Here I consider some of the time series aspects of longitudinal models in more detail.

In Section 9.2, I mentioned that the term *time series analysis* is typically used when one or, perhaps, a few sequences are involved. For this reason, time series analyses often involve long series. After all, if we have only one series, it had better be long enough to show something. The distinguishing feature of longitudinal studies is the consideration of multiple sample units that are followed through time. Repeated observations on the same sample unit often cannot be treated as independent (Gurevitch and Chester 1986). Most often, the individual units are assumed independent from one another, although this need not be the case, particularly with random group effects or with spatiotemporal phenomena. As in previous models, individual differences may be accommodated in the deterministic (process) part of the model (e.g., treatment effects or covariates), in the stochastic part (random effects, distance relationships), or in both. The series themselves are often short. They may come from a controlled experiment, but, particularly with Bayesian structures, we can accommodate many types of information. Series are often nonstationary. The flexibility of Bayes is valuable, because sampling distributions are often non-Gaussian (generalized linear mixed models or worse), data sets are often unbalanced, and covariates are partially known.

Time is an independent variable, although we may treat it implicitly, by way of changes in covariates over time. From the large literature on longitudinal data structures, I focus on two features of particular relevance for ecological data, autoregression and heterogeneity among sample units. I build on the basic structure for linear mixed models introduced in Chapter 8. As before, I do not dwell on classical approaches, which are available in many software packages. I focus on essentials of the Bayesian approach, as these concepts are readily implemented as part of more complex models.

9.14.1 Autoregression

Serial correlation in observations can be modeled with an autoregressive structure. Because studies are often short, it is common to use the AR(1) model of Section 9.3.5, but now in the context of multiple sequences, that is, one for each observational unit. Consider a linear model with normally distributed error. Let $y_{i,t}$ be the tth observation on individual or sample unit i. Here is an example containing an explanatory variable x that can change through time,

$$y_{i,t} = \mu_{i,t} + \varepsilon_{i,t}, \quad i = 1, \ldots, n, \ t = 1, \ldots, T_i \qquad 9.43$$

$$\mu_{i,t} = \mathbf{x}_{i,t}\,\boldsymbol{\alpha} = \alpha_0 + \alpha_1 x_{i,t}$$

where $\mathbf{x}_{it} = [1 \ x_{it}]$ and $\boldsymbol{\alpha} = [\alpha_0 \ \alpha_1]^{\text{T}}$. The error term contains serial correlation from one time step to the next with a parameter ρ for the lag-one correlation and a variance of σ^2,

$$\varepsilon_{i,t} \sim N(\rho\varepsilon_{i,t-1}, \sigma^2)$$

The mean response depends on t through the explanatory variable $x_{i,t}$. The error term depends on t through autocorrelation. The assumption that measurements are multivariate normal with AR(1) covariance structure means that the distribution for the first measurement is assumed to be

$$y_{i,1} \sim N\left(\mathbf{x}_{i,1}\boldsymbol{\alpha}, \frac{\sigma^2}{1-\rho^2}\right)$$

The mean of this distribution is the first row of the design matrix multiplied by the column vector $\boldsymbol{\alpha}$, that is, $\alpha_0 + \alpha_1 x_{i,1}$.

Now consider the full time series for one individual. Longitudinal models can have a covariance matrix for each sample unit. Let $\mathbf{y}_i$ be the vector of T_i observations on the ith individual. The history of measurements on this individual is $\mathbf{y}_i^T = [y_{i,1}, \ldots, y_{i,T_i}]$. We now have a covariance matrix from Equation 9.9 for the measurements on each individual i, having dimensions T_i y T_i. Recall that stationarity requires $|\rho| < 1$.

The linear model has design matrix with covariates $\mathbf{X}_i$ and parameters for fixed effects α. Because there is a sequence of measurements for individual i, there is a design matrix for each individual. For an intercept and a covariate x_i, which also changes with time, the design matrix for the model 9.43 can be written as

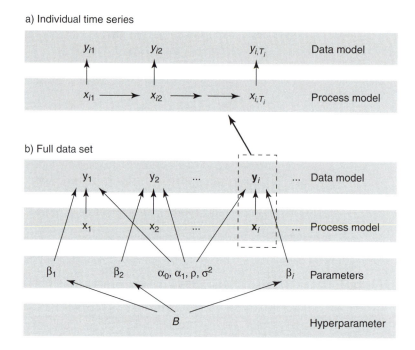

a) Individual time series

b) Full data set

FIGURE 9.16. A longitudinal model with fixed and random effects. Each individual $\mathbf{y}_i$ consists of a sequence of measurements that may be related to covariate (fixed) effects (*above*). Parameters are associated with the whole population (fixed effects) and with individuals (random effects). The random effects are drawn from a distribution (in this case, with covariance matrix $\mathbf{B}$), which represents an additional stage.

$$\mathbf{X}_i = \begin{bmatrix} 1 & x_{i,1} \\ \vdots & \vdots \\ 1 & x_{i,Ti} \end{bmatrix}$$

with covariates $\alpha = [\alpha_0, \alpha_1]^T$ that apply to the entire population.

I develop the model beginning with the distribution of one measurement on one individual, moving to the distribution of a sequence of measurements, and then to the distribution of a data set (multiple individuals). The distribution for the sequence of measurements on individual i can be written as

$$\mathbf{y}_i \sim N_{T_i}(\mathbf{X}_i \alpha, \Sigma_i)$$

The mean vector is

$$\mathbf{X}_i \alpha = \begin{bmatrix} \alpha_0 + \alpha_1 x_{i1} \\ \vdots \\ \alpha_0 + \alpha_1 x_{iT_i} \end{bmatrix}$$

and the covariance is given by 9.9. This model can be analyzed in classical and Bayesian frameworks. Lairde and Ware (1982) is a useful reference for the traditional treatment of such models. Standard texts include Lindsey (1999) and Diggle et al. (1996). To allow for unequal sample intervals, Jones and Ackerson (1990) recast this as a continuous time model, and they compare results with a continuous time Kalman filter. Due to its flexibility and the opportunity to build from the Gibbs sampling framework, I focus on the Bayesian approach.

A Bayesian model with standard distributions for regression parameters, variance, and serial autocorrelation extends Example 9.1, with the model

$$p(\alpha,\sigma^2,\rho|\mathbf{y},\mathbf{X},\ldots) \propto \prod_{i=1}^{n} N_{T_i}(\mathbf{y}_i|\mathbf{X}_i\alpha,\Sigma_i)N_2(\alpha|\mathbf{a},\mathbf{V}_\alpha)IG(\sigma^2|s_1,s_2)Unif(\rho|-1,1)$$

A sampling-based approach entails no new methods. The conditional posteriors follow.

1. The regression parameters have conditional posterior

$$p(\alpha|\mathbf{y},\mathbf{X},\sigma^2,\rho\ldots) \propto \prod_{i=1}^{n} N_{T_i}(\mathbf{y}_i|\mathbf{X}_i\alpha,\Sigma_i)N_2(\alpha|\mathbf{a},\mathbf{V}_\alpha)$$

that is normally distributed with mean and variance

$$\mathbf{Vv} = \mathbf{V}\left(\sum_{i=1}^{n}\mathbf{X}_i^T\Sigma_i^{-1}\mathbf{y}_i + \mathbf{V}_\alpha^{-1}\mathbf{a}\right)$$

$$\mathbf{V}^{-1} = \sum_{i=1}^{n}\mathbf{X}_i^T\Sigma_i^{-1}\mathbf{X}_i + \mathbf{V}_\alpha^{-1}$$

2. For the variance, sample from

$$p(\sigma^2|\mathbf{y},\mathbf{X},\alpha,\rho\ldots) \propto \prod_{i=1}^{n} N_{T_i}(\mathbf{y}_i|\mathbf{X}_i\alpha,\Sigma_i)IG(\sigma^2|s_1,s_2)$$

$$= IG\left(\sigma^2\bigg|s_1 + \frac{1}{2}\sum_{i=1}^{m}T_i,s_2 + \frac{1}{2}\sum_{i=1}^{m}(\mathbf{y}_i - \mathbf{X}_i\alpha)^T\mathbf{R}_i^{-1}(\mathbf{y}_i - \mathbf{X}_i\alpha)\right)$$

where $\mathbf{R}_i$ is given by Equation 9.9.
3. Sample the AR(1) parameter ρ indirectly, using, say, Metropolis-Hastings or inverse distribution sampling.

The only change from Example 9.1 is the likelihood, which is here taken over multiple sequences.

9.14.2 Additional Sources of Heterogeneity

In addition to variability associated with time, there may be variance among sample units. This variability has a different structure than $\varepsilon_{i,t}$, of the last model which describes variation associated with each individual and time increment. A model that allows for persistent differences among sample units is needed. Such differences might be evident if multiple measurements are made on an individual or group over time. I follow the general approach outlined by Lairde and Ware (1982) with a Bayesian implementation (Zeger and Karim 1991). This is a straightforward extension of the mixed model of Section 9.2.3.

Heterogeneity among individuals is implemented with a term for random effects. From Section 8.2.3, the basic model individual i at time t becomes

$$y_{i,t} = \mathbf{x}_{i,t}\alpha + \mathbf{w}_{i,t}\beta_i + \varepsilon_{i,t} \qquad 9.44$$

As before, the random effects $\mathbf{w}_{it}\beta_i$ might include as many variables as there are fixed effects. The random effects describe how individual responses may differ from the population mean responses contained in $\mathbf{X}_{i,t}$. The T_i measurements on individual y_i are distributed as

$$\mathbf{y}_i|\beta_i \sim N_{T_i}(\mathbf{X}_i\alpha + \mathbf{W}_i\beta_i, \Sigma_i)$$

There are two design matrixes and two parameter vectors. The mean response for the ith individual is the sum of two components, the first being deterministic (fixed effect) and the second being stochastic (random effect), in the sense that the different vectors for each individual, β_i, will not be assigned to specific causes. The individual contributions are modeled stochastically, because we do not know why their responses differ. If we could know the causes for this variability and we could measure them, we would include those additional explanatory variables as fixed effects, thus increasing the size of design matrix $\mathbf{X}_i$ (Section 8.2.3). Because the mean response is contained in the first term $\mathbf{X}_i\alpha$, the random effects are assumed to have mean zero. We are interested in the variance of these random effects as a summary of how much of the response is tied up in individual differences. We can include a random effect in the design matrix $\mathbf{W}_i$ corresponding to every fixed effect in design matrix $\mathbf{X}_i$, but we do not have to. We might decide that only a subset of the covariates requires random effects. The random effects are usually assumed to be normally distributed,

$$\beta_i \sim N(0, \mathbf{V}_b)$$

A full Bayesian analysis requires priors for α, $\mathbf{V}_b$, σ^2, and ρ. The conditional posterior for α (conditioned on $\mathbf{Y}$, β, σ^2, and $\mathbf{V}_b$) is now $N(\alpha|\mathbf{Vv}, \mathbf{V})$ with

$$\mathbf{V}^{-1} = \sum_{i=1}^{n}\mathbf{X}_i^T\Sigma_i^{-1}\mathbf{X}_i + \mathbf{V}_\alpha^{-1}$$

$$\mathbf{v} = \sum_{i=1}^{n}\mathbf{X}_i^T\Sigma_i^{-1}(\mathbf{Y}_i - \mathbf{W}_i\beta_i) + \mathbf{V}_\alpha^{-1}\mathbf{a}$$

For more efficient sampling integrate over the random effects

$$\mathbf{V}^{-1} = \sum_{i=1}^{n} \mathbf{X}_i^T (\Sigma_i + \mathbf{W}_i \mathbf{V}_b \mathbf{W}_i^T)^{-1} \mathbf{X}_i + \mathbf{V}_\alpha^{-1}$$

$$\mathbf{v} = \sum_{i=1}^{n} \mathbf{X}_i^T (\Sigma_i + \mathbf{W}_i \mathbf{V}_b \mathbf{W}_i^T)^{-1} \mathbf{Y}_i + \mathbf{V}_\alpha^{-1} \mathbf{a}$$

For random effects the conditional posterior has the form $N(\beta_i | \mathbf{V}_i \mathbf{v}_i, \mathbf{V}_i)$ with

$$\mathbf{V}_i^{-1} = \mathbf{W}_i^T \Sigma_i^{-1} \mathbf{W}_i + \mathbf{V}_b^{-1}$$

$$\mathbf{v}_i = \mathbf{W}_i^T \Sigma_i^{-1} (\mathbf{Y}_i - \mathbf{X}_i \alpha)$$

MCMC algorithms for random effects models can be slow to converge. When parameters are sampled one at a time, there is strong autocorrelation in the chain. Chib and Carlin (1999) describe block sampling of groups of parameters.

9.14.3 Application: Fecundity of Trees

Seed production of trees was modeled using an allometric function of diameter, AR(1) error distribution, and random effects to allow for differences among trees in how seed production changes with diameter (Clark et al. 2004). Because seeds cannot be counted on trees, Clark et al. (2004) embedded this fecundity model within a structure that allows for inference based on seed rain and observations of tree status (reproductive or not, Section 8.5). Here I consider just the longitudinal portion of that full model. An extension to multivariate responses is discussed in Section 9.18.4. For this example, $x_{i,t}$ is the log diameter of tree i in year t, $y_{i,t}$ is the corresponding log seed production, and β_i is the extent to which the diameter response of the ith tree departs from the population average. Random effects were limited to a single parameter (diameter effect), because this was sufficient to capture individual differences. This hierarchical Bayes approach for the longitudinal effects is similar to that of Lange et al. (1992). Then $\mathbf{W}_i$ is a vector of ones, and random effects are

$$\beta_i \sim N(0, \tau^2)$$

with an inverse gamma prior for the variance making up the lower stage. The MCMC algorithm for this model included direct sampling of conjugate distributions (regression parameters and variances) with a Metropolis-Hasting step for the autocorrelation.

Together with data models for seed traps and status observations (Chapter 10), Clark et al. (2004) estimated model parameters and seed production of every tree in the study area every year. Seed production was estimated to have low values in most years and periodic eruptions, with the degree of correlation among individuals being mostly positive but highly variable (Figure 9.17). This masting effect means that most of the variance is contained in the year-to-year fluctuations, that is, large σ. Diameter effect ($\alpha_1 x$) and individual

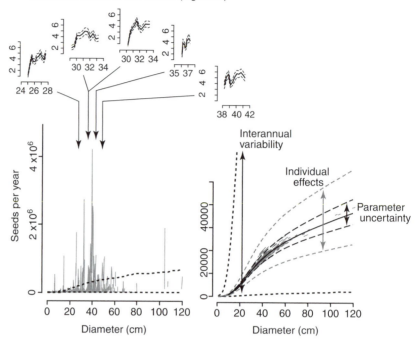

Individual estimates with 95%CI (log scale)

FIGURE 9.17. Estimates of seed production by *Liriodendron* trees from the southern Appalachians and Piedmont of North Carolina. Individual trees are represented by sequences from three to eleven years in length (*lower left*). Estimates show high fecundity in mast years with intervening low values. Selected trees, plotted on a log scale, show example year-to-year estimates with 95 percent confidence intervals (*upper left*). Variance is partitioned at *lower right*, showing contributions from parameter uncertainty, random individual effects (β_i with total variance τ^2), and temporal variability (σ^2). The model includes a maturation schedule, which accounts for the accelerating part of the curve (Section 8.5). From Clark et al. (2004).

differences (τ) were of similar magnitude (Figure 9.18). Lag-one autocorrelations within series varied among species, some positive, some negative. Thus, although tree size is the factor typically used to predict fecundity, the analysis that allows for process error, autocorrelation, and random effects, interannual variability can be large (Figure 9.18) and, for many species, overshadows the tree size effect. These interannual effects are represented by mast years with low production in intervening years. For most species, size effects and random individual effects accounted for much less of the variability.

9.14.4 Application: Harbor Seals in Prince William Sound

Ver Hoef and Frost's (2003) analysis of harbor seal abundance in Prince William Sound demonstrates an alternative treatment of time. They analyze a model for changes in abundance at twenty-five sites over ten years, based on

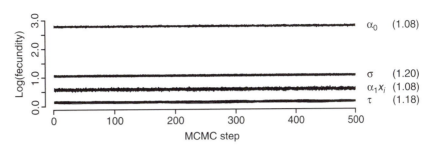

FIGURE 9.18. Twenty postconvergence MCMC sequences for the red maple data set, with parameters for the mixed effects model placed on a common scale. These include the intercept term α_0, the slope term $\alpha_1 x_i$, where diameter x_i is used for the mean tree diameter (28.6), random effects τ, and process error σ. At right are Gelman and Rubins' potential scale reduction factors (Section 7.5). From Clark et al. (2004).

counts from aerial surveys. A number of covariates were included that are known to affect when seals are most likely to be observed. Basic elements of the model can be summarized as

$$z_{ik,t} \sim Pois(y_{ik,t})$$

$$\ln(y_{ik,t}) = \mathbf{x}_{ik,t}\beta_i + \theta_{i,t} + \varepsilon_{ki,t}$$

$$\theta_{i,t} \sim N(\tau_{0i} + \tau_{it}t, \delta^2)$$

$$\tau_{0i}, \tau_{1i} \sim N_2(\eta, \gamma^2)$$

for the kth observation from location i at sample date t. There is a length-p vector of covariates $\mathbf{x}_{ik,t}$ with parameter vector β_i, with random site effects and error $\varepsilon_{ki,t}$ (distributions not shown) and random effects in the trend over time $\theta_{i,t}$

This approach accommodated the many factors that affect data (through covariates, treated as fixed effects), together with changes over time (random site effects in trend). The posterior site-specific trends for a standardized set of covariate values are shown as $e^{\theta_{it}}$ in Figure 9.19. Most are decreasing, but there is substantial heterogeneity. The overall decreasing trend dominates across sites (Figure 9.20).

9.15 Intervention and Treatment Effects

Time series data sets can be observational (monitoring) or experimental (with controlled treatments). Identification of factors responsible for dynamics depends on natural variability or on interventions. In the absence of substantial natural variability, even long time series data can be frustratingly uninformative.

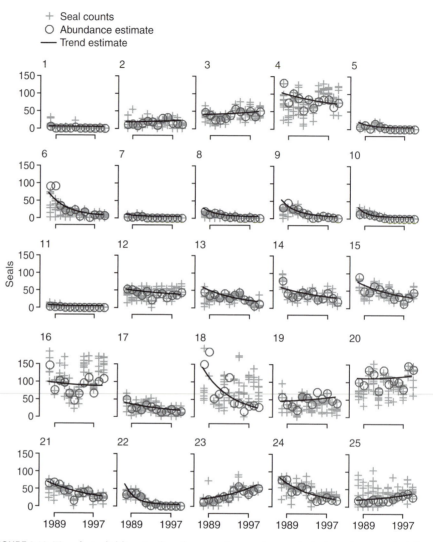

FIGURE 9.19. Trends (solid lines), abundance estimates (open circles), and counts (+) for twenty-five sites in Prince William Sound. From Ver Hoef and Frost (2003).

Some of the most basic relationships, such as density dependence, can be difficult to extract from time series data (e.g., Figure 9.13). A series of density estimates that remain near some carrying capacity may not be sufficient to identify the factors that regulate growth (Murdoch 1994)—effects of a vari-able can only be estimated if it changes over the course of the study.

The conflicting demands for sufficient duration, replication, and controlled or known changes in the important variables are less daunting for the Bayesian approach, because uncertainties in latent variables can be readily accommodated. This allows for a fuller exploitation of data sets that derive from uncontrolled or partially controlled variables. Many effects that cannot

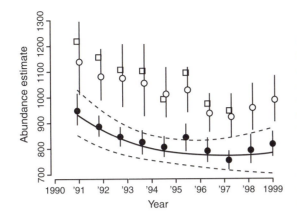

FIGURE 9.20. Trends for the full harbor seal population from Figure 9.20. Solid symbols show the estimates with covariates **x** set equal to zero, thus isolating the trend in time. Open symbols are for a different set of covariate values and show a comparison of the same model (open circles) with a classical Poisson regression model (open square). From Ver Hoef and Frost (2003).

be implemented as controlled experiments can be analyzed using observational data, provided we allow that they are not precisely known (Carpenter et al. 1994). In this section I focus on interventions, but the basic approach follows from longitudinal models discussed previously.

9.15.1 Design Considerations

Intervention involves manipulations of variables and observations of response. The design is longitudinal if observations are made repeatedly on the same units. The design is balanced, if all units are observed at the same times. It is most powerful when there are pretreatment data and there is replication. In the simplest design, samples obtained before and after intervention can be used to test whether the difference between treatments is significant (e.g., Stewart-Oaten et al. 1986). It will often be the case that there is more than one potential design for the treatment and for the responses. Some generic ones are summarized in Figure 9.21. Treatments might involve step changes, pulses, or a change in trend. Many of the responses to treatments might look more like combinations of the examples in Figure 9.21, and it may be difficult to anticipate how the response might play out over time. Yet each of the different patterns in Figure 9.21 implies a different design matrix.

There are at least three ways in which we can treat time in such an analysis; they are not mutually exclusive:

1. Time as a covariate—either time since an intervention is implemented on the ith individual, t_i^*, or specific times can be covariates. Time since an intervention is written here as $t_i - t_i^*$. If the treatment is imposed simultaneously on all sample units, I simply write $t - t^*$. Here we are typically examining the hypothesis that a treatment has an effect or not, intrepreted as a trend. Time might be a covariate for the analysis of any of the examples in Figure 9.21.

2. Implicit time—if the response variable $y_{i,t}$ is modeled in terms of a variable that affects it, $x_{i,t}$, then $y_{i,t}$ can depend implicitly on time.

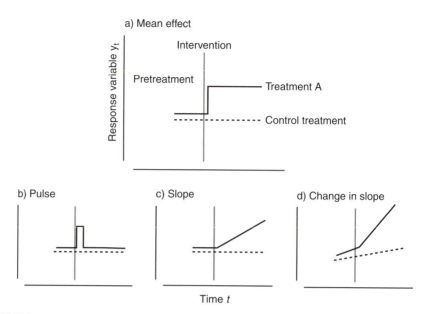

FIGURE 9.21. Four examples of ways in which an intervention might occur. The trend in the solid line "Treatment A" might not match that of the response to the treatment. For example, a step change in a treatment x (a) might elicit a trajectory of change in a response variable y.

Here we would monitor changes in $x_{i,t}$ or impose a treatment sequence $\mathbf{x}_i$ and model the response. Even if administered as a treatment, we may need to allow for variability and uncertainty in $\mathbf{x}_i$.

3. Time as autocorrelation—time may enter the stochastic term. In this case, we allow for influences that carry over from year to year (Section 9.4.5). This might be in addition to a model where covariate effects also change with time.

I have already considered autocorrelation in longitudinal studies. Those considerations apply to intervention analyses. Here I provide examples of the first two approaches.

9.15.2 Time as a Covariate

One of the most common approaches to intervention analysis entails comparisons of a treatment and control before and after an intervention. This usually takes the form of a hypothesis test. There is an effect, or not. Despite apparent simplicity, there are many design considerations. Is the intervention acute (a pulse) or chronic (persistent)? Is the response immediate, lagged, short-term, or persistent? If persistent, what is the expected form of the response? We might expect the response variable will immediately shift to a new level. Or we might expect the treatment to change the trend in the response variable.

There might be a perturbation followed by return to the baseline (Figure 9.21). Here I consider several designs for an intervention analysis imposed at time t^*.

For a pulse treatment, the effects associated with a short-term treatment might dissipate over time (Figure 9.21b). For $T_i = 3$ observations on individual i, where the treatment applies to the second observation, the design matrix might be constructed as

$$\mathbf{W}_i = \begin{bmatrix} 1 & 0 \\ 1 & 1 \\ 1 & 0 \end{bmatrix} \begin{array}{l} \text{pretreatment} \\ \text{treatment} \\ \text{post-treatment} \end{array}$$

Alternatively, we might assume that a new mean level applies after the treatment, for example, if the treatment is persistent (Figure 9.21a). Then

$$\mathbf{W}_i = \begin{bmatrix} 1 & 0 \\ 1 & 1 \\ 1 & 1 \end{bmatrix} \begin{array}{l} \text{pretreatment} \\ \text{treatment} \\ \text{post-treatment} \end{array}$$

Whereas in the first case, the treatment effect is limited to a single sample date, in the second case, it persisted across post-treatment observations.

For a third alternative, we might look for a treatment effect that sets a new trajectory in time (Figure 9.21c). Suppose there are two pretreatment observations followed by administration of a treatment to a subset of the subjects at time t_i^*. In this case we might have $s_{it} = \max(0, t_i - t_i^*)$ such that observations prior to intervention are all handled as pretreatment. For time t, we have $\mathbf{w}_{it} = [1\ s_{it}]$, with the full matrix for individual i being

$$\mathbf{W}_i = \begin{bmatrix} 1 & s_{i1} = 0 \\ 1 & s_{i2} = 0 \\ 1 & s_{i3} = 1 \\ 1 & s_{i4} = 2 \end{bmatrix}$$

For the full model I assume that treatment effects manifest as population-level differences, with random individual effects allowing for heterogeneity. Treatment and control groups may differ (for reasons that may go unrecognized), both before and after treatment. By allowing for these potential differences, we increase confidence in the inference on treatment effects. I introduce an indicator variable δ_i, which is equal to zero for a control treatment and one for a Treatment A group. The design matrix for fixed effects includes some or all of the random effects. I construct the design matrix with columns for the intercept and the slope, followed by columns for the treatment effects on those coefficients,

$$\mathbf{X}_i = \begin{bmatrix} \mathbf{W}_i & \delta_i \mathbf{W}_i \end{bmatrix} \qquad\qquad 9.45$$

$$= \begin{bmatrix} 1 & 0 & 0 & 0 \\ 1 & 0 & 0 & 0 \\ 1 & s_{i3} & 0 & 0 \\ 1 & s_{i4} & 0 & 0 \end{bmatrix} \text{(control)}$$

and

$$\mathbf{X}_i = \begin{bmatrix} 1 & 0 & 1 & 0 \\ 1 & 0 & 1 & 0 \\ 1 & s_{i3} & 1 & s_{i3} \\ 1 & s_{i4} & 1 & s_{i4} \end{bmatrix} \text{(Treatment A)}$$

(Carlin and Louis 2000). Regression parameters are $\alpha = [\alpha_0, \alpha_1, \alpha_2, \alpha_3]^T$ and $\beta_i = [\beta_{i0}, \beta_{i1}]^T$.

How do we interpret the coefficients? It may help to consider the four elements of this design. For the control group prior to treatment the expected response is

$$E[y_{it} | \beta_t, \delta_i = 0, t < t^*] = \alpha_0 + \beta_{0i}$$

For the treatment group, the expected response is

$$E[y_{it} | \beta_t, \delta_i = 1, t < t^*] = \alpha_0 + \beta_{0i} + \alpha_2$$

The two fixed effects describe pretreatment differences between treatment and control populations. If α_2 does not differ from zero, then the two intervention groups are similar. They need not be similar in order to draw inference on treatment effects, but we might still need them in the model to obtain proper estimates of treatment effects.

Following treatment, the control group is

$$E[y_{it} | \beta_i, \delta_i = 0, t > t^*] = \alpha_0 + \beta_{0i} + (\alpha_1 + \beta_{1i})(t - t^*)$$

and the treatment group is

$$E[y_{it} | \beta_i, \delta_i = 1, t > t^*] = \alpha_0 + \beta_{0i} + \alpha_2 + (\alpha_1 + \alpha_3 + \beta_{1i})(t - t^*)$$

The population-level response to the treatment is given by the slope parameters, with the treatment effect on slope being α_3. If α_1 is different from zero, there is a trend following t^*, regardless of treatment. The additional effect of treatment is α_3. α_1 need not be zero in order to draw inference on the treatment effect. Again, we may still need to include it in the model.

In the foregoing model, I make no allowance for a temporal trend before t^*. If there is a trend in data before treatment, I could ask whether or not the treatment changes the trend (Figure 9.21d). Instead, I might examine the treatment time as a changepoint in a preexisting trend that may change at t^*. The design matrix could then have a third covariate, $w_{it} = [1, t, \max(0, t - t^*)]$. Now there is a slope term and a change in slope term. Lange et al. (1992) used this model to determine how lymphocyte counts of AIDS-infected patients changed following drug treatment.

9.15.3 Implicit Time

An advantage of the Bayesian approach is the capacity to allow for variability and uncertainty in covariates. Whereas the need for complete knowledge of covariates restricts classical approaches to highly controlled experiments, a Bayesian treatment of observational or experimental data might not be too different. Within the Bayesian framework, we specify the model for observations y_{it} conditional on x_{it} and then include the model for x_{it} at a different stage.

Explicit modeling of covariates in the context of intervention can be viewed as a way of extending the range of natural variability. If there is a causal relationship between x and y, but natural variability is limited at the times of observation, intervention may help quantify this relationship. Moreover, interventions often allow only partial control of variables, and quantification of the relationship between x and y may be more informative and more realistic than would be a test of how y responds to t. For example, interventions involving exclusion of predators (Dial and Roughgarden 1995) or prey (Sabo and Power 2002) change the density of the intended populations, but they often do not fully exclude them. Yet there may still be information on those densities before and after intervention that could be introduced in the model. A test of treatment effect (exclusion) might not provide the same insight as an explicit model involving relationships between predators and prey. Here is an example where an intervention involves explicit modeling of time dependence in x and y.

9.15.4 Application: Fecundity Responses to Elevated CO$_2$

LaDeau and Clark (2006; Clark and LaDeau 2006) extended the fecundity model of Section 9.15.2 to an experimental setting. The full design for this model included assimilation of multiple data types, but I focus here on that portion of the design associated with intervention. The experiment involved collecting data on cone production from trees prior to and then following the intervention, which entailed fumigation with CO$_2$ elevated to a concentration of 565 ppm.

The model builds on the random effects model

$$y_{i,t} = \mathbf{x}_{i,t}\alpha + \mathbf{w}_{i,t}\beta_i + \varepsilon_{i,t}$$

For the CO$_2$ experiment, there is a single random effect on diameter and fixed year effects, κ_t. Year effects are taken as fixed, because this analysis is part of a large number of studies involving climate variables associated with particular years. For our purposes, it does not make sense to treat years as exchangeable. This model is

$$y_{i,t} = \mathbf{x}_{i,t}\alpha + \beta_i + \kappa_t + \varepsilon_{i,t}$$

$$\varepsilon_{i,t} \sim N(0,\sigma^2)$$

$$\beta_i \sim N(0,\tau^2)$$

The design matrix for the ith tree includes columns for intercept, treatment (elevated/ambient), log diameter (D_{it}), treatment history (CO_2 level), and ice damage (some trees damaged in 2003). For a treatment tree this matrix is

$$\mathbf{X}_i = \begin{bmatrix} 1 & 1 & D_{i,1996} & 0 & 0 \\ 1 & 1 & D_{i,1997} & 0 & 0 \\ 1 & 1 & D_{i,1998} & 0.192 & 0 \\ \vdots & \vdots & \vdots & \vdots & \vdots \\ 1 & 1 & D_{i,2003} & 0.192 & 1 \end{bmatrix} \text{ice-damage year}$$

For a control tree,

$$\mathbf{X}_i = \begin{bmatrix} 1 & 0 & D_{i,1996} & 0 & 0 \\ 1 & 0 & D_{i,1997} & 0 & 0 \\ 1 & 0 & D_{i,1998} & 0 & 0 \\ \vdots & \vdots & \vdots & \vdots & \vdots \\ 1 & 0 & D_{i,2003} & 0 & 1 \end{bmatrix} \text{ice-damage year}$$

Of course, the year effects could be subsumed in the design matrix. This would be accomplished by including a column for each year, with zeros and ones included only in the year corresponding to an observation y_{it}. I write year effects separately here. The CO_2 variable is given as $C_{it} = \log(c_{it}/365)$, where $c_{it} = 365$ for ambient conditions and $c_{it} = 565$ for the elevated treatment. With this scaling, the CO_2 treatment is either zero (ambient) or 0.192 (elevated). We used this scaling as opposed to 0,1, because ecologists are interested in the proportionate increase in yield for a proportionate change in CO_2 concentration.

Note that column 2 identifies a tree as belonging to a particular treatment. We allowed for the fact that the assignment of ambient and elevated treatments may not have resulted in samples with identical fecundity potential. There can be inherent differences in individuals assigned to the two treatments. We may not care what these differences are, but we still want to allow for them to ensure better estimates of treatment effects. In the notation of Equation 9.43, this design matrix as

$$\mathbf{X}_i = [\mathbf{W}_i \; \delta_i\mathbf{W}_i \; \mathbf{D}_i \; \mathbf{C}_i \; \mathbf{I}_i]$$

where $\mathbf{W}_i = \beta_i$, and the last three vectors include only fixed effects, being diameter, CO_2, and ice damage.

Here is the interpretation. The expected response from a tree that will receive the ambient treatment is

$$E[y_{it}|\beta_i] = (\alpha_0 + \beta_i) + \alpha_2 D_{it}$$

The pretreatment $(t \leq t^*)$ model for those that will receive the elevated treatment is

$$E[y_{it}|\beta_i, t \leq t^*] = (\alpha_0 + \alpha_1 + \beta_i) + \alpha_2 D_{it}$$

whereas the post-treatment response is

$$E[y_{it}|\beta_i, t > t^*] = (\alpha_0 + \alpha_1 + \beta_i) + \alpha_2 D_{it} + \alpha_3 C_{it}$$

The mean intercept for treatment trees differs from that of ambient trees by an amount α_1. This is not a treatment effect. The treatment effect is α_3.

The full model for this experiment included different types of data and is fully described by LaDeau and Clark (2006). I simply point out that this portion of the model is linear and easily analyzed with a Gibbs sampler. Here is the full model for the regression part of this model:

$$p(\alpha,\beta,\kappa,\sigma^2,\tau^2|\mathbf{X},\mathbf{Y},\dots) \propto \prod_{i=1}^{n} N_{S_i}(Y_i|\mathbf{X}_i\alpha + \mathbf{1}\beta_i + \kappa_i,\sigma^2\mathbf{I}_{S_i})$$

$$\times\ N_5(\alpha|\mathbf{a}_\alpha,\mathbf{V}_\alpha) \prod_{t=1}^{T} N(\kappa_t|0,v_\kappa) \prod_{i=1}^{n} N(\beta_i|0,\tau^2)$$

$$\times\ IG(\sigma^2|a_\sigma,b_\sigma)IG(\tau^2|a_\tau,b_\tau),$$

where S_i is the number of years for which there are observations on individual i and κ_i is the vector of coefficients for those years. S_i and κ_i differ among individuals, because trees became reproductively mature in different years, and some died. Reproductive maturity is a partially hidden state and was estimated as part of the model.

The Gibbs sampler takes some time to converge, in part, due to further complexity in the model not described here. Figure 9.22 includes only a short MCMC sequence that follows convergence. Parameter estimates show that fecundity increases with diameter (positive α_2) and CO_2 (positive α_3), and it decreases with ice damage (negative α_4) (Figure 9.22). They further show that trees selected for inclusion in the elevated CO_2 treatments tend to have inherently lower fecundity than those in ambient treatments (negative α_1). Allowing for this inherent difference provides more confidence in the estimate of the CO_2 effect using parameter α_3. Note that random effects are substantial ($\tau \gg \sigma$). Year effects (κ) follow the stand level trends in log seed production, a phenomenon known as masting.

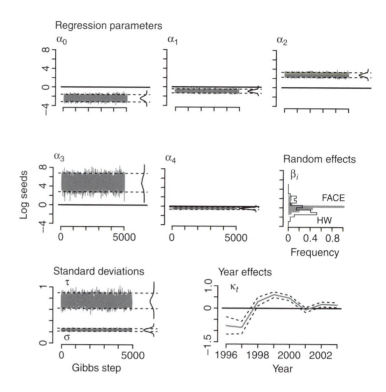

FIGURE 9.22. Examples of MCMC output for fixed effects parameters α and variance parameters σ and τ, together with a histogram of posterior means for random effects β_i, and mean and 95 percent credible intervals for year effects κ_t. There are trees from two data sets included in this analysis, and random effects for each are shown separately (FACE and HW) (see Clark and LaDeau 2006).

9.16 Capture-Recapture Studies

Longitudinal studies often document changes in variables that are best represented by discrete classes (Section 2.2.4). These might include binomial transitions, such as survival, Poisson transitions, such as births, or multinomial state transitions in structured populations (Chapter 2). They might involve changes in location, defined by discrete patches, jurisdictional units, and so forth. These are longitudinal data structures, in the sense that they consist of repeated measures on many individuals, and individual sequences are typically short.

Regardless of approach, analysis of capture-recapture data highlights two considerations that appear in many contexts, but play an especially prominent role here. These are missing data and stratification. *Missing data* refers to the fact that study individuals can be observed (recaptured) or not. The data model consists of a probability of recapture. For example, the Bayesian missing data representation includes both a vector of true states, which can be unknown, together with a vector for capture histories. *Stratification* refers to the use of different parameter values for different stages, groups, times, and so forth. Due to the many combinations of ways to stratify, challenging model selection issues emerge; model averaging can be relevant.

In Chapter 2, I introduced the stage-transition model as a Markov process, where transitions, summarized by the transition matrix **S**, depend only on the current state. Each column of the transition matrix represents the multinomial probabilities of transition to any other state. Observations are equally spaced in time. I considered the first-order Markov process, whereby transitions conditionally depend only on the current state, and not on previous states. Capture-recapture models are used to estimate state transitions, including survival, stage transitions, and movement.

The literature on classical capture-recapture applications is huge, including some valuable overviews (e.g., Williams et al. 2001; White and Burnham 1999). The early models were used to estimate population size and survival (Jolly 1965; Seber 1965; Lebreton et al. 1992). The approach has been extended to include changes in location and temporary emigration (Hestbeck et al. 1991; Schwartz et al. 1993; Kendall and Nichols 2002) and stage transitions (Arnason 1972; Nichols et al. 1992; Brownie et al. 1993; Caswell 2001; Fujiwara and Caswell 2002). There have been several classical models for transition rates that incorporate random effects (e.g., Burnham and White 2002). Estimating population size from capture-recapture data involves many of the same issues, including heterogeneity in recapture rates (Burnham and Overton 1978). Dorazio and Royle (2003) provide a recent review of the challenges, and analyze mixture models that accommodate variability in recapture rates. Link (2003) argues that detection variability among individuals may thwart efforts to estimate population size.

I provide most detail on Bayesian approaches, again motivated by their flexibility. The missing value approach that I summarize here follows from state space models of Section 9.6, with states now being categorical, and the data model is a Bernoulli likelihood of being observed or not. The missing value structure for capture-recapture data extends back to at least George and Robert (1992). Not all Bayesian approaches adopt the missing data representation. Royle and Link (2002) described a random effects model for recapture and survival probabilities within a more traditional framework. Sauer and Link (2002) illustrate a hierarchical application for a nonstructured model. Bayesian approaches can accommodate covariates (Catchpole et al. 2000), autocorrelation (Johnson and Hoeting 2003), and random effects in survival (Barry et al. 2003). My treatment leans most heavily on Dupuis (1995), who extended the missing value approach to populations structured by state. King and Brooks (2002) showed how reversible jump MCMC could be used for averaging models for stage-structured populations. Clark et al. (2005) included random effects in stage-structured models, applied predictive loss as a model selection tool, and showed how alternative data models are readily accommodated by the missing value structure.

9.16.1 Missing Data Structure

As an introduction, I consider the problem of estimating survival from recapture data. A population is sampled on $T = 5$ occasions, equally spaced in time. At each sample date, individuals are captured, marked (if not already marked in a previous census), and released. The state of individual i at time t is $z_{it} = 1$ (alive) or $z_{it} = 0$ (dead). An individual that dies between the fourth and fifth censuses has a history of states,

$$\mathbf{z}_i = [z_{i1}, \ldots, z_{iT}] = [1, 1, 1, 1, 0]^T$$

Note that a one can become a zero (death), but not vice versa.

Classical approaches to the analysis of marked individuals can be conveniently done with computer programs such as MARK (White and Burnham 1999). Because these methods are widely available, I immediately depart from the traditional approach by introducing the missing data structure (Dupuis 1995; King and Brooks 2002; Clark et al. 2005).

Because recapture of live individuals is not certain, I distinguish the recapture history of an individual from its history of states $\mathbf{z}_i$ (Figure 9.23). Individual i has recapture history $x_i = [x_{i1}, \ldots x_{iT}]^T$, where x_{it} indicates whether or not the individual was observed at sample date t. Suppose the recapture history for this individual is

$$x_i = [1, 0, 1, 0, 0]^T$$

where entries for each census indicate whether the individual was captured ($x_{it} = 1$) or not ($x_{it} = 0$). This recapture history indicates that individual i was marked at the first sample date and only captured again at the third sample date. I know this individual to be alive at date $t = 2$, but I do not know whether it was alive at date $t = 4$ or 5. Here are three state vectors $\mathbf{z}_i$ that are consistent with this recapture history:

$$\mathbf{z}_i = [1,1,1,1,1]$$

$$\mathbf{z}_i = [1,1,1,1,0]$$

$$\mathbf{z}_i = [1,1,1,0,0]$$

(Figure 9.24 shows one of these possibilities.) To infer which of these potential histories is most likely, we must pool information from a number of individuals

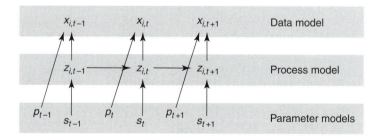

FIGURE 9.23. The missing data structure for capture-recapture analysis. The latent processes z_i are not observed, but must be estimated through the recapture histories x_i. Some of the latent states z_{it} are known (when any succeeding observation shows that individual to be alive), whereas others are not.

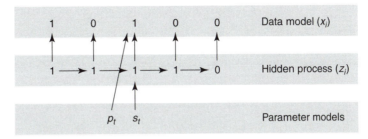

FIGURE 9.24. An example history of states z_i with a corresponding recapture history.

in the form of parameter estimates. These parameter estimates, derived from multiple recaptures and individuals, represent survival and recapture rates that are most consistent with the full data set. These parameter estimates, in turn, can inform posterior probabilities for each of the state histories z_i. Before pursuing the missing value approach further, consider maximum likelihood estimates for capture-recapture sequences. This interlude will provide some motivation for considering the missing value framework.

Maximum likelihood estimates for probability of survival s_t from date t to date $t + 1$ and recapture probability p_t come from the likelihood for the ith recapture history. There is a sampling distribution for the joint probability of zeros and ones that comprise each individual's recapture history. The probability that a live individual will be recaptured is

$$p(x_{it} = 1 | z_{i,t} = 1) = p_t$$

The probability of observing at date t an individual that was alive at date $t - 1$ is the joint probability of surviving and being captured,

$$p(x_{it} = 1 | z_{i,t-1} = 1) = s_{t-1}p_t$$

If an individual is not recaptured at date t but is captured at some date in the future, it must have been alive at date t. For the example at hand, the individual must have been alive at date 2, despite not having been observed at date 2, because it was alive at date 3.

If there are no subsequent recaptures, the true state is unknown. The classical approach focuses on the probability of observations x_{it}, as these are the basis for the likelihood. One step ahead of the last capture, we have the probability of no recapture,

$$
\begin{aligned}
p(x_{i,t+1} = 0 | x_{i,t} = 1) &= p(x_{i,t+1} = 0 | z_{i,t} = 1) \\
&= p(z_{i,t+1} = 1)p(x_{i,t+1} = 0 | z_{i,t+1} = 1) + p(z_{i,t+1} = 0)p(x_{i,t+1} = 0 | z_{i,t+1} = 0) \\
&= s_t \times (1 - p_t) + (1 - s_t) \times 1
\end{aligned}
$$

The terms represent the total probability that the individual survived, but was not recaptured, and the individual died. Two years since the last capture, we have a more complex probability. The enumeration of all possible recapture histories makes for complicated likelihoods. Here is one for the example $x_i = [1,0,1,0,0]^T$:

$$p(\mathbf{x}_t = [1,0,1,0,0]) =$$

$$\overbrace{s_1(1 - p_2)}^{2} \overbrace{s_2 p_3}^{3} \overbrace{s_3(1 - p_4)}^{4} \overbrace{s_4(1 - p_5)}^{5} \qquad \rightarrow p(\mathbf{z}_i = [1,1,1,1,1])$$

$$+ \ \overbrace{s_1(1 - p_2)}^{2} \overbrace{s_2 p_3}^{3} \overbrace{s_3(1 - p_4)}^{4} \overbrace{(1 - s_4)}^{5} \qquad \rightarrow p(\mathbf{z}_i = [1,1,1,1,0])$$

$$+ \ \overbrace{s_1(1 - p_2)}^{2} \overbrace{s_2 p_3}^{3} \overbrace{(1 - s_3)}^{4} \qquad \rightarrow p(\mathbf{z}_i = [1,1,1,0,0])$$

The numbers above coefficients indicate the sample date. At the right, I indicate the history of states to which each term belongs. Clearly, for long-term data sets, the complexity of the likelihood becomes substantial. And there is a different likelihood for each recapture history. An example considered below represents a subset of recapture data from $> 10^4$ animals over seventeen years.

Software programs, such as MARK, can analyze these data sets from a classical approach (White and Burnham 1999). If data or assumptions do not conform to options available in existing software, requiring new algorithm development, this classical approach can be daunting.

A missing value structure simplifies in the sense that we can write a likelihood for the latent states $z_{i,t}$ and observations $x_{i,t}$, based on local rather than global relationships. This was the same basic approach adopted with the state space model. Here again we can separate the process part of the problem from the data part of the problem (Figure 9.23). In this case, modeling locally means focus on the transitions that link z_{it} to x_{it} and to $z_{i,t-1}$. A sampling-based analysis can marginalize over the many possibilities. This approach allows estimation of posterior probabilities for the states themselves together with parameters.

With this local approach, construct the likelihood associated with a given state conditioned on the previous state and on the event that $x_{it} = 0$. The state transitions are

$$p(z_{i,t} = 0|z_{i,t-1} = 1)=1 - s_t$$

and

$$p(z_{i,t} = 1|z_{i,t-1} = 1) = s_t$$

Then proceed to the joint likelihood of x_{it}'s and z_{it}'s, where the z_{it}'s are either observations that individuals are alive ($x_{it} = 1$) or simulated states (if $x_{it} = 0$, the true state could be zero or one). Assuming that animals may be first marked at different sample dates, let the sample dates for individual i extend from $t = t_i, \ldots, T$, where t_i is date of first capture and T is the termination date of the study. The full likelihood can be written as

$$p(\mathbf{x},\mathbf{z}_{(-)}|s,\mathbf{p},\mathbf{x}_0) = \overbrace{\prod_{i=1}^{n}\prod_{t=t_i+1}^{T} p(x_{i,t}|z_{i,t},p_t)}^{A} \overbrace{\prod_{i=1}^{n}\prod_{t=t_i}^{T-1} p(z_{i,t+1}|z_{i,t},s_t)}^{B}$$

where n is number of individuals, $\{z_{i,t_i}\} = \mathbf{x}_0 = [x_{t_1},\ldots,x_{t_n}]$ is the vector of initial states at the first capture, and $\mathbf{z}_{(-)}$ represents states at all sample dates exclusive of the first. Hereafter I simplify notation and let $\mathbf{z} = \mathbf{z}_{(-)}$. The first component, A, represents observations conditioned on true states, which depend on recapture probability p_t. The second component, B, represents the process itself, which is survival. Both components might be binomial probabilities. Component A is

$$\prod_{i=1}^{n}\prod_{t=t_i+1}^{T} p(x_{it}|z_{it},p_t) = \prod_{i=1}^{n}\prod_{t=t_i+1}^{T} p_t^{u_{i,t}}(1-p_t)^{v_{i,t}}$$

where

$$u_{i,t} = I(x_{i,t} = 1)$$

indicates the event that an individual is observed and

$$v_{i,t} = I(x_{i,t} = 0, z_{i,t} = 1) = 1 - u_{i,t}$$

indicates not having observed an individual. (Recall that $I(\ )$ is the indicator function and assumes a value of one when its argument is true, and zero otherwise.)

The second component, B, is the partially observed survival process,

$$\prod_{i=1}^{n}\prod_{t=t_i}^{T-1} p(z_{i,t+1}|z_{i,t},s_t) = \prod_{i=1}^{n}\prod_{t=t_i}^{T-1} s_t^{w_{i,1,t}}(1-s_t)^{w_{i,0,t}}$$

where

$$w_{i,1,t} = I(z_{i,t} = 1, z_{i,t+1} = 1)$$

indicates survival and

$$w_{i,0,t} = I(z_{i,t} = 1, z_{i,t+1} = 0) = 1 - w_{i,1,t}$$

indicates death.

Substituting these distributions in the likelihood and including priors for survival and recapture results in the full model

$$p(\mathbf{x},\mathbf{z}|s,\mathbf{p}) \propto \prod_{i=1}^{n}\prod_{t=t_i+1}^{T} p_t^{u_{i,t}}(1-p_t)^{v_{i,t}} \prod_{i=1}^{n}\prod_{t=t_i}^{T-1} s_t^{w_{i,1,t}}(1-s_t)^{w_{i,0,t}}$$

$$\times \prod_{t=2}^{T} Beta(p_t|\alpha_{p_1},\alpha_{p_2}) \prod_{t=1}^{T-1} Beta(s_t|\alpha_{s_1},\alpha_{s_2})$$

the beta priors being conjugate for the binomial likelihoods. To simplify notation, I have assumed identical prior parameter values for all sample intervals, but there could be reasons to adopt different priors for each sample interval. This would be valuable if capture efforts or information on survival risks varied during the study (Clark et al. 2005). Note that there may be different numbers of animals contributing to the estimates of survival and recapture for each sample date.

9.16.2 A Gibbs Sampler

The latent states must be simulated. By definition, the state of an individual is $z_{it} = 1$ if the current or any later sample shows it to be alive. Thus, there is no need to simulate these states. For the remaining states, the probability of being in state 0 or 1 is conditioned on the preceding state. At the last capture, individual i is in state 1. At the next date, the probability of being alive is

$$p(z_{i,t} = 1 | z_{i,t-1} = 1, x_{i,t} = 0, s_t, p_t) = \frac{s_t(1 - p_t)}{s_t(1 - p_t) + (1 - s_t)} \qquad 9.46$$

and the probability of being dead is

$$p(z_{i,t} = 0 | z_{i,t-1} = 1, x_{i,t} = 0, s_t, p_t) = \frac{1 - s_t}{s_t(1 - p_t) + (1 - s_t)}$$

For the Gibbs sampler, initialize a state vector for each individual in the data set. This vector will have elements $z_{it} = 1$ for all sample dates before the last capture, and $z_{it} = 0$ or 1 for the sample dates thereafter. For the capture history $\mathbf{x}_i = [1,0,1,0,0]^T$, the initialization could be $\mathbf{z}_i^{(0)} = [1,1,1,0,0]^T$ or $\mathbf{z}_i^{(0)} = [1,1,1,1,0]^T$ or $\mathbf{z}_i^{(0)} = [1,1,1,1,1]^T$. All of these are consistent with the recapture history.

The Gibbs sampler draws new values for the parameters and then updates the states $\mathbf{z}$. Conditional posteriors can be directly sampled. If the Gibbs sampler begins with recapture probability, sample from

$$p(p_t | z, \alpha_{p_1}, \alpha_{p_2}) = Beta\left(p_t | \alpha_{p_1} + \sum_{i \in \{t\}} u_{i,t}, \alpha_{p_2} + \sum_{i \in \{t\}} v_{i,t} \right)$$

where $\{t\}$ is the set of individuals for which z_{it} is imputed to be in state 1 at time t. Then, for survival, sample from

$$p(s_t | z, \alpha_{s_1}, \alpha_{s_2}) \sim Beta\left(s_t | \alpha_{s_1} + \sum_{i \in \{t\}} w_{i,1,t}, \alpha_{s_2} + \sum_{i \in \{t\}} w_{i,0,t} \right)$$

Now update the $\mathbf{z}$'s. Do this sequentially for each individual. Because dead individuals cannot become alive, all sample dates before the last one do not need updating. If I initialized individual i with the state vector $z_i^{(0)} = [1,1,1,0,0]^T$, I would begin updating at the fourth sample date. At the fourth date, I determine the probabilities of remaining alive from Equation 9.44 and draw a new state from the Bernoulli

$$p(z_{i,t}|z_{i,t-1} = 1, s_t, p_t) = Bernoulli(z_{i,t}|\theta_t)$$

where

$$\theta_t = \frac{s_t(1 - p_t)}{s_t(1 - p_t) + (1 - s_t)}$$

This is the updated state for this individual and sample date. If a zero is drawn, this individual is estimated to be dead for all remaining sample dates, and the algorithm proceeds to the next individual. If a one is drawn, the algorithm proceeds to the next sample date and makes a draw to update $z_{i,t+1}$. After updating all individuals, the algorithm proceeds to the next Gibbs step. If the chain of estimates from the Gibbs sampler for a state z_t contains, say, 80 percent zeros, then the most probable posterior estimate that the individual was alive at sample date t is 0.2.

9.16.3 Stratification

The foregoing example represents but one of many ways to stratify transitions. This model is based on the assumption that survival and recapture probabilities at one sample date are different from (and independent of) those for all other sample dates. This assumption would be used if there is reason to believe that rates change from one date to the next. Of course, reasonable parameter estimates require that there are enough individuals represented at each sample date. With the assumption of independent estimates, there is no way to pool information from other sample dates to inform parameters that apply only to a given sample date. This assumption (different and independent parameters at each sample date that apply to all individuals) is but one of many ways in which we might stratify the data. I consider differences among groups later. For now simply note that I might assume that transition probabilities remain constant over all or a subset of sample dates. For example, if survival rates are assumed to be constant, then the model becomes

$$p(\mathbf{x}, \mathbf{z}|s, \mathbf{p}) \propto \prod_{i=1}^{n} \prod_{t=t_i+1}^{T} p_t^{u_{it}} (1 - p_t)^{v_{it}} \prod_{i=1}^{n} \prod_{t=t_i}^{T-1} s^{w_{i,1,t}} (1 - s)^{w_{i,0,t}}$$

$$\times \prod_{t=2}^{T} Beta(p_t|\alpha_{p1}, \alpha_{p2}) \, Beta(s|\alpha_{s_1}, \alpha_{s_2})$$

with conditional posterior for survival being

$$p(s|z, \alpha_{s_1}, \alpha_{s_2}) \sim Beta\left(s \left| \alpha_{s1} + \sum_{i=1}^{n} \sum_{t=t_i}^{T-1} w_{i,1,t}, \alpha_{s_2} + \sum_{i=1}^{n} \sum_{t=t_i}^{T-1} w_{I,0,t} \right. \right)$$

Similar modifications allow for assumptions for subsets of sample dates.

The question of how many strata to apply is a model selection problem. In a classical setting, AIC can be used to compare models. A likelihood ratio test would typically not be used, because many of the stratifications under consideration will not be nested. In this Bayesian framework, we might use Bayesian P values (King and Brooks 2002). Because each potential stratification represents a different model, the large number of potential models means that it will not be possible to undertake all possible comparisons. This is particularly true when we take up stage-structured models.

King and Brooks (2002) present a model-averaging approach that is accomplished together with posterior simulation using a reversible jump MCMC simulation. The algorithm provides parameter posteriors that average over the potential stratifications, thus incorporating the uncertainty in stratification.

9.16.4 Covariates and Higher-Order Effects

We can extend this structure to include covariate effects. Here survival or recapture probabilities could depend on differences among individuals that are linked to covariates and might be measured. Survival probability can be a GLM structure (Section 8.2), with a logit link (Royle and Link 2002). Extrabinomial variation can include random effects (e.g., Barry et al. 2003) and autocorrelation (Johnson and Hoeting 2003). For a GLM of survival, we could specify

$$Bernoulli(s_{i,t})$$

where

$$logit(s_{i,t}) = \mathbf{x}_i \beta$$

and where $\mathbf{x}_i \beta$ are covariate effects. Within the missing value representation, we would include a prior for regression parameters β and σ^2. The AR(1) model has the error discussed in Section 9.4.5.

As with the state space model, transitions could also depend on individual state at some time in the past. These lagged effects could affect recapture probability (component A), survival probability (component B), or both. Johnson and Hoeting reanalyzed with AR(2) effects in survival for a data set on northern pintails (*Anas acuta*) that had previously been analyzed without lag terms. They found a substantial lag-2 influence on survival, which also appeared to influence the estimate of trends over time. Given that lags can often come from age or size structure, explicit modeling of such structure is an alternative (Section 9.17).

9.16.5 Random Effects

The foregoing approach assumes parameters are unrelated across strata. A model-averaging scheme integrates the uncertainty associated with different assumptions about stratification, but it does not change the fundamental assumption for a given model—that parameters are independently applied. Differences among individuals can be included as covariates, but there will often be large individual differences, even when the causes for those differences cannot be measured (Clark 2003).

Random effects can be incorporated by adding an additional stage for one or more parameters. Here I illustrate the approach using a data set consisting of seventeen years of recaptures of common terns (Clark et al. 2005). There appeared to be variability in recapture probability, but that variability was associated with individuals rather than sample date. We have individual-level information, because an animal is captured repeatedly. Traditional treatment of this problem would be to stratify the data and estimate an independent recapture probability for each individual. Figure 9.25 shows a random effects model that allows for individual differences, but assumes individuals are drawn at random from a population. Note that each recapture history (upper boxes) has an associated probability of recapture. The data model (likelihood for the x_i's) is

$$p(\mathbf{x}|\mathbf{z},\mathbf{p}) = \prod_{i=1}^{n} \prod_{t=t_i+1}^{T} (p_i)^{u_{i,t}} (1 - p_i)^{v_{i,t}}$$

Note that recapture probability is assumed to be constant over time, but to vary among individuals—p has an i subscript, rather than t. The conditional probability for a recapture parameter is

$$p(p_i|x,z,\alpha_{p1},\alpha_{p2}) \propto \prod_{t=t_i+1}^{T} (p_i)^{u_{i,t}} (1 - p_i)^{v_{i,t}} Beta(p_i|\alpha_{p1}, \alpha_{p2})$$

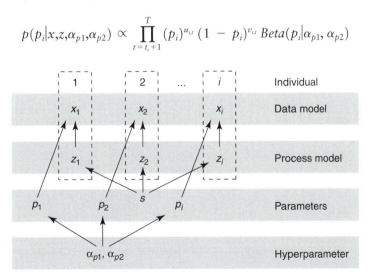

FIGURE 9.25. A random effects model for recapture probability. The hierarchical structure is achieved with the inclusion of an additional stage. Each of the recapture histories (labeled 1, 2, ... i) has an associated recapture probability p_i. Probabilities are drawn from a population distribution.

$$= Beta\left(\alpha_{p1} + \sum_{t=t_i+1}^{T} u_{i,t}, \alpha_{p2} + \sum_{t=t_i+1}^{T} v_{i,t}\right)$$

where $u_{it} = 1$, if the individual is recaptured at sample date t, and zero otherwise, and $v_{it} = 1$, if the individual is not recaptured at sample date t, and zero otherwise. α_{p1} and α_{p2} are gamma distributed hyperpriors. For example, the first parameter would be sampled from

$$p(\alpha_{p1}|p,\alpha_{p2}) \propto \prod_{i=1}^{n} Beta(p_i|\alpha_{p1},\alpha_{p2})Gam(\alpha_{p1}|a_1,a_2)$$

This would be sampled indirectly, using, say, Metropolis-Hastings (Clark et al. 2005).

Analysis shows posterior probabilities of being alive or dead (z_{it}) and individual recapture p_i, together with observed states (Figure 9.26). The posteriors

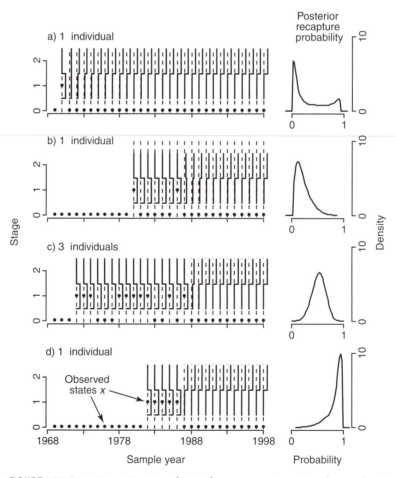

FIGURE 9.26. Posterior estimates of state for common terns (unobserved = 0, alive = 1, dead = 2). Right are posterior estimates of recapture probability for each recapture history. From Clark et al. (2005).

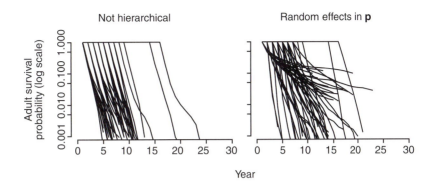

FIGURE 9.27. Posterior probability of being in the live class with time since first capture. Without random effects (*left*), transition to the dead class proceeds at the same rate for all individuals. The random effects model allows that the transition depends on recapture probability. From Clark et al. (2005).

for recapture range from low (Figure 9.26a) to high (Figure 9.26d). These posteriors reflect the borrowing of information across the full data set. For example, an individual that is never recaptured does not have a zero probability of recapture (Figure 9.26a). And an individual that is recaptured repeatedly may still be missed. In Figure 9.26d, the model assigns nonzero probability to the possibility that the individual is still alive the year after it was last recaptured.

The allowance for random effects in recapture probability has a large impact on estimates of survival. The probability of being in the alive state is plotted over years since first capture in Figure 9.27 for models with and without random effects. The long life estimated for the random effects model comes from its allowance for individuals having substantial probability of still being alive years after last being recaptured. Without random effects, all individuals are assumed to have the population mortality rate, and move to the dead stage at the same pace since the last recapture.

9.17 Structured Models as Matrixes

Structure is introduced into models in a number of ways (Section 2.7). For models framed in continuous time and continuous state, partial differential equations are the most common way of introducing structure. For discrete time and continuous state, we might use integral projection (Section 2.13) or integro-difference equations. I defer discussion of (discretized) partial differential equations and integro-difference equations until Chapter 10, where I consider migration with continuous space. For discrete classes, systems of ordinary differential equations can be used, but, in the context of data, we often discretize time.

Here I consider examples of discrete-time models for state variables that are structured in terms of discrete categories, such as age, size, gender, or developmental stage (Section 2.7). These include matrix models (this section) followed by systems of difference equations (Section 2.17). The former is simply a multinomial extension of the binomial model for two states, alive and dead (previous

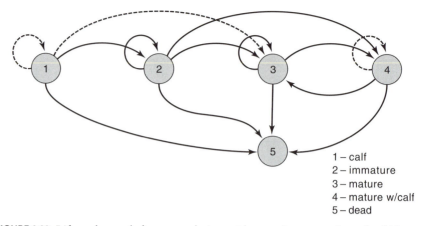

1 – calf
2 – immature
3 – mature
4 – mature w/calf
5 – dead

FIGURE 9.28. Life cycle graph for a population with $m = 5$ stages, where the fifth stage is "dead." This graph expands on Fugiwara and Caswell's (2002) right whale example: the three dashed lines have been added to make simulation experiments more challenging. This graph does not yet include parameter or data models.

section). In this section I combine the demographic methods for matrixes of Chapter 2 and Appendix E with the missing value representation for capture-recapure data of Section 9.16. As pointed out there, a number of studies have addressed this problem using classical capture-recapture models, and there are several Bayesian studies. I focus on the Bayesian approach.

To allow for more states than alive and dead I incorporate additional stages and transition probabilities. Again, states might be associated with a host of different attributes, such as size, life history, or location. For example, consider the transition matrix for five states, where the last state is dead. Clark et al. (2005) considered the right whale examples of Fugiwara and Caswell (2002), but arbitrarily added some additional transitions to make the model more challenging (Figure 9.28). Although biologically unreasonable, the increased complexity helps demonstrate the approach using simulated data. I follow with several real data sets.

I begin with a transition matrix (Section 2.2.4) that can be decomposed as $\mathbf{A} = \mathbf{F} + \mathbf{S}$, where $\mathbf{F}$ and $\mathbf{S}$ are matrixes consisting of fecundity and growth/survival matrixes, respectively (Section E.3.1). The matrix $\mathbf{S}$ for Figure 9.28 looks like this:

$$\mathbf{S} = \begin{pmatrix} s_1 g_{11} & 0 & 0 & 0 \\ s_1 g_{21} & s_2 g_{22} & 0 & 0 \\ s_1 g_{21} & s_2 g_{32} & s_3 g_{33} & s_4 g_{34} \\ 0 & s_2 g_{42} & s_3 g_{43} & s_4 g_{44} \end{pmatrix}$$

For simplicity, I assumed that transitions do not depend on time, although time-dependent transitions could be readily included. Let S_k represent the stages that can be reached from stage k. For example $S_2 = [2, 3, 4]$. The transition probabilities among states must sum to one,

$$\sum_{j \epsilon S_k} g_{jk} = 1$$

This is the sum of g's in the kth column.

I proceed as in Section 9.19, with the allowance that observed states $\mathbf{x}$ and hidden states $\mathbf{y}$ can have values other than zero and one, there are transitions g, and there can be survival and recapture probabilities associated with each stage k. For example, the true states for individual i might be $\mathbf{z}_i = [2, 2, 3, 5, 5, 5]$, meaning that the individual was first captured in state 2, between the second and third sample dates it made the transition to state 3, and it died (state 5) by sample date 4. An observation vector for this individual might be $\mathbf{x}_t 5 [2, 0, 3, 0, 0, 0]$. This individual was missed at the second sample date. Based on the life cycle graph (Figure 9.28), I know that it could have been in state 2 or state 3 at sample date 2. I know this, because only these states can both be accessed from state 2 and are accessible by state 3. In other words, these two possibilities for sample dates 1 to 3 are $2 \rightarrow 2 \rightarrow 3$ or $2 \rightarrow 3 \rightarrow 3$.

For the stage-structured model, I introduce a corresponding vector for the unknown states $\mathbf{y}_i = [0, 2, 0, 5, 5, 5]$ (Dupuis 1995). These unknown states, together with $\mathbf{x}_i$, fully determine the vector of true states $\mathbf{z}_i$. I now define the indicators in terms of the multiple states k,

$u_{ik,t} = I(x_{i,t} = k)$ (observed in state k)
$v_{ik,t} = I(y_{i,t} = k)$ (in state k, but unobserved)
$w_{ijk,t} = I(z_{i,t} = k, z_{i,t+1} = j)$ (made the transition from j to k)

Furthermore, I indicate survival in any state as $w_{i(-m)k, t}$ (any state other dead state m) and death as $w_{imk,t}$.

A Bayesian model where transitions do not depend on sample date but do depend on state is

$$p(\mathbf{x}, \mathbf{z}_{(-)}|s, \mathbf{g}, \mathbf{p}, \mathbf{x}_0 \; priors) \propto \prod_{i=1}^{n} \prod_{t=t_i+1}^{T} p_k^{u_{ik,t}} (1-p_k)^{v_{ik,t}}$$

$$\times \prod_{i=1}^{n} \prod_{t=t_i}^{T-1} (s_k)^{w_{i(-m)k,t}} (1 - s_k)^{w_{imk,t}} g_{jk}^{w_{ijk,t}}$$

$$\times \prod_{k=1}^{m-1} Beta(p_k|\alpha_{p_1}, \alpha_{p_2}) \; Beta(s_k|\alpha_{s_1}, \alpha_{s_2}) \; Dir(\mathbf{g}_k|\alpha_k)$$

The Dirichlet prior for the transitions consists of a parameter vector $\boldsymbol{\alpha}_k$ having the same length as the vector $\mathbf{g}_k$.

Analysis of this model is described in Clark et al. (2005). The Gibbs sampler is structured as for the survival model, with additional sampling required for parameters in each stage k,

$$p(p_k|\mathbf{x}, \mathbf{y}, \alpha_{p_1}, \alpha_{p_2}) = Beta\left(\alpha_{p1} + \sum_{i=1}^{n} \sum_{t=t_i+1}^{T} u_{ik,t}, \alpha_{p2} + \sum_{i=1}^{n} \sum_{t=t_i+1}^{T} v_{ik,t}\right)$$

$$p(s_k|\mathbf{x},\mathbf{y},\alpha_{s_1},\alpha_{s_2}) = Beta\left(\alpha_{s_1} + \sum_{i=1}^{n}\sum_{t=t_i}^{T-1} w_{i(-m)_{k,t}}\alpha_{s_2} + \sum_{i=1}^{n}\sum_{t=t_i}^{T-1} w_{imk,t}\right)$$

The vector of growth transitions $\mathbf{g}_k$ comes from the conditional Dirichlet posterior,

$$p(\mathbf{g}_k|x,y,\alpha_k) = Dir\left(\alpha_{1k} + \sum_{i=1}^{n}\sum_{t=1}^{T-1} w_{i1k,t}, \ldots, \alpha_{(m-1)_k} + \sum_{i=1}^{n}\sum_{t=1}^{T-1} w_{i(m-1)k,t}\right).$$

Sampling from the Dirichlet is described in Box 7.1.

Sampling of unknown states is multinomial, rather than Bernoulli,

$$p(z_{i,t}|z_{i,t-1},z_{i,t+1},s,g,p)=Multinom(z_{i,t}|1,\theta_{i1,t},\ldots,\theta_{i(m-1),t})$$

(Box 7.1), where elements of the parameter vector are

$$\theta_{ij,t} = p(z_{i,t} = j|z_{i,t-1} = k, z_{i,t+1} = l, s, g, p) = \frac{\overbrace{s_k\,g_{jk}}^{B_1}\ \overbrace{(1 - p_j)}^{A}\ \overbrace{s_j\,g_{lj}}^{B_2}}{\displaystyle\sum_j s_k g_{jk}\,(1 - p_j) s_j\, g_{lj}}$$

The factors identified in the numerator are, respectively, B_1, the probability of making the transition from k to j, A, the probability of being unobserved in state j, and B_2, the probability of making the transition from state j to state l.

For the example in Figure 9.29, Clark et al. (2005) simulated data sets and fitted models with and without random effects among groups in growth transitions $\mathbf{g}_k$. The correct model in this case has ten different groups with the true parameter values indicated as vertical dashed lines. Parameter estimates are shown as posterior densities, together with group-level estimates for the $\mathbf{g}_k$. Predictive loss (Section 7.6) identifies the correct model (that with random effects in $\mathbf{g}_{,k}$ but not in $\mathbf{s}$ or $\mathbf{p}$). The incorrect, nonhierarchical model provides an inaccurate estimate of p_4, and fails to capture the variability among g_{jk}.

An example applied to soft-furred rats (*Praomys delectorum*) from East Africa includes two stages, subadults and adults (Figure 9.30). Transitions include g_{11} (remain subadult) and $g_{21} = 1 - g_{11}$ (subadult to adult). The analysis involved six sites, each of which might have different transition and survival rates. It was not possible to stratify by site (i.e., assume different parameters for each site), because at some sites there were not enough transitions to estimate all parameters. A random effects model allows us to borrow strength among sites, an alternative to the assumption that sites are independent. An analysis of these data allowing for random site effects in survival, recapture probability, and growth transitions indicated large variability in both $\mathbf{s}$ and $\mathbf{p}$ that is masked in the nonstratified analysis (dashed lines). There were important differences in transition rates, with the sites from Ngangao showing especially low adult survival. Hyperparameters describe this among-site heterogeneity (horizontal dashed lines are posteriors for each of six sites). The narrow densities for the

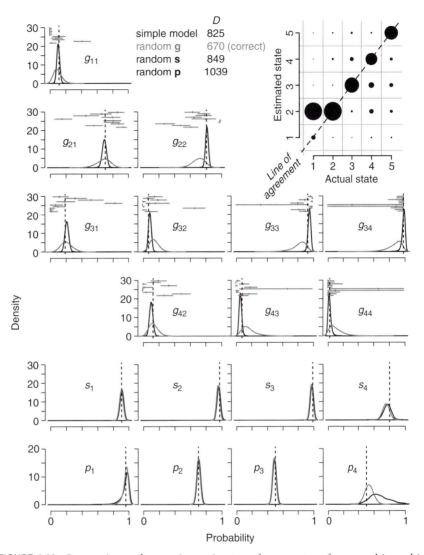

FIGURE 9.29. Comparison of posterior estimates of parameters from nonhierarchical (black lines) and the correct model allowing heterogeneity in **g** (red lines) for the life cycle diagram shown in Figure 9.28. Above posteriors for population estimates of **g** are ten group-level estimates (95 percent confidence interval with posterior mean) from the hierarchical model. Correct values for each parameter are shown as dashed vertical lines. Estimates of states are shown in *upper right* with line of agreement. Predictive loss for four models is shown above (*D*). From Clark et al. (2005).

nonstratified analysis largely reflect sample size, whereas the hierarchical model accommodates and describes variability (Clark 2003).

How important is the heterogeneity described by the hierarchical analysis? Several recent studies demonstrate that these random effects have a large impact on the estimates of life history (Carothers 1973, 1979; Cam et al.

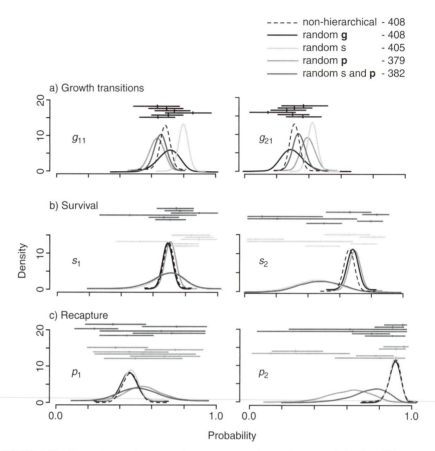

FIGURE 9.30. Posterior estimates of parameters from four models for Taitu rats. Posteriors are identified by color with predictive loss at *upper right*. Horizontal lines show posterior mean and 95 percent credible interval for each of six sites. For example, the model with random effects in **g** includes estimates for each site in black above the estimates for the full population in *a*. From Clark et al. (2005).

2002; Royle and Link 2002) and population growth (Clark 2003) that are estimated from census data. Consider simple estimates of survival and maturation age. Clark et al. (2005) used the posterior estimates from the hierarchical model in Figure 9.30 to construct predictive distributions for survival (Equation E.15) and maturation age (Equation E.16), using matrix methods, outlined in Appendix E.3. Both individual survival and population extinction depend on the variability in these quantities. Without stratification, that variability is predicted to be small (Figure 9.31, left-hand side). The variability among sites is large and it translates to large variability in life history (Figure 9.31, right-hand side). That variability provides a buffering capacity for the species that is not apparent from the model that assumes homogeneity.

In contrast to plugging estimates into a matrix that derives from a number of studies as the basis for predictive intervals of population growth and life

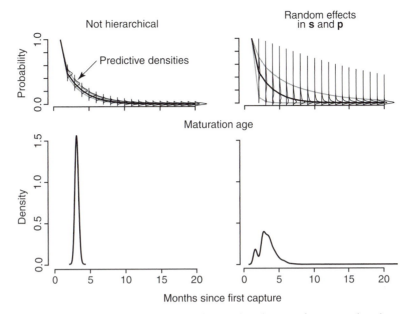

FIGURE 9.31. Comparison of posterior predictive distributions for survival and matura-tion age for rats using the nonstratified and the hierarchical random effects model. From Clark et al. (2005).

history (e.g., Box 5.3, Section 5.7), with this approach we are simultaneously estimating all parameters. Data for transition matrixes often are not collected in a way that can provide confident estimates of how vital rates are correlated with one another. It is reasonable to expect that allocation to fecundity may come at the cost of reduced growth, increased mortality risk, or both. Estimates of correlations must come from observations of vital rates from the same individuals. If such correlations are available, they can be used to calcu-late approximate population growth rates (Section 5.7). If such approxima-tions are not appropriate, standard methods for drawing from a joint parameter distribution can be challenging (e.g., Doak et al. 1994; Caswell 2001). On the other hand, if the joint posterior is obtained by MCMC meth-ods described here, the growth rate prediction is a simple extension that can be incorporated directly into the MCMC loop, just as we predicted life history schedules in the example from Section 9.19. The predictive distributions of survival and maturation in Figure 9.31 incorporate the full covariance struc-ture of posterior estimates.

There are a large number of variations on the mark-recapture theme. Fugiwara and Caswell (2002) specify an uncertainty associated with stage recognition, which also can be estimated (Clark et al. 2005). Kauffman et al. (2003) include dead recoveries in their model for survival of American pere-grine falcons (*Falco peregrinus anatum*).

9.18 Structure as Systems of Difference Equations

Matrix models can be written equivalently as a system of difference equations. In ecological applications, state transitions are often expected to be, at least in part, nonlinear. For a given situation, we can test for nonlinearity (Ives et al. 1999). Often, not all states are observable. It may be desirable to incorporate process uncertainty in models based on different equations in flexible ways. In this section, I include examples that span a range of options in terms of data, process, and parameter models.

9.18.1 A Lake Food Web

To understand how top-down and bottom-up forces affect trophic interactions, Ives et al. (2003) developed a model for application in lakes having different nutrient loadings and levels of predation. This study is unique in its integration of data modeling with a sophisticated analysis of model behavior, extending concepts introduced in Chapter 2 and Appendix B. Nutrient loading, particularly phosphorus, increases phytoplankton productivity, whereas grazing by zooplankton, especially large *Daphnia*, reduces phytoplankton density. Small fish feed on *Daphnia*, leading to dominance by the small zooplankton species that have only limited impact on phytoplankton. The small planktivorous fish that feed on *Daphnia* tend to be rare in the presence of large fish. Thus, by manipulating combinations of nutrients and large fish populations, one can expect contrasting effects on lake productivity (Carpenter et al. 2001). Either manipulation might be expected to destabilize the food web (Appendix B).

Ives et al. (2003) used a discrete Gompertz equation model with process and observation errors to infer food web consequences of nutrient addition and predation. The basic model is

$$\mathbf{x}_t = \mathbf{a} + \mathbf{B}\mathbf{x}_{t-1} + \mathbf{C}\mathbf{u}_t + \varepsilon_t \qquad \text{\textit{process equation}}$$

$$\mathbf{y}_t = \mathbf{x}_t + \mathbf{g}_t \qquad \text{\textit{observation equations}}$$

$$\mathbf{z}_t = \mathbf{u}_t + \mathbf{h}_t$$

$$\varepsilon_t \sim N_p(\mathbf{0}, \Sigma)$$

$$\mathbf{g}_t \sim N_p(\mathbf{0}, \mathbf{V}_g)$$

$$\mathbf{h}_t \sim N_q(\mathbf{0}, \mathbf{V}_h)$$

$$\mathbf{V}_g = Diag[0.04, 0.04, 0.16, 0.16]$$

$$\mathbf{V}_h = Diag[0, 0.36]$$

where $\mathbf{x}_t$ is the length-p vector of log species abundances, $\mathbf{a}$ is a vector of intercepts, $\mathbf{B}$ is the p by p matrix of interaction terms, $\mathbf{u}_t$ is a length-q vector of covariates, **C** is a p by q matrix of coefficients, and ε_t is the length-p vector of process errors. Observed abundances and covariates are $\mathbf{y}_t$ and $\mathbf{z}_t$, respectively. In this example, $p = 4$, with the vector $\mathbf{x}_t$ containing abundances for large phytoplankton, small phytoplankton, *Daphnia*, and non-*Daphnia* zooplankton. The zero variance for the first element of $\mathbf{V}_h$ is intended simply as a notational device to indicate that there is assumed to be no observation error in the first covariate. In fact, the design matrix **B** was supplemented to include additional columns for lake differences, which I have omitted to simplify notation. There are $q = 2$ covariates, with $\mathbf{u}_t$ containing levels of phosphorus, which was added to two of three lakes, and to planktivory, which differed among lakes due to predation by largemouth bass on planktivorous minnows.

Because the model is linear in state variables and regression parameters, substantial progress can be made using classical approaches. Ives et al. (2003) applied several techniques to estimate parameters. The variances for observation errors $\mathbf{V}_g$ and $\mathbf{V}_h$ were set at values that were larger than those estimated from separate analyses of similar data. They constrained some parameter estimates to specific ranges of values to assure that interactions between zooplankton and phytoplankton were estimated to have the appropriate signs. Because the study involved several lakes and manipulations over time (intervention), they were able to estimate effects of nutrients and predation.

Ives et al. (2003) further explored the multivariate extension of stability properties outlined in Appendix B. The stationary distribution for this model is Gaussian with a covariance matrix that depends on the process error Σ and the interaction coefficients **B**. Ignoring covariates $\mathbf{u}_t$, the covariance is updated as

$$\mathbf{V}_t = \sum_{s=0}^{t-1} \mathbf{B}^s \Sigma (\mathbf{B}^s)^T$$

Stability depends on **B**: if the eigenvalues of **B** all lie within the unit sphere (May 1974), then $\lim_{t\to\infty} \mathbf{B}^t \to 0$. They found the high-planktivory lake to be least stable. They attributed much of this effect to *Daphnia* grazing and, secondarily, to self-regulation by small phytoplankton. Other estimates were consistent with expected interactions involving nutrient limitation and predation.

9.18.2 Application: Skagerrak Cod

Bjørnstad et al. (1999) considered cod populations from survey data on the Norwegian Skagerrak coast. They used an age-structured model to determine if age structure could account for the two-to three-year oscillation in the observed time series. The model for four cohorts can be written as the nonlinear difference equations

$$x_{t+1} = (z_t + w_t)e^{\alpha + \varepsilon_t}$$
$$y_{t+1} = cx_t^\phi y_t^\gamma$$
$$z_{t+1} = 0.4 \times y_t$$
$$w_{t+1} = 0.4 \times z_t$$

where x_t and y_t are two juvenile cohort densities, and z_t and w_t are two adult cohort densities. The parameters ϕ and γ represent within-cohort (density dependence: $\phi < 1$) and between-cohort (cannabalism: $\gamma < 0$) effects, respectively. Adult survival is density-independent, assumed to be the proportion 0.4. The reproductive rate is α. For one version of the model there was no process stochasticity in reproduction ($\varepsilon_t = 0$). A second version included process variability in reproduction, with

$$\varepsilon_t \sim N(0,\sigma^2)$$

Observations of juvenile densities were assumed to be unbiased Poisson variates

$$x_t^{(o)} \sim Pois(x_t)$$

$$y_t^{(o)} \sim Pois(y_t)$$

These are observation errors. Adult cohorts are unobserved and treated deterministically. Finally, noninformative priors were placed on C, γ, α, ϕ, and σ^2. The joint posterior was simulated using MCMC in BUGS. For the model with process variability, posteriors were centered near $C = 0.6$, $\gamma = -0.06$, $\alpha = 2.5$, $\phi = 0.7$, and $\sigma^2 = 1$.

In simulation, Bjørnstad et al. (1999) found that limit cycles occurred for values of $\gamma < 0.4\phi - 1$, that is, if intracohort dynamics are substantially stronger than intercohort dynamics. Their estimates of ϕ were too low to predict stable limit cycles in the deterministic model, but they did predict damped oscillations. Stochastic simulations (i.e., including ε_t) produced the high-frequency oscillations and low-frequency variation. They concluded that age structure was an important contribution to the high-frequency variation and that stochastic reproduction contributed to the low-frequency variation.

9.18.3 Application: London Measles

There is a large literature on the application of modern statistical methods for epidemiological data. Methods can be found in the public health and ecological literature, as well as in standard statistical journals. I do not attempt an overview of the many approaches used for modeling disease. Hierarchical Bayes can provide some powerful advantages for epidemiological data. This example introduces a basic approach. Although disease models deal with interacting species, many epidemic models can be framed as structured models, because the disease population is modeled indirectly, through numbers of infected individuals (Anderson and May 1991). Disease reproduction involves transmission from infected to susceptible individuals. The model can simply track changes in state. For example, the Susceptible-Infected-Recovered (SIR) model for measles assumes transitions among these states that, indirectly, result from transmission, infection, and recovery (Figure 9.32). This example concerns measles reported from the prevaccination era (1944–1964) from London (Bjørnstad et al. 2002; Grenfell et al. 2002) (Figure 9.33).

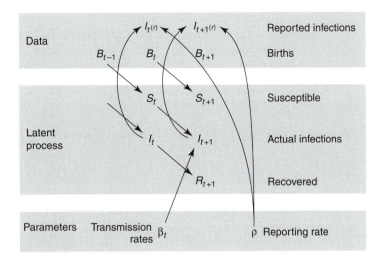

FIGURE 9.32. Graph of the SIR model used by Clark and Bjørnstad (2004) to model London measles data.

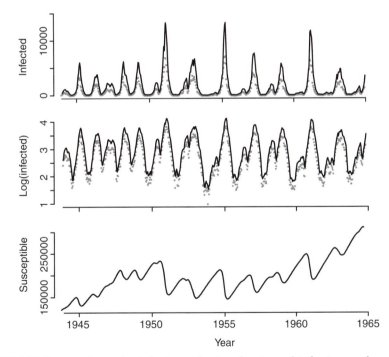

FIGURE 9.33. Reported measles infections (dots) and estimated infections and susceptibles for the London measles data. From Clark and Bjørnstad (2004).

The process model is modified from Bjørnstad et al. (2002) and Grenfell et al. (2002). Change in the susceptible class occurs with addition of new births, B, and losses as the disease runs its course.

Newborns become susceptible within several months, and before vaccination began in 1966, about 95 to 99 percent of all individuals in urban areas contracted the disease during their lifetimes. Because recovery occurs in two weeks, this period is taken as the time step for the model. Data are available at this time scale. Infection occurs with probability $\emptyset$,

$$I_t \sim Bin(S_{t-1}, \emptyset_{t-1})$$

where

$$\phi = \beta_w \times \frac{I_{t-1}}{N_{t-1}}$$

where N_t is total population size. In other words, susceptible individuals become infected in proportion to the fraction of infected already in the population. The transmission parameter β_w has subscript $w = 1, \dots 26$, referring to the wth two-week increment of the year. In reality, the process will be more complex than a simple proportionality, due to heterogeneity in the transmission process (Bjørnstad et al. 2002). I do not explicitly model recovery, as all infected individuals are assumed recovered by the next two-week time increment.

Infection I_t is not directly observed, but rather only those cases reported,

$$I_t^{(o)} \sim Bin(I_t, \rho)$$

where ρ is the fraction of cases reported. The full model used by Clark and Bjørnstad (2004) is

$$p(\mathbf{I}, \beta | \mathbf{I}^{(o)}, \mathbf{B}, \rho, priors) \propto \prod_{t=1}^{T} Bin(I_t | S_{t-1}, \phi_{t-1}(I_{t-1}, S_{t-1}, \beta_w)) \prod_{t=1}^{T} Bin(I_t^{(o)} | I_t, \rho)$$

$$\times \prod_{w=1}^{26} Unif(\beta_w | 0, 1000) Beta(\rho | 45, 53)$$

The informed prior on ρ comes from the knowledge that nearly all children contract the disease, allowing for confident estimates of reporting rates from the long-term birth data. Of particular interest are the transmission parameters β, but we must simultaneously model infections $\mathbf{I}$. Conditioned on $\mathbf{I}$, susceptibles $\mathbf{S}$ are deterministic. Conditionals needed for the Gibbs sampler include infected

$$p(I_t | I_{t-1}, I_{t+1}, I_t^{(o)}, S_{t-1}, \beta_w, \rho, \dots) \propto$$

$$Bin(I_t | S_{t-1}, \phi(I_{t-1}, S_{t-1}, \beta_w)) Bin(I_{t+1} | S_t, \phi(I_t, S_t, \beta_{w+1})) Bin(I_t^{(o)} | I_t, \rho)$$

sampled with a Metropolis step, and transmission parameters β_w, each of which is conditionally dependent on the two week period w, taken over all years,

$$p(\beta_w | I_t \epsilon w, S_{(t-1)\epsilon w}, \dots) \propto \prod_{t \epsilon w} Bin(I_t | S_{t-1}, \phi(I_{t-1}, S_{t-1}, \beta_w)) Unif(\beta_w | 0, 1000)$$

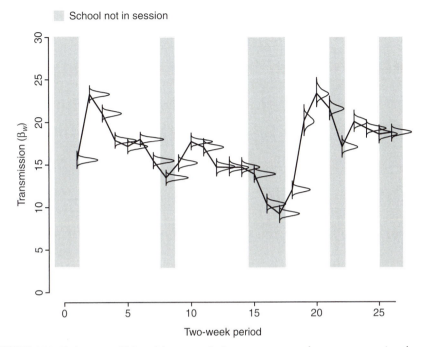

School not in session

FIGURE 9.34. Estimates of biweekly transmission parameters, shown as posterior densities, increase sharply when schools open and subsequently decline. From Clark and Bjørnstad (2004).

A Metropolis step was used with the normal proposal density.

Among the more striking features of the data is the contrast between the prominent annual to biennial cycle in infections, which results from buildup of susceptibles through new births and removal by infection (Figure 9.33), and the seasonal progression of transmission parameter estimates (Figure 9.34). The seasonal changes in transmission rate are largely masked by the cycles in infections, but can be estimated together with the unobserved infected class (Figure 9.34). Bjørnstad et al. (2002) and Grenfell et al. (2002) describe dynamics in detail.

9.18.4 Allocation Trade-offs

Ecologists have long attempted to understand trade-offs within individuals (allocation to one activity leaves less for another) (Mangel 2003), among individuals (individuals that do well at one activity may be less or more successful at another) (Charnov 1991; Carey 2001), and among species (species that compete well for one resource may compete less well for another) (Niklas and Enquist 2002; Bonsall and Mangel 2004). In the case of tree allocation to growth verses reproduction, the effort to identify trade-offs within individuals (over time) and among individuals has been frustrated by the need for detailed observations that span long time periods and include a large number of individuals. Using the approach to estimate fecundity, summarized in Section 8.5.1

(Figure 9.17), combined with growth estimated from thousands of individuals over thirteen years, Clark et al. (in review) used a multivariate model to determine relationships through time and among individuals. Define the length $k = 2$ response vector

$$\mathbf{y}_{i,t} = [\ln(d_{i,t})\ \ln(f_{i,t})] \qquad \forall Q_{i,t} = 1$$

where $d_{i,t}$ is the diameter growth increment form year $t - 1$ to year t and $f_{i,t}$ is the number of seeds produced by the tree i in year t. As in Section 8.5, the qualifier $Q_{i,t} = 1$ says that this response vector exists only for tree years in which the individual is reproductively mature. Immature trees ($Q_{i,t} = 0$) have a model for growth, but not fecundity. The goal is to determine how the joint response is influenced by tree size and light available and to determine how sources of variation are related.

The process model is

$$\underset{(1 \times k)}{\mathbf{y}_{i,t}} = \underset{(1 \times p)}{\mathbf{x}_{i,t}} \underset{(p \times k)}{\alpha} + \underset{(1 \times k)}{\kappa_t} + \underset{(1 \times k)}{\beta_i} + \underset{(1 \times k)}{\varepsilon_{i,t}}$$

with residual variance and random effects

$$\varepsilon_{i,t} \sim N_k (\mathbf{0}, \Sigma)$$

$$\beta_{i,t} \sim N_k (\mathbf{0}, \mathbf{V}_\beta)$$

Σ is the k by k covariance matrix that holds the covariances among the k different responses in $\mathbf{y}_{ij,t}$. $\mathbf{V}_\beta$ is the covariance matrix for random effects. κ_t is the vector of fixed effects for year t, included because the thirteen-year series for the updated analysis is sufficiently long to explored masting. (The previous analysis of Clark et al. 2004 used AR(1) error.) For this particular model, $k = 2$ responses in the vector $\mathbf{y}_{ij,t}$.

The length p vector of covariates,

$$\mathbf{x}_{i,t} = [1\ \ln(D_{i,t})\ [\ln(D_{i,\,t=1})]^2\ \ln(\lambda_{i,\,t-1})\ \ln(d_{i,t})]$$

has an intercept, linear and quadratic effects of tree size, represented by diameter $D_{i,t}$, canopy status $\lambda_{i,t}$ (Section 8.5.2), and the diameter increment from the previous year $d_{i,\,t-1}$. This last element represents and AR(1) term for diameter increment and the effect of previous year diameter growth for fecundity. Note that the response of an individual over T yr can be written as

$$\underset{(T \times k)}{\mathbf{Y}_i} = \underset{(T \times p)}{\mathbf{X}_i} \underset{(p \times k)}{\alpha} + \underset{(T \times k)}{\kappa} + \underset{(T \times 1)(1 \times k)}{\mathbf{1}_T\ \beta_i} + \underset{(T \times k)}{\mathbf{E}_i}$$

$$\mathbf{E}_i \sim N_{Tk}(\mathbf{0}, \mathbf{S})$$

where the block diagonal covariance matrix is $_{(TkS \,\times\, Tk)} = \mathbf{I}_T \otimes \Sigma_k$ and κ is the matrix having rows κ_t.

Direct sampling is done from the conditional posteriors. For fixed effects, first stack the matrixes,

$$\mathbf{X} = \begin{bmatrix} \mathbf{X}_1 \\ \vdots \\ \mathbf{X}_n \end{bmatrix}, \mathbf{Y} = \begin{bmatrix} \mathbf{Y}_1 \\ \vdots \\ \mathbf{Y}_n \end{bmatrix}, \text{and } \mathbf{Z} = \begin{bmatrix} \mathbf{Z}_1 \\ \vdots \\ \mathbf{Z}_n \end{bmatrix}$$

where $\mathbf{Z}_i = \mathbf{Y}_i - \kappa - \mathbf{1}_T \beta_i$. Include only ij for which the tree is mature. For the vectorized parameter matrix of fixed effects, the conditional posterior is

$$vec(\alpha)|\mathbf{X},\mathbf{Y},\Sigma,\ldots \sim N_{kp}(\mathbf{Vv},(\Sigma+\mathbf{V}_\beta) \otimes \mathbf{V})$$

where $\mathbf{V}^{-1} = \mathbf{X}^T \mathbf{X} (1 + w_\alpha)$ and $\mathbf{v} = \mathbf{X}^T\mathbf{Z} + w_\alpha\mathbf{X}^T\mathbf{X}\alpha_0$. The design-based prior here is taken to be $vec(\alpha) \sim N_{kp}(vec(\alpha_0), \Sigma \otimes (w_\alpha\mathbf{X}^T\mathbf{X}^{-1})$, for p by k matrix of prior values for α. Large prior weight is specified with large scalar value w_α.

There is a pair of fixed year effects for each year, the prior for all pairs taken to be $N(\kappa_t|\mathbf{0},\mathbf{V}_\kappa)$. For the tth year, the conditional posterior is

$$\kappa_t \sim N_k (\mathbf{V}_t\mathbf{v}_t,\mathbf{V}_t)$$

where $\quad \mathbf{V}_t^{-1} = \Sigma^{-1}\sum_i I(Q_{i,t} = 1) + \mathbf{V}_\kappa^{-1} \quad$ and $\quad \mathbf{v}_t = \Sigma^{-1} \text{ is} \sum_{\forall(Q_{it} = 1)}$ $(\mathbf{y}_{i,t} - \mathbf{x}_{i,t}\alpha - \beta_t)$

Based on the prior $IW(\mathbf{V}_\Sigma|\mathbf{R}_\Sigma,r_\Sigma)$ process errors are sampled from the inverse Wishart,

$$\Sigma^{-1} \sim W\left(\left[\mathbf{Z}^T\mathbf{Z} + r_\Sigma\mathbf{V}_\Sigma \right]^{-1n}, \sum_{i=1}^{n} I(\max(Q_{i,t}) = 1) + r_\Sigma \right)$$

where $\mathbf{Z}$ is now the stacked matrix of sub matrixes $\mathbf{Z}_t = \mathbf{Y}_{i,t}\alpha - \kappa - \mathbf{I}_T\beta_i$. Using the prior $IW(\mathbf{V}_\beta|\mathbf{R}_\beta,r_\beta)$ random effects covariance is sampled from

$$\mathbf{V}_\beta^{-1} \sim W\left(\left[\sum_{i=1}^{n} I(\max_t(Q_{i,t}) = 1)\beta_i^T\beta_i + r_\beta\mathbf{R}_\beta \right]^{-1}, \sum_{i=1}^{n} I(\max_t(Q_{i,t}) = 1) + r_\beta \right)$$

The max( ) function in the indicator functions means that if the tree is estimated to be mature in any year, I include a random effect for that individual.

From the ensemble of species fitted with this model, *Liriodendron tulipifera* serves as an example. Interactions over time enter the analysis in several ways. Because trees grow, time enters through the design matrix, with effects of ln diameter, the square of ln diameter, and last year's diameter increment. These effects on the current diameter increment are α_{21}, α_{31}, and α_{51}, respectively. Effects on the current fecundity are α_{22}, α_{32}, and α_{52}, respectively (Table 9.1). Variation in time that is shared across the population is contained in year effects κ (Figure 9.35a and b). Growth rates were low during the relatively dry years, and fecundity was especially low since 2000. *Liriodendron* shows masting, but not as pronounced as in many other species, an effect that would be picked up

TABLE 9.1.
Posterior Means, Standard Errors, and 95 percent Credible Intervals for the Fixed Effects α

Parameter*	Covariate	Estimate	Standard Error	0.025	0.975
		Response: in diameter growth increment			
α_{11}	Intercept	−2.93	0.0445	−3.01	−2.84
α_{21}	$\ln(D_{i,t-1})$	−0.0654	0.0264	−0.118	−0.0137
α_{31}	$[\ln(D_{i,t-1})]^2$	0.00962	0.00415	0.00158	0.0180
α_{41}	$\ln(\lambda_{i,t-1})$	0.101	0.00354	0.0937	0.107
α_{51}	$\ln(d_{i,t-1})$	0.00859	0.00377	0.00123	0.0158
		Response: in fecundity			
α_{12}	Intercept	2.19	0.145	1.91	2.48
α_{22}	$\ln(D_{i,t-1})$	1.101	0.0993	0.901	1.29
α_{32}	$[\ln(D_{i,t-1})]^2$	−0.207	0.0169	−0.239	−0.172
α_{42}	$\ln(\lambda_{i,t-1})$	0.146	0.00929	0.128	0.164
α_{52}	$\ln(d_{i,t-1})$	0.0136	0.0105	−0.00701	0.0348

*Subscripts indicate row and column number for the matrix of fixed effects α.

in the second column of κ and plotted in Figure 9.35b (Clark et al. 2004). The lag-1 autocorrelation for diameter increment is positive (Fig 9.35d) in agreement with the positive estimate of α_{51} (Table 9.1). For this analysis, there is no evidence that a large growth increment in one year results in lower growth the next year (α_{s1}).

There is strong evidence to suggest that fecundity is not sacrificed for rapid growth, neither within individuals over time nor among individuals. The estimate of $\alpha_{52} = 0.0136$ with 95 percent confidence interval straddling zero indicates that high-growth years show no tendency to result in lower fecundity the next. The year effects have not taken up obvious correlations between the two response variables (Figure 9.35a, and b). The correlations between fecundity versus previous diameter increment and the next diameter increment are slightly positive and negative, respectively. Clearly, within individuals, growth increment and fecundity do not show trade-offs for this species.

Individuals that grow rapidly on average produce more seeds on average (Figure 9.35c). Increased light availability has a positive effect on both variables (positive estimates of α_{41} and α_{42}).

9.18.5 Further Examples Based on Structured Populations

An increasing number of studies involve creative applications in structured population settings. As a sequel to the analysis of Soay sheep using the SETAR model (see Section 9.13), Coulson et al. (2001) applied a stage-structured

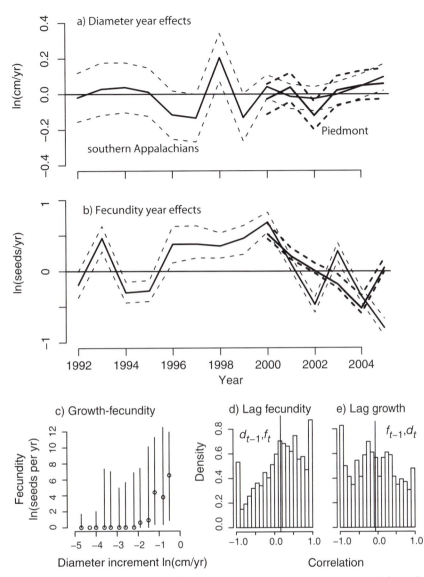

FIGURE 9.35. Posterior estimates of your effects $\mathbf{b}_t$ for diameter growth rate (*a*) and fecundity (*b*) with 95 percent confidence intervals. Relationships between the two are represented by scatter plots of predictive means for in fecundity versus in diameter (*c*), the same with a one-year lag on fecundity (fecundity in yr t versus diameter in yr $t - 1$) (*d*) and the random effects β_i (*e*), indicating the relationship between individuals in terms of fecundity versus diameter growth taken spanning the full duration of the data sets.

model, where covariate effects, such as climate variables and density dependence, were independently estimated for transitions (e.g., survival). The authors applied these estimates in a model with process noise that apparently was not fitted to data (described in an on-line supplement), but found that estimates from the model time series did describe the data well. Modeling experiments involving different assumptions about climate variability highlighted the importance of nonlinear responses to climate together with age structure and sex ratio. For example, prolonged severe winter reduces population size and, thus, density dependence. With better average weather, weather fluctuations conspire with density-dependence to cause population crashes. If the weather is most favorable, crashes do not occur.

Several studies have addressed the problem of modeling structured populations, when there is a capacity to identify some age classes, but not all. For a population model not unlike that described in Section 9.20.1, Link (2003) worked from a system of linear difference equations to estimate survival of an age-structured population of whooping cranes (*Grus americana*). The example was similar to that of the NSO in Section 2.8 in that individuals might remain in the oldest class for more than one year. Link et al. could readily identify young of the year on the basis of color, but older birds could not be confidently aged. Birds live approximately seven years, and they breed primarily in years 6 and 7. Link et al. used a model that treated all age classes other than first year as latent states. The likelihood consisted of a Poisson production of newborns by age class 7 and annual survival rates for all age classes. Estimated year and age effects on survival rates were products of the analysis.

The problem of inference for a structured population where we have only counts of total population size is not uncommon. Lavine et al. (2002) addressed the challenge of modeling a structured population, where seedlings fall in two classes. As in the whooping crane example, seedlings can be identified only as first (new) and > first year (old) seedlings. The number of individuals in these two classes in year t is $y_{1,t}$ and $y_{2,t}$, respectively. The goal was to estimate survival rates s_1 and s_2 for these two classes. The number of new seedlings that survive to become old seedlings is $Bin(y_{1,t-1}, s_1)$, and the number of old seedlings that survive to remain in the old class is $Bin(y_{2,t-1}, s_2)$. The probability of obtaining $y_{2,t}$ old seedlings in year t comes from the total probability of all combinations of survival from the two classes. This convolution is

$$p(y_{2,t}|y_{1,t-1}y_{2,t-1},s_1,s_2) = \sum_{x=\max(0,y_{2,t}-y_{1,t-1})}^{\min(y_{2,t-1},y_{2,t})} Bin(y_{2,t} - x|y_{1,t-1},s_1)Bin(x|y_{2,t-1},s_2)$$

This is the likelihood. It says that x of the survivors come from old seedlings and $y_{2,t} - x$ come from first year seedlings. The summation gives the probability taken over all combinations of first and second year survival. The summation limits bound the number of survivors that must have come from old seedlings last year. Lavine et al. (2002) coupled the process model for survival with a data model for the probability that a seedling would, in fact, be observed, and they extended the process model to include covariates. They were able to demonstrate when it would be more efficient to simply count the number of individuals in a plot as opposed to the more laborious tagging of each seedling. Wakefield and Shaddick (2005) point out that, for survival probabilities near zero, this convolution simplifies to a Poisson.

A similar model could be developed for the case where there is reproduction and survival both contributing to the number of individuals y_t at time t. For birth rate b and survival s, we could have

$$p(y_t|y_{t-1},b,s) = \sum_{x=0}^{\min(y_{t-1},y_t)} Pois(y_t - x|by_{t-1})Bin(x|y_{t-1},s)$$

As with the previous case, we have a number of ways of arriving at any particular observation, given that we only observe the sum. Informative priors may be critical in such cases.

The combination of population structure and partial observation can lead to likelihood functions even more complex than those already encountered in preceding examples (e.g., He 2003). These complex likelihoods result from the fact that there may be many combinations of probabilistic events that could give rise to a particular observation. The missing value structure used for capture-recapture data (Section 9.18) represents one approach for avoiding likelihood functions with complex convolutions.

9.19 Time Series, Population Regulation, and Stochasticity

Complications associated with temporal dependence highlight some of the advantages of a Bayesian approach, including flexibility to accommodate process and observation errors. Because modeling and computation focuses on conditional relationships, dynamics models do not require anything fundamentally different from static ones. Methods are not overly reliant on distributional assumptions. Carlin et al. (1992) describe extensions for non-Gaussian errors. For example, a *scale mixture* refers to a constant variance parameter scaled by time-dependent parameters that accommodate broad assumptions concerning error structure. Stroud et al. (2003) address the case where the variance depends on the state variable itself. In this chapter I have summarized ways to address some of the challenges associated with sampling and heterogeneity.

Despite the importance of time series data and the promise of new methods, the limitations can be underappreciated. Regardless of how careful (and technically sound) the modeling effort, a single time series may not provide much information on relationships of interest. The time series of U.S. population size (Figure 2.1) is unusually long and derives from a large sample size. Yet hindsight shows that it can be difficult to extract even the simplest relationships between population size and growth rate (Section 2.1). For humans, time series of total population size is supported by detailed demographic data. Ecological data are more limited and lack the supporting information that could help identify the sources of variability in time series (e.g., Krebs et al. 1995). For example, nearly a century of annual censuses of the BPF moose population does not provide clear indication of density dependence (Figure 9.36). Allee effects may be common, but they apply when densities are low and, thus, are particularly difficult to infer (Figure 9.37). Parameter estimates obtained from populations at high abundance may not describe the processes that dominate at low

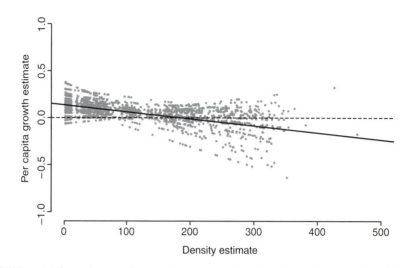

FIGURE 9.36. The estimate of growth rate versus density for BPF moose data allowing for the uncertainty in population size and in growth rate. This plot does not show strong evidence for population regulation (a decline in growth rate with population size—compare with Figure 2.9a).

abundance. Because ecological data are typically short-term, access to complementary sources of information, including total abundance data, demographic data (e.g., Korpimaki and Norrdahl 1991; Korpimaki et al. 1991), and covariates, could greatly improve insight. Intervention can provide a powerful means for extending an analysis beyond the range of variability likely to occur in a limited observational study (Section 9.14).

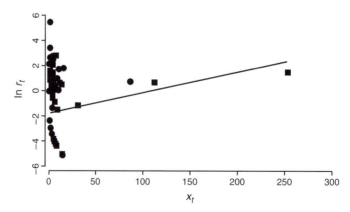

FIGURE 9.37. Plot of log population growth rate of gypsy moths in small outlying populations following detection of the new colony of size x_t. Although many of the populations went extinct, the regression line (and, thus, C) is difficult to identify due to the low densities and consequent high variability in growth rate. From Liebhold and Bascompte (2003).

This chapter has provided only a brief introduction to some of the ways in which the treatment of complexity, variability, and unknowns in models affects inference and prediction. These considerations are relevant to the efforts aimed at separating complex dynamics that result from deterministic relationships (including chaos) from "stochastic" change using time series data. Among other things, spectral analysis helps identify superimposed cycles that can obscure one another or that may be hidden by aperiodic variation. Although obvious cycles may be identified from power spectra (or even raw time series), sensitivity to initial conditions (a definition of chaos) cannot. Chaotic dynamics may be explored through Lyapunov exponents (Ellner and Turchin 1995). Chaotic dynamics may trace out geometric patterns in phase space, whereas stochastic variation does not (Appendix B), but chaos is typically not identified in this way. Because stochasticity takes up everything left out of the deterministic component of the model, the distinction between chaotic versus stochastic is somewhat arbitrary. The fact that stochasticity is commonly used to describe chaos in deterministic models (e.g., ensemble analysis [Section 11.2.3]) raises further doubt concerning the utility of this dichotomy. The answer to the question, "What is the stochastic part of this process?" could be "All of the influences not included in the deterministic specification of the model." Because processes are left out of models for many reasons, and any one of them could be nonlinear, the distinction may not mean much—stochastic elements can soak up a lot of chaos simply because it is not well described by the specific formulation used for deterministic relationships.

This chapter has summarized only a few of the ways in which stochastic elements of a time series model can be treated in an analysis. To explore that portion of the data that can be accounted for with deterministic relationships, Grenfell et al. (2002) compare simulation models of measles dynamics with time series of measurements that make allowance for underreporting of infections. Alternatively, in forecasting applications it is common to forecast data, rather than an underlying process, because the data are observable (and thus can be compared with predictions), whereas an underlying process is not observable (West and Harrison 1997). Either approach is fine, provided it is clear what is being shown, and both are preferable to the more common practice of comparing simulations of a process model against data, ignoring the data-generating mechanism. Moreover, stability analyses of time series data (Ellner and Turchin 1995) can be applied to estimates of the underlying process, including the process error, rather than to the data generated thereof (Ives et al. 2003).

When detailed information is available, more complex models lend deeper insight. As illustrated by a number of examples in this chapter, heterogeneity can be taken up with some form of structure, covariates, or both. In the case of population dynamics, structure accommodates the fact that demographic rates can depend on age or stage. There can be heterogeneity in space, in time, and among individuals. Here again, the importance of assumptions behind demographic models that contain structure can be unappreciated.

There has been substantial effort in the last decade to understand how variability in time affects population growth (Caswell 2001; Lande 2003).

Variability among individuals can also be substantial, but it has received much less attention (Cam et al. 2002; Clark et al. 2003a, 2004). Interactions involving time and population heterogeneity are widespread. In some regions, year-to-year variation in seed production is linked to weather, nonlinear physiological changes, and interactions among individuals (Wright et al. 1999; Kelly and Sork 2002). Interactions involving temporal variation and mortality risk can likewise be substantial. Selective mortality during extreme events, such as fires, floods, high winds, and droughts (Foster 1988; Condit et al. 1995; Clark 1996; Langtimm and Beck 2003) affects demographic rates in subsequent years. The apparent decreased survival of Florida manatees during hurricanes means that the individuals susceptible to hurricane risk (the precise cause of death is not known) are preferentially culled and not exposed to risk during hurricanes that follow shortly thereafter (Figure 9.38). Differential risk can result from variability in physiology, behavior, size, and location. For trees exposed to hurricane winds, large individuals are most susceptible. Size can be admitted as a covariate (Section 8.2.2), but the change in representation of size classes before and after the hurricane means that inference will depend on when a study is conducted (e.g., whether or not the last hurricane removed susceptible individuals).

There are more subtle influences that impact inference in demographic studies. For example, variability in capture rates among individuals can influence the demographic parameters that are estimated from them (Carothers 1973, 1979; Clark et al. 2005). Under the standard assumption of a homogeneous population, a survivor function describes an individual's probability of surviving to a particular age. Even if the model allows that mortality risk can change with age, there is an underlying assumption that all individuals in the population flip the same coin in a given year with a probability s_x of surviving to the next year. The interacting effects of variability in time and within

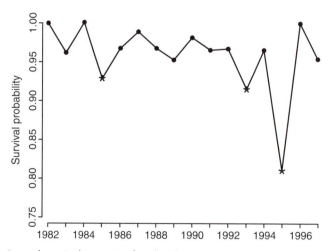

FIGURE 9.38. Annual survival estimates for Florida manatees are low during hurricane years. (From Langtimm and Beck (2003). One of the highest survival rates occurs the year after the lowest rate (1995).

populations make the interpretation more complicated than this. As the study advances susceptibles are progressively lost, and the data set is dominated by increasingly robust elements of the population. This effect is not trivial, having both ecological (Cam et al. 2002; Clark 2003) and evolutionary (Carey et al. 1998) consequences. It is so obvious as to be the basic dynamic addressed by epidemiology models—how average risk changes as individuals contract, die from, or possibly recover from infection. Development of immune responses to disease depends on it. Yet it is rarely admitted by ecological models of demography.

Decisions regarding whether to stratify, allow for random effects, or some combination thereof can have large impact on inference (Section 9.16). Stratification has long been viewed as the only option. In some cases, there is clear motivation for stratification. For example, males and females may differ in behavior or be subject to different risks, and it may be easy to distinguish sexes in field studies. There might be reasons simply to model the sexes independently (e.g., Grant and Grant 1992). More typically, stratification schemes are highly subjective. For example, there may be many ways to define size classes. Estimates of elasticities and parameter variances depend not only on growth rates, but also on class definition (Vandermeer 1978; Moloney 1986; Enright et al. 1995). Integral projection models provide one alternative (Easterling et al. 2000), where transitions are described by a function that is continuous in state. As discussed in this Section 9.17, the assumption that stage transitions are independent is often unrealistic. Random effects models may often meet the needs of an analysis better than the assumption of independent parameter values for each class. We typically expect correlations over time within the same individuals and interdependence among individuals or groups. The capture-recapture methods for joint estimation of demographic rates (Section 9.16) address some of the concerns that growth rates calculated from matrixes of independently estimated parameter values miss parameter correlations.

The seeming contradiction between the view that linear models do not well describe most populations, on the one hand, and widespread use of matrix population models, on the other, bears a bit more discussion. The standard demographic models have traditionally been linear, deterministic, and based on parameter estimates that derive from short-term observations. Natural populations are nonlinear, continually responding to environmental changes (Coulson et al. 2001), and sufficiently complex to require one or more stochastic elements in models (Bjørnstad and Grenfell 2001; Lande et al., 2003; Clark et al. 2004). Published demographic studies rarely include more than a few years of data (e.g., review of Clark et al. 1999), and demographic rates of most natural populations fluctuate dramatically from year to year. We would not want to use the model

$$n_t = n_0 \lambda^t$$

for long-term extrapolation, and the same concern applies to matrix projection. The dominant eigenvalue and corresponding eigenvectors only apply to a population that has achieved equilibrium stage structure (Fox and Gurevitch

2000). These two assumptions of linearity (generally safe only in the short term) and equilibrium stage structure (only true if the process has been free of perturbations for an extended interval) are typically at odds and suggest caution in the interpretation of results. Likewise, integral projection models are linear in density. Observed stage structure reflects recent environmental changes (e.g., Doak and Morris 1999). Nonlinear models replace matrix transition elements with functions of population density (e.g., Grant and Benton 2000; Caswell 2001), but they require data for the range of relevant densities, which may be hard to obtain. Perhaps the biggest advantage of modern methods lies not in the promise that they can provide a quick fix for some or all of the challenges of modeling time series, but rather in the opportunity they provide to exploit available information fully and to accommodate complexity in the underlying processes.

10 Space-Time

Declines in biodiversity, responses to rapidly changing climate and land cover, the spread of introduced species—it is hard to think of an ecological process that is uninfluenced in dramatic ways by its location and the time at which it is observed. Ecological processes are inherently and unavoidable spatio-temporal. There has been tremendous progress in recent decades concerning how space and time can influence concepts about ecosystems. These advances range from the implicitly spatial aspects of disturbance, migration, and colonization to the explicit effects of local topography, soils, and land cover on resources and the populations and processes that depend on them.

I make no attempt to summarize the explosion of important results that have appeared in the environmental literature. As with other topics in this book, the reader is referred to this extensive literature as a basis for understanding implications. Okubo (1980) and Murray (1989) are important references on process models for populations. Chapters in Tilman and Kareiva (1997) consider some of the implications from models that treat space in implicit versus explicit ways. A growing landscape ecology literature is beginning to connect empirical patterns of heterogeneity that are understood from map data and remote sensing with basic ecological process models (Urban 2000; Turner et al. 2001). Connections to statistical models have improved dramatically in the last decade, from the seminal work of Cressie (1993) to the proliferation of spatial approaches in modern Bayes (Bannerjee et al. 2004).

The organization of this chapter progresses from implicitly spatial and deterministic to explicitly spatial, with integration of stochastic and deterministic elements. I consider spatio-temporal processes in terms of implicit and explicit relationships. At the most implicit level, space can be considered as simply an effect. The fluctuating recruitment studied in lottery models can be viewed as a process that plays out spatially (Warner and Chesson 1985; Chesson 2000a, 2000b). (I do not take up lottery models here, but note that they have been explored in the context of data [e.g., Pfister 1995].) Metapopulation models sometimes treat space in this way (Section 10.3). Somewhat more explicit are models that summarize spatial effects in terms of distance relationships (Ribbens et al. 1994; Clark et al. 1999; Fraser et al. 2001). We might consider how distances among objects influence dynamics without ever explicitly referencing location. Models that summarize spatial effects in terms of distance include migration models, which concern how far a population spreads in a given time interval. Examples of models for dispersal are included in Sections 10.4–10.5 and several for migration are included in Sections 10.1–10.2.

Explicitly spatial models reference location and, as a consequence, might consider influences that vary in space that are extrinsic to the process itself. For example, models are used to examine how migration rates, habitat selection, and species interactions might be affected by landscape heterogeneity (Harrison 1989; Lewis and Murray 1993; Grünbaum 1998; Morales and Ellner 2002; Boyce et al. 2003; Jonsen et al. 2003; Fortin et al. 2005). Estimation of those effects often requires a model that treats space explicitly. With location reference, such as Cartesian coordinates (x, y), we can consider not only distance relationships, but also spatial covariates (e.g., spatial process) and unexplained stochasticity (process error) that is location-specific. Examples of explicitly spatial models are included in Table 10.1. Some of these models are considered in Sections 10.6–10.10.

Specification of deterministic versus stochastic components is especially challenging, because one or both may be spatial. In statistical terms, traditional ecological models can be viewed as fully deterministic (e.g., partial differential equations) or fully stochastic (e.g., probabilistic dispersal). The current trend in the statistical literature is to combine deterministic and stochastic elements to allow for inference on the process of interest, with full accommodation of unknowns. An extended example is discussed in Section 10.10.

10.1 A Deterministic Model for a Stochastic Spatial Process

The most basic stochastic model of spread, the so-called random walk, can be the starting point for a model of movement, the diffusion equation. The summary of this model follows Levin (1986). The transition from stochastic to deterministic treatment comes when we take the limit to an infinitely large population size. This section serves to introduce this connection between fully stochastic and fully deterministic models. This is followed by classical inference for diffusion (Section 10.2).

Assume that space is one-dimensional with a population centered at location $x = 0$ (Figure. 10.1a, $t = 0$). At each time increment Δt individuals step distance Δx to the left or right, each with probability $\lambda = \frac{1}{2}$. For a large population, roughly half move to the right and half to the left. After the next time increment ($t = 2\Delta t$), about half of the right population and half of the left $\frac{1}{4}$ population have moved back to the center. The other half of the left population ($\frac{1}{4}$ of the total) is at location $-2\Delta x$ and the other half of the right is at location at location $2\Delta x$. After n time increments (elapsed time $= n\Delta t$) the most distant individuals may have reached locations as far as $-n\Delta x$ and $n\Delta x$. The probability that an individual finds itself m steps to the right after n time steps depends on the number of steps it has taken to the right and to the left,

$$r - l = m.$$

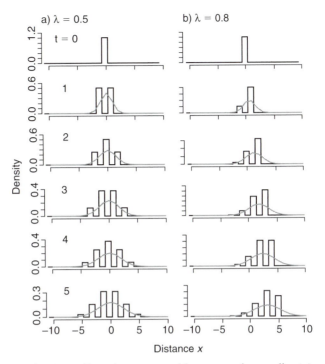

FIGURE 10.1. Distributions of location generated by two random walks, (a) unbiased and (b) biased. The normal approximation (Equation 10.4) is overlaid for comparison. The process begins at $t = 0$ and continues for five time increments (bottom panels).

The total number of steps is

$$r + l = n.$$

Thus, the number of steps to the right is $r = m + n - r = (n + m)/2$, and the number to the left is $l = (n - m)/2$.

The number of different paths by which the individual might arrive at location m is described by the binomial coefficient

$$\binom{n}{r} = \frac{n!}{r!l!} = \frac{n!}{r!(n-r)!} = \frac{n!}{\left(\dfrac{n+m}{2}\right)!\left(\dfrac{n-m}{2}\right)!}$$

The probability then is the binomial distribution,

$$p(m) = \frac{n!}{\left(\dfrac{n+m}{2}\right)!\left(\dfrac{n-m}{2}\right)!}\left(\frac{1}{2}\right)^{n}$$

This is the probability that r of n total steps will be to the right, given an unbiased probability of $\lambda = \frac{1}{2}$.

$$p(m) = Bin(r|n, \lambda) = \binom{n}{(n+m)/2}\lambda^{(n+m)/2}(1 - \lambda)^{(n-m)/2}$$

Provided that the transition probability is $\frac{1}{2}$, this binomial distribution is centered on (has the mean location) $x = 0$,
and the binomial variance $n\lambda(1 - \lambda)$ increases with the number of steps taken (Figure. 10.1a).

Over time (assuming large n), the binomial looks increasingly Gaussian (Figure. 10.1a). Stirling's formula can be used to show that the binomial converges to the Gaussian distribution with large n,

$$\lim_{n\to\infty} p(m) \to (2/\pi n)^{1/2}\exp\left[-\frac{m^2}{2n}\right]$$

Now assume that m and n are continuous. Distance $x = m\Delta x$, and time $t = n\Delta t$. In the limit as λ and λt become small, define the diffusion coefficient

$$D \equiv \frac{1}{2} \times \frac{(\Delta x)^2}{\Delta t}$$

This is one-half times the squared step size in a time increment λt. By substitution obtain the density at time t,

$$p(x) = \frac{1}{2(\pi Dt)^{1/2}}\exp\left[-\frac{x^2}{4Dt}\right] \qquad 10.1$$

Classical diffusion can be related to the random walk using a Taylor series (Appendix A). Consider a concentration of particles at location x at time t, $n_{x,t}$. Suppose particles move at random, to the left a distance Δx with probability λ_l and to the right distance Δx with probability λ_r over the time interval Δt. Note that $\lambda_l + \lambda_r = 1$. We can write the following expressions for the concentrations of particles at $x - \Delta x$, x, and $x + \Delta x$. First is the change in particles at location x,

$$n_{x,t+\Delta t} = \lambda_r n_{x-\Delta x,t} + \lambda_l n_{x+\Delta x,t} \qquad 10.2$$

the sum of terms emanating from the right and the left, respectively. This expression is not yet useful, because it contains the arbitrary quantities Δx and Δt. To eliminate them, expand and then take limits. Expanding n about x and t (Appendix A), we have

$$n_{x,t+\Delta t} \approx n_{x,t} + \Delta t\frac{\partial n}{\partial t}$$

$$n_{x-\Delta x,t} \approx n_{x,t} - \Delta x\frac{\partial n}{\partial x} + \frac{\Delta x^2}{2}\frac{\partial^2 n}{\partial x^2}$$

$$n_{x+\Delta x,t} \approx n_{x,t} + \Delta x\frac{\partial n}{\partial x} + \frac{\Delta x^2}{2}\frac{\partial^2 n}{\partial x^2}$$

Substitution back into equation 10.2 gives

$$n_{x,t} + \Delta t \frac{\partial n}{\partial t} = \lambda_r \left[n_{x,t} - \Delta x \frac{\partial n}{\partial x} + \frac{\Delta x^2}{2} \frac{\partial^2 n}{\partial x^2} \right] + \lambda_l \left[n_{x,t} + \Delta x \frac{\partial n}{\partial x} + \frac{\Delta x^2}{2} \frac{\partial^2 n}{\partial x^2} \right]$$

Now collect the terms containing $n_{x,t}$ and its first and second derivatives,

$$n_{x,t}(1 - \lambda_r - \lambda_l) + \Delta t \frac{\partial n}{\partial t} = (\lambda_l - \lambda_r)\Delta x \frac{\partial n}{\partial x} + (\lambda_r + \lambda_l)\frac{\Delta x^2}{2} \frac{\partial^2 n}{\partial x^2}$$

Because the λ's sum to unity, simplify to obtain

$$\frac{\partial n}{\partial t} = \frac{(\lambda_r - \lambda_l)}{\Delta t}\Delta x \frac{\partial n}{\partial x} + \frac{\Delta x^2}{2\Delta t} \frac{\partial^2 n}{\partial x^2}$$

In the limit as Δx and Δt tend to zero, define advection and diffusion coefficients to be

$$\lim_{\Delta t, \Delta x \to 0} \frac{(\lambda_r - \lambda_l)\Delta x}{\Delta t} \to v$$

$$\lim_{\Delta t, \Delta x \to 0} \frac{\Delta x^2}{2\Delta t} \to D$$

We now have the advection diffusion equation

$$\frac{\partial n}{\partial t} = v \frac{\partial n}{\partial x} + D \frac{\partial^2 n}{\partial x^2} \qquad 10.3$$

Diffusion or advection alone can be obtained by simply including only the relevant term. This is the Fokker-Planck diffusion equation.

An advecting/diffusing population expands from an initial location at a rate given by the solution to the diffusion equation. For an initial condition of $n(x,0) = n_0\delta(x)$, where $\delta(x)$ is the Dirac delta function,[1] the solution to 10.3 is a normal distribution scaled by the initial number n_0,

$$n(x,t) = \frac{n_0}{2\sqrt{\pi Dt}} \exp\left[-\frac{(x - vt)^2}{4Dt} \right] \qquad 10.4$$

with mean vt and variance $\sigma^2 = 2Dt$ (Crank 1975). This is the normal approximation shown in Figure 10.1b. Thus, the mean is determined by advection, and both mean and variance increase linearly with time. If there is no advection, that is,

---

[1]The Dirac delta function is a conceptual device that is defined to be everywhere zero except at $x = 0$, where it is infinity large. It is further defined to integrate to one.

$$\frac{\partial n}{\partial t} = D \frac{\partial^2 n}{\partial x^2} \qquad\qquad 10.5$$

then the mean remains at the initial location, but the variance causes the population to spread out over time. If all individuals start at the same initial location, the solution to equation 10.5 is equation 10.1 multiplied by n_0; it is compared with an advection model in Figure 10.2. Advection affects the location of the distribution, but not its shape (compare Figure 10.2a, b). Advection translates the distribution downstream, but diffusion is responsible for the spread. Overviews of diffusion with ecological examples include Okubo (1980) and Murray (1989). Advection diffusion models may be used to understand the critical patch size needed for a growing population that experiences losses due to diffusion across the patch boundary (Kierstead and Slobodkin 1953) and to the drift paradox, or the rate at which a population must grow to offset downstream losses due to advection (Lutscher et al. 2005).

In two dimensions (x,y), for constant D, the diffusion equation is

$$\frac{\partial n}{\partial t} = D \frac{\partial^2 n}{\partial x^2} + D \frac{\partial^2 n}{\partial y^2}$$

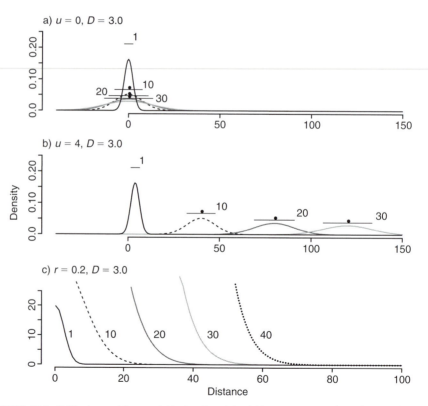

FIGURE 10.2. Diffusion with no drift (*top*), with drift (*center*), and with population growth (*bottom*). The upper two panels show the mean and standard deviation for each distribution.

For the same initial condition, the solution is analogous is to 10.1 times n_0,

$$n(x,y,t) = \frac{n_0}{4\pi Dt}\exp\left[-\frac{x^2 + y^2}{4Dt}\right]$$

10.2 Classical Inference on Population Movement

The diffusion equation is a deterministic summary of a process that is viewed as stochastic at the level of individual movement. The applications of classical inference that follow treat this deterministic equation as a sampling distribution for individual movement (Section 10.2.1) or as the deterministic mean of a sampling distribution (Section 10.2.2). Remaining sections consider some methods for estimating aspects of dispersal based on models of population spread.

10.2.1 Estimating D and V

The consequence of diffusion and advection can be written in the form of a density function (e.g., equation. 10.4). If the assumptions are appropriate, we could treat this as a sampling distribution and derive estimates directly. Suppose we release individuals at location $x = 0$ and we wish to estimate diffusion and advection coefficients based on their distribution at time t. Assume that movement is one-dimensional and recovery is unbiased with respect to distance. The likelihood function for the advection-diffusion equation, still in one dimension x, is

$$L(\mathbf{x};D,v) = \prod_{i=1}^{n}\frac{1}{2\sqrt{\pi Dt}}\exp\left[-\frac{(x_i-vt)^2}{4Dt}\right] \propto D^{-\frac{n}{2}}\exp\left[-\frac{1}{4Dt}\sum_{i=1}^{n}(x_i-vt)^2\right]$$

with log likelihood

$$\ln L(\mathbf{x};D,v) \propto -\frac{n}{2}\ln D - \frac{1}{4Dt}\sum_{i=1}^{n}(x_i - vt)^2 \qquad 10.6$$

To find the ML estimate of the advection parameter we differentiate

$$\frac{\partial}{\partial v}\ln L = \frac{2t}{4Dt}\sum_{i=1}^{n}(x_i - vt) = \frac{\sum_{i=1}^{n}x_i}{2D} - \frac{nvt}{2D}$$

Setting this equal to zero we obtain the ML estimate

$$v_{ML} = \frac{\overline{x}}{t} \qquad 10.7$$

This makes sense as the average displacement for elapsed time t.

For the diffusion coefficient

$$\frac{\partial}{\partial D} \ln L = -\frac{n}{2D} + \frac{1}{4D^2 t} \sum_{i=1}^{n} (x_i - vt)^2$$

Setting this equal to zero and solving for D yields

$$D = \frac{1}{2nt} \sum_{i=1}^{n} (x_i - vt)^2$$

We can now substitute v_{ML} to obtain

$$D_{ML} = \frac{\sum_{i=1}^{n} (x_i - \bar{x})^2}{2nt} = \frac{\sigma_x^2}{2t} \qquad 10.8$$

This is one-half of the variance σ_x^2 divided by elapsed time. To obtain standard errors determine Fisher Information. The second derivatives are

$$\frac{\partial^2}{\partial v^2} \ln L = -\frac{nt}{2D} = -\frac{nt^2}{2\sigma_x^2}$$

and

$$\frac{\partial^2}{\partial D^2} \ln L = \frac{n}{2D^2} - \frac{1}{2D^3 t} \sum_{i=1}^{n} (x_i - \bar{x})^2 = \frac{n}{2D^2} - \frac{n\sigma_x^2}{2D^3 t}$$

Now substitute for D_{ML},

$$\frac{\partial^2}{\partial D^2} \ln L = \frac{2t^2 n}{\sigma_x^4} - \frac{4t^2 n}{\sigma_x^4} = -\frac{2t^2 n}{\sigma_x^4}$$

Recall that the standard error is one divided by the square root of the negative Fisher Information,

$$se_v = \sqrt{2/n} \frac{\sigma_x}{t} \qquad 10.9$$

and

$$se_D = \frac{\sigma_x^2}{\sqrt{2nt}} \qquad 10.10$$

Note that the diffusion coefficient has units of distance squared over time. It is half the mean-squared deviation per unit time. The diffusion coefficient would best be estimated by observation of individuals. It would be equally straightforward to solve for different numbers of individuals captured at different times and to extend this to two dimensions.

There are a number of reasons why we might not want to model spread data this way. First, the treatment of the partial differential equation obtained from the diffusion model as a sampling distribution implies that we have an unbiased (and complete!) sample of individual movement. Each individual is an observation. In practice, the nature of the data would require that we overlay this process model with a data model. For example, if collections come from traps (e.g., insects), the likelihood might be Poisson (Section 10.2.2). If the individuals were marked and released, we need a capture-recapture model that allows for incomplete recovery. If recapture rates decline with distance, it may be difficult to identify parameters, or estimates may be biased. Finally, spread may be more variable than diffusion. Some alternative models are described later in this chapter. A different sampling design is illustrated in the next example. Failing direct observation, D might be estimated as one-half times the slope of the regression of mean squared deviation against time (e.g., Kareiva 1983).

Example 10.1. Estimate of the diffusion coefficient in two dimensions
Assume individuals are released and recovered at distances r_i at sample times t_i. There is no advection, and no directional bias in diffusion. From the two-dimensional likelihood,

$$L(\mathbf{r},\mathbf{t};D) = \prod_{i=1}^{n} \frac{1}{4\pi Dt_i} \exp\left[-\frac{r_i^2}{4Dt_i}\right]$$

where $r^2 = x^2 + y^2$, we have the log likelihood

$$\ln L \propto -n\ln D - \frac{1}{4D}\sum_{i=1}^{n}\frac{r_i^2}{t_i}$$

Setting the derivative equal to zero and solving for D, we have the ML estimate

$$D_{ML} = \frac{1}{4}\sum_{i=1}^{n}\frac{r_i^2}{t_i}$$

This is the mean of the squared displacement of each individual divided by the elapsed time for that individual, divided by 4. This result differs from 10.8 in being two-dimensional and in allowing for a different recovery time for each individual.

Example 10.2. Flea beetles on collards
Kareiva (1982) considered a slightly different model to analyze movement of flea beetles. The experiment involved releasing flea beetles on collard patches on lush and stunted collard patches. He recorded their locations one hour after release. Most had moved no more than one patch to the right or left.

In this case, beetles might move to the right, move to the left, or remain on the initial patch. Then

$$\lambda_0 + \lambda_1 + \lambda_r = 1$$

where λ_0 is the probability of not moving. Then the equation for concentration of beetles contains an additional term for those that stay put,

$$n_{x,t+\Delta t} = \lambda_0 n_{x,t} + \lambda_r n_{x-\Delta x,t} + \lambda_l n_{x+\Delta x,t}$$

You can work through this problem to arrive at an equation similar to the answer given above. The difference is that movement is slowed, because

$$\lambda_1 + \lambda_r < 1$$

From the equation,

$$\frac{\partial n}{\partial t} = \frac{(\lambda_r - \lambda_l)}{\Delta t} \Delta x \frac{\partial n}{\partial x} + \frac{(\lambda_r + \lambda_l)}{2\Delta t} \Delta x^2 \frac{\partial^2 n}{\partial x^2}$$

we see that the diffusion coefficient is now

$$D = \frac{(\lambda_l + \lambda_r)(\Delta x)^2}{2\Delta t}$$

Kareiva estimated diffusion using recapture data. For lush patches he found that total fraction that moved was 0.303. Because patches were $\Delta x = 2$ m apart, and $\Delta t = 1$ hr, then $D = 0.61$ m^2 hr^{-1}. The value for stunted patches, $D = 1.63$ m^2 hr^{-1}, was substantially higher.

10.2.2 Incorporating Population Growth

The foregoing models assume a fixed number of individuals—the population does not grow, it only spreads. Now consider the model that combines population growth with diffusion (Skellam 1951; Okubo 1980; Andow et al. 1990; Van den Bosch et al. 1990). The model includes a first term, describing exponential population growth, and a second diffusion term from equation 10.5,

$$\frac{\partial n}{\partial t} = \alpha n + D \frac{\partial^2 n}{\partial x^2} \qquad 10.11$$

where x is linear distance for this one-dimensional model. This means that the population grows exponentially and individuals subsequently diffuse from their point of origin in an unbiased fashion (there is no advection term in the model) away from the source in one dimension. The solution to this equation is

$$n(x,t) = \frac{n_0}{2\sqrt{\pi Dt}} \exp\left[\alpha t - \frac{x^2}{4Dt}\right] \qquad 10.12$$

Note the relationship to equation 10.1. The diffusion component allows the population to spread (Figure. 10.2c). Population growth affects the rate of spread, because reproduction at each new location contributes offspring (by diffusion) to other locations. The rate of spread of the population front approaches a traveling wave solution

$$x/_t = \sqrt[2]{\alpha D} \qquad 10.13$$

This result means that the population front expands as a wave with a velocity that depends on the rate of population growth and on the rate of diffusion. Note how the population profiles in Figure 10.2c advance to the right at a constant rate. Derivations of this result are available in a number of excellent references, including Okubo (1980). Skellam (1951) derived the result and used it to interpret muskrat expansion in Europe. He also applied the model to oak invasion of Europe following the last ice age. Here the model did a poor job, because invasion was faster than could be explained by diffusion. Lubina and Levin (1988) used it to interpret twentieth-century recovery of sea otter populations along the California coast (Example 10.3). Turchin (1998) reviews application to a number of other examples.

Most applications to data do not involve formal inference. Rough calculations based on model predictions (e.g., area occupied by the population) provide valuable insight into not only population spread, but also the life history characteristics and potential barriers to dispersal that determine it. I consider formal inference later in this chapter. For now, notice that a formal classical approach could involve population counts z_j for $j = 1, \ldots, m$, where samples are obtained at locations and times (x_j, t_j), and they span width X_j and duration T_j as

$$L(\mathbf{z}, \mathbf{x}, \mathbf{t}; n_0, D, \alpha) = \prod_{j=1}^{m} Pois(z_j; X_j T_j g_j)$$

$$g_j = \frac{n_0}{2\sqrt{\pi D t_j}} \exp\left[\alpha t_j - \frac{x_j^2}{4 D t_j}\right]$$

The function g_j is the expected density for sample j, accounting for the initial density, diffusion, and advection parameters. It has units of distance^{-1}time^{-1} and thus is scaled by sample width (distance) and duration (time) to obtain the Poisson mean. Maximum likelihood estimators are not available in closed form, but they can be obtained numerically (provided sufficient information in data). The assumption of a spatial sample of counts is but one of many sampling distributions that might be applied. Treatments of this type of model are discussed in Sections 10.4 and 10.5.

10.2.3 Nonlinear Population Growth

Interestingly, adding nonlinear population growth does not change the asymptotic wave speed. In general, for a nonlinear growth function $f(n)$,

$$\frac{\partial n}{\partial t} = f(n) - D\frac{\partial^2 n}{\partial x^2} \qquad 10.14$$

The asymptotic wave speed is given by

$$x/t = 2\sqrt{\left.\frac{df}{dn}\right|_{n=0} D}$$ 10.15

(Kolmogorov et al. 1937; Fisher 1937). Note that this result is consistent with the result for linear population growth, that is, if α is the density-independent growth rate, then

$$\left.\frac{df}{dn}\right|_{n\to 0} \to \alpha$$

This result is handy, because it permits solutions for the rate of advance despite the fact that we could not solve for population density in the interior.

Example 10.3. Rebound of the California sea otter
Data on population spread provide insight into population life history within the context of expansion. Lubina and Levin (1988) calculated a rate of California sea otter expansion following near extinction before 1914 as

$$x/t = 2\sqrt{\alpha D}$$

where x/t is the rate of expansion (km yr^{-1}) to the north or south, α is the intrinsic rate of increase (yr^{-1}), and D is a diffusion coefficient or the mean squared rate of individual displacement (km^{2}yr^{-1}). Their estimates of population size come from samples of the total otter population (Figure. 10.3b). A rough estimate of α from the regression of change in log density with time, when density is low, is obtained from

$$\ln n_t = \ln n_0 + \alpha t$$

A simple least-squares fit for the values from early in the twentieth century is

$$\ln n_t = -92.9 + 0.051t$$

($r^2 = 0.97$) indicating a rate of increase of roughly 5 percent per year. The range along the coast occupied by otters from the same surveys is plotted in Figure 10.3a. The velocity of spread is roughly given by a regression of the increasing range,

$$x = -10,824 + 5.61t$$

($r^2 = 0.87$). A rough estimate of the diffusion coefficient comes from $5.61 = 2\sqrt{0.0509D}$, or $D = 150$ km^{2}yr^{-1}. If the generation time for otters is five years, then the net reproductive rate is obtained from $\alpha = \ln R_0/T$. Then $0.0509 = \ln R_0/5$, or $R_0 = 1.3$ (Sections E.1, E.2). This is the reproductive output required to sustain a 5 percentage annual growth rate for a population having a generation time of five years.

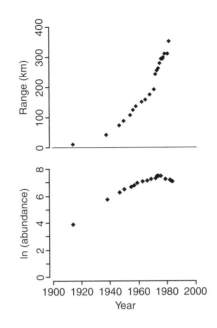

FIGURE 10.3. Plots of California sea otter range and size since the population began to rebound early in the twentieth century. Data from Lubina and Levin (1988).

10.2.4 Population Spread in Discrete Time

Wave speed can also be solved using a model in discrete time and continuous space. Here I follow Kot et al. (1996). Consider again the linear difference equation model for population growth,

$$n_{t+T}(x) = R_0 n_t(x)$$

where R_0 is the net reproductive rate, and T is the generation time (Section E.1). On average, individuals produce R_0 offspring. Some fraction of offspring produced at location y disperse to location x. The density function describing this probability of dispersing a distance $r = |x - y|$ is termed the dispersal kernel, $K(r)$. Assume a one-dimensional environment, and dispersal can occur to the right or left with equal probability. Population growth for a population with generation time $T = 1$ is

$$n_{t+1}(x) = R_0 \int_{-\infty}^{\infty} k(x - y)n_t(y)dy \qquad 10.16$$

The model says that population growth at location x is the sum of offspring deriving from all other locations y.

A traveling wave moves forward at a constant rate, so it must satisfy

$$n_{t+1}(x) = n_t(x - c)$$

where c is the velocity of spread (Kot et al. 1996). The wave maintains the same shape from one time step to the next, and the rate of spread is constant. By substitution into Equation 10.16, we obtain

$$n_t(x - c) = R_0 \int_{-\infty}^{\infty} k(x - y)n_t(y)dy$$

Equation 10.16 is linear in n and so has an exponential solution (refer back to Appendix B). Consider one that satisfies

$$n_t(x) \propto e^{-sx}$$

Substituting into equation 10.16 obtain

$$e^{-sx}e^{sc} = R_0 \int_{-\infty}^{\infty} k(x - y)e^{-sy}dy$$

Make the variable change $r = x - y$

$$e^{-sx}e^{sc} = R_0 \int_{-\infty}^{\infty} k(r)e^{-s(x-r)}dr$$

Now cancel the factor e^{-sx} to obtain

$$e^{sc} = R_0 \int_{-\infty}^{\infty} k(r)e^{sr}dr$$

The integral is the *moment-generating function* of the dispersal kernel. Let $M(s)$ represent this integral, that is,

$$M(s) = \int_{-\infty}^{\infty} k(r)e^{sr}dr \qquad 10.17$$

Recognize this as the expected value of e^{sr}. The solution must satisfy

$$e^{sc} = R_0 M(s)$$

or

$$c = \frac{1}{s}\ln(R_0 M(s)) \qquad 10.18$$

which has unknowns c and s. The value of c we seek is the minimum, which can be obtained by differentiating and setting the result equal to zero,

$$\frac{d}{ds}\left[\frac{1}{s}\ln(R_0 M(s))\right] = \frac{-\ln(R_0 M(s))}{s^2} + \frac{1}{sM(s)}\frac{dM(s)}{ds} = 0 \quad 10.19$$

The value of s that satisfies this equation is plugged into 10.18. This result is due to Weinberger (1982). Kot et al. (1996) used this approach as a starting point for calculating spread when a kernel is fat-tailed.

Example 10.4. The Gaussian dispersal kernel
Consider a population that disperses according to the one-dimensional Gaussian dispersal kernel

$$k(x) = N(x|0,\sigma^2)$$

The moment-generating function for the normal distribution is

$$M(s) = e^{\sigma^2 s^2/2}$$

and Equation 10.17 is

$$c = \frac{1}{s}\left[\ln R_0 + \frac{\sigma^2 s^2}{2}\right]$$

Solving 10.18, the minimum value of s is

$$s^* = \frac{\sqrt{2\ln R_0}}{\sigma}$$

Now substitute for s in Equation 10.18 to obtain

$$c = \sigma\sqrt{2\ln R_0}$$

From the binned dispersal data in Figure 10.4a I calculate a value of $\sigma = 44.1$. For $R_0 = 10$, the wave speed is 95 m per generation. Equation 10.18 is drawn in Figure 10.4b.

Example 10.5. The exponential dispersal kernel
The one-dimensional exponential or Laplace kernel

$$k(x) = \frac{e^{-|x|/\alpha}}{2\alpha}$$

has moment-generating function

$$M(s) = \frac{1}{1 - \alpha^2 s^2}$$

The modeling strategy outlined above assumes we can represent the dispersal kernel with a parametric function $f(x)$. There are cases where we will choose to avoid assumptions concerning kernel shape and work directly from the data. A nonparametric dispersal estimator (Clark et al. 2001b) replaces the

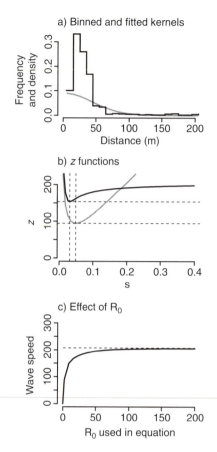

a) Binned and fitted kernels

b) z functions

c) Effect of R_0

FIGURE 10.4. Binned dispersal data compared with best-fitting Gaussian kernel (a) and z functions with wave speeds indicated by minimum values (b). (c) shows that, with large R_0, wave speed asymptotes at the maximum disperser in the data set.

parametric moment-generating function of equation 8.17 with a nonparametric one,

$$M(s) = E[e^{sX}] = \frac{1}{n} \sum_{i=1}^{n} e^{sX_i} \qquad 10.20$$

The nonparametric estimator proceeds as before, substituting this empirical estimator into Equation 10.18 and 10.19. It is solved numerically.

A variant is available for binned data, which has the empirical moment-generating function

$$M(s) = \sum_{j=1}^{m} k(x_j) e^{sx_j} \qquad 10.21$$

where $f(x_j)$ is the number of observations in the jth of m total bins (Figure 10.4a). Using the same approach, Equation 10.18 is drawn (Figure 10.4b) and the minimum is the wave speed. For large values of R_0 the predicted wave speed asymptotes at the extreme disperser in the data set (Figure 10.4c). This is a good indication that the estimate is an artifact of limited sample distance. I return to such considerations in Section 10.6.5.

For classical inference, Clark et al. (2001b) recommend a bootstrapping approach, whereby resamples are drawn from the empirical distribution and used to calculate a wave speed. The distribution of wave speeds generated in this fashion is conditional on the typically unknown constant R_0, as the total propagule production will be unobserved. Alternatively, one might hope to estimate both the dispersal kernel and R_0 directly from data on population spread.

10.2.5 Population Spread with Stage Structure

The foregoing model makes the assumption that all individuals are identical in terms of both growth and dispersal. Not all life history stages disperse, and calculations of spread rate may depend on dispersal in different life history stages. Neubert and Caswell (2000) extended the integro-difference equation model to accommodate the case where dispersal can potentially occur from any stage to any other stage. Moreover, each stage-to-stage dispersal process may be described by its own dispersal kernel. This is a deterministic approach that is reasonable when stages can be recognized as such and have transition parameters that are known.

Consider again the linear difference equation growth model with age structure

$$\mathbf{n}_{t+1} = \mathbf{A}\mathbf{n}_t$$

$\mathbf{A}$ is the transition matrix (Section 2.9.2). Combined with the Equation 10.16, we have the stage structured integro-difference equation model

$$\mathbf{n}_{t+1}(x) = \int_{-\infty}^{\infty} [k(x - y) \circ \mathbf{A}]\mathbf{n}_t(y)dy \qquad 10.22$$

where K is the matrix of dispersal kernels describing the probability of dispersing from stage j to stage i during a time step, and $\circ$ is the Hadamard product. The Hadamard product of two matrixes is a new matrix where each element ij is the product of the corresponding ij^{th} elements (Section C.6).

Example 10.6. Territorial species may be characterized by (nonmigratory) dispersal of juveniles from the natal area to establish new breeding territories. Turchin (1998) uses data from R. J. Gutierrez to analyze movement of northern spotted owl and suggests dispersal distances may average about 16 kilometer. A three-stage life history, such as that used for the northern spotted owl (Section 2.9.2) of juveniles, subadults, and adults, might be modeled with a projection matrix such as

$$\mathbf{A} = \begin{bmatrix} 0 & 0 & b \\ s_0 & 0 & 0 \\ 0 & s & s \end{bmatrix}$$

consisting of juveniles, subadults, and adults (Lande 1988; McKelvey et al. 1993; Franklin et al. 2000; Clark 2003). If dispersal to establish new territories occurs as juveniles become subadults according to a Gaussian dispersal kernel, then the Hadamard product is

$$k(x - y) \circ \mathbf{A} = \begin{bmatrix} \delta(x) & \delta(x) & \delta(x) \\ N(x|y,\sigma^2) & \delta(x) & \delta(x) \\ \delta(x) & \delta(x) & \delta(x) \end{bmatrix} \circ \begin{bmatrix} 0 & 0 & b \\ s_0 & 0 & 0 \\ 0 & s & s \end{bmatrix}$$

$$= \begin{bmatrix} 0 & 0 & \delta(x)b \\ N(x|y,\sigma^2)s_0 & 0 & 0 \\ 0 & \delta(x)s & \delta(x)s \end{bmatrix}$$

$\delta(x)$ is the Dirac delta function. This model could be analyzed in the context of spread using methods described above. An important caveat would be that the dispersal behavior summarized in the model must represent that which controls population spread. Influences of other owls (density-dependent dispersal) could make this assumption unrealistic.

Analysis of this model proceeds as before. A traveling wave must satisfy

$$\mathbf{n}_{t+1}(x) = \mathbf{n}_t(x - c)$$

By substitution into Equation 10.22, obtain

$$\mathbf{n}_t(x - c) = \int_{-\infty}^{\infty} [\mathbf{k}(x - y) \circ \mathbf{A}]\mathbf{n}_t(y)dy$$

The exponential solution has the form

$$\mathbf{n}_t(x) = \mathbf{w}e^{-sx},$$

where $\mathbf{w}$ is the distribution of stages in the traveling wave. Substitution gives

$$\mathbf{w}e^{-sx}e^{sc} = \left[\mathbf{A} \circ \int_{-\infty}^{\infty} \mathbf{k}(x - y)e^{-sy}dy\right]\mathbf{w}$$

The variable change $u = x - y$ gives

$$\mathbf{w}e^{-sx}e^{sc} = \left[\mathbf{A} \circ \int_{-\infty}^{\infty} \mathbf{k}(u)e^{-s(x-u)}du\right]\mathbf{w}$$

or

$$\mathbf{w}e^{sc} = \left[\mathbf{A} \circ \int_{-\infty}^{\infty} \mathbf{k}(u)e^{su}du\right]\mathbf{w}$$

As with the unstructured case, we arrive at an integral expression that is now a matrix of moment-generating functions,

$$\mathbf{M}(s) = \int_{-\infty}^{\infty} \mathbf{k}(u)e^{su}du$$

Let

$$\mathbf{H}(s) = \mathbf{A} \circ \mathbf{M}(s).$$

The resulting expression

$$e^{sc}\,\mathbf{w} = \mathbf{H}(s)\,\mathbf{w}$$

shows that e^{sc} must be an eigenvalue and $\mathbf{w}$ an eigenvector of $\mathbf{H}(s)$. To solve for the wave speed find the largest eigenvalue ρ_1 of $\mathbf{H}(s)$ and then determine the wave speed as

$$c = \frac{1}{s}\ln \rho_1 \qquad\qquad 10.23$$

As before, this requires that we minimize this function with respect to s.

Example 10.7. Neubert and Caswell (2000) use the example of teasel with transition matrix

1		2	3	4	5	6
1	0	0	0	0	0	402.6
2	0.974	0	0	0	0	0
3	0.017	0.011	0	0	0	8.25
4	0.004	0.002	0.077	0.212	0	69.2
5	0.003	0	0.038	0.281	0	3.81
6	0	0	0	0.063	1	0

with dispersal estimated as a double exponential and dispersal parameter $\alpha = 0.257$ m. Because seed dispersal only occurs during the final stage, the dispersal matrix has Dirac delta functions everywhere except the last column:

$$\mathbf{f}(x - y) = \begin{bmatrix} \delta(x) & \delta(x) & \delta(x) & \delta(x) & \delta(x) & Exp(x - y|\alpha) \\ \delta(x) & \delta(x) & \delta(x) & \delta(x) & \delta(x) & Exp(x - y|\alpha) \\ \delta(x) & \delta(x) & \delta(x) & \delta(x) & \delta(x) & Exp(x - y|\alpha) \\ \delta(x) & \delta(x) & \delta(x) & \delta(x) & \delta(x) & Exp(x - y|\alpha) \\ \delta(x) & \delta(x) & \delta(x) & \delta(x) & \delta(x) & Exp(x - y|\alpha) \\ \delta(x) & \delta(x) & \delta(x) & \delta(x) & \delta(x) & Exp(x - y|\alpha) \end{bmatrix}$$

The matrix of moment-generating functions has ones everywhere except the last column, which is filled with the moment-generating function for the exponential kernel

$$\mathbf{M}(s) = \begin{bmatrix} 1 & 1 & 1 & 1 & 1 & \dfrac{1}{1-\alpha^2 s^2} \\[2ex] 1 & 1 & 1 & 1 & 1 & \dfrac{1}{1-\alpha^2 s^2} \\[2ex] 1 & 1 & 1 & 1 & 1 & \dfrac{1}{1-\alpha^2 s^2} \\[2ex] 1 & 1 & 1 & 1 & 1 & \dfrac{1}{1-\alpha^2 s^2} \\[2ex] 1 & 1 & 1 & 1 & 1 & \dfrac{1}{1-\alpha^2 s^2} \\[2ex] 1 & 1 & 1 & 1 & 1 & \dfrac{1}{1-\alpha^2 s^2} \end{bmatrix}$$

The dominant eigenvalue of $\mathbf{H}(s)$ is $\rho_1 = 6.086$, and $c = 0.56$ m per year (Figure 10.5). The normal distribution results in less rapid spread.

Formal inference requires estimates of both matrix elements and the dispersal kernel. If possible, it is preferable to estimate parameters together, including parameter covariances. Capture-recapture methods, such as discussed in Section 9.19, are an option for the matrix elements. The methods there allow us to define states in terms of location, so it may be possible to estimate matrix elements and dispersal simultaneously. Caswell (2001) provides some options for classical inference where movement occurs between discrete patches. In many cases, parameter estimates may come from a number of different studies, in which case covariances among estimates will be unknown.

10.2.6 Spread with Long-Distance Dispersal

The examples thus far assume population movement occurs locally. Within a time increment Δt the population can only move small increments to the right or left. In many cases, dispersal may occur by more than one mechanism (Clark 1998;

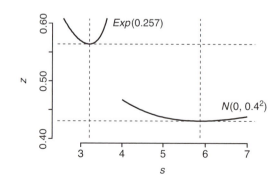

FIGURE 10.5. Predicted rates of spread (values of z) from two dispersal kernels with similar mean distances but different tail shapes with the teasel example.

Higgens and Richardson 1999). For example, acorns may be cached locally by squirrels and dispersed long distances by jays (Johnson and Webb 1989). The spread of exotic species often involves short-distance natural dispersal mechanisms and long-distance jumps by humans (Sharov and Leibold 1998; Suarez et al. 2001). Wind fields within and above canopies may result in heterogeneous dispersal patterns (Nathan et al. 2002; Soons et al. 2004; Katul et al. 2005).

There are a number of ways to represent multiple types of dispersal in models of population spread. Shigesada et al. (1995) consider outlying colonies established by long-distance dispersal that subsequently coalesce as a result of local population growth. Lewis (2000) uses a stochastic model with multiple dispersal modes to examine how stochasticity influences rate of spread. A number of studies explore multiple dispersal modes in simulation (Clark 1998; Higgins and Richardson 1999). Detailed models can be used where meteorological information is available (Sharpe and Fields 1982; Okubo and Levin 1989; Greene and Johnson 1989; Andersen 1991; Nathan et al. 2002). As formal inference is the topic, I consider one approach that has directly incorporated dispersal data.

Recent analytical results demonstrate that the traveling wave solutions described above break down when individuals can make long-distance leaps. The point of breakdown in the traveling wave approximation occurs when the dispersal kernel becomes more leptokurtic than exponential (Mollison 1972, 1977). Recall that the wave speed for the Gaussian dispersal kernel is

$$x/t = 2\sqrt{\ln R_0 D}$$

In other words the dispersal kernel $x \sim N(0, \sigma^2))$ implies a velocity of order $O(\sqrt{\ln R_0})$. Using the model of Kot et al. (1996) for fat-tailed kernels that fitted dispersal data of Clark (1998) (dispersal kernel $x^{1/4} \sim N(0, \sigma^2))$ the velocity has order $2t \cdot O[(\ln R_0)^2]$, where t is time in generations. Thus, the rate accelerates over time. This accelerating spread is consistent with some observations of invading populations and suggests an important role for the fat tail (Lewis and Pacala 2000; Kot et al. 2004). Obviously, velocity cannot increase indefinitely, so there must be a different way to model population spread that yields an asymptotic rate of spread, like the result we obtained for diffusion.

Clark et al. (2001a) developed a method to estimate spread based on extreme values derived from a stochastic approach that yields finite estimates for some kernels that are fatter than exponential. Distributions of extreme dispersal events are based on the dispersal kernel and on R_0. Fat-tailed dispersal exaggerates the importance of seed production and thus allows for rapid spread. Spread cannot be infinitely fast, because R_0 is finite, and there is always a "farthest" seed. Some insight on the rate of spread can be gained from knowing the density of extreme dispersal events.

Let $p(x)dx$ be the probability that the farthest-dispersing individual settles on the interval $(x, x + dx)$. The probability density function is the product of the events that a seed settles at x and all other seeds from the parent settle closer to the source,

$$p(x) = R_0 k(x) \left[\int_{-\infty}^{x} k(y)dy \right]^{R_0 - 1} \qquad 10.24$$

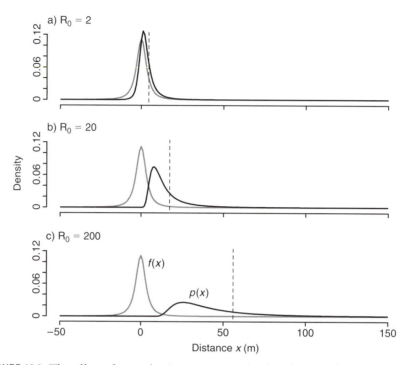

FIGURE 10.6. The effect of reproductive output on the distribution of extremes. From Clark et al. (2001b).

where $k(x)$ is the dispersal kernel (also a probability density function). The factor R_0 in Equation 10.24 is the number of ways in which we can obtain a given outcome (any one of R_0 total offspring can be the extreme case).

The density of extremes incorporates the contributions of both the dispersal kernel and the net reproductive rate R_0 (Figure 10.6). Net reproductive rate has a dramatic effect on the extremes. With $R_0 = 2$, extreme dispersal does not much exceed the kernel itself (Figure 10.6a). With $R_0 = 200$, the extreme extends well beyond the dispersal kernel (Figure 10.6c).

The velocity of spread is determined with a variable change (Appendix D). The density of velocity estimates is taken from the density of extreme values scaled by generation time

$$g(c) = p(x)\left|\frac{dx}{dc}\right| = p(cT) \cdot T$$

The expected rate of spread is

$$E[c] = \frac{1}{T}\int_{-\infty}^{\infty} xp(x)dx = \int_{-\infty}^{\infty} cg(c)dc$$

For tree populations having a two-dimensional Student's t dispersal kernel

$$k(x) = \frac{1}{2\sqrt{2u}\left(1 + \dfrac{x^2}{2u}\right)^{\frac{3}{2}}}$$

the expected velocity for the kernel used by Clark et al. (2001b) is roughly

$$E[c] \approx \frac{1}{T}\sqrt{\frac{\pi u R_0}{2}}$$

The pattern of spread is erratic, with occasional large leaps and many shorter steps (Figure 10.7). Overall, the expected velocity is in agreement with the analytical approximation, but any one realization can depart substantially from the analytical approximation. Rates of spread calculated from dispersal data show a broad range of rates that are influenced by seed production, dispersal, and maturation age. Clark et al. (2000b) also found that variable seed production can slow the rate of spread well below that expected from a point estimate of R_0.

Among the important insights gained from the study of fat-tailed dispersal kernels is the qualitative change that occurs in models when the kernel is

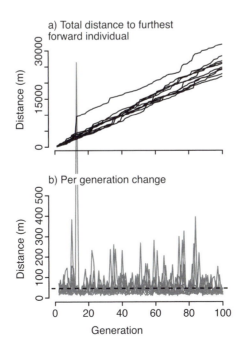

FIGURE 10.7. Simulated spread by fat-tailed dispersal compared with the analytical solution (dashed line in b). Ten realizations are shown (Clark et al. 2001b).

more leptokurtic than exponential. For diffusion, a stochastic model will yield the same rate of spread as a deterministic model. For stochastic models with density dependence, the situation becomes more complicated. Mollison (1991) pointed out that, with density dependence, a stochastic model of spread based on fat-tailed dispersal does not increase indefinitely. Kot et al.'s (1996) derivation of the accelerating spread with fat-tailed, deterministic spread provided a baseline for understanding transient acceleration. Applications to empirical dispersal kernels and with dispersal data (Clark et al. 2001a) reaffirm the importance of treating dispersal as a stochastic process for the leptokurtic dispersal commonly observed for seeds. Using a model for branching random walks, Kot et al. (2004) provided a critical link between stochastic dispersal and integro-difference equations. Their demonstration that linear, stochastic population growth still leads to accelerating spread sharpens our understanding of these relationships and emphasizes the need to consider carefully how stochasticity, kernel shape, and density dependence are handled in specific settings.

10.2.7. What Is R_0 for a Spreading Population?

The concept of R_0 in models of population spread is worth some elaboration. The concept is not applied in a uniform way. Many theorists think of R_0 as the density-independent potential for growth. This concept does not square with the fact that R_0 is typically calculated from data, where there is density dependence that is not modeled as such (e.g., life tables compiled from vital rate data). For a population that is not changing in density, $R_0 = 1$; each individual replaces itself with one offspring (Section E.1). Although individuals of many populations are capable of high fecundity, large production of seeds, eggs, and the like must be balanced by high mortality, such that the population does not continue to increase. In other words, if an individual that lives long might produce 10^6 offspring, then the fraction of offspring that survive to realize this potential could be roughly 10^{-6}. Large R_0 values are possible, but they cannot remain high for extended periods. Thus, population ecologists tend to think in terms of R_0 values near one, despite recognition that under optimum conditions R_0 might be much larger.

For a spreading population, one must consider R_0 in terms of survival in areas not now occupied by the population. If the unoccupied region allows for high survival of offspring, then the individual's contribution to the population R_0 is the sum of low survival of offspring that settle nearby and high survival for those landing far away. Clearly, a single number does not describe this situation. Moreover, parameter estimates often come from the place where the population is now, but are used in calculations of spread into some new environment. The value needed for a model of spread is that appropriate for the new environment. In most ecological models that assume values of R_0 near one, long-distance dispersal can have only a limited impact on expected spread rate, and rapid spread is hard to achieve.

By contrast, Skellam (1951) used a value of $R_0 = 9,000,000$ for his calculations of Holocene oak expansion into western Europe. This value of R_0 would seem to violate standard views about R_0 that come from nonspreading populations. It does not make sense to expect that each individual will be replaced by 9,000,000 new offspring next generation, each of which will beget 9,000,000 the next, and so forth. Clearly, Skellam (1951) had in mind the number of seeds that might be produced by an individual that survives to old age. This concept of R_0 focuses on the number of offspring that a parent might beget beyond the current range. If that area is unoccupied, a new colonist might realize high survival rates and contribute many new propagules beyond the frontier. And so forth.

So what value of R_0 is appropriate? Direct observations are difficult to come by, so formal inference is rarely possible. Reasonable values must take cognizance of the setting. Although Skellam's (1951) guess of nearly 10^7 acorns over the lifetime of an oak is reasonable (e.g., Clark et al. 2004), during the Holocene oaks were spreading into regions already occupied by forest. We expect survival rates in areas already occupied by forest to be not much greater than those where oaks were already present. So we might disagree with this notion of R_0. Yet, there are examples of contemporary invasions where the concept of R_0 as lifetime production from individuals that have unusually high survival is appropriate. For example, invading zebra mussels (*Dreissena polymorpha*) fully occupy a substrate (Ricciardi et al. 1998), at which point we expect R_0 to be roughly one. This low value of R_0 is not appropriate for the spread of zebra mussel throughout the upper Mississippi in the late twentieth century, because they were capable of high survival in areas not yet colonized. In the absence of estimates of R_0 taken from the region into which the population is spreading, careful thought must be given to what constitutes a reasonable value. In most cases, predictions of spread potential should involve sensitivity to R_0 (Clark et al. 2001b, 2003).

10.2.8. Predictive Capacity

The erratic spread possible when dynamics are controlled by long-distance dispersal has long been recognized in epidemiology (Mollison 1972, 1977). The establishment of outlying populations determines the rate of spread (e.g., Lewis 1997; Shigesada et al. 1995; Clark et al. 2001a), but it is still finite. For especially large data sets, it can be possible to gain some rough estimates of spread from simple plots of distances from population frontiers to newly formed outliers (Figure 10.8). Sharov and Leibhold (1998) used such estimates as basis for exploring the U.S. Forest Service's concept of a barrier zone that would be needed to slow the spread of gypsy moth invasion. These erratic patterns make predictions of a specific invasion inherently poor. Far more valuable in such instances is the understanding of dispersal vectors and survival potential in new habitats. This knowledge can provide insight as to where potential for rapid invasion is large (e.g., Kolar and Lodge 2002). Approaches for estimating dispersal potential are taken up in the following section.

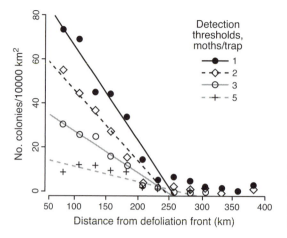

FIGURE 10.8. Gypsy moth colonies at distances from the expanding population frontier using detection thresholds of one, two, three, and five moths as detection thresholds. From Sharov and Leibhold (1998).

10.3. Island Biogeography and Metapopulations

Many spatial models used by ecologists involve discrete locations that may be isolated to varying degrees and connected by dispersal (Harrison 1989). Such models are used to explore how local dynamics and evolutionary change within a habitat combine with barriers to or losses during dispersal to describe abundance and extinction risk (Hanski 1998). Metapopulation models incorporate several aspects of such populations:

1. The metapopulation consists of an ensemble of local populations.
2. Local populations communicate by dispersal.
3. Interactions occur locally and, sometimes, during dispersal.

In this section I provide only a brief overview and several examples involving data. The data usually consist of abundance observations, which become the basis for inferring dynamics, represented by parameters in the model.

10.3.1 The Levins Model: A Single-Species Patch Occupancy Model

The earliest patch occupancy model is attributed to Levins (1969). A landscape consists of identical patches that are either occupied or not by members of the population of interest. Propagules emanate from occupied patches, and they enter a bath from which they are distributed evenly across the landscape. Residence time in the bath is zero—redistribution is immediate. Successful colonization occurs only on unoccupied patches. Patches have no explicit location relative to one another. There is a sufficiently large number of patches such that extinction can be treated as a deterministic rate rather than an episodic process.

   The basic model is a simple differential equation for change in occupancy over time. The rate of change in the fraction of patches occupied p is given by the rate at which new patches are colonized $C(p)$ minus the rate at which they go extinct,

$$\frac{dp}{dt} = p\left[C(p) - e\right]$$

where colonization rate is

$$C(p) = m(1 - p)$$

The parameter m is a rate of propagule production, a fraction $1 - p$ of which land on unoccupied patches. Thus,

$$\frac{dp}{dt} = mp(1 - p) - ep \qquad\qquad 10.25$$

For the population to persist, colonization rate $C(p)$ must exceed a threshold set by extinction, and vice versa. Thus, if patches are small (high extinction rate), or highly isolated (small colonization rate), a species is likely to go extinct. To see this, note that the model can be rearranged to look like a logistic equation,

$$\frac{dp}{dt} = (m - e)p\left[1 - \frac{p}{1 - e/m}\right],$$

so it will have the same behavior. The equilibrium at $dp/dt = 0$ is

$$p^* = 1 - e/m$$

and (trivially) $p^* = 0$. For the nontrivial equilibrium to exist, $e < m$. Stability analysis (Appendix B) proceeds as

$$\frac{\partial}{\partial p}\left(\frac{dp}{dt}\right)_{p^*} = m - 2mp^* - e$$

To evaluate this dependency at the equilibrium, substitute p^* for p to obtain

$$\frac{\partial}{\partial p}\left(\frac{dp}{dt}\right)_{p^*} = e - m$$

Because we already know $e < m$, *this result is negative, and thus the equilibrium is stable.*

The equilibrium patch occupancy fraction is usually equated with the probability of finding a patch to be occupied. Here I describe several ways to model occupancy based on the binomial. As a simple illustration, consider Hanski's (1991) modification of the Levins model under the assumption that migration rates decrease exponentially with distance from source (isolation) and extinction rates decrease exponentially with island area

$$m_i = m_0 e^{-\beta_1 D_i}$$
$$e_i = e_0 e^{-\beta_2 A_i}$$

where D_i and A_i are distance to and area of the ith island, and m_0, e_0, β_1, and β_2 are parameters. Then the equilibrium probability of occupancy for patch i is

$$p_i = 1 - e_i/m_i = 1 - \frac{e_0}{m_0}\exp[-\beta_1 A_i + \beta_2 D_i]$$

Clearly, this is the basis for a Bernoulli likelihood with parameter p_i.

We can determine the critical area A_c for a given degree of isolation D_i by solving for A_i at equilibrium for patch occupancy $p^* = 0$,

$$A_c > \frac{1}{\beta_1}[\beta_2 D + \ln(e_0/m_0)]$$

At increasing levels of isolation, this model says the area per patch must increase commensurately. Hanksi (1991) and Peltonen and Hanski (1991) show that there is a tendency for occupied islands to exceed this value, and vice versa.

10.3.2 The MacArthur/Wilson Island Biogeographic Model

Classic biogeography theory assumes that the number of species present on an island represents a balance between immigration and extinction rates (MacArthur and Wilson 1967). The MacArthur/Wilson Island Biogeographic Model (MWIB) model can be viewed as a simplified metapopulation model, where all colonists are viewed as immigrants that derive from outside the ensemble of islands. The landscape consists of identical islands within an uninhabitable matrix, existing in one of two states, occupied versus unoccupied. Propagule production is typically taken to be constant and external to the island system: propagules come from a mainland. Colonization occurs only on unoccupied islands (others are already colonized). There is assumed to be a sufficiently large number of islands such that extinction can be treated as a deterministic rate rather than an episodic process.

The MWIB model differs from the Levins model (see below) in having a mainland source of propagules that is independent of the islands themselves.

$$\frac{dp}{dt} = colonization - extinction$$

$$\frac{dp}{dt} = C(p) - ep = m(1 - p) - ep$$

For this model, the equilibrium occurs when the two rates are equal,

$$p^* = \frac{m}{e + m}$$

The effects of isolation and distance on colonization and immigration can be applied to this model. For example, Hanski (1991) replaced the extinction rate e with a function that depends on island area

$$e = e' \, A^{-\beta_1}$$

for parameters e' and β_1. To show that this could be treated as a general linear model, I modify this slightly to

$$e = e' e^{-\beta_1 A}$$

The probability that an island i is occupied (at equilibrium) can be written as

$$p_i = \frac{m}{m + e' e^{-\beta_1 A_i}} = \frac{e^{x_i \beta}}{1 + e^{x_i \beta}}$$

or

$$\text{logit}(p_i) = \mathbf{x}_i \beta$$

where the vector of covariates, in this case is $\mathbf{x}_i = [1 \ A_i]$ and the parameter vector would be interpreted as $\beta^T = [\ln(m/e') \ \beta_1]$. Peltonen and Hanski's (1991) closely related model quantified the incidence data of shrews on islands. They found that the estimate of the effect of island area on incidence was related to the body sizes of different shrew species.

10.3.3 Application: Immigration/Extinction Curves

Manne et al. (1998) used long-term census data of terrestrial birds (i.e., neither aquatic nor marine) from thirteen British islands to estimate the form of the immigration and extinction curves. They wanted to know whether or not these curves were concave, as originally suggested by MacArthur and Wilson. Because censuses were conducted annually, we can draw inference on immigration (absent, then present) and extinction (present, then absent). I use this example to demonstrate application of a parametric bootstrap (Section 5.3.3).

The approach differs from the patch occupancy approach in several ways. Rather than model presence or absence of a single species on an island, Manne et al. (1998) modeled the number of species on an island at time t. Species number changes with immigration and extinction. Change in species number could be modeled as

$$S_{t+1} = S_t + I_t - E_t$$

for the number of species on an island that remain from the preceding year S_t, new immigrants I_t, and species lost through extinction E_t. Numbers of immigrants and emigrants are treated as binomial random variates. First consider immigration. There are species on the mainland that are not now on the island and can immigrate. The number of such species is $= P - S_t$. The probability of immigration is taken to be a decreasing function of the number of species already present, and it could be nonlinear. I use a function similar to that of Manne et al. (1998),

$$\gamma_t = \beta_1(1 - S_t/P)^{\beta_2}$$

where β_1 and β_2 are fitted parameters. Thus, the distribution of immigrants is $I_t \sim Bin(P - S_t, \gamma_t)$. The probability of emigration is taken to be an increasing function of species present,

$$\theta_t = \beta_3(S_t/P)^{\beta_4}$$

where β_3 and β_4 are also fitted parameters. The distribution of emigrants is

$$E_t \sim Bin(S_t, \theta_t)$$

The joint probability of immigrants and emigrants is

$$L(\mathbf{I},\mathbf{E};\beta) = \prod_{t=1}^{T} Bin(I_t|P - S_t,\gamma_t) \prod_{t=1}^{T} Bin(E_t|S_t,\theta_t)$$

$$= \prod_{t=1}^{T} \gamma_t^{I_t}(1 - \gamma_t)^{P-S_t-I_t}\ \theta_t^{E_t}(1 - \theta_t)^{S_t-E_t}$$

Manne et al. (1998) assumed that all states (P, I_t, E_t, and S_t) are known. With these assumptions we might fit this model directly to data, that is, ignore the time series character of data. This is possible, because, conditioned on known P and S_t, immigrations and extinctions are independent, as assumed by the likelihood function, which is written as a product for all observations I_t and E_t. Note that, under these assumptions, the estimates of parameters β_1 and β_2 are independent of those for β_3 and β_4; there is no need to fit the two binomial likelihoods simultaneously. For a number of islands, Manne et al. (1998) estimated the shapes of the curves to be concave.

In the event that P and S_t were not precisely known, we might allow for the time series character, which could be accommodated by a parametric bootstrap (Section 9.5). For this example, I consider a time series with S_t being unknown. I do not have the original data, but simulated data to demonstrate the idea, with the advantage that we can check the algorithm (because parameters used to generate data are known).

To demonstrate I assume a pool of $P = 100$ species available for colonization from the mainland and observations for eighty consecutive years. I retain the assumption that there are no observation errors in I_t and E_t, but I now suppose that S_t is unknown. For the parametric bootstrap, I begin with estimates of parameters. I used maximum likelihood to estimate a vector of four parameters $\hat{\beta} = [\hat{\beta}_1, \hat{\beta}_2, \hat{\beta}_3, \hat{\beta}_4]$ fitted to data generated as a time series using the values [0.05, 2.0, 0.4, 1.5] and the binomial probabilities. The bootstrap loop consists of generating a full time series of values followed by estimation of four new parameter values. The first value cannot be generated in this way, so we might simply condition on the observed first value. In this case, I conditioned on the first value S_1 and simulated the following:

$$I_t^{(b)} \sim Bin(P - S_t^{(b)}, \gamma(S_t^{(b)}; \hat{\beta}_1, \hat{\beta}_2))$$
$$E_t^{(b)} \sim Bin(S_t^{(b)}, \theta(S_t^{(b)}; \hat{\beta}_3, \hat{\beta}_4))$$
$$S_{t+1}^{(b)} = S_t^{(b)} + I_t^{(b)} - E_t^{(b)}$$

The superscript indicates that there will be a full time series of simulated data for each bootstrap resample b. At each time step, the immigration and emigration probabilities γ_t and θ_t were determined, based on the current S_t, and binomial draws generated the updated values of S_{t+1} (Figure 10.9, upper panel). From each full sequence, I obtained estimates of the vector of $\boldsymbol{\beta}(b)$. This is the bth resample estimate and consists of a vector of four parameter values. Repeating

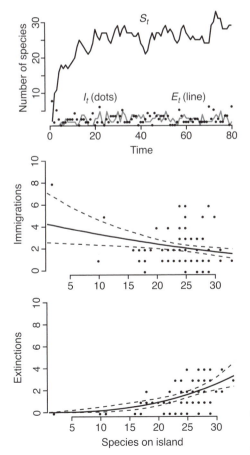

FIGURE 10.9. A simulated data set of immigrations and extinctions based on the model of binomial probabilities.

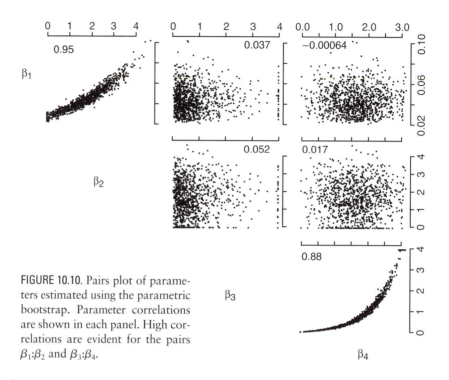

FIGURE 10.10. Pairs plot of parameters estimated using the parametric bootstrap. Parameter correlations are shown in each panel. High correlations are evident for the pairs $\beta_1:\beta_2$ and $\beta_3:\beta_4$.

this process B times results in a B by 4 matrix of parameter estimates, with each of the four columns being a list of estimates for one parameter.

From the bootstrapped parameter estimates I plot each of the combinations of parameter pairs to discover strong correlations between those that describe immigration and extinction functions (Figure 10.10). The scatter plots describe confidence intervals obtained by bounding the central 95 percent of the estimates for each parameter. The bootstrap also provides parameter correlations, shown as correlation coefficients and illustrated by the orientation of scatter in Figure 10.10. The difficulty identifying both β_1 and β_2 and both β_3 and β_4 suggests a simpler model. Note that the parameter estimates describing immigration are independent of those describing extinction.

10.3.4 Other Models of Patch Occupancy

Models of patch occupancy can be framed in many ways. The simple differential or difference equation models of previous sections are among the many examples. Other approaches include partial differential equation models (e.g., Hastings and Wolin 1989; Clark 1991), nonlinear matrix models (Caswell and Cohen 1991), systems of ordinary differential equations (Hanski 1991), and cellular automata (Molofsky et al. 2001; Turcotte et al. 2002). Although often not advertised as such, many models in the literature concern the basic elements of metapopulation models: internal dynamics coupled with dispersal among suitable sites. Here are a few examples.

The analysis of the 2001 foot-and-mouth epidemic in Britain involved farms as infection sites, with spread of the disease among sites (Keeling et al.

2001). Several livestock species contract the disease, and each has a susceptibility S_k and transmission β_k, assembled as vectors **s** and **β**. Let θ_{jt} be the probability that farm j becomes infected at time t and $\{n_t\}$ be the farms already infected at time t. Using the structure outlined by Keeling et al. (2001), we could write this model as

$$y_{j,t} \sim Bernoulli(\theta_{j,t}(k(\mathbf{r}_j;u))$$

$$\theta_{j,t} = 1 - \exp\left[-\mathbf{s}^T\mathbf{n}_{j,t} \sum_{i\epsilon\{n_j\}} \beta^T\mathbf{n}_{i,t}k(r_{ij};u) \right]$$

where $\mathbf{n}_{i,t}$ is a vector containing the number of livestock of each susceptible species on farm i at time t. The highly leptokurtic shape of the estimated kernel (Figure 10.11) comes from the fact that local transmission is likely to occur between neighboring farms combined with long-distance infections that occur from a variety of mechanisms. The resulting pattern of spread included clusters of infection dispersed patchily over the island. An analysis by Ferguson et al. (2001) approached dispersal models somewhat differently from Keeling et al. (2001), apparently modeling directly dispersal distances between farms, perhaps limiting the data to include only those farms for which the source of the infection could be confidently identified.

Metapopulation models can be used to describe populations of aquatic organisms that occupy lakes. As in the Keeling et al. (2001) model, Havel et al. (2002) used a Bernoulli sampling distribution to describe distance effect of source areas on the probability that the zooplankter *Daphnia lumholtzi* would invade lakes in Missouri. Their model could be written as

$$y_j \sim Bernoulli(\theta_j(x_j\beta,k(\mathbf{r}_j;u))$$

$$logit(\theta_j) = x_j\beta + \beta_f \sum_{i=1}^{n} k(r_{ij};u)$$

$$k(r_{ij};u) = e^{-ur_{ij}}$$

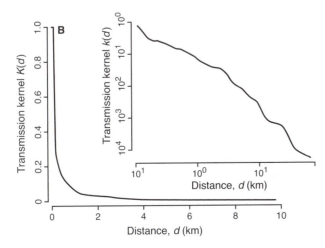

FIGURE 10.11. Estimated dispersal kernel for the transmission of foot-and-mouth disease across Britain in 2001. From Keeling et al. (2001).

where the vector of covariates $\mathbf{x}_j$ included factors that might affect the capacity of *Daphnia* to colonize the lake, such as size and epilimnetic temperature, and there are n total lakes in the study. Note that, in this case, k is not strictly a dispersal kernel, but rather acts more like a distance effect, and β_f is the regression coefficient for this effect. They found that invaded lakes tended to be large and had warm epilimnetic waters and the scale parameter u was estimated to be in the range of 20 to 30 kilometer.

To accommodate landscape heterogeneity, Ovaskainen (2004) used a diffusion model to describe movement within patches, with boundary conditions between patches that allow for biased movement toward or away from patches of differing suitability. Several studies have quantified how boundaries affect movement—emigration rates can be influenced by the nature of nearby patches and characteristics of the boundaries themselves (e.g., Haddad 1999; Ries and Debinski 2001). In Ovaskainen's model the landscape consists of a series of patches $i = 1, \ldots, n$, each potentially having a unique rate of diffusion,

$$\frac{\partial u}{\partial t} = -\rho u + \frac{\partial}{\partial x}\left(D_i(t)\frac{\partial u}{\partial x}\right) + \frac{\partial}{\partial y}\left(D_i(t)\frac{\partial u}{\partial y}\right)$$

where $D_i(t)$ is the diffusion coefficient in the ith patch at time t and ρ is the mortality rate. To incorporate bias, the boundary between two patches i and j is described by the velocities of movement on each side of the boundary and flux across the boundary (Ovaskainen and Cornell 2003). Exterior boundaries were reflecting. At any given time, an individual is somewhere on the landscape or dead; there is assumed to be no reproduction. From this basic model, Ovaskainen (2004) determined probabilities that an individual is in a particular patch, probability that an individual is dead, and so forth.

A simplified version of this model was fitted to recapture data for the false heath fritillary butterfly (*Melitaea diamina*) within a landscape of open, habitable, patches and closed forest. Data consist of recaptures on marked individuals; there is an estimated recapture probability p. Time and location of initial capture and recaptures is the basis for model parameterization. This model assumed identical diffusion coefficients D and mortality rates ρ, a single parameter describing the relative difference in velocity of individual movement on each side of the boundary, and recapture probability. The likelihood function was maximized numerically, with bootstrapped confidence intervals. Ovaskainen (2004) estimated a recapture probability of $\frac{1}{3}$ to $\frac{1}{2}$, an average lifespan of $1/\rho = 10$ days, and an average lifetime movement of 1.5 kilometer. He further estimated a density 115 times higher in open patches than in closed forest. For a more complex model that included different parameter values for open and forest areas, parameter identification was difficult.

Moorcroft et al. (2005) used a pair of partial differential equations to describe movement of wolf packs, one for the wolves, including random motion, avoidance of scent marks from neighboring packs, and responses to steep terrain and prey, and another for scent marks, including deposition and decay. Their modeling approach differs from others discussed here in that the equations were simulated to achieve an equilibrium distribution for the two state variables (wolf packs and scent marks), which was then used to fit the model based on observed locations. In other words, the temporal aspect of data is not included. They obtained good fits to the spatial location data, and parameter estimates that helped explain key aspects of population dynamics.

Bayesian approaches have also been applied to metapopulation models. Drechsler et al. (2003) used a Bayesian framework to estimate the influence of patch size and distance on colonization and extinction of the Glanville fritillary, *Melitaea cinxia*, which occupies small meadows on the Åland Islands of southwestern Finland. They wrote a model for colonization much like those described for seed dispersal and disease transmission and a second equation for extinction, with parameters to describe an exponential kernel and the effects of island size. The joint posterior was then used in simulations of different landscape scenarios to consider management options most likely to ensure survival. The combination of Bayesian inference and forward simulation meshes nicely, because parameter uncertainty can be directly mixed through use of MCMC output for Monte Carlo simulation. Together these studies demonstrate complementary approaches to modeling populations that are spatially heterogeneous. All depend on assumptions that heterogeneity can be simplified in different ways.

Molecular tools are becoming increasingly popular in the effort to understand the migration that occurs among suitable habitats. Bayesian methods show promise for understanding metapopulation dynamics where there may be a need to combine molecular evidence of population source with additional information. There is a rapidly growing literature on this topic. I mention just a few examples. Rannala and Mountain (1997) used multilocus genotype information to assess immigrant ancestry of human populations in northern Australia. They began with a multinomial likelihood for the numbers of copies of each allele at a locus in each of the populations that might exchange migrants. The number of categories in the multinomial is set at the number of alleles observed in the population. Of course, the true number is unknown, but the observed number is used as a substitute. The parameters of the multinomial are the underlying allele frequencies to be estimated. With a conjugate Dirichlet prior, they determined posterior frequency estimates. Under the assumption of random mating, they write the probabilities of an individual genotype conditional on parents from different source populations. The marginal probability of the genotype for each potential source population is then obtained by mixing over the Dirichlet posterior. The relative probabilities for different source populations were then compared.

The basic approach was extended by Gaggiotti et al. (2004) to include genotypes of a new (second generation) Orkney gray seal colony, combined with the likelihood of Rannala and Mountain (1997) for source colonies. Gaggiotti et al.'s likelihood for the newly colonized population is a probability for an individual genotype obtained by weighting the contributions of different parental genotypes from different source populations by (1) the relative proportion of immigrants from those sources and (2) the degree to which parents tend to mate with individuals from the same source population. In turn, the relative proportion of immigrants from each source was expressed in terms of factors likely to affect migration, such as population density and migration distance. A model-averaging approach was used to integrate over model uncertainty. They obtained posterior estimates of the contribution of gray seal source populations to a new colony in the Orkney Islands and were able to relate those contributions to covariates. The degree of assortative mating was estimated simultaneously.

As an additional example, Okuyama and Bolker (2005) showed that ecological covariates could contribute geographical detail to patterns inferred from haplotype data of sea turtle origins.

Molecular evidence will vastly extend our understanding of dispersal, particularly within a formal inferential framework, but we do not yet know the full potential. Sequence data are complementary to ecological evidence on dispersal, making hierarchical Bayes a logical tool for analysis. Yet applications to date have entailed sophisticated models applied to relatively simple settings. For example, the Orkney gray seal study assumed that sample individuals were all F_1 individuals derived from only two source populations, which were discrete (they came from different islands). Still, the posterior density for assortative mating was essentially flat, and those for covariates, source distance and density, broadly spanned zero. In most ecological settings, sources will not be identifiable with discrete islands, but will be broad and diffuse or consist of a combination of potentially many islands and/or mainland populations. Most target populations will consist of overlapping generations. It seems clear that ecological insight will continue to play a large role in such analyses. More comprehensive characterization of potential sources may be possible as molecular methods are further refined. In any event, there will remain strong impetus to increase the amount of ecological information that can be combined with molecular evidence to better refine estimates.

10.4 Estimation of Passive Dispersal

Understanding migration and the importance of space in competition models depends on estimates of dispersal. Estimation can be complicated, because the process is rarely observed directly. Instead, inference depends on capture-mark-recapture or on spatial relationships among offspring and adults. For capture-recapture data, dispersal distances can be estimated if location is taken as a state. Individuals can have a location at a given sample date in the same way that they have a stage assignment (age, size, and so forth) (classical approaches are reviewed in Williams et al. 2001, p.455). The capture-recapture techniques of Section 9.19 are appropriate. Here I mention some techniques for dispersal estimation based on data obtained by indirect methods.

Passive dispersal can be estimated using process models having degrees of complexity that depend, in large part, on the amount of information available. Propagules can derive from point or dispersed sources. The source can be continuous, intermittent, or active only once. A dispersed source can be one-dimensional (e.g., smoke from a flame front), two-dimensional (pollen from a patch of flowering plants), or three-dimensional (seeds from the crown of a large tree). Movement is subject to gravity and drag (properties of the dispersing entity), winds that can be described by three vectors (including vertical updrafts), and turbulence. There is a huge literature from many disciplines on movement that is controlled by these forces from the military (Sutton 1953), earth and atmospheric sciences (Bagnold 1941; Okubo 1971), and plant (Gregory 1968; Okubo and Levin 1989; Greene and Johnson 1989) and animal (Turchin 1998) sciences. In ecology, models have been applied to cases where detailed information is available on source location and transport (Sharpe and Fields 1982; Okubo and Levin 1989; Andersen 1991). For example, Nathan et al. (2002) coupled detailed meteorological evidence with three-dimensional information on canopy structure to conduct simulations of individual seeds in a closed forest to show the importance of updrafts for seed dispersal. In cases where detailed information is not available, low-dimensional dispersal kernels are used to summarize movement (Ribbens et al. 1994; Clark et al. 1998, 1999, 2004; Dalling et al. 2002). Because this type of model has been parameterized directly from observations it is relevant to the present discussion of data modeling.

10.4.1 Seed Shadows

Seed production (fecundity) and seed movement are often considered together, because they may be estimated simultaneously, as suggested in Section 10.2.2. In Chapter 9, I discussed a model for longitudinal effects to estimate individual, size, and year effects on fecundity of trees (Figure 9.13). In fact, these estimates come from a process model, containing fixed and random effects, with data models for several types of observations. Data models for this problem depend on how observations are made. Some examples were used to demonstrate the flexibility of the approach in Chapter 8. The distribution of seeds can be used to estimate fecundity (Figure 10.12), provided we connect them with an appropriate data model. Then the inference problem can be represented as in Figure 10.9. An overview of the classical approach is provided in Clark et al. (1999), including parameter estimation, model selection, and predictive distributions for seed shadows. Subsequent application of Bayesian approaches, which accommodated additional sources of uncertainty, showed that that classical models overestimated fecundity and dispersal distance (Clark et al. 2004).

10.4.2 Kernel Shape

The shape of the dispersal kernel is determined by the height and shape of the crown, characteristics of the stand that affect wind profiles at the time when seeds are released, and characteristics of the seed, including settling velocity and dispersal vectors. Distributed and elevated sources (such as tree crowns) tend to have areas of high seed deposition at least as wide as the crown, and rapid decline with increasing radius. The shape of the tail depends on the dispersal vector (winds, birds, water) and on secondary dispersal (e.g., birch seeds blowing across snow, caching by rodents).

Among the most popular functional forms for seed shadows are power exponentials,

$$k(r) = \frac{1}{C}\exp\left[-\left(\frac{r}{\alpha}\right)^c\right] \qquad 10.26$$

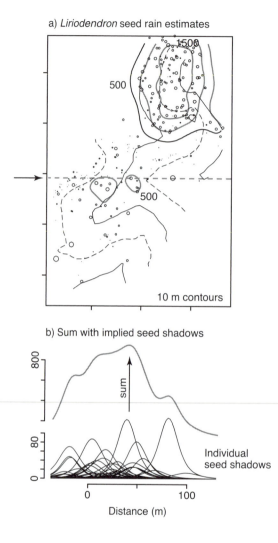

a) *Liriodendron* seed rain estimates

1500

500

500

10 m contours

b) Sum with implied seed shadows

800

sum

80

Individual
seed shadows

0

0   100

Distance (m)

FIGURE 10.12. Seed rain map (*a*) and the summed seed shadow model (*b*) for the transect indicated in (*a*). The estimate for the sum line in (*b*) is an explicit function of the individual seed shadows shown below it.

where C is a normalization constant and α and c are dispersion and shape parameters, respectively. To normalize the density, integrate over both distance r and arc angle. If dispersal is assumed to be rotationally symmetric, then arc angle does not appear in the kernel, but we still have to normalize in all directions. For the power exponential, the two integrations give

$$C = \int_0^\infty \oint_{2\pi} \exp\left[-\left(\frac{r}{\alpha}\right)^c\right]d\phi dr = 2\pi \int_0^\infty r \exp\left[-\left(\frac{r}{\alpha}\right)^c\right]dr = \frac{2\pi\alpha^2\Gamma(2/c)}{c}$$

The inner arc-wise integration introduces the factor $2\pi r$. We then integrate over distance. The normalized density is

$$k(r) = \frac{c}{2\pi\alpha^2\Gamma(2/c)}\exp\left[-\left(\frac{r}{\alpha}\right)^c\right] \qquad 10.27$$

The power exponential accommodates a broad area of high seed density beneath a distributed and elevated crown. This shape is obtained for values of $c > 2$ (e.g., Ribbens et al. 1994 use $c = 3$) (Figure 10.9). A bivariate Gaussian density is obtained for $c = 2$ (recall that $\Gamma(1) = 1$). Fatter-tailed kernels are obtained with values of $c < 1$. The exponential density, which is commonly applied to dispersal data, has $c = 1$ (note that $\Gamma(2) = 1$) (Figure 10.13). Still fatter tails, with $c = 1/2$ ($\Gamma(4) = 3! = 6$), were used by Kot et al. (1996) and Clark (1998) to explore how long-distance dispersal influence population spread.

Katul et al. (2005) recently proposed an inverse Gaussian distribution for wind dispersal of seeds, which is similar in shape and complexity to this power exponential model. For a single point source, it has the added advantage that the two parameters of the model can be related to summaries of wind velocity. Of course, application depends on knowledge of wind conditions at the time of seed release (seed release varies with wind condition), and tree canopies represent distributed sources, thus requiring some type of spatial integration over source volume. Moreover, the solution applies to the downwind direction.

Anisotropic kernels typically require more parameters than do those that are rotationally symmetric. If one is interested only in directional bias in a process, as opposed to identifying the specific relationship between sources and settling locations, then bivariate normal or Student's t distributions can be used, the directional bias being summarized in terms of the correlation matrix. Calder et al. (2004) used an anisotropic kernel to model spread of particulates. These kernels make no distinction between upwind and downwind. Morales et al. (2004) apply a wrapped Cauchy distribution for turn data, which does allow for a flexible treatment of directional data.

For cases where there is a prevailing direction to advection, the situation requires an upwind/downwind distinction. For advection in streams, Lutscher et al. (2005) use an advection-diffusion model to derive a one-dimensional kernel that is Laplace (Section F.3.4) in the absence of diffusion, but is otherwise skewed downstream. A number of models have been used to explicitly include effects of winds on dispersal of seed (e.g., Okubo and Levin 1989; Katul et al. 2005).

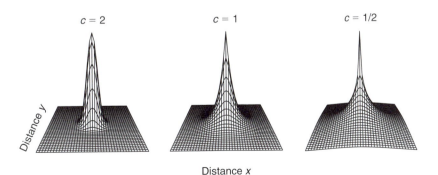

FIGURE 10.13. Seed shadows for the power exponential. Examples are fat-tailed ($c = \frac{1}{2}$), exponential ($c = 1$), and Gaussian ($c = 2$).

10.4.3 Mixtures

Often seeds may be dispersed by more than one mechanism, each with a characteristic scale. For instance, many conifers shed a portion of the seeds from cones still attached to trees, while others remain in the cones. *Cercus canadensis* and *Robinia pseudoacacia* release a portion of their leguminous seeds from pods still on the trees, others remain in pods, while still others are consumed and dispersed by vertebrates. Fleshy-fruited species often have some seeds fall to the ground and others transported by birds, bats, foxes, deer, or bears (e.g., Flinn et al. 2005). In each of these cases we might consider the kernel to be a mixture of two or more processes. For the simplest case of two components, represent this discrete mixture with the fraction p characterized by one dispersal vector and the remaining $1 - p$ by the second vector (Figure 10.14). This mixed kernel has the formula,

$$k(r) = pk_1(r) + (1 - p)k_2(r)$$

where $k_i(r)$ correspond to kernels describing the two components. Ferguson et al. (2001) used such a mixture for spread of foot-and-mouth disease, where one component was independent of distance, representing human transport of infected animals, essentially independent of distance. The two-part mixture is a special case of that with an arbitrary number of components,

$$k(r) = \sum_{i=1}^{n} p_i k_i(r)$$

In practice, discrete mixtures can face identifiability problems. Clark et al. (2003b) use prior information to constrain the components on a discrete mixture for dispersal data.

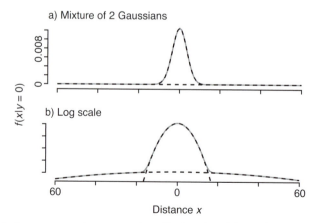

FIGURE 10.14. A two-part mixture, to describe local and long-distance dispersal. This kernel is Gaussian ($c = 2$) and has one percent in a tail with $\alpha = 50$ and 99 percent in the local component with $\alpha = 5$. This conditional representation at $y = 0$ is a cross-section through the two-dimensional kernel.

An alternative to this discrete approach is a *continuous mixture*, which assumes a full range of variability in the dispersal process (Clark et al. 1999). A continuous mixture well describes the case of seed dispersal by many tree species, because seed release over extended time periods means that seeds are subjected to a range of unknown conditions and, in some cases, vectors. A straightforward approach to the problem is to assume that the power exponential is a reasonable description of dispersal for a given instant, but that the dispersion parameter is itself a random variable. Suppose that for a given value of the dispersion parameter α the kernel is Gaussian ($c = 2$),

$$k(r|\alpha) = \frac{1}{\pi\alpha^2}\exp\left[-\left(\frac{r}{\alpha}\right)^2\right]$$

There are mechanistic reasons why this would never be precisely correct, but it works as an approximation. Now if α varies at random according to density $p(\alpha)$, then the kernel actually observed will be more variable than Gaussian. The continuous mixture is a marginal density obtained by integrating over the variability in α,

$$k(r) = \int_0^\infty f(r|\alpha)p(\alpha)d\alpha$$

For conjugate pairs, such as the Gaussian-inverse gamma, we can solve this integral (Box 10.1). This two-dimensional Student's t distribution, or 2Dt, fits seed shadows better than the traditional power exponential, because it better accommodates the tail of the distribution (Clark et al. 1999). The Gaussian model predicts too many seeds at intermediate distances, because data tend to be leptokurtic, with more seeds close to and distant from the source (Figure 10.15). The density of α has a mode at low values, but is also leptokurtic. Other wind-dispersed species show similar shapes. By contrast, the animal-dispersed *Carya* and *Nyssa* have most dispersal clustered near the parent and an especially long tail.

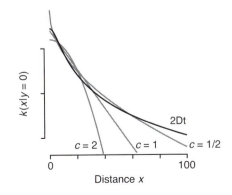

FIGURE 10.15. Comparison of the power exponential (dashed lines) with the 2Dt model for the case where all seeds have the mean dispersal distance of 10. The 2Dt has a fat tail $\left(\text{like } c = \frac{1}{2}\right)$ but also a convex form near the source (like $c = 2$). From Clark et al. (1999).

Box 10.1 The 2Dt dispersal kernel

The parameterization for the two-dimensional Student's t distribution that has been used in several dispersal studies is

$$k(r) = \frac{1}{\pi u \left(1 + \dfrac{r^2}{u} \right)^2}$$

(Clark et al. 2004). This parameterization implies two degrees of freedom and a fat tail. The dispersal distance r is obtained as $r^2 = x^2 + y^2$. The density of dispersal distances for the source at location (0,0) is obtained by first integrating arc-wise,

$$K(r) \equiv \oint_{2\pi} k(r)d\phi = 2\pi r k(r)$$

To obtain the mean dispersal distance, find the first moment from the integral over distance,

$$\mu_1 = \int_0^\infty rK(r)dr = 2\pi \int_0^\infty r^2 k(r)dr$$

We can solve this, beginning with the substitution, $v = r^2/u$, giving us

$$\sqrt{u} \int_0^\infty \frac{v^{1/2}}{(1 + v)^2} dv$$

Recognizing the integral as a beta function, we obtain (see Box D.1)

$$\sqrt{u}B(3/2,1/2) = \sqrt{u}\frac{\Gamma(3/2)\Gamma(1/2)}{\Gamma(2)} \sqrt{u}\frac{\sqrt{\pi}/2 \cdot \sqrt{\pi}}{1}$$

or

$$\mu_1 = \frac{\pi\sqrt{u}}{2}$$

(Clark et al. 1999). With this parameterization, the variance is not finite. The probability of dispersal beyond a distance K is obtained from the cumulative distribution function of the arc-wise integrated kernel $\int_R^\infty K(r)dr$. With the substitution $v = 1 + \frac{r^2}{u}$, we obtain

$$G(R) \equiv \int_R^\infty K(r)dr = \int_R^\infty v^{-2}dv = -v^{-1}\Big|_R^\infty$$

Substituting for v and evaluating limits we have

$$G(R) = \left(1 + \frac{R^2}{u} \right)^{-1}$$

10.4.4 Inference

There are two common types of dispersal data that are fitted to dispersal models. These are dispersal distances and density data. I begin with a classical approach for distance data. This summary follows Clark et al. (1999).

Density Data—Consider a data set consisting of observed dispersal distances that are recorded without regard to direction. These data arise when all propagules are recovered. Each propagule represents an observation with a corresponding distance from the point of release. A sampling distribution for these data is the arc-wise integral of the two-dimensional kernel. For the exponential family this sampling distribution for the ith seed comes from Equation 10.26,

$$L(r_i;\alpha,c) = \oint_{2\pi} \frac{c}{2\pi\alpha^2\Gamma(2/c)} \exp\left[-\left(\frac{r_i}{\alpha}\right)^c\right] d\phi = \frac{r_i c}{\alpha^2\Gamma(2/c)} \exp\left[-\left(\frac{r_i}{\alpha}\right)^c\right]$$

Example 10.8. Estimate a two-dimensional Gaussian kernel from distance data Assume that n propagules are released, and all are recovered. The likelihood function for the Gaussian kernel is

$$L(\mathbf{r};\alpha) = \prod_{i=1}^{n} \frac{2r}{\alpha^2} \exp\left[-\left(\frac{r_i}{\alpha}\right)^2\right] \propto \alpha^{-2n} \exp\left[-\frac{1}{\alpha^2}\sum_{i=1}^{n} r_i^2\right]$$

The MLE is the root mean square distance

$$\alpha_{ML} = \sqrt{\overline{r^2}}$$

Note that this sampling distribution has a mode $r_{mode} = \frac{\alpha}{c^{1/c}}$ that is displaced from the source, due to the increased sampling area with distance. Using the same approach we can show the MLE for exponential parameter (i.e., c is fixed at one in Equation 10.26) to be the mean dispersal distance $\alpha_{ML} = \bar{r}$.

Example 10.9. Estimate a 2Dt kernel with censoring
The 2Dt dispersal kernel with two degrees of freedom has the form given in Box 10.1. Full recovery of released individuals may not be possible, because the area that must be searched increases with the square of distance.

Suppose that n propagules are released, all m propagules that settle within distance R are recovered, and none of the $n-m$ that settle beyond R are recovered. The elements of the likelihood are found in Box 10.1,

$$L(\mathbf{r};u) = \prod_{i=1}^{m} K(r_i;u) \times G(R;u)^{n-m}$$

The second component is the cumulative distribution function that applies to nonrecovered individuals. Of course, we can differentiate the log likelihood with respect to parameter u, but we cannot obtain a closed form solution for the MLE of u. It can be found numerically. This likelihood assumes that we know how many propagules were lost and that all those lost traveled at least distance R.

Indirect Estimation with Density Data—The foregoing likelihood assumes settling distances of propagules are known, at least within some interval. In Example 10.9, distances are known for m individuals, and $n-m$ land in the interval (R, ∞). The kernel itself is the basis for the likelihood of observed travel distances. More often, the process of dispersal is not directly observed, and inference comes from indirect information. We might assume the same underlying process, but require a different sampling distribution and a more complex model, depending on how the data are collected. As discussed in Chapter 8, seed traps are a common method for obtaining seed rain data. A number of traps are deployed, often such that a given trap may collect seed from multiple trees. The seed shadows of individual trees may overlap, making it impossible to say which tree is the source of a given seed. In this case fecundity and dispersal can be estimated using a summed seed shadow model (Figure 10.9). The sampling distribution applies not to the distances themselves, because dispersal distances are unknown. Instead, the sampling distribution applies to numbers of seeds in a trap. Poisson or negative binomial distributions are reasonable candidates. For a Poisson, the likelihood for seeds in the jth seed trap has the form

$$s_j \sim Pois(A_j T_j g_j \, (\mathbf{y}, k(\mathbf{r}_j; u)))$$

where s_j is the density of seed recovered from the jth trap, A_j is the area of the jth trap, T_j is time span for the collection, and $g_j(\ )$ is the model of seed production and dispersal, which depends on a m-length vector of seed production from trees $\mathbf{y}$ over the interval T_j and the vector of distances $\mathbf{r}_j$ from all trees to trap j. The model predicts an expected value for each seed trap, and the Poisson assigns a probability to the observed datum based on this expectation. This is a variant on the examples from Chapter 8. Classical estimation involves maximizing the likelihood function over all parameters (Ribbens et al. 1994; Clark et al. 1998). This is done numerically. Different kernels can be compared using LR tests (nested models) or AIC (Clark et al. 1999).

10.4.5. Some Challenges of Dispersal Estimation

Many processes of ecological interest involve long-distance dispersal. For example, the longest dispersal events can, in principle, have an overwhelming impact on the rate of population spread (Sections 10.3 and 10.5). When R_0 is large, nonparametric kernels are dominated by the most distant observed event (Figure 10.4). Unfortunately, there is no good method for estimating rare events. Parametric dispersal kernels are insensitive to them. The shape of a fitted kernel is overwhelmingly influenced by the abundant, short dispersal distances. Of course, if events are rare, the longest event cannot be observed, as the area of the annulus increases with the square of distance.

"Mechanistic" models that explicitly consider wind fields and behavior of dispersal vectors provide valuable insight into the process of population spread. They do not alleviate the problems of estimating extreme events; rather, they shift the problem to one of constructing scenarios for the extreme wind fields and extreme behavior of vectors. One must make assumptions of, say, how many hurricanes might occur or extreme fluctuations in vertebrate vector abundance and movement. Despite the challenges of estimating and predicting rare events, methods presented here work well for estimation of movement where data are abundant. Where much is unknown, a Bayesian framework can help by allowing for uncertainty and data that may come from multiple sources (Clark et al. 2003b; Gaggiotti et al. 2003; Okuyama and Bolker 2005). Still, extreme events highlight the limitations to inference when data are limited.

10.5 A Bayesian Framework

A Bayesian framework allows for inference for more complex problems. Consider the metapopulation problem, where population survival depends on the capacity to recolonize sites at least as fast as local populations go extinct. Here we are faced with the need to estimate not only the local demographic rates that determine persistence on different size islands, but also the spatio-temporal relationships among islands and movement among them. I mentioned several examples of metapopulation studies in Section 10.3. In this section I extend the Bayesian approach mentioned in Section 8.5 for estimation of seed dispersal based on the same underlying process discussed in Section 10.2.

Many of standard limitations with the classical approach cause severe bias for dispersal estimation of seeds. Inflexibility precludes incorporation of the appropriate structure for fecundity schedules, dispersal, and sampling variability and leads to large overestimates in fecundity and in dispersal distance. To allow for the stochasticity in these factors Clark et al. (2004) combined data models described in Section 8.5 with the longitudinal mixed model of Section 9.17 as outlined in Figure 10.16. The full model includes data models for two data sets: (1) seed traps and (2) status observations that are connected to the fecundity process f and maturation status Q. Estimates were obtained for dispersal $k(r;u)$, with distance r and dispersal parameter u, female fraction ϕ (for species that are dioecious), covariates $\mathbf{x}_i$ with parameters $\boldsymbol{\beta}$, autocorrelation ρ, heterogeneity among individuals τ^2, and prediction of seed production for individual i in year t. The fixed year effects of Section 9.18.4 are an alternative to the autocorrelation approach. Stochasticity brings in additional parameters, including observation errors on tree status ν and process error in log fecundity, σ^2. Note that the connection between data sets for maturation status and seed traps comes with the maturation function describing the probability $\phi_{i,t}$ of being in the mature state $Q_{i,t}$ (Figure 10.16). The observations of tree status, $q_{i,t}$, help constrain the parameter estimates for all other parts of the

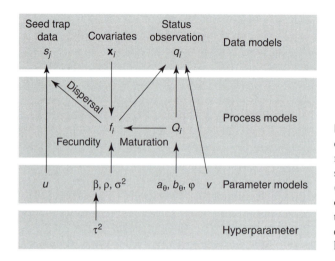

FIGURE 10.16. The model used to estimate fecundity from a combination of status observations and seed trap counts by Clark et al. (2004). This model extends that of Figure 8.11 including an additional stage that allows for heterogeneity among trees, described by variance τ^2.

model, because different status observations (mature or not) have different implications for seed production.

Fecundity $f_{i,t}$ may depend on the size and vigor of the parent plant. It has units of log seeds per year. Fecundity generally increases with tree size (a covariate in $\mathbf{x}_i$), but it also depends on other factors (Figure 9.17). The kernel $k(r)$ is a probability density function describing the density of individuals expected to land per unit area as a function of radius r from the parent. The seed shadow for the ith tree in year t is $g_{ki,t} = f_{i,t} k(r_{ki})$ the number of seeds that land at a location that is distance r (Section 8.5). The predictive distribution for the seed shadow (seed density at distance r) with covariates $\mathbf{x}_i$ mixes over the posteriors for parameters,

$$p(g(f_i, r_i)|\mathbf{x}, \mathbf{s}, q_i) = \int p(g(f_i, r_i)|\mathbf{p}, \mathbf{x}, \mathbf{s}, q_i)p_i(\mathbf{p}|\mathbf{x}, \mathbf{s}, \mathbf{q})d\mathbf{p},$$

where $\mathbf{p}$ is the vector of parameters and latent variables (Section 8.5). As usual, this integration is accomplished directly within the MCMC simulation.

Posterior estimates for some of the parameters related to the maturation schedule and observation error are shown in Figure 10.17. The uncertainty in these parameter estimates is integrated in the predictive distributions of seed dispersal in Figure 10.18. Due to large sample sizes, parameter uncertainty is a small part of the total stochasticity in the model. Most of the spread comes from variability within the population (among individuals) and over years (masting). Figure 10.9 shows how the individual seed shadows combine to produce a seed surface for the landscape.

The shapes of the fecundity schedules and their dependency on resources are important for fitness, because different functional forms imply very different net reproductive rates (Chapter 2). The size at reproductive maturity varies among species (Figure 10.14c) and with resources. LaDeau and Clark (2001)

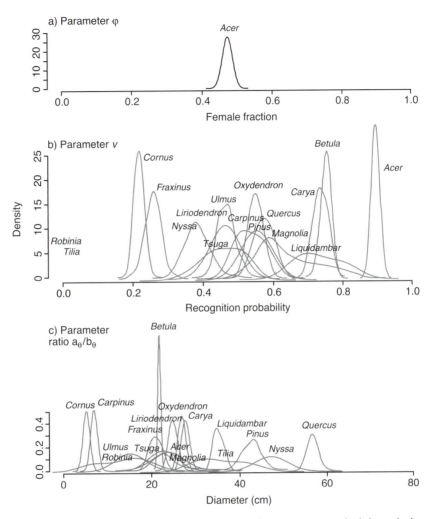

FIGURE 10.17. Posteriors for parameters related to the maturation schedule and observation error. *Acer rubrum* is dioecious and estimated to be approximately half female (*a*). Species vary considerably in terms of our capacity to identify mature individuals from the ground (*b*). The diameter at which about half the population is reproductive also varies widely among species (*c*). From Clark et al. (2004).

modeled reproductive maturity in the context of size and CO_2 content of the atmosphere to show that elevated CO_2 decreases the diameter at which individuals become reproductive (Section 9.17.4).

The model that allows for different sources of stochasticity produced predictions of seed rain that differed substantially from classical methods used by Clark et al. (1999) and that were more consistent with alternative evidence. For all species, the hierarchical Bayes framework (HB in Figure 10.19) yielded much lower estimates of fecundity than were obtained by traditional maximum likelihood approach (ML in Figure 10.19). Greenberg and Parrasol (2002) estimated fecundity of oaks from the same region in situations where crowns were

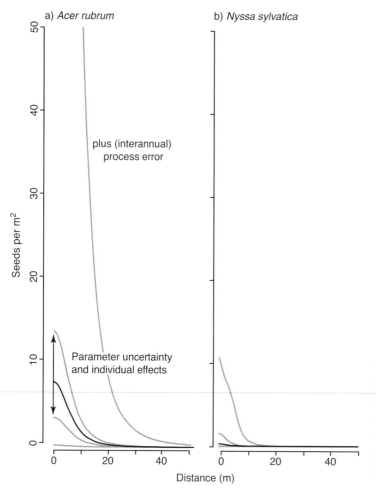

a) *Acer rubrum*

b) *Nyssa sylvatica*

plus (interannual) process error

Parameter uncertainty and individual effects

Seeds per m²

Distance (m)

FIGURE 10.18. Seed shadows predicted from the model in Figure 10.13. The 95 percent predictive intervals integrate parameter uncertainty and random individual effects (inner dashed curves) and process error, which is primarily interannual variability. From Clark et al. (2004).

sufficiently separated from one another such that rough estimates of seed production could be obtained from seed traps (GP in Figure 10.19). Their estimates constitute independent evidence and agree with the hierarchical Bayes approach.

Why do the classical and hierarchical Bayes approaches yield such different answers? There are probably several reasons related to flexibility that allows for more realistic assumptions. With a classical approach we can estimate only one fecundity parameter. The many sources of variability that contribute to fecundity cannot be identified with a classical model, so all stochasticity is shoe-horned into the sampling distribution for seed trap data. This distribution does not capture the variability associated with the fecundity process. The model of seed production is nonlinear and, thus, translates process error and random effects in ways that affect parameter estimates. The flexibility of hierarchical Bayes allows us to organize the deterministic and stochastic elements in ways that make ecological sense and are much less constrained by modeling limitations.

Two of the challenges involved in direct dispersal estimation involve the fact that recovery rates of propagules can be low and correlated with the dispersal

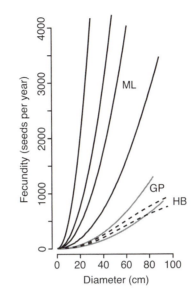

FIGURE 10.19. Estimates of *Quercus* fecundity schedules using classical maximum likelihood (ML), hierarchical Bayes (HB), and an independent study of isolated trees by Greenberg and Parrasol (2002).

process itself. The collinearity results from the fact that area (and thus search effort) expands with the square of distance. For this reason, methods involving density data at a range of distances provide a valuable option. Nonetheless, it is usually not possible to estimate rare, long-distance dispersal events; if they can be observed routinely, they are not rare. If they are truly rare, they will have almost no impact on the shape of a parametric kernel, and nonparametric kernels will be of little use, being overly sensitive to sampling at distance (Clark et al. 2003b; see previous section). Simulation studies can provide valuable insight concerning different transport mechanisms at these scales (Nathan et al. 2002).

Dispersal in complex landscapes, such as those resulting from land cover change, bring in scale-dependent effects that are not always conducive to modeling with simple dispersal kernels. For example, wind- and bird-dispersed seed can tend to accumulate near windbreaks in pastures and around isolated trees (Holthuijzen and Sharik 1985; Guevara and Laborde 1993; Harvey 2000). Clearly, such processes require models that include landscape characteristics.

10.6 Models for Explicit Space

Until now, I have considered spatial processes for which explicit location has not played an important role. In the models discussed thus far, location has mostly been relative, for example, the location of offspring relative to the location of the parent. I have often used a distance variable r_{ij} to stand in for location variables, say, the distance between locations $[x_i, y_i]$ and $[x_j, y_j]$.

It will often be efficient and even necessary to consider location explicitly. It can be efficient, because the distances between n objects occupy an n by n symmetric matrix. For multivariate problems, this becomes cumbersome. In Euclidean space, their locations only require an *n* by 2 matrix of paired coordinates. It can be necessary, because spatial influences may be nonstationary, anisotropic, or both. If so, knowledge of distance is not enough—we also need location (stationarity assumption) from which we can determine direction (isotropy assumption). For weak or second-order stationarity we require a finite mean and covariance, where the latter does not depend on location. (The stationarionarity assumption arises in space, much like it does for time [Section 9.3.2].) Processes within flowing waters are anisotropic, because they depend on whether one location is upstream or downstream of another. Although I focus on two-dimensional processes, much of the discussion generalizes to three dimensions. I begin with stationary, isotropic models, followed by nonstationary models.

For the remainder of this chapter, I reference the location of an observation *i* by its Cartesian coordinates $\mathbf{s}_i = [x_i, y_i]$. If distance is Euclidean (a straight line from object *i* to object *j*), then location is related to distance as

$$r_{ij} = \|\mathbf{s}_i - \mathbf{s}_j\| = \sqrt{(x_i - x_j)^2 + (y_i - y_j)^2} \qquad 10.28$$

where $\mathbf{s}_i = [x_i, y_i]$ and $\mathbf{s}_j = [x_j, y_j]$. Following tradition, in the remaining sections of this chapter, I use *x* and *y* to identify locations, despite the fact that they are also used to designate observations or state variables. Context should make the meanings clear.

Spatial models typically deal with one of three types of data. Let $z(\mathbf{s})$ represent the value of a state variable observed at the set of locations $\mathbf{s} = [\mathbf{s}_1, \ldots, \mathbf{s}_n]$. For the two-dimensional setting a specific location has Cartesian coordinates $\mathbf{s}_i = [x_i, y_i]$. Locations *x* and *y* represent continuous spatial variables. For the first two types of data, there is no stochasticity associated with *x* and *y*. For the third type, location is stochastic.

1. *Point-referenced data*—This term, used by Gelfand et al. (2001), corresponds to Cressie's (1993) term *geostatistical data* and involves observations that derive from fixed locations **s**. Carlin and Louis (2000) use the term *point source data models*. Observations of a process $z(\mathbf{s})$ that is continuous in space at predetermined locations can include air- and water-quality monitoring, variables that determine risk of infection or mortality, and so forth. Although sample locations are fixed, we view the underlying process as being spatially continuous. I consider examples for which $z(\mathbf{s})$ is either continuous or discrete in state.

2. *Block-referenced data*—Observations come from fixed (but not necessarily regular) areas, such as the pixels on digital images and census information that is summarized at the level of jurisdictional units or physiographic provinces. Carlin and Louis (2001) refer to counts or averages obtained for different regions as regional summary data models. Cressie (1993) uses the term *lattice data*, again, without requiring regular spacing. Here $z(B)$ refers to an attribute associated with a sample area or block *B*. It may be viewed as a spatial integral, average, aggregate, and so forth. Because aggregation is involved, within-block heterogeneity and scale- and position-related effects have important consequences for

inference and prediction. As with point-referenced data, sample locations are known.

3. *Point patterns*—For point patterns, location itself is a random process as might apply to spatial patterns of individuals in a population or the locations of events. Of course, point patterns can be spatio-temporal, such as the locations of tree falls across a landscape over time. With such data, one is often asking whether events are, in some sense, random in space, time, or both.

Although point patterns are commonly used for descriptive purposes, they have fewer applications than do data sets with known location. They are addressed by some widely available references (Schabenberger and Gotway [2005] is a recent example). I provide an introduction to the first two of these model types, in Sections 10.7 and 10.8, respectively.

10.7 Point-Referenced Data

10.7.1 A Basic Model

Spatial models can take up space in the process, in the error, or both. There can be differing opinions on this, depending on application and scale, but my view is that it is generally preferable to include spatial interactions in the mean structure of the model. Spatial covariance is treated as error or a random effect when it cannot be explained by the mean structure or it is deemed undesirable to do so. There might be spatial relationships among observations that are so well described by the process model that there is nothing spatial left over for the error term. If so, this happy situation would allow us to treat observations as independent (conditioned on the process). In practice, models often include both, because the process model does not fully accommodate the spatial relationships in data. In fact, many models include space only in the stochastic term(s). Experimental designs, where attention focuses on fixed treatment effects, might address spatial effects through randomization. Coarse-scale spatial variation might be accommodated by a block design, with randomization within blocks to address fine-scale variation. Even if not accommodated by randomization, it might be undesirable to include spatial autocorrelation in the mean structure. This stochastic treatment of space and its relationship to spatio-temporal processes is the topic of the remainder of this chapter. I begin with point-referenced, geostatistical data, where the covariance function or semivariogram is used.

These concepts are related to the distinction between design-based inference and model-based inference. The former is based on random sampling, whereas the latter is based on the assumption that the underlying process is uncertain (Varhoef 2002).

Consider a variable $z(\mathbf{s})$ measured at locations $\mathbf{s}_1, \ldots, \mathbf{s}_n$. Due to environmental heterogeneity, locations close to one another might be more similar than locations distant from one another. At fine spatial scales, they covary, but this correlation of an attribute z with itself decays with distance. Obvious examples include soil properties (Robertson 1987). If we design an experiment that requires the assumption of independent samples, we might simply overstep intermediate scale spatial covariance, locating samples at random, with the constraint that they be sufficiently distant from one another to minimize spatial covariance. Otherwise, there will be redundancy in samples: close samples do not bring much new information, and they violate the assumption of independence. If goals include inference on the scale of spatial variation or on a spatial process, then close sampling may be critical, and leftover spatial error may be unavoidable. Blocking is used to allow for variation at coarse scales. For example, to estimate seed germination within and among stands, Ibáñez et al. (2007) modeled random stand effects and random plot effects to account for variation not explained by covariates at scales of landscapes (10 ha) and microsites (10 m^2), respectively.

Suppose that we cannot assume spatial independence at the scales of interest. The spatial model $z(\mathbf{s})$ in two or more dimensions is often referred to as a *random field*. A *Gaussian random field* refers to the case where $\mathbf{z}(\mathbf{s})$ is assumed to be multivariate normal with a vector of means and a covariance matrix. The interactions among locations are assumed to go both directions, although they could be anisotropic.

The covariance function is central to spatial modeling and enters through the covariance matrix. For a simple linear model with Gaussian error, we might move from the standard covariance matrix used for simple regression models, with zero off-diagonal elements, to one that admits a nonzero element for every element of the covariance matrix. This symmetric matrix contains $n(n + 1)/2$ unique elements, that is, a large number of estimates. Unlike in multivariate methods, where there is true replication to support estimates of these elements, few spatial applications entail true replication—each sample comes from a different location (see Sampson and Guttorp [1992] for a method that applies when true replication is available). If samples are obtained repeatedly from the same locations, we often view them as samples from a spatio-temporal process and, thus, still not as true replicates of the same location and instant. So we typically cannot expect to estimate all of the elements of a covariance matrix for spatial or spatio-temporal models.

Many of the modeling strategies for spatial data involve ways to limit the dimensionality of this covariance matrix. This is often possible, because the variation in data may be concentrated in a small number of dimensions. This is the case when, say, covariance declines with distance. The covariance matrix has nonzero values concentrated near the diagonal and zero or near zero values far from the diagonal. For block-referenced or lattice data, a model for this situation could allow for a neighborhood of interactions, as when the observation in a patch conditionally depends only on the patches that border it. Here, spatial interactions are summarized by a connectivity matrix. I discuss block-referenced data in Section 10.8. For point-referenced data, we might specify a parametric form for the covariance matrix as a function of distance. Although all elements of the covariance matrix may be nonzero, they can all come from a function of distance that includes only a few estimated parameters. Parametric representations of spatial covariance can assure that the covariance matrix will be positive definite.

A basic model for a spatial process might look like this,

$$z(\mathbf{s}) = \mu(\mathbf{s}) + w(\mathbf{s}) + \varepsilon(\mathbf{s}) \qquad 10.29$$

where $\mu(\mathbf{s})$ is a (deterministic) trend surface, $w(\mathbf{s})$ are random spatial effects, and $\varepsilon(\mathbf{s})$ are observation errors (Cressie 1993). To limit notation, I have omitted a subscript i on $\mathbf{s}$ in each of the terms in Equation 10.28. We might think of these three terms as process or coarse-scale variation, additional finer-scale spatial relationships not already contained in the process term, and error, respectively. For now, I assume this to be a second-order stationary process, given that the mean is described by $\mu(\mathbf{s})$. In other words, the full process need not be stationary, but the stochastic terms

$$w(\mathbf{s}) + \varepsilon(\mathbf{s}) = z(\mathbf{s}) - \mu(\mathbf{s})$$

are stationary. For these terms, the mean is everywhere zero, the covariance is finite, and the covariance between two points depends only on their distance and not on their location. This two-dimensional model applies to map data. I say more about stationarity later in this section as part of the discussion of variograms.

Clearly, each of these terms can be spatial, potentially accommodating variation at different scales. At the coarsest scale, the mean response surface can be spatial, and might involve explanatory covariates. A simple linear form could be

$$\mu = \mathbf{X}\beta$$

where $\mu = [\mu(\mathbf{s}_1), \ldots, \mu(\mathbf{s}_n)]$, $\mathbf{X}$ is the n by p design matrix, and β is the length-p vector of parameters. Spatial covariates can enter through design matrix $\mathbf{X}$. A row of this design matrix could be as simple as, say, an intercept with Cartesian coordinates,

$$\mathbf{x}_i = \begin{bmatrix} 1 & \mathbf{s}_i^T \end{bmatrix} = \begin{bmatrix} 1 & x_i & y_i \end{bmatrix}$$

This would result in a planar surface. Higher-order (polynomial) terms can be used to allow for hills and valleys. Alternatively, covariates in the design matrix could be spatial. This mean structure is, of course, deterministic (unless random effects enter through, say, β—more on this in Section 10.9). Alternatively, μ could be nonspatial.

There are two stochastic terms, introduced to describe spatial and non-spatial errors, respectively. By nonspatial, I mean that the last term does not have spatial covariance—of course, there is a different value for this term at each location. It combines variation at scales too fine to be resolved by sampling and measurement error (Cressie 1993). These two sources of variation could only be resolved through true replicate sampling. The distribution might be normal,

$$\varepsilon(\mathbf{s}_i) \sim N(0, \sigma^2)$$

with $cov[\varepsilon(s_i),\varepsilon(s_j)] = 0$ for all $i \neq j$. The covariance matrix is diagonal, $\Sigma = \sigma^2 I_n$. If measurement error is independent of fine-scale spatial variation, it makes sense to view σ^2 as simply the sum of these two variances.

The central term $w(s)$ takes up spatial variation at intermediate scales. Said another way, for all $i \neq j$ $cov[z(s_i),z(s_j)] = cov[w(s_i),w(s_j)]$, because spatial covariance is limited to the second term. For this term, we consider stochastic relationships involving the covariance function and related representations, such as the semivariogram. To simplify notation, let C_{ij} represent an interaction or relationship between location s_i and s_j. Initially, assume that $C_{ij} \equiv cov[w(s_i),w(s_j)]$, the ij^{th} element of a covariance matrix for observations $z(s)$. Spatial correlation is related to the covariance as

$$\rho(r_{ij}) = \frac{C_{ij}}{C_{ii}}$$

(again, still assuming second-order stationarity). The subscript in the denominator represents the nonspatial variance, which, for this stationary process, is taken to be everywhere the same. If the variance in the w term at zero distance is $var(w(s)) = cov[w(s_i),w(s_j)] = C_{ij} = \sigma_w^2$, then the covariance function can be written as

$$C_{ij} = \sigma_w^2 \rho(r_{ij})$$

where $\rho(r)$ is the correlation function, which has a value of one at distance $r = 0$ (Section 9.3.7). Note that the total variance of $z(s)$ is

$$var(z(s)) = var(w(s)) + var(\varepsilon(s)) = C_{ii} + \sigma^2 = \sigma_w^2 + \sigma^2$$

Still under the assumption of second-order stationarity (Section 9.3.2), the covariance between locations i and j only depends on the distance between them, r_{ij}, not on their actual locations. Much like the assumption of independent and identically distributed (i.i.d.) random variables simplifies nonspatial analysis, this stationarity assumption greatly facilitates spatial analysis.

A traditional way of expressing spatial relationships in the geosciences involves the mean squared difference in z with increasing distance r, or

$$2\gamma(r_{ij}) \equiv var[z(s_i) - z(s_j)] \qquad 10.30$$

This is the variance of a difference, and can be solved using the same approach applied to the variance of a sum in Section D.5.3,

$$var[z(s_i) - z(s_j)] = var[z(s_i)] + var[z(s_j)] - 2cov[z(s_i),z(s_j)]$$

With second-order stationarity, the first two terms on the right-hand side both equal the sum of the total nonspatial variance, $\sigma_w^2 + \sigma^2$. With appropriate substitutions we have the *semivariogram*

$$\gamma(r_{ij}) = \sigma^2 + \sigma_w^2(1 - \rho(r_{ij}))$$

This is equivalent to

$$\gamma(r_{ij}) = C_{ii} - C_{ij} = \sigma^2 + \sigma_w^2 - \sigma_w^2 \rho(r_{ij})$$

For spatial modeling, the correlation function is assumed to have a parametric form, such as an exponential,

$$\rho(r_{ij};\phi) = e^{-\phi r_{ij}}$$

The parameter ϕ must be estimated along with other parameters in the model. A number of parametric forms are used to describe distance relationships.

There are several terms associated with semivariograms that reflect the fact that spatial covariance decreases with distance. As distance between locations approaches zero, we have the *nugget effect*, that is, measurement error and fine-scale variation. This is the value assigned to $\gamma(0) = \sigma^2$. Note that if there is no variation at this fine scale, then $\lim_{r_{ij} \to 0} (C_{ii} - C_{ij})$ and $\gamma(r_{ij}) = 0$.

With increasing distance between locations, the semivariogram may increase and level off at a *sill*. This value applies to $\lim_{r_{ij} \to \infty} \gamma(r_{ij}) \to \sigma^2 + \sigma_w^2$. This value is reached at an approximate distance termed the *range*, which might have an explicit parameter associated with it in the function ρ.

The semivariogram is an alternative way to represent the covariance function. So why might we choose to work with a semivariogram? One advantage of the semivariogram comes from that fact that it may be used when the process $z(\mathbf{s})$ is not second-order stationary. Recall from Section 9.3.2 that a time series is second-order stationary if the mean and covariance do not depend on when the process is observed. A random walk is not stationary, having mean and variance given by Equation 9.2, but the differenced series is stationary (Example 9.1). We have the same idea applied to spatial processes—for the covariance function to be valid, the covariance function must be positive-definite. If the process is not second-order stationary, the differenced process $z(\mathbf{s}_i) - z(\mathbf{s}_j)$ may still be stationary. It is easy to show that all second-order stationary processes satisfy this weaker restriction:

$$\begin{aligned}
\mathrm{var}[z(\mathbf{s}_i) - z(\mathbf{s}_j)] &= \mathrm{var}[z(\mathbf{s}_i)] + \mathrm{var}[z(\mathbf{s}_j)] - 2\mathrm{cov}[z(\mathbf{s}_i),z(\mathbf{s}_j)] \\
&= 2\mathrm{var}[z(\mathbf{s}_i)] - 2C_{ij} \\
&= 2(C_{ii} - C_{jj}) = 2\gamma(r_{ij})
\end{aligned}$$

If the process is not second-order stationary, then the C_{ij}'s are not valid covariances, but one can still calculate a semivariogram.

Returning to the model based on Equation 10.28, because terms are independent, the full covariance matrix is obtained by summing the covariance matrices that apply to the second and third terms,

$$Var[z(\mathbf{S})] = \mathbf{C}_\phi + \Sigma$$

where $\Sigma = \sigma^2 \mathbf{I}_n$. This can also be expressed as a two-stage model

$$\mathbf{z}|\mathbf{W}_\phi \sim N_n(\mathbf{X}\beta + \mathbf{W}_\phi, \Sigma)$$

$$\mathbf{W}_\phi \sim N(0, \mathbf{C}_\phi)$$

where $\mathbf{C}_\phi$ is the covariance matrix for spatial effects with elements $C_{ij} = \sigma_w^2 \rho(r_{ij};\phi)$.

Integrating over random spatial effects yields the marginal distribution

$$\mathbf{z} \sim N_n(\mathbf{X}\beta,\mathbf{\Sigma}_\theta)$$

where the covariance matrix $\mathbf{\Sigma}_\theta = \mathbf{C}_\phi + \mathbf{\Sigma}$ contains parameters $\theta = [\sigma_w^2,\sigma^2,\phi]$. For known θ, we have the generalized least-squares estimates of regression parameters,

$$\hat{\beta}|\theta = (\mathbf{X}^T\mathbf{\Sigma}_\theta^{-1}\mathbf{X})^{-1}\mathbf{X}^T\mathbf{\Sigma}_\theta^{-1}\mathbf{y}$$

Of course, θ will typically be unknown.

The spatial relationships introduce some new issues for such estimators. I mentioned earlier that positive autocorrelation reduces the amount of information in the data. To illustrate the effect on the standard error in estimates, consider just the variance for a simple least-squares estimate of the mean surface μ having variance σ^2 and AR(1) correlation ρ (Equation 9.9),

$$Var[\hat{\mu}] = (\mathbf{1}^T\mathbf{\Sigma}_\rho^{-1}\mathbf{1})^{-1} = \frac{\sigma^2(1 + \rho)}{n(1 - \rho) + 2\rho}$$

In the absence of autocorrelation ($\rho = 0$) we recover the parameter variance seen previously (Equation 5.3). If there is complete correlation ($\rho = 1$), then we have the information equivalent to one sample, that is, $Var[\hat{\mu}] = \sigma^2$; repeated sampling does nothing to improve the estimate. In other words, we are just observing the same thing, not repeated samples thereof.

Moreover, because parameters in the covariance matrix θ are unknown, regression parameter estimates are typically obtained by plugging in estimates for variance parameters obtained by other means. MLE's for parameters contained in the covariance matrix are biased. Restricted maximum likelihood (REML) results in less bias (e.g., Ver Hoef et al. 2001). In general, most methods rely on asymptotic properties. Because the Bayesian approach readily accommodates parameter uncertainty, I focus on hierarchical formulations in a Bayesian context that can be simulated using MCMC.

The covariance function and semivariogram represent ways to simplify the covariance structure of spatial data. With these tools, we can move to inference and prediction. First I review Bayesian kriging, a common approach that focuses on spatial prediction.

10.7.2 Bayesian Kriging

Kriging has traditionally emphasized predicting spatial pattern in a variable that has been sampled at known locations. It makes explicit use of the spatial covariance to obtain a predictive interval for new locations. From observations $z(\mathbf{s})$ at locations $\mathbf{s} = [\mathbf{s}_1, \ldots, \mathbf{s}_n]$ we wish to predict $z(\mathbf{s}')$ at locations $\mathbf{s}' = [\mathbf{s}'_1, \ldots, \mathbf{s}'_m]$. I refer to the latter as a prediction grid. Note that the number of sample locations

(n) need not equal the number of locations (m) at which we will predict $z(\mathbf{s}')$. For example, the concentration of a nutrient $z(\mathbf{s})$ might be sampled at random or irregularly spaced locations, but we wish to produce a regular grid of predictions $z(\mathbf{s}')$, that is, a map. The most common approach to this problem is termed *kriging*. Unfortunately, the prediction intervals obtained by traditional means do not include the uncertainty in the parameter estimates. The classical approach estimates variance conditioned on parameters, but does not provide a straightforward mechanism for integrating parameter uncertainty into predictions. I focus on Bayesian kriging, because it has the advantage of providing estimates of parameter uncertainty and incorporating it in predictions of $z(\mathbf{s}')$, again relying on the wonders of MCMC. Carlin and Louis (2000) and Banerjee et al. (2004) provide descriptions of Bayesian kriging, which involves a parametric model for spatial covariance to obtain the predictive distribution $p(z(\mathbf{s}')|z(\mathbf{s}))$. Both data and inference are point-referenced; they just refer to different points. A brief summary of kriging brings us immediately to the change of support problem and block-referenced data.

Consider the stationary Gaussian process, disregarding for now observation error ε,

$$z(\mathbf{s}_i) = \mu(\mathbf{s}_i;\beta) + w(\mathbf{s}_i;\phi),$$

where $\mu(\mathbf{s}_i)$ is the underlying trend surface with parameters β, and $w(\mathbf{s})$ accounts for spatial covariance. As above, the parameter ϕ determines the scale of the spatial correlation—we will estimate it together with β. Let $\mathbf{z}_s = [z(\mathbf{s}_1), \ldots ,z(\mathbf{s}_n)]$ represent the set of n measurements at sample locations and $\mu_\beta = [\mu(\mathbf{s}_1;\beta), \ldots , \mu(\mathbf{s}_n;\beta)]$ be the vector of mean values. The spatial component, $\mathbf{w}_s$ and, therefore, $\mathbf{z}_s$, has the n by n covariance matrix $\mathbf{C}_\phi$ with elements

$$C_{ij} = \text{cov}(z(\mathbf{s}_i),z(\mathbf{s}_j)) = \sigma^2 \rho(r_{ij};\phi),$$

where the autocorrelation function might be a simple exponential, $\rho(r_{ij};\phi) = e^{-\phi r_{ij}}$ to simplify notation, $\mathbf{z}_s \equiv \mathbf{z}(\mathbf{s})$. The parameter ϕ determines the range of the process, with large ϕ indicating a short range. The stationary Gaussian model has the likelihood

$$p(\mathbf{z}_s|\beta,\sigma^2\phi) = N_n(\mathbf{z}_s|\mu_s,\mathbf{C}_\phi)$$

$$= (2\pi)^{-n/2}|\mathbf{C}_\phi|^{-1/2} \exp\left(-\frac{1}{2}(\mathbf{z}_s - \mu_s)^T\mathbf{C}_\phi^{-1}(\mathbf{z}_s - \mu_s) \right)$$

where μ_s is the mean at location s and includes parameters β. Here are the basic kriging steps for predicting points from point-referenced data in a Bayesian framework (lab moment):

1. Specify priors and construct an MCMC algorithm for simulation. A trend surface that is linear in regression parameters (e.g., a polynomial) can have normal priors for β and an *IG* prior for σ^2, which allows for direct sampling (Chapter 5). The scale parameter ϕ must be

sampled indirectly (e.g., Metropolis-Hastings). At this point the algorithm will produce chains of estimates, the gth sample being $[\beta^{(g)}, \phi^{(g)}]$.

2. We desire a predictive distribution for the m new points $\mathbf{s}'$, $p(\mathbf{z}_{s'}|\mathbf{z}_s)$. Begin with the joint distribution of $\mathbf{z}_s$ and $\mathbf{z}_{s'}$ (conditioned on parameter values),

$$p\left(\begin{bmatrix} \mathbf{z}_s \\ \mathbf{z}_{s'} \end{bmatrix}\Big|\beta,\phi,\sigma^2\right) = N_{n+m}\left(\begin{bmatrix} \mu_s \\ \mu_{s'} \end{bmatrix}, \begin{bmatrix} \mathbf{C}_s & \mathbf{C}_{s,s'} \\ \mathbf{C}_{s,s'}^T & \mathbf{C}_{s'} \end{bmatrix}\right),$$

where $\mathbf{C}_{s,s'} \equiv \text{cov}(\mathbf{z}_s,\mathbf{z}_{s'})$, each element of which would depend on the distance between the respective locations in $\mathbf{s}$ and $\mathbf{s}'$. This is an application of prediction using linear models from Chapter 8. From this joint distribution of observations and predictions, obtain the conditional distribution of predictions from the m dimensional normal distribution with conditional mean and variance

$$E(\mathbf{z}_{s'}|\mathbf{z}_s,\beta,\phi) = \mu_{s'} + \mathbf{C}_{s,s'}^T\mathbf{C}_s^{-1}(\mathbf{z}_s - \mu_s)$$

$$\text{var}(\mathbf{z}_{s'}|\mathbf{z}_s,\beta,\phi) = \mathbf{C}_{s'} - \mathbf{C}_{s,s'}^T\mathbf{C}_s^{-1}\mathbf{C}_{s,s'}$$

In other words, draw values for the prediction grid $\mathbf{z}_{s'}$ directly from the multivariate normal having this mean and variance.

3. We now have the distribution for predictions that is conditioned on parameter values. Predictions that integrate parameter error must marginalize over the posterior uncertainty in β and ϕ. This is easy, because at each Gibbs step g, draw values for $\mathbf{z}_s^{(g)}$ from the conditional distribution in Step 2. In other words, simulate the predictive distribution of $\mathbf{z}_{s'}$ simultaneously with the parameters. Parameter uncertainty is mixed together with the spatial effects.

It can be noted that the exponential covariance model

$$\text{cov}(z(\mathbf{s}_i),z(\mathbf{s}_j)) = \sigma^2 e^{-\phi r_{ij}}$$

is a special case of the Matern class

$$\text{cov}(z(\mathbf{s}_i),z(\mathbf{s}_j)) = \frac{\sigma^2 \nu^{1/2}\phi r_{ij}}{2^\nu \Gamma(\nu)}K_\nu(2\phi\nu^{1/2}r_{ij}) \qquad 10.31$$

where ν determines the smoothness of the function and K_ν is the modified Bessel function, which is available in many software packages due to its engineering applications. This is exponential for $\nu = \frac{1}{2}$, and it approaches the Gaussian as ν becomes large. This kernel comes up in later sections of this chapter.

10.7.3 Noninformative Priors

The spatial covariance structure will often be unknown, if for no other reason than its inclusion in the model derives from concerns that spatial correlation may exist (or not). If there is no insight as to the nature of this correlation, we would like to specify a noninformative prior for it. Unfortunately, if the mean of the random field is unknown, the likelihood for correlation parameters is bounded away from zero and, thus, the posterior is improper. This problem was identified and explored by Berger et al. (2001) and led to recommendation of a reference, nonuniform prior that results in a proper posterior. They provide general results. I illustrate using a specific example.

Consider the random field having the mean value at location i given by

$$\mu(\mathbf{s}_i;\beta) = \mathbf{x}_i\beta$$

with covariance matrix having elements

$$C_{ij} = \sigma^2 e^{-\phi r_{ij}}$$

the full matrix being represented as the product

$$\mathbf{C} = \sigma^2 \mathbf{R}_\phi$$

where $\mathbf{R}_\phi$ is the symmetric matrix containing correlations. The vector $\mathbf{x}_i$ will typically have a leading one, corresponding to the intercept β_0. Remaining columns may involve locations x_i and y_i, or they may be values of covariates at location $\mathbf{s}_i$. The likelihood is

$$p(\mathbf{z}|\beta,\sigma^2,\phi) = \frac{1}{(\sqrt{2\pi}\sigma)^2|\mathbf{R}_\phi|^{1/2}}\exp\left[-\frac{1}{2\sigma^2}(\mathbf{z} - \mathbf{X}\beta)^T\mathbf{R}_\phi^{-1}(\mathbf{z} - \mathbf{X}\beta)\right]$$

Berger et al.'s reference prior is

$$p(\beta,\sigma^2,\phi) = \frac{1}{\sigma^2}\sqrt{tr(\mathbf{V}^2) - \frac{1}{n-p}tr(\mathbf{V})^2}$$

where

$$\mathbf{V} = \left(\frac{\partial}{\partial\phi}\mathbf{R}_\phi\right)\mathbf{R}_\phi^{-1}\left[\mathbf{I} - \mathbf{X}(\mathbf{X}^T\mathbf{R}_\phi^{-1}\mathbf{X})^{-1}\mathbf{X}^T\mathbf{R}_\phi^{-1}\right]$$

and p is number of parameters. The first factor on the right-hand side is the matrix of derivatives of each element, in this case having elements

$$\left(\frac{\partial}{\partial\phi}\mathbf{R}_{\phi,ij}\right) = -r_{ij}\mathbf{R}_{\phi,ij}$$

This prior is noninformative in regression parameters β and is proportional to σ^{-2}.

Should you use this prior? Perhaps, but I think the important point is to be careful with noninformative priors. Throughout this book I have recommended that proper priors help to head off some hard-to-identify problems. In the case of spatial covariance it can be worthwhile to adopt a strategy of examining sensitivity to different proper priors, rather than specifying an improper one. In the case of spatial autocorrelation one may often have some basis for prior specification derived from considerations of scales of interest and so forth.

10.8 Block-Referenced Data and Misalignment

10.8.1. A Basic Model

The inconvenient referencing of spatial information can be one of the most challenging and time-consuming aspects of spatial models. Ecological problems often come with several data sets that are referenced in different ways. We may wish to draw inference at a different reference or to predict underlying state variables that vary continuously based on data that come from discrete areas or blocks. As with many aspects of public health, climate, and geosciences, ecologists cannot avoid the change of support problem, which concerns the fact that the spatial reference of data sets and of inference can differ. For example, Gelfand et al. (2001) consider the relationship between ozone levels (point-referenced measurements) and human health (block-referenced records organized by zip code). Fuentes and Raftery (2005) combine weekly averages of SO_2 concentrations from an irregular point-referenced data set with block-referenced model predictions based on emissions data and local weather. Wikle et al. (2001) consider weather predictions at one scale together with satellite imagery at another. In the latter two cases, one of the data types is model output. There may be nested grids, where observations at one scale might be fully contained within a single block at the next (e.g., counties within states, watersheds or life zones within continents) (Wikle and Berliner 2005). Block-referenced data sets may be misaligned, say, if one data set comes from counties and another from precincts or from satellite imagery (Agarwal et al. 2005). Climate impacts on population growth can involve matching data sets with different reference (e.g., Stenseth et al. 1999). How do we draw inference and make predictions from these different data types? I outline the general approach of Gelfand et al. (2001), leaving several examples for later.

The example of Bayesian kriging from Section 10.7, that is, points to points ($\mathbf{z}_s$ to $\mathbf{z}_{s'}$), is one of four possibilities (Gelfand et al. 2001). A typical notation for block-referenced data, where the observation represents an average, looks like this:

$$z(B_j) = |B_j|^{-1} \int_{B_i} z(\mathbf{s}) d\mathbf{s}$$

Here, $|B_j|$ is taken to be the area of block B_j, and the integral represents the spatial average of the process over the block (Cressie 1993), that is, the locations $\mathbf{s}$ that lie within B_j. Again, it can be useful to think of $z(\mathbf{s})$ as a surface. Because we are dealing with random samples, this is a stochastic integral.

This integral cannot be used for population abundance, which is discrete at any given location s_i. Likewise, it could not be used for a binary variable, such as mortality, but it could be used for the underlying mortality risk. Such problems involving non-Gaussian sampling distributions build on this basic framework (Diggle et al. 1998), as in Section 10.4. Here I discuss how Bayesian kriging can be extended to the change of support problem (Gelfand et al. 2001).

Let $\mathbf{z}_B = [z(B_1), \ldots, z(B_m)]$ represent the set of all block averages, that is, one for each map unit. We may have point-referenced data as the basis for block inference ($\mathbf{z}_s$ to $\mathbf{z}_B$), or vice versa ($\mathbf{z}_B$ to $\mathbf{z}_s$). Or, we may have data from one set of blocks and we desire inference at some different set of blocks ($\mathbf{z}_B$ to $\mathbf{z}_{B'}$). This is still a kriging problem, but it involves a few more considerations than the points to points case. In ecology, the typical approach involves simply averaging over samples from locations $\mathbf{s}$ that fall within block B_j. This can be problematic, because there may be no samples within a given block. Moreover, averaging block by block ignores spatial relationships that span blocks.

We first need the covariances involving blocks. For random points in blocks j and j' the covariance matrix has elements

$$C_{B_{jj'}} = |B_j|^{-1}|B_{j'}|^{-1}\sigma^2 \int\limits_{B_j} \int\limits_{B_{j'}} \rho(\mathbf{s},\mathbf{s}';\phi)d\mathbf{s}d\mathbf{s}'$$

For a location $\mathbf{s}_i$ and a random point in block j the covariance matrix has elements

$$C_{s_i,B_j} = |B_j|^{-1}\sigma^2 \int\limits_{B_j} \rho(\mathbf{s}_i,\mathbf{s}';\phi)d\mathbf{s}'$$

Blocks have irregular shapes, so Gelfand et al. (2001) recommend evaluating these integrals numerically under the assumption of uniform distributions within blocks.

To predict blocks from points, use the same kriging steps with the addition of some estimates for the elements of Step 2. For the points-to-points case, the $\mu(\beta^{(g)})$ and $\mathbf{C}_{s,s'}(\sigma^{2(g)},\phi^{(g)})$ are directly available in the MCMC simulation—draw samples for the parameters, which, in turn, are the basis for sampling the predictive distribution. For points-to-blocks, this joint distribution is needed:

$$p\left(\begin{bmatrix} \mathbf{z}_s \\ \mathbf{z}_B \end{bmatrix}\middle|\beta,\phi,\sigma^2\right) = N_{n+m}\left(\begin{bmatrix} \mu_s \\ \mu_B \end{bmatrix},\begin{bmatrix} \mathbf{C}_s & \mathbf{C}_{s,B} \\ \mathbf{C}_{s,B}^T & \mathbf{C}_B \end{bmatrix}\right)$$

I have simply replaced the $\mathbf{s}'$ subscripts with B subscripts. Unlike the points-to-points case, these elements are not directly available. They can be obtained by spatial averaging, taking account of covariances. Gelfand et al. (2001) suggest taking these averages from within blocks as part of the MCMC algorithm. To obtain this average, establish $l = 1, \ldots, L_k$ sample locations $\mathbf{s}_{k,l}$ in the kth block. The more sample points, the better will be the representation of this average. For the gth Gibbs step and the kth block we would obtain the estimate of the surface (i.e., the kth element of the vector μ_B) as

$$\mu_{B_k}{}^{(g)} = L_k^{-1} \sum_l \mu(\mathbf{s}_{k,l}; \beta^{(g)})$$

For the j,k^{th} element of the covariance matrix $\mathbf{C}_B$, we have

$$\mathbf{C}_{B_{j,k}}{}^{(g)} = L_j^{-1} L_k^{-1} (\sigma^2)^{(g)} \sum_l \sum_{l'} \rho(r_{jl,kl'}; \phi^{(g)})$$

where $r_{jl,kl'}$ is the distance between sample points located at $\mathbf{s}_{jl}$ and $\mathbf{s}_{kl}$. $\mathbf{s}_{jl}$ is the location of the lth sample point in the jth block. Both of these sets of sample points are not data; they are sample points established for the purposes of estimating block averages. For the i,k^{th} element of the cross covariance $\mathbf{C}_{s,B}$ we have

$$\mathbf{C}_{s_i,B_k}{}^{(g)} = L_k^{-1} (\sigma^2)^{(g)} \sum_l \rho(r_{i,kl}; \phi^{(g)})$$

$r_{i,kl}$ is the distance between the observation located at $\mathbf{s}_i$ and the sample point established at $\mathbf{s}_{kl}$ for purposes of estimating the surface in block B_k. $\mathbf{s}_{jl}$ is the location of the lth sample point in the jth block. We can use the same set of sample locations $\mathbf{s}_{jl}$ for each MCMC step. We now have all of the elements needed to sample from the conditional distribution, which follows directly from Step 2. The other cases (blocks-to-points, blocks-to-blocks) follow directly from Gelfand et al.'s (2001) paper, which includes an extension to spatio-temporal data.

10.8.2 Conditional Autoregression

One of the most common treatments of spatial covariance for block-referenced data involves means and variances for block j or point location $\mathbf{s}_i$ conditional on (1) distances from other samples, (2) proximity to neighboring blocks, or (3) a combination thereof. Here I consider conditional autoregression, where neighboring blocks share some connectivity, and this spatial relationship is used as basis for representing autocorrelation. The approach begins with a specification of the expected response conditioned on nearby blocks. For block j, we need to specify

$$z_j | \{z_k, k \neq j\}$$

Define a neighborhood for a block j that might consist of all blocks that are, say, adjacent to it. If z_j conditionally depends only on neighbors, then we have a spatial Markov process, termed a *Markov random field*. This neighborhood is designated $\{n_j\}$ and consists of n_j neighbors. If block j conditionally depends only on its neighborhood, then there is need to specify only those relationships,

$$z_j | \{z_k, k \in \{n_j\}\}$$

The interactions among locations are represented by elements of a matrix **C**, which is no longer the covariance matrix, but rather contributes to it. It defines connectivity of the landscape. In effect, the elements of **C** are weights. If the variance is constant σ^2, then the joint distribution is

$$\mathbf{z} \sim N_n \left(\boldsymbol{\mu}, (\mathbf{I}_n - \mathbf{C})^{-1}\sigma^2\mathbf{I}\right)$$

where the expectation for a given block depends on contributions from neighbors

$$E[z_j | \{z_k, k \neq j\}] = \mu_j + \sum_k C_{jk}(z_k - \mu_j)$$

There are some constraints on the covariances and, thus, on **C**. The matrix $(\mathbf{I} - \mathbf{C})^{-1}$ must be symmetric and positive definite. The diagonal elements $C_{jj} = 0$. There must be symmetry in the sense that $C_{jk}\sigma^2 = C_{kj}\sigma^2$. The weights in **C** can depend on distances among locations. If so, the weights decrease with distance.

The neighborhood might consist only of adjacent blocks or of blocks within a certain distance. This is known as a conditional autoregressive (CAR) model. In the following section I consider such covariance structures in the context of some general spatial approaches.

10.9 Hierarchical Treatment of Space

10.9.1. Elements of the Model

As with nonspatial models, it can often be efficient to specify relationships conditionally and work from a hierarchical structure. Consider how the elements of the model contribute to spatial inference. Block-referenced data typically entail numbers of events. Those counts may derive from a process that is viewed as more or less spatially continuous. Such data models are non-Gaussian, in that a binomial or Poisson sampling distribution may describe observations for the block, but it can be underlain by a Gaussian structure viewed as continuous not only in space, but also in state (Diggle et al. 1998). The approach can be likened to generalized linear models (Chapter 8), in which a linear model lies beneath a non-Gaussian sampling distribution with, say, a log or logit link function, although other assumptions are possible (e.g., Henderson et al. 2002; Kelsall and Wakefield 2002).

As for count data in general, the commonly used likelihoods are binomial, Poisson, or (less often) multinomial. A Poisson likelihood for the number of events for group j in block i at time t is

$$z_{ijt} \sim Pois(A_{ijt} y_{ijt})$$

where the constants A_{ijt} incorporate factors affecting the sample size, including sample duration and area. Conditional on y, observations are independent.

We take up spatial effects at the next stage. In disease risk models, the Poisson parameter may be reparameterized to a relative risk (Carlin and Louis 2001; Thogmartin et al. 2004).

At the lower stage, effects might be described by a linear model, with a suitable link function (e.g., logit for the binomial, log for the Poisson). This model will consist of block effects, subgroup effects, and errors. At this stage, I proceed as I did for a nonspatial generalized linear model (Chapter 8). For the Poisson, a simple specification for the underlying process might be

$$\ln y_{ijt} = b_i + g_j + \varepsilon_t$$

including block, group, and time. There can be different types of effects at the block and the group levels. As an example, consider fixed effects (including main effects and interactions), random effects, and spatial effects for blocks,

$$b_i = \mathbf{x}_i\beta + w_i + \phi_i$$

The fixed effects in the first term would not include an intercept if that would be redundant with other parameters in the model. The second and third terms, θ_i and ϕ_i, include random effects associated with blocks and spatial relationships, respectively.

As with any regression problem, decisions regarding which terms to include depend on how much information is available, model use, and patterns in the data themselves. The first term corresponding to fixed effects would be used only if covariates are available. On the other hand, if covariates fully account for block differences, we might not require the second and third terms. If fixed effects are insufficient, block differences might involve random effects, spatial effects, or both. The random effects are block differences that are exchangeable and not described by proximity to other blocks. This term would take up block differences unexplained by any fixed effects. If spatial effects are included in the model, the random effects should be specified with mean zero,

$$w_i \sim N(0,\nu)$$

with proper hyperprior on variance ν (Gelfand and Smith 1990). As in nonspatial applications, random effects allow us to borrow strength across sample units, which is especially important for small sample sizes.

The spatial dependence is contained in the third term. A CAR specification for spatial dependence could have

$$\varphi_i|\varphi_{k \neq i} \sim N\left(\overline{\varphi}_i, \sigma^2/n_i\right)$$

where the conditional mean for block i is estimated from the neighborhood,

$$\overline{\varphi}_i = \left.\sum_k c_{ik}\varphi_k \middle/ \sum_k c_{ik}\right. = \frac{1}{n_i}\sum_{k \in \{n_i\}} \varphi_k$$

with weights $c_{ik} = 1$ for blocks within the neighborhood of i and zero otherwise. In other words, the neighborhood determines the mean for block i, and the larger the neighborhood, the smaller the variance (the more smoothing). We further need to impose a "sum to zero" constraint on the spatial effects, $\sum_{i=1}^{n} \varphi_i$, which ensures a well-behaved posterior (Besag et al. 1995). This can be done at the end of each MCMC step by subtracting the mean taken over all blocks,

$$\varphi_i^{(g)} - \frac{1}{n} \sum_{i=1}^{n} \varphi_i^{(g)}$$

Finally, if error terms for time ε_t are included, there must be a constraint such as $\varepsilon_1 = 0$ or $\sum_t \varepsilon_t = 0$. As with the spatial effects, the ε_t can be centered after each MCMC step (Gelfand et al. 1998).

The prior specification is especially important if both random and spatial effects of blocks are included in the model, determining whether the analysis will "tell primarily a spatial story or primarily a heterogeneity story" (Gelfand et al. 1998, p. 315). The zero mean of the w_i prior is used to avoid redundancy between w_i and ϕ_i. We can potentially fit both, because priors impose different constraints (exchangeable effects w_i and spatial effects ϕ_i). But if we tightly constrain the random effects (hyperpriors that limit the range of v), then spatial effects will dominate. If priors on both are noninformative, we may identify neither. Thus, we might often choose to include one or the other. For example, we might seek covariates that account for some of the random and spatial effects, thus allowing us to simplify. Any information on the scale of variation to be expected can be helpful for prior specification.

There may be further information available at different scales. At the group level, we may incorporate additional effects g_j. For understanding seedling dynamics, Ibáñez et al. (2007) incorporate climate variation in time, soil moisture variation at the scale of stands, and light availability at the scale of plots within stands.

The full model involves priors, which might be Gaussian for regression parameters. Priors for variances should be proper, again exploiting knowledge of the range of variation to be expected, where possible. This would allow for direct sampling of parameters for the regression. Due to the Poisson sampling distribution, we must also sample the y's, which requires an indirect step, such as M-H.

10.9.2 Continuous Variation from Discrete Data

Block-level inference may not be sufficient to reveal the character of an underlying continuous surface, call it $y(\mathbf{s})$. Blocks may vary in size and shape, the partition being imposed by factors having little to do with the spatial scale of the process. In such cases, there may be no ecological justification for deciding which blocks should be regarded as neighbors. There may be no reason to expect that changes in the surface $y(\mathbf{s})$ should correspond with block boundaries. The data may overemphasize the random aspect of spatial effects simply as a result of how a map is partitioned (Figure 10.17a).

Kelsall and Wakefield (2002) consider this underlying risk surface to be a Gaussian random field, when the data are spatially aggregated within blocks. This is much like the blocks-to-points case of Gelfand et al. (2001), but they introduce the notion of a spatial density function across a block. For an underlying continuous surface $y(\mathbf{s})$ they use the concept of a block average that is related to the continuous surface by the integral

$$y_i = \int_{B_i} y(\mathbf{s})f_i(\mathbf{s})d\mathbf{s}$$

where $f_i(\mathbf{s})$ is the probability density of the population across block i. The spatial covariances between random points in two blocks are contained in matrix $\mathbf{C}_B$ having elements

$$\mathrm{cov}(\ln y_i, \ln y_j) = \sigma^2 \rho(B_i, B_j; \varphi)$$

where $\rho(B_i, B_j; \varphi)$ is the average correlation between points selected at random from within B_i and B_j.

The prior specification for the log relative risks associated with blocks $\mathbf{y} = [\ln y_1, \ldots \ln y_n]$ is Gaussian, with prior mean μ and n by n covariance matrix $\mathbf{C}_B$ (Kelsall and Wakefield modify this slightly for efficiency purposes). The correlation model they use is a cubic function with a range parameter φ. For convenience φ is defined for discrete values with prior multinomial parameters $\mathbf{D}$.

To predict the trend surface for log relative risk across a grid of locations $\mathbf{s}'$, they use the covariance between a block and a point $\mathbf{C}_{B,s}$,

$$\mathrm{cov}(\ln y_i, \ln y(\mathbf{s}_j)) \approx \sigma^2 \int_{B_i} \rho(\mathbf{s}_i, \mathbf{s}'; \varphi)f_i(\mathbf{s}')d\mathbf{s}'$$

$$= \sigma^2 \rho(B_i, \mathbf{s}_j; \varphi)$$

the average covariance between the log risk at a random point in B_i a location $\mathbf{s}_j$. Then the conditional mean and variance of the surface of m points $\mathbf{s}'$ is

$$E[\ln y(\mathbf{s}')|\mathbf{y}] = \mu \mathbf{1}_m + \mathbf{C}_{Bs}^T \mathbf{C}_B^{-1}(\mathbf{y} - \mu \mathbf{1}_n)$$

$$\mathrm{var}[\ln y(\mathbf{s}')|\mathbf{y}] = \sigma^2 (\mathbf{C}_s - \mathbf{C}_{Bs}^T \mathbf{C}_B^{-1} \mathbf{C}_{Bs})$$

where $\mathbf{C}_s$ is the m by m matrix with elements $\sigma^2 \rho(\mathbf{s}'_i, \mathbf{s}'_j)$ and $\mathbf{C}_{Bs}$ is a n by m matrix with elements $\sigma^2 \rho(B_i, \mathbf{s}'_j)$.

From Figure 10.20b, the within-block variation is apparent. The disjointed map of mortality ratios in Figure 10.20a seems to mask some of the important spatial features apparent with the allowance for within block smoothing. On the down side, this method is computationally intensive and perhaps impractical for many applications.

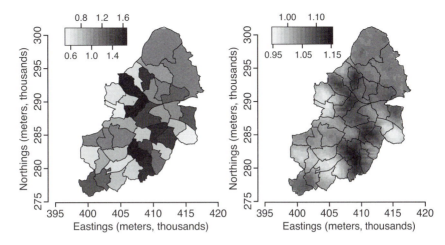

FIGURE 10.20. Standardized mortality ratios for thirty-nine wards in Birmingham, England, calculated as observed versus expected cases (*left*), and posterior median relative risk $y(\mathbf{s})$ (*right*). From Kelsall and Wakefield (2002).

10.9.3 Spatial Effects "in the Process"

In some sense, we could regard the foregoing treatment of space as something of a nuisance. If our primary interest is in prediction (e.g., $\mathbf{z}$), then random and distance effects discussed thus far may suffice. However, spatial relationships may enter as a more explicit aspect of the process of interest. In fact, this is preferable—we would rather understand the process as opposed to simply allowing for spatially correlated error. For the general model

$$z(\mathbf{s}) = \mu(\mathbf{s}) + w(\mathbf{s}) + \varepsilon(\mathbf{s})$$

we might use the first term to represent a process about which we wish to draw inference. If this process can be linearized, we could then build on the structure outlined above. A very simple regression relationship might involve coefficients that are spatial. Thus, the first two terms can be used to define a mixed model

$$z(\mathbf{s}) = \mathbf{x}(\mathbf{s})\alpha + \mathbf{w}(\mathbf{s})\beta(\mathbf{s}) + \varepsilon(\mathbf{s})$$

where $\mathbf{w}(\mathbf{s})$ might be a subset of the covariates in $\mathbf{x}(\mathbf{s})$, the parameter vector α represents fixed effects, and $\beta(\mathbf{s})$ contains spatial relationships (Gelfand et al. 2003). This could be a spatial version of the random effects models discussed in Chapter 8. Wikle and Anderson (2003) used this approach to assess how tornado reporting depended on population density and, thus, probability of detection. In this case the spatial random effects describe how the coefficients vary in space, in much the same way as they previously varied among individuals.

Although regression models can be flexible, they may not suffice for many spatial processes. Such models accommodate space in terms of its effects on the residual covariance structure and on regression parameters in a linear model. Hierarchical structures allow us to write a process model that may be more complex, overlay the process with a data model, and establish context with parameter model. Several examples already mentioned incorporate physical models of a spatial process. For instance, to model tropical winds, Wikle et al. (2001) assume conditionally independent observations of wind speed linked to an underlying model that treats wind components in a more realistic way. I summarize some of the key elements with an extended example in Section 10.12.

10.9.4. Dimensionality Reduction and Nonstationarity

Among the greatest challenges of spatial analysis is the high dimensionality associated with map data. For example, Gaussian models require the inverse and determinant of a covariance matrix. Two-dimensional spatial data sets generated from satellite imagery are far too large for standard manipulations. A model may involve a prediction grid that does not coincide with the locations of observations. The prediction grid itself for climate and weather models may include 10^6 grid locations (Bengtsson et al. 2003). Each update of a numerical weather prediction model involves large numbers of observations to be ingested in a large prediction grid (Patil et al. 2001).

In addition to the size of spatial problems is the unavoidable complication that processes are often nonstationary. Nonstationary covariance structure poses problems for parameter estimation. If spatial dependence varies from place to place, the foregoing treatments do not apply. A number of new approaches are being developed for the problems of high-dimensional, nonstationary spatial models. This is an area of active research, with a comprehensive treatment being beyond the scope of this text. Here I simply mention some promising directions that can be pursued further through cited papers.

Nonstationarity and dimension reduction may be addressed through use of basis functions. Various transformations of space can produce stationarity, while at the same time allowing for simplification. Returning to the basic model

$$z(\mathbf{s}) = \mu(\mathbf{s}) + w(\mathbf{s}) + \varepsilon(\mathbf{s})$$

take the second spatial term to be a random effect, which, for the full data set, is the n by 1 vector

$$\mathbf{w} \sim N_n(0, \Sigma_w)$$

This can be modeled as

$$\mathbf{w} = \mathbf{W}\beta$$

where $\mathbf{W}$ is the n by m matrix of basis functions and $\boldsymbol{\beta}$ is the length-m vector of coefficients. For location i, this term takes the form

$$w(\mathbf{s}_i) = \mathbf{w}(\mathbf{s}_i)\beta = \sum_{k=1}^{m} w_k(\mathbf{s}_i)\beta_k$$

Now the coefficients have distribution

$$\beta \sim N_m(0, \Sigma_\beta)$$

If basis functions are orthogonal, then $\beta = \mathbf{W}^T\mathbf{w}$ and $\Sigma_\beta = \mathbf{W}^T\Sigma_w\mathbf{W}$ (Wikle and Royle 2004). The dimension of the model can be lower than the original spatial domain, $m \leq n$. The covariance matrix Σ_β is m by m, which may be smaller than the original n by n, and it may have simpler structure.

Basis functions can come from any of several approaches. In Fourier analysis, the basis functions are sines and cosines. The coefficients are amplitudes assigned to each of m frequencies. Each basis function has k maxima and k minima—they are waves. Wavelets can also serve as basis functions (Nychka et al. 2002).

The climate literature uses the term *empirical orthogonal functions* (EOF's) to refer to basis functions obtained from a principal components analysis. Specifically, one conducts an eigenanalysis, with each basis function being orthogonal to previous ones, and maximizes the residual variance. The matrix $\mathbf{W}$ holds the first m eigenvectors, and **?** holds the corresponding eigenvalues. To determine the fraction of the total variance contained in each eigenvalue, divide by the sum over all eigenvalues. The dimensionality is reduced to the first m eigenvectors, deemed sufficient to capture most of the important variation. Often a substantial fraction of the variance can be captured with small m. In the case of EOF's, the basis functions are not prespecified functions, but projections of the data onto new dimensions that capture most of the variance. Wikle and Royle (2004) use this approach for inference and prediction on migrating bird densities (Chapter 11).

Smoothing kernels can also be used as basis functions. In this case, the random spatial term might look like this:

$$w(\mathbf{s}) = \mathbf{f}(\mathbf{s}',\mathbf{s};\theta)\mathbf{w}(\mathbf{s}') = \sum_{k=1}^{m} f(\mathbf{s}_k' - \mathbf{s};\theta)w(\mathbf{s}_k')$$

where θ are kernel parameters and $w(\mathbf{s})$ is a Gaussian noise process (Higdon 2002). In Calder et al.'s (2004) model, the Gaussian process is taken at $k = 1, \ldots, m$ locations on a prediction grid, which does not correspond to the n sample locations. The model convolves discrete measurements to create a smooth spatial field. The kernel can be anisotropic. They model the full process $\mathbf{z}$ on the locations where measurements are obtained (Section 10.9.5). These kernels are sometimes termed *radial basis functions*.

In a related approach, Fuentes (2001, 2002) describes how to model a nonstationary process as a mixture of stationary random fields $z_k(\mathbf{s})$. The locations $\mathbf{s}_k$ for $k = 1, \ldots, m$ are taken to be the centers for kernel functions The density of the grid must be sufficient to estimate the covariance, with m selected by, say, AIC or BIC. The model looks like this:

$$z(\mathbf{s}) = \mathbf{f}(\mathbf{s})\mathbf{z}_\theta(\mathbf{s}) = \frac{1}{m}\sum_{k=1}^{m} f(\mathbf{s}_k - \mathbf{s})z_{\theta_k(\mathbf{s}_k)}(\mathbf{s})$$

The m different processes $z_{\theta_k(s_k)}(\mathbf{s})$ are stationary and uncorrelated. Yet the covariance for the full process $z(\mathbf{s})$ is nonstationary. The covariance function has m terms, involving the m kernels and m covariance functions. From this, Fuentes (2002) provides the corresponding spectral density, which is the Fourier transform of the covariance function. The Fourier transform is estimated from the periodogram. The periodogram is the discrete Fourier transform of the sample covariance. In the case of this mixture, the full periodogram consists of m terms. We end up with m sets of parameter estimates, one set for each of the m kernels. In her application to air pollution data, Fuentes (2002) uses a sample grid of $m = 9$ with Matern kernels. Rather than the 7209 by 7209 matrix represented by the physical map, which could not be inverted, this model has a highly simplified covariance matrix.

Fuentes and Raftery (2005) provide an example for prediction of sulfate concentrations in the atmosphere, where two data sets bear on the same process, but at different resolutions (Section 8.4). This example combines a number of issues that have appeared in Chapters 8 and 10. Measurements of atmospheric SO_2 concentrations are available from a network of sites across eastern North America. Complex terrain and fine-scale spatial variation means that we cannot simply interpolate from this coarse network to predict spatial trends. The second source of information combines emissions estimates with weather models to predict the spatial pattern of SO_2. Thus, there is space in the mean structure, in the sense that the simulation model of SO_2 concentrations involves atmospheric transport. Both measurements and model output can be combined, provided we can write an appropriate data model that links both to the underlying SO_2 concentrations. Fuentes and Raftery's (2005) data models allow for measurement error in concentrations and bias in model output. All errors are spatial, and there is a change-of-support issue: measurements are from points, model output might be for blocks, and inference should be at points (a prediction grid). Prediction entails blocks-to-points. The predictive distributions for concentrations of sulfate that came from a model with nonstationary covariance revealed some important limitations of both types of data and demonstrated the value of combining them within a consistent framework. Together, they combined different types of data, allowed for error and bias in both, and addressed the change-of-support and nonstationarity issues.

It should be emphasized that basis functions are a valuable tool for dimension reduction, even for stationary covariance. In the following section, I return to nonstationarity in the context of space-time covariance.

10.9.5. Spatio-Temporal Covariance

Time brings some added complications to spatial models. The combined treatment of space and time has been approached in two general ways. The first involves adding one more dimension to spatial models. The second involves incorporating space into a process model that looks much like the state space models of Chapter 9. Both are discussed in this section, followed by an extended example of the latter in Section 10.10.

Building on the spatial models of the last section involves incorporating a time variable,

$$z(\mathbf{s},t) = \mu(\mathbf{s},t) + w(\mathbf{s},t) + \varepsilon(\mathbf{s},t)$$

This requires a spatio-temporal covariance function

$$\text{cov}[z(\mathbf{s}_i,t_k),z(\mathbf{s}_j,t_l)]$$

for a sample obtained at location $\mathbf{s}_i$ at time t_k and one obtained at location $\mathbf{s}_j$ at time t_l. The simplest covariance functions are *separable*, being a simple product of spatial covariance and temporal covariance,

$$\text{cov}[z(\mathbf{s}_i,t_k),z(\mathbf{s}_j,t_l)] = C_{ik,jl} = \sigma^2 \rho_s(r_{ij};\phi)\rho_t(\Delta_{kl};\theta)$$

where the correlation functions have subscripts denoting space and time and Δ_{kl} is the elapsed time between sample times t_k and t_l. This form is convenient, because it allows us to treat space and time separately. We might use some common parametric forms for correlation functions, such as

$$\rho_s(r_{ij};\phi) = e^{-\phi r_{ij}}$$

and

$$\rho_t(\Delta_{kl};\theta) = e^{-\theta \Delta_{kl}}$$

In other words, the total (nonspatial) variance is σ^2, and the covariance decays with distance at proportionate rate ϕ and over time at proportionate rate θ. Of course, this replaces the nonspatial assumption that the covariance matrix has zeros everywhere except along the diagonal. Chen et al. (2004) describe how the vectorized form for $z(\mathbf{s},t)$ has a spatio-temporal covariance matrix that is a Kronecker product of spatial and temporal covariance matrixes. This form has the advantage that inverse and determinant of the spatio-temporal covariance matrix, needed for computation, is obtained from those of the two separate covariance matrixes, which have much lower dimension.

There are several ways to approach nonseparable spatio-temporal covariances. Cressie and Huang (1999) use a Fourier transform representation to derive several functional forms (see also Gneiting 2002). Chen et al. (2006) envision a landscape where different scales of spatial covariance are associated with partitions of the full landscape and over time. The full set of such partitions subdivides the area of interest over the duration of the study. For example, if there are two spatial dimensions, then the partition is taken over the three-dimensional space-time domain. There is a kernel function associated with each region, which places maximum weight on nearby influences and diminishing weight with distance and over time according to a kernel function. There are centroids that shift in importance over time. Model selection could be used to determine the number of regions. One can also test for separability of spatial and temporal covariances, which simplifies the problem.

For many processes, spatial relationships change over time, and temporal relationships depend on distance or location. There are no simple solutions, but there are some options. In a hierarchical Bayes framework, there can be an observation model, assuming conditional independence, atop a spatio-temporal model with process and leftover spatio-temporal covariance, with additional stages for priors and hyperpriors (Wikle et al. 2001). As we have seen, the approach is flexible, but spatio-temporal covariances can be challenging. More promising in these situations is the approach that may involve a process that evolves over time, much like the state space models of Chapter 9.

Calder et al.'s (2004) model introduced in Section 10.9.4 applies a state space framework for change in particulate transport over time. There are two components, each of which is modeled with the process and observation equations

$$z(\mathbf{s},t) = \mathbf{x}(\mathbf{s},t)\beta + w(\mathbf{s},t) + \varepsilon(\mathbf{s},t)$$

$$w(\mathbf{s},t) = \mathbf{w}(\mathbf{s}',t)\mathbf{f}(\mathbf{s},\mathbf{s}',\theta_t) \qquad \text{random spatial term at sample locations } \mathbf{s}$$

$$\mathbf{w}(\mathbf{s}',t) = \mathbf{w}(\mathbf{s}',t-1) + \mathbf{v}(\mathbf{s}',t) \qquad \text{process at lattice points } \mathbf{s}'$$

where design vector $\mathbf{x}(\mathbf{s},t)$ holds covariates, β is a parameter vector, $\mathbf{w}(\mathbf{s}',t)$ is the underlying process at lattice locations with covariance matrix

$$\mathbf{v}(\mathbf{s}',t) \sim N_m(0,\sigma_{vt}^2\mathbf{I}_m)$$

and observation error

$$\varepsilon(\mathbf{s},t) \sim N_n(0,\sigma_{\varepsilon t}^2\mathbf{I}_n)$$

Here the set of kernel functions has parameters that vary over time. The full process $z(\mathbf{s},t)$ involves effects of covariates in the first term, spatial smoothing in the second term, and observation errors in the third term. Depending on how space is treated in the process component of the model, error terms may depend only on time (Calder et al. 2004).

Further considerations include a dynamic trend surface $\mu(\mathbf{s},t)$, which can involve covariates, some or all of which are also spatio-temporal. Like the covariance matrix, the trend may be represented as a mixture taken over partitions of space-time. Chen et al. (2006) combine this technique with data models for model output and observations that are misaligned to obtain predictive distributions of wind fields over Chesapeake Bay. Observations are sparse, so the model of atmospheric circulation has an important role.

10.10 Application: A Spatio-Temporal Model of Population Spread

The house finch (*Carpodacus mexicanus*) was introduced in Long Island, New York, in 1940 and subsequently spread throughout eastern North America (it is native to the western United States and Mexico). The expansion is documented

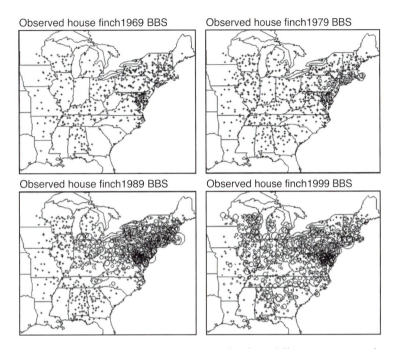

FIGURE 10.21. The set of observation locations for four different time periods. In our notation these locations are $\mathbf{s} = [\mathbf{s}_{1t}, \ldots, \mathbf{s}_{nt}]$. From Wikle (2003a).

from the North American Breeding Bird Survey (BBS), which is conducted by volunteers who travel along prescribed routes, stopping to detect birds (Figure 10.21). The data are nonuniformly distributed in space and time and contain observer bias (Link and Sauer 1998). This example demonstrates how space may enter the model in several ways.

Wikle (2003a) used a diffusion model to predict changing densities of the house finch over time and to estimate population growth rates, spatially varying diffusion, and errors resulting from model misspecification and observations. He begins with the assumption that observations are conditionally Poisson, with an underlying process λ that varies in space and time,

$$z_t(\mathbf{s}_i) \sim Pois(\lambda_t(\mathbf{s}_i))$$
$$\ln \lambda_{it} = \mu_t + \mathbf{k}_{it}^T \mathbf{u}_t + \eta_{it} \qquad 10.32$$

μ_t is the trend in log abundance that applies to the entire population. The second term describes population spread and relates it to the observations—it is spatiotemporal. The third term is observation error. I consider each of these in turn.

10.10.1 Temporal Trend

On average, the density of a population can increase or decrease (Veit and Lewis 1996). Population growth is modeled locally, because it should depend on the density of finches in the area, and there may be spatial random effects.

Because these effects are local, the overall change in density can be as simple as a random walk. This is the assumption used by Wikle,

$$\mu_{t + \Delta t} = \mu_t + w_t$$

$$w_t \sim N(0,\sigma^2)$$

The spatially local dynamics in the second term describe departures from this population average.

10.10.2 Population Spread

The two components of the second term in Equation 10.30 describe (1) a diffusion process $\mathbf{u}_t$ at regularly spaced grid points laid over a map of eastern North America (Figure 10.22) and (2) the relationship between those grid points and the observations at $\mathbf{s}_i$, $i = 1, \ldots, n_t$ (Figure 10. 20) contained in the length m column vector $\mathbf{k}_{it}$. Suppose the prediction grid is rectangular, and there are $m = m_x m_y$ total grid points, m_x being the number in the east-west (x) direction and m_y in the (y) north-south direction. Then $\mathbf{u}_t$ is length-m vector of log densities at each grid point (x,y) and could be stacked row by row as $\mathbf{u}_t = [u_{1,1,t}, u_{1,2,t}, \ldots, u_{1,n_y,t}, \ldots]^T$. These grid points must be related to observations, which are scattered over the eastern United States (Figure 10.21). Wikle uses a length-m vector $\mathbf{k}_{it}$ to assign the ith observation to the closest grid point. Each observation i has a m by 1 vector that has a one in the position for its associated grid point and zero everywhere else. Note that this vector provides a means for weighting observations by distance from grid points. Then the second term in Equation 10.30,

$$\underset{(1 \times m)(m \times 1)}{\mathbf{k}_{it}^T \quad \mathbf{u}_t}$$

describes how the entire mapped process at time t affects the ith observation.

Prediction Grid

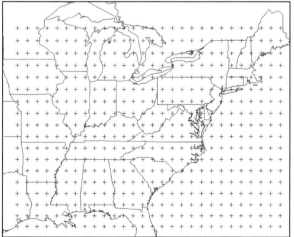

FIGURE 10.22. The prediction grid, consisting of m total grid locations, being m_x points east-west and m_y points north-south. From Wikle (2003a).

Spread is taken to be a reaction-diffusion process. The simple model from Section 10.2 for two dimensions is

$$\frac{\partial u}{\partial t} = ru + \frac{\partial}{\partial x}\left(D(x,y)\frac{\partial u}{\partial x}\right) + \frac{\partial}{\partial y}\left(D(x,y)\frac{\partial u}{\partial y}\right)$$

Note that the diffusion coefficient D itself is taken to be location-dependent— it is a function of (x,y). At this point, you may wonder why there is no density dependence—the first term describes unbounded growth. That will be discussed later. Taking the derivatives of the products we have

$$\frac{\partial u}{\partial t} = ru + \frac{\partial D}{\partial x}\frac{\partial u}{\partial x} + D\frac{\partial^2 u}{\partial x^2} + \frac{\partial D}{\partial y}\frac{\partial u}{\partial y} + D\frac{\partial^2 u}{\partial y^2}$$

To apply this model on a grid requires discretization. The forward difference approximation is used for the time derivative on the left,

$$\frac{\partial u}{\partial t} \approx \frac{u_{x,y,t+\Delta t} - u_{x,y,t}}{\Delta t}$$

Forward difference refers to the fact that the derivative is approximated over the interval $(t, t + \Delta t)$ as opposed to being, say, centered on a time point t. This is consistent with the way I have treated time throughout this book. For the spatial derivatives, use the centered difference approximation. In the x direction, this derivative is centered on x and, thus, averaged over $(x - \Delta x, x + \Delta x)$. This centering avoids a directional bias that would result if we used a forward difference or a backward difference in space. Then

$$\frac{\partial u}{\partial x} \approx \frac{u_{x+\Delta x,y,t} - u_{x-\Delta x,y,t}}{2\Delta x}$$

$$\frac{\partial^2 u}{\partial x^2} \approx \frac{1}{2\Delta x}\left[\frac{u_{x+\Delta x,y,t} - u_{x,y,t}}{\Delta x} - \frac{u_{x,y,t} - u_{x-\Delta x,y,t}}{\Delta x}\right]$$

$$= \frac{u_{x+\Delta x,y,t} - 2u_{x,y,t} + u_{x-\Delta x,y,t}}{2\Delta x^2}$$

(the difference of the differences), and

$$\frac{\partial D}{\partial x} \approx \frac{D_{x+\Delta x,y} - D_{x-\Delta x,y}}{2\Delta x}$$

We have the same relationships in the y direction. Now substitution in the continuous model gives

$$\frac{u_{x,y,t+\Delta t} - u_{x,y,t}}{\Delta t} = ru_{x,y,t}$$

$$+ \left[\frac{D_{x+\Delta x,y} - D_{x-\Delta x,y}}{2\Delta x}\right]\left[\frac{u_{x+\Delta x,y,t} - u_{x-\Delta x,y,t}}{2\Delta x}\right]$$

$$+ D_{x,y}\left[\frac{u_{x+\Delta x,y,t} - 2u_{x,y,t} + u_{x-\Delta x,y,t}}{2\Delta x^2}\right]$$

$$+ \left[\frac{D_{x,y+\Delta y} - D_{x,y-\Delta y}}{2\Delta y}\right]\left[\frac{u_{x,y+\Delta y,t} - u_{x,y-\Delta y,t}}{2\Delta y}\right]$$

$$+ D_{x,y}\left[\frac{u_{x,y+\Delta y,t} - 2u_{x,y,t} + u_{x,y-\Delta y,t}}{2\Delta y^2}\right]$$

$$+ \gamma_{x,y,t}$$

It is now apparent that this discretized model of diffusion is linear both in the latent variables $\mathbf{u}_t$ and in the diffusion coefficients $\mathbf{D}$. And the error is normal. This is probably not accidental. Wikle's (2003a) algorithm exploits this linear structure to allow for direct sampling of each within the MCMC. To arrive at normal distributions, simply organize the diffusion model in two different ways, one for sampling $\mathbf{u}_t$ and a second for sampling $\mathbf{D}$.

To sample the log densities, rearrange the model in terms of the densities $u_{x,y,t}$ and their corresponding coefficients. Note that in the following, we have five coefficients, one for each of densities that lie within the neighborhood of $u_{x,y,t}$

$$u_{x,y,t+\Delta t} = u_{x,y,t}\left[1 + r\Delta t - D_{x,y}\left(\frac{1}{\Delta x^2} + \frac{1}{\Delta y^2}\right)\Delta t\right]$$

$$+ u_{x-\Delta x,y,t}\left[(-D_{x+\Delta x,y} + 2D_{x,y} + D_{x-\Delta x,y})\frac{\Delta t}{4\Delta x^2}\right]$$

$$+ u_{x-\Delta x,y,t}\left[(D_{x+\Delta x,y} + 2D_{x,y} - D_{x-\Delta x,y})\frac{\Delta t}{4\Delta x^2}\right]$$

$$+ u_{x,y-\Delta y,t}\left[(-D_{x,y+\Delta y} + 2D_{x,y} + D_{x,y-\Delta y})\frac{\Delta t}{4\Delta y^2}\right]$$

$$+ u_{x,y+\Delta y,t}\left[(D_{x,y+\Delta y} + 2D_{x,y} - D_{x,y-\Delta y})\frac{\Delta t}{4\Delta y^2}\right]$$

$$+ \Delta t\gamma_{x,y,t}$$

The first five terms describe contributions from (1) the same location [population growth $(1 + r\Delta t)$ minus those that move to one of the neighboring grid points], and (2 through 5) those that move to location (x,y) from one the neighboring grid points over the course of one time step.

To facilitate computation, use matrixes. First, vectorize this discrete model for the neighborhood of (x,y),

$$u_{x,y,t+\Delta t} = \mathbf{h}(\mathbf{D}_{x,y},r)\mathbf{u}_{(x,y)t}$$

where $\mathbf{u}_{(x,y),t}$ is a column vector of densities at the five locations that constitute the neighborhood of $u_{x,y,t}$, $\mathbf{h}(\mathbf{D}_{(x,y)},r)$ is the 1 by 5 row vector of coefficients in brackets, and $\mathbf{D}_{(x,y)}$ is the set of five diffusion coefficients. For example, if $\mathbf{u}_{i,j,t}$ is the density at the jth grid point in the x direction and the kth grid point in the y direction, then $\mathbf{u}_{(j,k),t} = [u_{j,k,t}, u_{j-1,k,t}, u_{j+1,k,t}, u_{j,k-1,t}, u_{j,k+1,t}]^{\text{T}}$. The first element of $\mathbf{h}($) would be

$$h_1(D_{jk}) = 1 + r\Delta t - D_{j,k}\left[\frac{1}{\Delta x^2} + \frac{1}{\Delta y^2}\right]\Delta t$$

and so on.

To simplify notation, let the qth element of $\mathbf{h}(\mathbf{D}_{j,k},r)$ be h_{qjk}. These vectors are used to assemble a matrix, so we can write an expression for the entire grid. Place densities for all grid points in a single length-$m_x m_y$ vector and then organize an $m_x m_y$ by $m_x m_y$ matrix $\mathbf{H}$ such that the appropriate coefficients are multiplied by the neighborhood densities for each grid point. Here is a structure that could be used for a grid of $m_x = m_y = 5$:

$$\begin{bmatrix} u_{11} \\ u_{12} \\ u_{13} \\ u_{14} \\ u_{15} \\ u_{21} \\ u_{22} \\ \vdots \end{bmatrix}_{t+\Delta t} = \begin{bmatrix} h_{111} & h_{511} & 0 & 0 & 0 & h_{311} & 0 & \cdots \\ h_{412} & h_{112} & h_{512} & 0 & 0 & 0 & h_{312} \\ 0 & h_{413} & h_{113} & h_{513} & 0 & 0 & 0 \\ 0 & 0 & h_{414} & h_{114} & h_{514} & 0 & 0 \\ 0 & 0 & 0 & h_{415} & h_{115} & h_{515} & 0 \\ h_{221} & 0 & 0 & 0 & \underline{0} & h_{121} & h_{521} \\ 0 & h_{222} & 0 & 0 & 0 & h_{422} & h_{122} \\ \vdots & & & & & & & \ddots \end{bmatrix}\begin{bmatrix} u_{11} \\ u_{12} \\ u_{13} \\ u_{14} \\ u_{15} \\ u_{21} \\ u_{22} \\ \vdots \end{bmatrix}_{t}$$

You will immediately appreciate the diagonal structure of the matrix. An exception is the zero in row 6, column 5 (I have underlined it). This is a boundary condition. The map has boundaries that can be treated in a number of ways. In this example, I have fixed them at zero, because the grid includes areas that the finch population never invades. The position 6,5 corresponds to the contribution from grid point 1,5 to grid point 2,1. Because grid point 1,5 is located on a boundary I define its contribution to be zero. To treat boundaries differently from the rest of the grid, include them as a separate term in the model (Wikle 2003a).

Spatio-temporal noise in the diffusion process (model misspecification) enters as

$$\gamma_{x,y,t} \sim N(0, \sigma_\gamma^2 \mathbf{I})$$

and, thus,

$$\Delta t \gamma_{x,y,t} \sim N(0, \Delta t^2 \sigma_\gamma^2 \mathbf{I})$$

(Chapter 9). This is not to be confused with the term for stochasticity associated with sampling, or observation error.

10.10.3 Observation Error and Parameter Variability

Observation error contrasts with process error, in that it is associated with observations at locations s_i (not grid points j,k), and it does not affect the process itself. Wikle uses the simple assumption of

$$\eta_{it} \sim N(0,\sigma_\eta^2)$$

Diffusion might vary systematically with location, so Wikle used the parameterization

$$\mathbf{D}|\mathbf{d}_0,\sigma_D^2,\phi \sim N(\mathbf{d}_0,\mathbf{C}(\sigma_D^2,\phi))$$

with the covariance matrix having elements

$$\text{cov}(D_{jk},D_{j'k'}) = \sigma_D^2\rho_s(r_{ij};\phi)$$

and distance decay

$$\rho(r_{ij};\phi) = e^{-\phi r_{ij}}$$

(Section 10.7.1). There will be hyperpriors on the prior mean d_0 and covariance parameters σ_D^2 and ϕ

10.10.4 Sampling the Latent Variables and Diffusion Coefficients

To fill out the model, Wikle uses normal priors for the initial densities, inverse gammas for variance parameters, a uniform prior for ϕ, and normal hyperpriors for the d_0. Gibbs sampling involves some straightforward extensions of methods from Chapter 5 (Wikle 2003a). Here I focus on the sampling of latent states and diffusion coefficients. Obviously the grid itself poses the biggest challenge. The conditional posterior for the latent densities at a given grid point j,k at time t includes the five grid points contained in its neighborhood at time $t - \Delta t$ and at $t + \Delta t$ and all of the observations that directly depend on density at grid location j,k at these two time intervals. Assemble the vectors $\mathbf{k}_{it}$ for each observation i into a m by n matrix that includes all samples, $\mathbf{K}_t = [\mathbf{k}_{1t}, \ldots, \mathbf{k}_{nt}]^T$. Likewise, place the log trends in a length-n vector $\Gamma_t = [\ln\lambda_1, \ldots, \ln\lambda_n]^T$. The conditional posterior for the density grid at time t is

$$p(\mathbf{u}_t|\ldots) \propto N_n(H_t|\mu_t \, \mathbf{1}_n + \mathbf{K}_t^T\mathbf{u}_t,\sigma_\eta^2\mathbf{I}_n)N_m(\mathbf{u}_t|\mathbf{H}\mathbf{u}_{t-\Delta t},\Delta t^2\sigma_\gamma^2\mathbf{I}_m)$$

$$N_m(\mathbf{u}_{t+\Delta t}|\mathbf{H}\mathbf{u}_t,\Delta t^2\sigma_y^2\mathbf{I}_m)$$

$$= N(\mathbf{V}_t\mathbf{v}_t,\mathbf{V}_t)$$

where

$$\mathbf{V}_t^{-1} = \frac{\mathbf{K}_t^T\mathbf{K}_t}{\sigma_\eta^2} + \frac{1}{\Delta t^2\sigma_\gamma^2}(\mathbf{I}_m + \mathbf{H}^T\mathbf{H})$$

$$\mathbf{v}_t = \frac{1}{\sigma_\eta^2}\mathbf{K}_t(\Gamma_t - \mu_t\mathbf{1}_n) + \frac{1}{\Delta t^2\sigma_\gamma^2}(\mathbf{H}\mathbf{u}_{t-\Delta t} + \mathbf{H}^{\mathsf{T}}\mathbf{u}_{t+\Delta t})$$

We now have a simple rule for updating the entire grid for time t. There may be additional considerations that complicate implementation (e.g., due to its size, matrixes may be hard to invert). Regardless, the framework is an important starting point.

Wikle used the same approach to sample the diffusion coefficients. If we collect coefficients in terms of the D's we have

$$u_{x,t,t+\Delta t} = u_{x,y,t}(1 + r\Delta t)$$

$$+ D_{x,y}\Delta t\left[\frac{u_{x+\Delta x,y,t} - 2u_{x,y,t} + u_{x-\Delta x,y,t}}{2\Delta x^2} + \frac{u_{x,y+\Delta y,t} - 2u_{x,y,t} + u_{x,y-\Delta y,t}}{2\Delta y^2}\right]$$

$$+ D_{x-\Delta x,y}\Delta t\left[\frac{-(u_{x+\Delta x,y,t} - u_{x-\Delta x,y,t})}{4\Delta x^2}\right]$$

$$+ D_{x+\Delta x,y}\Delta t\left[\frac{u_{x+\Delta x,y,t} - u_{x-\Delta x,y,t}}{4\Delta x^2}\right]$$

$$+ D_{x,y-\Delta y}\Delta t\left[\frac{-(u_{x,y+\Delta y,t} - u_{x,y-\Delta xy,t})}{4\Delta y^2}\right]$$

$$+ D_{x,y+\Delta y}\Delta t\left[\frac{u_{x,y+\Delta y,t} - u_{x,y-\Delta y,t}}{4\Delta y^2}\right]$$

$$+ \Delta t\gamma_{x,y,t}$$

If we now place these coefficients of the $D_{x,y}$ in a m by m matrix $\mathbf{F}_t$, such that

$$\mathbf{u}_{t+\Delta t} = (1 + r\Delta t)\mathbf{u}_t + \mathbf{F}_t\mathbf{D} + \gamma_t$$

then the second term would look like this:

$$\begin{bmatrix} f_1 & f_3 & 0 & 0 & 0 & f_5 & 0 & 0 & \cdots & & & & \\ f_2 & f_1 & f_3 & 0 & 0 & 0 & f_5 & 0 & & & & & \\ 0 & f_2 & f_1 & f_3 & 0 & 0 & 0 & f_5 & & & & & \\ \vdots & & & & & & & & & {\scriptstyle m_x+1} & {\scriptstyle m_x+2} & {\scriptstyle m_x+3} & \\ f_4 & 0 & 0 & 0 & 0 & 0 & 0 & 0 & \cdots & f_2 & f_1 & f_3 & \cdots \\ 0 & f_4 & 0 & 0 & 0 & 0 & 0 & 0 & & 0 & f_2 & f_1 & \\ \vdots & & & & & & & & & & & & \ddots \end{bmatrix}_t$$

$$\begin{bmatrix} D_{11} \\ D_{12} \\ D_{13} \\ \vdots \\ D_{22} \\ D_{23} \\ \vdots \end{bmatrix}\begin{matrix} 1 \\ 2 \\ 3 \\ \vdots \\ m_x + 2 \\ m_x + 3 \\ \vdots \end{matrix}$$

The coefficients f in each row are understood to be referenced for the appropriate grid points for the neighborhood in question and to include Δt. We now have the conditional posterior

$$p(\mathbf{D}|\dots) \propto \prod_{t=1}^{T} N_m(\mathbf{u}_{t+\Delta t}|(1 + r\Delta t)\mathbf{u}_t + \mathbf{F}_t\mathbf{D}, \Delta t^2\sigma_\gamma^2\mathbf{I}_m)N_m(\mathbf{D}|\mathbf{d}_0, \mathbf{C}(\sigma_D^2,\phi))$$

$$= N_m(\mathbf{D}|\mathbf{V}\mathbf{v},\mathbf{V})$$

where

$$\mathbf{V}^{-1} = \frac{1}{\Delta t^2\sigma_\gamma^2}\sum_{t=1}^{T}\mathbf{F}_t^{\mathsf{T}}\mathbf{F}_t + \mathbf{C}^{-1}$$

$$\mathbf{v} = \frac{1}{\Delta t^2\sigma_\gamma^2}\sum_{t=1}^{T}\mathbf{F}_t^{\mathsf{T}}(\mathbf{u}_{t+\Delta t} - (1 - r\Delta t)\mathbf{u}_t) + \mathbf{C}^{-1}\mathbf{d}_0$$

10.10.5 Finch Diffusion

Unlike previous efforts, where heterogeneity in spread had to be neglected, Wikle (2003a) found that both the mean diffusion process (Figure 10.23a) and its variability (Figure 10.23b) changed over the course of the invasion process. The process began slowly, with low population growth rates and low diffusion rates in the Northeast. The process was most rapid in the Midwest, where expansion occurred primarily in the 1980s and 1990s. Spread was also most variable in the Midwest. Some of the heterogeneity appears related to human population densities.

10.11 How to Handle Space

We have seen that space makes models complicated. In most cases there will not be one objective approach to spatial effects. Rather, there will be options that require decisions, many of which are contingent on computation. Should space enter the model as part of a spatial process? Should it be included as covariance to ensure conditional independence of samples? Are there spatial covariates, random spatial effects, or distance-related dependencies? It would be nice if a simple model selection protocol could sort this out for us, but model selection for hierarchical models remains a point of discussion (Chapter 8).

To emphasize the subjectivity, consider again the simple linear model used for kriging of Section 10.7 with a trend surface and spatial covariance structure,

$$z(\mathbf{s}_i) = \mu(\mathbf{s}_i;\beta) + w(\mathbf{s}_i;\phi)$$

$$\mathbf{W}_\phi \sim N_n(0,\mathbf{C}_\phi)$$

where $\mathbf{W}_\phi$ is the matrix of spatial random effects, i.e., the set of all $w(\mathbf{s}_i; \phi)$

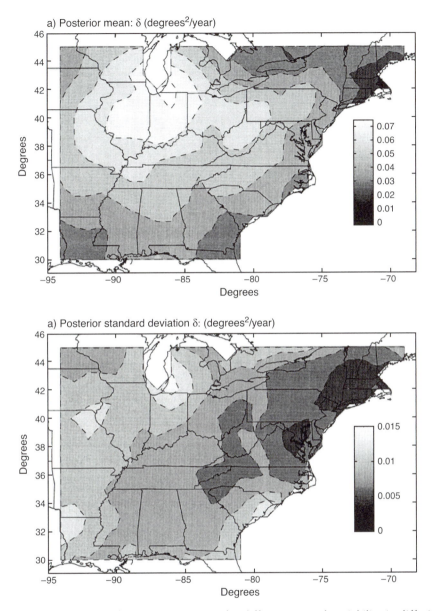

FIGURE 10.23. Maps of posterior estimates for diffusion (*a*) and variability in diffusion (*b*) for the house finch example. From Wikle (2003a).

At one extreme, we could put all space in the mean structure,

$$\mu(\mathbf{s}_i;\beta) = \mathbf{x}(\mathbf{s}_i)\beta + w(\mathbf{s}_i)$$

$$\mathbf{W} \sim N_n(0,\sigma^2\mathbf{I})$$

with a high-degree polynomial to ensure that the trend surface captures the local scale variation. In other words, $\mathbf{x}(\mathbf{s}_i)\beta$ must be flexible enough such that residual autocorrelation can be ignored. Note that the stochastic term is now nonspatial. At the other extreme, we could place all space in the stochastic term,

$$z(\mathbf{s}_i) = \mu + w(\mathbf{s}_i;\phi)$$

Of course, these two extremes are nested within the more general model that accepts space in both terms. Yet the two extremes might fit the data equally well. Modeling decisions must include consideration of how the state variables and parameters are to be interpreted as well as efficient computation. For instance, Schabenberger and Gotway (2005) suggest that, when trends are imposed by external forces, such as topography, space might be included in the mean structure, whereas intrinsic interactions (e.g., competition between plants) should enter through the autocorrelation structure. Obviously, such simple dichotomies will often be unavailable.

Computation is a serious consideration. Can the model be analyzed? The size of the complex spatial problem will often constrain modeling options, let alone extensive model selection efforts. As discussed in Chapter 1, formal model selection should often take a back seat to other considerations. A recent overview of selection for hierarchical models is provided by Spiegelhalter et al. (2002).

This and related issues mean that spatial models are an active area of research by Bayesian statisticians. From the foregoing discussion one can appreciate that there are many interesting challenges and room for creative solutions. We can expect this field to develop rapidly, with plenty of relevance for ecological applications.

11 Some Concluding Perspectives

11.1 Models, Data, and Decision

On 19 February 2001, foot-and-mouth disease (FMD) was discovered in Essex, England. Within two weeks it spanned Britain. The National Farmers Union told BBC News that something went "badly wrong" on Waugh's farm, the supplier of hogs to the Essex packing plant.[1] FMD is one of the most contagious diseases known, infecting all cloven-hoofed animals, including pigs, cattle, and sheep. Infected animals shed large amounts of the virus before symptoms are apparent. The virus survives on clothes, shoes, and car tires. It is borne by winds. Fifty-seven farms were infected before the first case was recognized—a crisis in progress before there was much awareness of its existence.

The human element was central not only to spread of the disease, but also to how model predictions were eventually used in control efforts. Government was blamed for failing to anticipate the FMD crisis. On 21 June 2002 the BBC reported that "the government was warned two years before the foot-and-mouth crisis that its vets could be overwhelmed by rapid spread of the disease." A Brigadier Birtwistle claimed that there was "a startling level of incompetence at every level." The authorities had been "slow off the ground" and apparently ignorant of the basic ecology of spread. A government's advisor, Mark Woolhouse, believed that an earlier ban on the movement of animals could have reduced the number of cases by half: "In a sense the epidemic in the UK was not one epidemic, it was lots of little epidemics some of which ran into each other, all started by individual movements from one particular market or another." Corruption brought additional distraction. Some officials faced prosecution. Farmers were charged with illegal movement of livestock after the ban

---

[1] Information on FMD came from the BBC web sites. Waugh was charged with such crimes as "failing to notify authorities," "causing unnecessary suffering to pigs," "feeding unprocessed waste to animals," and "failing to properly dispose of animal by-products." Waugh had "no regrets," claiming ignorance of the disease. During the trial the judge was shown a video of Waugh's pigs twitching on the floor.

was imposed. According to the BBC, "the costs of dealing with the outbreak spiraled out of control, with the government having to spend money hand over fist to get things done quickly." Six million animals were slaughtered in the crisis, which the NAO says cost the government £3bn and tourism and the rural economies more than £5bn.

Ecological models and data figured prominently in the FMD crisis. David King, the United Kingdom's chief science advisor, turned to two research teams in March 2002, when the disease was still on the rise.[2] Models were constructed. Data included farm locations, estimated number of livestock, disease spread, and the culling process. Scenarios were developed based on model predictions to assist decision. According to the Imperial College model, if the government had succeeded in culling every infected farm within twenty four hours and every adjacent farm within forty eight hours, the number of cases would have been cut by 66 percent and the number of farms culled by 62 percent. The other team put those numbers at 43 percent and 46 percent, respectively (these models are discussed in Section 10.10.4). Although crude, both models indicated that the government needed to get serious about its policy to cull infected farms within twenty four hours; in addition, livestock at all adjacent properties needed to be destroyed.

The impact depended not only on appropriate models, but also on non-science elements of the problem. Imperial College's Neil Ferguson argued that scientific advice helped break "an air of defeatism" at the Department for Environment, Food & Rural Affairs, and led the government to adopt the massive culling that eventually helped reduce the number of cases. The recommendations for increased surveillance and development of plans for coping with new outbreaks had impact for several reasons. This was a crisis that demanded response. It affected all of society. There were political consequences for inaction. Several labs immediately responded to the call for new model development, and they worked fast. The science was ready for this problem, and not because researchers anticipated the specific case of FMD. Results were largely in agreement, because the science was in place. The government was desperate. Recommendations were rapidly adopted, and took government off the hook. Implementation was both possible and successful.

Such examples bring us to the broader topic of where models and data can contribute to relevant scientific understanding and policy. In this final chapter I return to some of the issues raised in Chapter 1. These fall under two headings. The first (Section 11.2) concerns the interesting crossroads we find ourselves at, in terms of large, heterogeneous data sets and exciting new machinery for exploiting them. With some of these techniques behind us, there is more to be said about issues first encountered in Chapter 1. I offer some concluding thoughts on the frameworks for inference, extending the data-process-parameter view to some of the apparent future directions on networks that will be increasingly important in the environmental sciences.

The second topic concerns the uses of models, specifically for understanding, prediction, scenarios, and decision. The FMD experience emphasizes that models have context. For contrast, one might point to climate change, a scientific concern

---

[2] Roy Anderson from Imperial College and Bryan Grenfell, Cambridge University.

emphasized by model analyses that until recently failed to grab the attention of the public at large (at least in the United States). This broader context is relevant to this book only as a reminder that models are more than probability statements. Decision, based on scenarios and predictions, is a human endeavor. The widespread availability of powerful new tools does not foreshadow an age of reliable ecological forecasts. My introductory comments on learning and decision based on models (Sections 1.2–1.6) preface some further remarks on the practice of prediction (Section 11.3).

11.2 The Promise of Graphical Models, Improved Algorithms, and Faster Computers

I began in Chapter 1 by pointing out the flexibility that comes with powerful new computational tools. The graphical approach, with Bayesian implementation, represents one type of framework that can be molded to diverse applications. By lifting many of the constraints that have traditionally tied formal inference to rigid designs, environmental scientists can develop models with more attention to the science and less obsession with whether or not the model can be analyzed.

In fact, the preoccupation with computation has not diminished, largely because the sophistication of models has expanded to exploit and push the limits of new tools fully. Modern computation is concerned with algorithm development and data structures needed for efficient analysis. Now that it is possible, in principle, to draw inference on complex networks and large spatial domains, computational challenges that were previously absent, or at least latent, come to the fore. The upshot is that computation plays as large a role as ever in modeling of complex problems; even at the model development stage, computational issues must be considered.

I have covered some basic techniques related to efficient computation, focusing on a few methods that have broad application. This summary only scratches the surface of an evolving, interdisciplinary field that involves statisticians, computer scientists, and their collaborators in a range of physical and social sciences. The modeling and computational methods covered in this book represent an introduction, at best. This introduction is sufficient to motivate some additional remarks on some of the directions applications may take over the next few years.

11.2.1 From Rigid Design to Flexible Networks: Promise and Limitations

As we have progressed through increasingly complex problems, models have grown from graphs containing a small number of boxes to sometimes rather cluttered networks. Graphical approaches are becoming an important tool in many disciplines, including physics, chemistry, the Internet, pilot-assisted flight, analysis of terrorist networks, genomics, atmospheric circulation, finance, and biomedical applications. The goal is learning, where complex relationships can be modeled simply at a local level, with computational machinery to marginalize over interactions. Data may enter such models in flexible ways at a range of scales. Information and uncertainty propagate throughout the network, providing learning across the full model. It is not just a forward simulation tool, but rather inferential and predictive.

As pointed out in Chapter 1, the increasing capacity to accommodate complexity does not imply that ecological process models should necessarily be complicated. For sure, nature is complicated, and models can allow for complexity and uncertainty that comes from variables that are unknown, not measured, and so forth. Yet the underlying process of interest may be relatively simple. It should be remembered that the stochastic elements, often emphasized here, have the important role of *decreasing* complexity, standing in for deterministic relationships that might involve unobserved variables, nonlinear relationships, and interactions that are poorly understood, or just more complex than we care to model. The goal is not to complicate the problem, but rather to accommodate complexity in the simplest way possible.

As ecologists increasingly confront high-dimensional problems, the challenges will continue to evolve. Issues that will fuel research include the need to increase efficiency with large data sets that reflect complex interactions. Some types of environmental data (climate, remote sensing) are already accumulating faster than they can be analyzed. Storage alone can present a serious challenge. Wireless networks, consisting of self-organizing nodes that require no infrastructure, are beginning to add new layers of highly unbalanced data streams. Adaptive sampling strategies, controlled by software, promise to provide dense but highly uneven coverage of environmental variation. For many spatio-temporal problems, exact answers are not an option. Algorithms and data structures will increasingly stress the trade-off between precision and efficiency. We can expect growing emphasis on how approximate our answers can be in order to allow for efficient computation (Govindarajan et al. 2004).

In view of the complexity and obscurity in ecological systems, there is need to reevaluate continually how much can be known for realistic time horizons. The graphical approach exploits modularization to help simplify modeling and computation. The approach to computation discussed in this book provides inroads to complexity, through modeling locally. This reductionist approach to science is pervasive, having been around long before statisticians began thinking graphically and computer scientists began to concentrate on formal methods for machine learning. Reductionist methods can facilitate inference and prediction in complex settings, but they do not settle the modeling challenges. Computers and modularization allow us to build models as complex as we please. But progress may depend on the extent to which nature is dominated by these local connections. If sparsely connected, there is potential to learn much about a large network. We may identify many of the important relationships, even for a network with many variables, provided the network is sparse. Conversely, a network containing the same number of variables, but densely connected, may remain intractable.

The limitations posed by complexity are apparent from practical applications of dense networks. Innovations such as Microsoft Office and computer-assisted flight anticipate actions based on previous decisions for a given situation. These uses of machine learning are based on a Bayesian network that is updated by methods like those described in this text. Appearance of the paperclip on the computer screen is evoked by prior assumptions and experience with what the user is likely to do, conditioned on what she or he has done recently. If these local interactions dominate, there may be no need to involve the full network for such decisions. Although the network itself may involve many types of decisions that are all marginally interdependent, if sparse (i.e., a limited number of connections), it may be sufficient to condition only on those that most directly affect the decisions likely for the situation at hand. On the other hand, if the user is as likely to do one thing as anything else, then the network is densely connected, parameterization is difficult, prediction is uncertain, and conditioning on recent actions may not provide much guidance. Not surprisingly, predictions from models of dense networks do not deal well with the unexpected.

The environmental sciences provide particular challenges associated with the fact that model structure is especially difficult to define. The issues of network size and connectedness are highly relevant for environmental problems, such as food webs and linked biogeochemical cycles. At any one point in time a network may be sparse. But that structure changes continually, as connections come and go, a fact long appreciated by community ecologists (e.g., Elton 1927; Paine 1988). Models of spatio-temporal relationships usually involve replicating the network over space and time (Chapter 10). The mix of spatial and temporal scales and continuous state variables with connections that come and go may not be efficiently modeled with any approach currently in use. Models involving nonstationary covariances (Section 10.9.8) represent creative, practical solutions to some specific challenges. Yet no minimal model may capture the key features of many important processes.

Bayes is not the only way to analyze large complex problems. Before discussing predictions based on complex models in Section 11.3, which I do from a Bayesian perspective, I return to several issues that concern the apparent classical/Bayes dichotomy.

11.2.2 Which Framework When?

This introduction has focused on common frameworks for inference, frequentist and Bayes, with some limited excursions into related topics. Because Bayes is so flexible, I have suggested that ad hoc approaches may become less important than was true in the past. In an effort to maintain focus and generality, I have not attempted to summarize the many eclectic methods that have appeared in the literature. This is not to say that ad hoc approaches are never warranted—just that it is a good idea to start with a consistent and well-understood framework.

Having now introduced both classical and Bayes frameworks, I add a few pragmatic suggestions, combined with a modest level of personal philosophy. New tools for modeling and computation methods provide not only flexibility, but also opportunity for ruinous blunders. There is no reason every analysis must include fancy statistics. If an analysis is contemplated for data already in hand, unnecessary pain may be avoided with a careful examination of scatter plots. Students can often be dissuaded from pointless modeling exercises simply by showing that data clearly provide limited insight on the question at hand. Visual inspection may be sufficient.

Should all analyses be Bayesian? For simple problems, results of classical and Bayesian approaches will yield nearly identical confidence envelopes. A simple problem is one for which the likelihood is low-dimensional (perhaps a few parameters), and there is no additional information to enter by way of priors or additional data models. It can be hard to see the advantage in recent Bayesian applications from the ecological literature that make no effort to exploit modern machinery. A classical framework with off-the-shelf hardware can be most efficient (Clark 2005). Classical confidence envelopes are not controversial (Section 5.2). Of course, P values are not to be interpreted as error probabilities (Section 6.4), but they might still be considered as a rough index of how well things are going.

On the other hand, this book has made the case that models should often be more complex than those covered in graduate statistics courses for environmental sciences. While I agree with the standard advice that models should often start simple, when dealing with a complex problem it can be efficient to attack it directly. If the key elements of the problem are known to be important, it is often best to design the model to capture those elements. A model can be unavoidably complex simply due to the indirect nature of data, mismatch of scales, and so on. The overfitting that comes from unnecessarily complex models is bad. But I have suggested that too much emphasis can be placed on formal model selection (Chapter 6). Simple indexes that boil an analysis down to a single number should be but one of many considerations that contribute to model selection.

At several points, I have discussed the alternatives of posterior inference versus simply integrating a likelihood (i.e., without a prior). In light of modern tools there may be declining interest in integrating likelihoods. Some will continue to adopt this approach based on a philosophical view that it avoids being explicitly Bayes (specifying a prior). This rationale seems to overlook that it is not frequentist either. The likelihoodist interpretation may appeal to some. There remain a surprising number of cases in the ecological literature that report simulated posteriors where simpler methods would be easier and more accurate. Gibbs sampling is a powerful innovation, because it extends our reach to more complex problems. It is also a method of desperation. Use of MCMC for low-dimensional, classical inference, where bootstrapping or even simple asymptotic approximations are appropriate, adds not only work but also opportunity for disaster.

Much of the lively debate on philosophy found in recent ecological writings can leave the impression that the statistical world is still highly polarized over frequentist versus Bayes. It is hard to generalize, but clearly Bayes and classical statistics coexist. For example, both are well-represented in the leading statistical journals. While highly polarized philosophical arguments have proven unproductive (e.g., they do not produce consensus among statisticians), philosophy remains important (e.g., Dawid 2004). In most circles, there is no longer a stigma associated with Bayes. If anything, the pendulum has swung in the other direction. It is important to distinguish when the differences are important.

Increasingly, one might acknowledge that the pragmatic approach may depend on the setting (e.g., Clark 2005). I certainly lean toward the Bayesian notion of a prior-data-posterior update cycle. A frequentist view of probability can also be reasonable in some applications. Some of the recent complaints about Bayes in the ecological literature stem from an inaccurate view of priors as a subversive way to bend the truth (i.e., the data), effectively lying with statistics. This position assumes that the analyst would not be forthcoming about the priors used, sensitivity to them, or both. To me, this issue seems more ethical than philosophical or scientific. If the goal is to deceive, there are far more effective ways to hide the evidence than with an informative prior. Sensitivity to the prior can be made explicit. There is no reason for prior weight to be hidden. Often one does not dwell on this issue, because the experienced practitioner can recognize many situations where the prior is obviously overwhelmed by data. Yet, it should be explicit. It also should be clear by now that an informative prior is not a subversive way to improve the fit. The best fit to the data comes with a noninformative prior—the informative prior tends to pulls the posterior away from the likelihood.

Finally, while some of us have emphasized the pragmatic advantages of a Bayesian approach, as a reaction to the persistent hand wringing over philosophy (Clark 2005), Bayes is more than just a tool. It can be tempting to exploit new machinery, without really buying into the prior-data-posterior concept by, say, relying on vague priors. It seems to me that the consistent framework provided by Bayes for accommodating complexity and uncertainty is also the justification for updating prior knowledge with the data at hand. If you drift into Bayes because it allows you to address problems you could not before, it may eventually occur to you that informative priors are consistent with that probability framework. One can take each analysis in isolation, or exploit the opportunity to integrate them formally.

11.2.3 Techniques That Are Neither Classical Nor Bayesian

My general endorsement of both classical and Bayesian approaches for estimation of environmental relationships should not be taken to mean that anything goes. Recent years have seen the emergence of many ad hoc methods that are neither classical nor Bayes and that do not yield confidence envelopes with clear interpretation. These often involve optimization or integration techniques applied to a cost function that is appended to a deterministic simulation model. This cost function might be squared error loss and is minimized over one or more data sets. The terms *inverse analysis, data fusion*, and *data assimilation* are sometimes loosely attached to these methods. Terms from classical and Bayesian statistics often appear, contributing to the confusion. I have made no effort to synthesize these various views. I mention a few examples here, in the interest of illustrating how important it is to understand the model assumptions and how they are being applied.

In the fisheries literature there is increasing use of the term *integrated analysis*, a method that applies when multiple sources of data might inform the same process. Integrated analysis dates at least to a clever hierarchical model of Fournier and Archibald (1982). They considered the combined information that could be obtained from observations of fish age structure and total fish caught in a given year. A lower-stage process model for demography links these two data sets. With some specific distributional assumptions and functional forms they were able to reparameterize the model in such a way that a likelihood function could be written and evaluated by simulation. The analysis predates modern computation methods and demonstrates how creativity can yield valuable inference. It is a classical approach.

More recently, one encounters in the fisheries literature statements to the effect that integrated analysis makes Bayes redundant. This belief apparently comes from the view that data additional to a simple likelihood can enter either through a second likelihood (termed *integrated analysis*) or through a prior (Bayes), but not both. One might take issue with this belief on both semantic and theoretical grounds. As we have seen on many occasions, Bayes can admit multiple data sets, coming either from more than one data model or through a prior. The constraints on the classical treatment of such problems are substantial. Samples must be independent. If we have multiple data sets $\mathbf{z}_1, \ldots \mathbf{z}_n$ that bear on the same quantity $\mathbf{y}$, then we need to model $\mathbf{y}$. As discussed in Section 8.4, the $\mathbf{z}$'s cannot possibly be independent, all coming from the same $\mathbf{y}$. Instead of this

$$\prod_{i=1}^{n} p(z_i|y)$$

(y does not vary) or this

$$\prod_{i=1}^{n} p(z_i)p(y_i)$$

(z and y are independent) we need this

$$\prod_{i=1}^{n} p(z_i|y_i)p(y_i)$$

that is, the joint probability of $\mathbf{y}$ and $\mathbf{z}$. Point estimates obtained from the former, which ignores stochasticity in $\mathbf{y}$, may be reasonable. But we have no estimate of uncertainty. The joint distribution of quantities that depend on one another cannot be modeled as independent. Likewise, suppose we have a demographic parameter θ we wish to estimate on the basis of several data sets $\{\mathbf{z}\}$. It would be a stretch to assume that populations sampled in different times and locations are described by the same demographic rate. By modeling θ, that is,

$$\prod_{i=1}^{n} p(z)_i|\theta_i)p(\theta_i)$$

we account for the variability. Interest may focus on individual θ_i, on the density for θ constructed from its posterior hyperparameter estimates, or both (Section 8.1). In other words, we do not assume that the joint distribution of $\mathbf{z}_i$ is the same as the product of the marginals, unless they are marginally independent.

This most fundamental assumption of likelihood-based methods is not even mentioned in recent promotions of integrated analysis. If one can fully accommodate the uncertainty into a classical setting, fine. Through some creative modeling, Fournier and Archibald (1982) were able to build multiple sources of stochasticity directly into the likelihood. In practice, attempting to ingest everything by way of the likelihood will involve constraining distribution assumptions and approximations. Hierarchical Bayes is specifically designed to accommodate marginal relationships through use of conditioning. It provides a transparent and flexible approach to data assimilation. In short, Bayes is not redundant in the sense that this literature suggests; it does everything that integrated analysis does, and more, not less.

In the climate community, recognition of the need for estimates of uncertainty led to use of *ensemble analysis*. With few recent exceptions, both weather and climate models are deterministic. A forecast of "*X* percent chance of rain" is, in fact, the result of a deterministic model. These models are not inferential (parameter values are assumed or estimated independently), and, traditionally, they have not accommodated uncertainty in the way that an inferential model would. One type of uncertainty estimate comes from ensembles of model runs, where the physical conditions at the start of the run are random. There is a logical motivation for ensemble analysis: atmospheric circulation is famously sensitive to initial conditions, being the reason why one cannot forecast the weather more than a week or two in advance. Unlike most environmental models, which converge to something,[3] numerical weather prediction models display chaotic behavior (Appendix B). Ensemble analysis summarizes complex behavior in the form of distributions. The stochastic initial conditions lead to stochastic outputs. The predictive intervals that result from ensemble analysis simply state the distribution of outcomes for variable initial conditions. Of course, there are many other sources of uncertainty that recent models are beginning to incorporate, such as process error, observation error, and parameter error (e.g., Berliner et al. 1999; Wikle et al. 2001). Moreover, numerical weather prediction is becoming increasingly sophisticated, with continuing substantial effort devoted to data assimilation and dimensionality reduction (e.g., Nychka et al. 2002; Miliff et al. 2003; Bengtsson et al. 2003; Gilleland et al. 2006).

Special considerations are needed when multiple data sets are available to inform a fully deterministic simulation model. This is a particular concern with various ecosystem models, which are now being applied to multiple data sets and prior distributions. As with any complex model, data may be available for different parts of the model. However, if the model is deterministic, the outputs (Poole and Raftery 2000) and intermediates (Bates et al. 2003) are implicitly determined by the inputs. These have been termed *induced outputs* and are ingested using an approach termed *Bayesian melding* (Poole and Raftery 2000). Consider a prior distribution for inputs x and outputs $y = M(x)$, where M is the deterministic model. There may be data models for

---

[3] Ecologists who focus on complex dynamics (Appendix B) may disagree with this statement. I would simply point out that models that exhibit such behavior are interesting precisely for that reason. Such models are a small subset of the environmental models used for a vast range of applications.

both x and y. Calculation of the posterior requires recognition of the fact that y is just a deterministic transformation of x. Wolpert (1995) and Poole and Raftery (2000) address a number of issues that pertain to melding of two data sets within a consistent framework. Although this book has emphasized the importance of stochastic modeling to account for the uncertainty, many environmental models (including weather, climate, and pollution transport) are deterministic. Bayesian melding can provide for data assimilation in such models, but it requires some careful treatment of data.

When simple cost functions are minimized over multiple data sets for purposes of fitting a deterministic simulation, the foregoing and other concerns arise. The confidence intervals have no probabilistic interpretation; they do not derive from a probability model. If the justification would be that, say, a squared error loss amounts to the assumption of a Gaussian likelihood, then the statistician would say that data sets cannot be treated as independent—they derive from the same process. Some cost functions have appended an additional term, sometimes termed a *prior*, intended to constrain the estimate. Although sounding Bayesian, we still have not modeled the joint distribution of nonindependent data sets. A Bayesian would further express concerns about the Borel paradox (Wolpert 1995; Poole and Raftery 2000). The point estimates may be reasonable, but we have no estimate of uncertainty. Moreover, there is no place for process error in such models, a serious disadvantage for modeling high-dimensional systems. The Bayesian alternative is valuable, because it results in a joint distribution for process and data that is coherent. The posteriors have a straightforward interpretation.

Such concerns do not imply that methods such as inverse analysis have no meaningful role. For example, inverse analysis might be used to sharpen a fuzzy image. In this case the concept of a true model (the view represented by the image) has some utility, and we may only care about finding the best possible reconstruction of the image. A confidence interval is not the objective. Likewise, it can provide best estimates of sources and sinks in ecosystem models based on concentration and flux data. On the other hand, the concept of a true environmental model rarely has any utility. The confidence and predictive intervals are usually more important than the point estimates.

These are just a few of the ways in which the proliferation of Monte Carlo and various optimization techniques has been adapted to specific situations. It is always worth a careful look at the distributional assumptions, which can be obscure. The emphasis on hierarchical models in this book stems from their transparency. In the following section, I consider the types of predictions that come from models of the environment and how they can be interpreted.

11.3 Predictions and What to Do with Them

11.3.1 Environmental Predictions

Simulation models are arguably the universal tool for environmental prediction. I began this book by suggesting that ecological predictions rarely engender much confidence from scientists and decision makers. On the one hand, it is important not to overstate the potential that technical advances can change this perception. Many important environmental phenomena will remain inherently unpredictable for the foreseeable future (e.g., Clark et al. 2001a; Carpenter 2002). Rigorous application of the best computational tools does not yet promise much guidance for densely connected networks, where the key components are unobservable (Section 11.2.1), especially when highly nonlinear. On the other hand, a number of the issues raised in these chapters indicate ways to increase the predictability of many important processes and to improve the accuracy of predictive intervals.

We have seen examples of how prediction can follow directly from the models used for Bayesian inference. Monte Carlo methods allow us to mix over parameter error as part of a predictive distribution of variables at the same time that we are simulating the posterior (Section 7.5). Obviously, in such cases we are using the same model for both inference and prediction (Figure 1.7). When prediction is done in this way, there is no question about whether parameter uncertainty is fully represented in the predictive distribution. There are still limitations that I get to shortly. First I simply point out some implications of using parameter values out of context.

Ecological simulation studies show increasing awareness of uncertainty. Traditionally, a simulation model would be constructed with a particular process in mind. The practitioner would then forage for parameter values that might be reported in the literature. If there is some estimation error reported with that value, there might be an effort to mix over that parameter error in the course of simulation. Ecological examples of error analysis include Lande (1987), Pacala et al. (1996), Clark et al. (1998), and Feiberg and Ellner (2003). There are many others. Yet, more commonly, point estimates are used.

Although often overlooked, parameter error is more widely recognized than are most source of uncertainty in models. And it is often not the most important source (Clark 2003), typically being swamped by other uncertainties (Figures 1.3, 1.5). Perhaps the most fundamental consideration comes from the fact that parameter estimates typically do not come from the process models in which they are being used for simulation. Parameter estimates depend on the data and on the model. If there is an important inconsistency, error propagation will not solve the problem.

There is often a mismatch in scale between models used for estimation and models used for predictions. Parameter estimates obtained at the level of plant cells and leaves often find their way into simulation models of global carbon cycling. There are scale-dependent processes that determine tree allocation and growth that do not operate at the leaf level (Jarvis et al. 1997). For example, there are superficial similarities in the leaf and tree responses to light and CO_2 (shown in Figure 8.7). Some of the important differences are summarized in the figure caption. Among the most obvious is the huge individual and interannual differences in the growth responses of trees that overwhelm the modest CO_2 enhancement. This contrasts with the leaf-level instantaneous gas exchange, for which the CO_2 effect is comparatively large.

Simulation studies rarely carry forward the stochasticity in data. The sampling distribution for data z represents the observation error that may surround the actual process of interest, **y**. Why include the error associated with **z** in a predictive distribution? In fact, prediction of **y′** and **z′** may both be of interest. We predict **z**, because **y′** is invisible, whereas **z′** can be compared with new data. The predictive interval for **y′** will typically look better (it is narrower than that for **z′**), but it is not expected to describe new observations, even if the model describes the process well. Proper predictive distributions ideally include all sources of stochasticity that can be recognized. The following summarizes how they collectively contribute to a prediction.

11.3.2 Prediction Protocol

The limitations of prediction are best appreciated through a formal treatment that identifies each of the uncertain elements. In the context of inference, I have discussed stochasticity associated with *parameter uncertainty, process error, observation error* (including latent variables), and *model uncertainty*. Inference involves at least some observables, call them **X**. Typically we assume that at least some of the components in **X** are known precisely. Prediction brings an additional source of uncertainty, because few of the variables may be known in advance. Then the prediction becomes conditional on a scenario for **X**, call it **X′**. This introduces *scenario uncertainty* (Draper et al. 1999). Not all predictions involve uncertainty in **X′**. For example, kriging (Section 11.9.3) produces a prediction for known locations **X′** based on observations of a response **y** at some other known locations **X**. Here I consider scenarios that are uncertain, as a future climate scenario **X′** that might affect vegetation **y′**. I discuss some practical applications of the prediction process in the context of techniques discussed in this book, by extending Section 7.5 to include the predictive density of **y′** in terms of data, which might involve predictors **X**, responses **y**, a model M, and a scenario **X′**.

I begin by drawing inference on model-dependent parameters θ_M, which can take the form of a posterior distribution $p(\theta_M|\mathbf{X},\mathbf{y},M)$. This posterior distribution for model-dependent parameters is proportional to the product of the likelihood $p(\mathbf{y}|\mathbf{X},M,\theta_M)$ and prior

$$p(\theta_M|\mathbf{X},\mathbf{y},M) \propto p(\mathbf{y}|\mathbf{X},M,\theta_M)p(\theta_M)$$

A model- and scenario-specific predictive distribution is represented as

$$p(\mathbf{y'}|\mathbf{X},\mathbf{y}M,\mathbf{X'}) = \int p(\mathbf{y'}|M,\theta_m,\mathbf{X'})|p(\theta_m|\mathbf{X},\mathbf{y},M)d\theta_M$$

This is the predictive density of Section 9.5.2, obtained by integrating over the joint posterior distribution of parameters.

If there is a set of models under consideration $\{M\}$, there could be a posterior for each model M. This is a model-averaged predictive distribution,

$$p(\mathbf{y'}|\mathbf{X},\mathbf{y},\{M\},\mathbf{X'}) = \int\limits_{\{M\}} \int p(\mathbf{y'}|M,\theta_m,\mathbf{X'})p(\theta_m|\mathbf{X},\mathbf{y},M)$$

$$p(M|\mathbf{X},\mathbf{y})d\theta_m dM$$

On the right-hand side, the posterior for the model comes from a prior-likelihood structure

$$p(M|\mathbf{X}, \mathbf{y}) \propto p(\mathbf{y}|\mathbf{X},M)p(M)$$

(Section 6.5). We can further integrate over the uncertainty attached to a range of scenarios $\{\mathbf{X}'\}$, represented as

$$p(\mathbf{y}'|\mathbf{X},\mathbf{y},\{M\}', \{\mathbf{X}'\}) = \int_{\{\mathbf{X}'\}} \int_{\{M\}} \int p(\mathbf{y}'|\mathbf{X}',M,\theta_m)p(\theta_m|\mathbf{X},\mathbf{y},M)$$

$$p(M|\mathbf{X},\mathbf{y})p(\mathbf{X}')d\theta_m dMd\mathbf{X}'$$

At this point, we have mixed over scenarios, models, and parameters. This is Draper et al.'s (1999) *composite predictive distribution*.

As discussed in Section 7.5.2, predictive intervals will typically be generated by Monte Carlo simulation. In an ideal world, this might occur as follows (Figure 11.2). Draw a scenario at random from $\{\mathbf{X}'\}$ as

$$\mathbf{X}'_k \sim p(\mathbf{X}')$$

where $p(\mathbf{X}')$ is the prior distribution of scenarios. If there is more than one model available, draw a model

$$M_m \sim p(M|\mathbf{X},\mathbf{y})$$

and a parameter set

$$\theta_{jm} \sim p(\theta_m|\mathbf{X},\mathbf{y},M_m)$$

from the respective posteriors. We can now generate a prediction based on the scenario, model, and parameter set

$$\mathbf{y}'_{jkm} \sim p(\mathbf{y}'|\mathbf{X}'_k,M_m,\theta_{jm})$$

The last step in this process might be a draw from the sampling distribution, or likelihood. This is a predicted observation. By repeating this process and assembling a histogram of predictions $\mathbf{y}'$, we have a basis for a predictive interval, defined for, say, the central 95 percent of predictions $\mathbf{y}'$.

Due to issues that are both technical and related to data availability, this ideal process can usually only be approximated. Nonetheless, it provides a useful framework for organizing the unknowns and formalizing the process of prediction. I consider some of the challenges for each of the steps summarized in Figure 11.1.

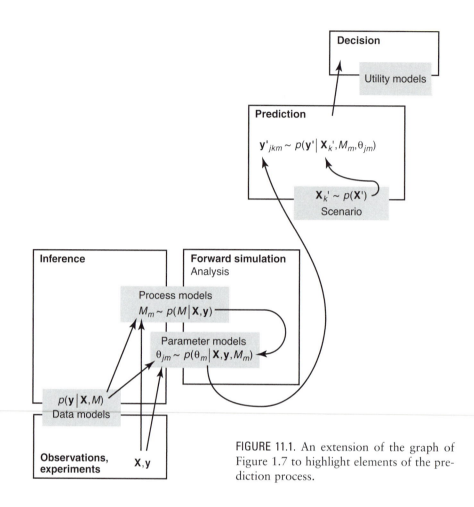

FIGURE 11.1. An extension of the graph of Figure 1.7 to highlight elements of the prediction process.

11.3.3 Scenario Uncertainty

Averaging over the scenarios that come to mind $\{\mathbf{X}'\}$ is a poor substitute for conditioning on a known set of conditions $\mathbf{X}$. As a result, formal predictive distributions based on the uncertainties should be usually viewed as too narrow. For the nuclear waste disposal risk assessment of Draper et al. (1999), the long life of radionuclides means that scenarios must include geological events, glaciation, and simple forgetfulness: drilling activity millennia from today could inadvertently encounter stored waste. None of these events have been observed, so we have no empirical basis for specification of $p(\mathbf{X}')$. The large uncertainties associated with future scenarios can overwhelm all aspects of the model that are supported by data. If so, predictions may be so broad as to provide little guidance. We can still hope to learn from the analysis where new data can be informative, and where additional data will not provide additional insight.

Scenario uncertainty also emphasizes the fact that predictions are conditioned not only on models, but also on the data used to fit the model. The data available here and now may not represent the conditions where predictions are desired. There is no real substitute for actual data. A scenario is a weak but often inevitable substitute.

11.3.4 Model Uncertainty

Prediction may entail a more careful examination of model uncertainty than is typical for inference. We often take the model M as fixed, in the sense that we condition everything on the one that, say, fits best. Of course, nothing in set $\{M\}$ is correct. Even if model averaging is used (Section 6.5.3), relatively few models are usually considered. We may average over more than one model, but the set of such models $\{M\}$ is necessarily subjective and limited. Model averaging alleviates some of the burden of defending M, but we still have to rationalize $\{M\}$.

Model averaging is not always feasible or even desirable. Still, there is usually more motivation for a model-averaged prediction than there is for a model-averaged parameter estimate (Section 6.5.3). Even for complex problems, several models may be available. For example, there are multiple models for numerical weather prediction and for general circulation of the atmosphere. They have

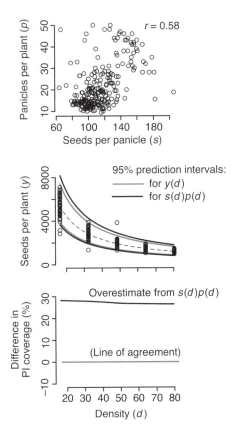

FIGURE 11.2. Overestimate of prediction interval (PI) width based on estimates obtained independently. (*a*) The correlation between two variables that can be used to estimate seeds plant. (*b*) Comparison of 95 percent prediction intervals obtained by direct estimation $y(d)$, which fully accounts for correlation in (*a*) and by propagating errors from independent fits of $s(d)$ and $p(d)$. (c) The width of the prediction interval is overestimated by almost 30 percent. Data from Reid and Fiscus (in preparation.)

traditionally been deterministic models, but they may be fed stochastic initial conditions, so-called *ensemble analysis* (Section 11.2.2). These are typically not implemented in an inferential mode, but one can still construct model-averaged predictive intervals (Raftery et al. 2005). This range of models exists only because governments are willing to invest heavily in weather and climate. For many large ecological problems, resources will limit the number of models that can be constructed for a given process.

11.3.5 Parameter Uncertainty

Even if simulation studies can exploit data obtained from relevant scales, the need to extract parameters from different studies can have important impacts on prediction, simply because relationships among components cannot be observed. Such relationships may be captured by parameter correlations (e.g., Lande 1987; Pacala et al. 1996). It is rarely appreciated that parameter correlations can have important impact. To illustrate with a simple transparent example, I compare an analysis that incorporates parameter correlation with one that does not. Consider predictions of seed yield y for rice plants grown under different densities d. In the first case, we have seed yield $y = sp = f_0(d)$ at density d as the product of seeds per panicle s and panicles per plant p, and we can observe y. In the second case, we analyze separately the two processes that make up seed yield, therefore, $s = f_1(d)$ and $p = f_2(d)$. We then predict seed yield y based on independent parameterizations of the components. To illustrate this effect, I use the same data sets for both approaches, although typically this problem is an issue when parameters (in this case, for s and p) come from different data sets. In other words, I am isolating the impacts that come from unknown relationships, in this case, the correlation between s and p.

In the first case, regression for total seed using log variables is

$$\ln y_i = a_0 + a_1 \ln d_t + \varepsilon_i$$

where ε_i is normally distributed. The data and prediction interval (dashed lines) are shown in Figure 11.4a, b.

In the second example we must proceed with parameter estimates obtained from two separate models, one for seeds per panicle s

$$\ln s_t = b_0 + b_1 \ln d_i + u_i$$

and one for panicles per plant p

$$\ln p_i = c_0 + c_1 \ln d_i + v_i$$

where u_i and v_i are normally distributed. Because I assume for this second example that y is unavailable, I predict its response to density based on estimates obtained from relationships involving quantities observed,

$$y(d) = s(d)p(d)$$

I do this in simulation: draw samples from the parameter covariance matrixes, draw errors u and v, calculate values of $s(d)$ and $p(d)$, then multiply those values to obtain a prediction of $y(d)$. Percentiles define the confidence interval shown as solid lines in Figure 11.2. The prediction interval for this second example is almost 30 percent wider than that for the direct method (Figure 11.2c).

The disagreement between prediction intervals is transparent in this simple example. There is a modest, positive correlation between s and p. Apparently, rice plants with access to greater resources at low density have not only more panicles, but also more seeds per panicle. Lacking knowledge of this correlation (obtaining estimates for each element of s and p independently) overestimates the predictive interval, because we overestimate the probability that s is low when p is high, and vice versa. Whether or not a 30 percent overestimate is an important problem depends on the question at hand. This example simply illustrates a problem that increases with the magnitude of any correlations that are ignored. Large simulation models with multiple parameters will manifest these errors in intractable ways.

What is the solution to this dilemma? If parameters can be estimated together, then correlations can be directly propagated to predictions. Ideally a parameter set is drawn from the joint posterior distribution. In principle, we can draw parameter sets directly from MCMC chains, which must be retained for this purpose (Section 7.5.2). There are some practical considerations here. First, the analysis may involve $> 10^5$ Gibbs steps, but only a fraction need be retained. Several thousand is often sufficient. If there are large numbers of random effects, the individual chains can often be discarded. There may be limited interest in the specific individuals or groups included in the study. The contribution of random effects may be obtained by drawing from the relevant distribution, using chains for hyperparameters.

In practice, some distributional assumptions may alleviate the need to retain MCMC results. For example, if the posterior can be approximated by few moments, it might be summarized by a normal distribution with a vector of means and a covariance matrix. For some parameters, it will help to save the covariance matrix, where some parameters are log-transformed. In cases where this simplification is possible, one can draw parameter sets θ_m from the multivariate normal.

If parameters are not fitted simultaneously, then parameter relationships are unavailable. For a Gaussian distribution, one might be tempted to simply make up covariances (i.e., assume a joint distribution). If there are many parameters, it may be difficult to construct a Gaussian approximation with a positive definite covariance matrix. If marginal posteriors are available, but the full joint posterior is not, it makes sense to begin with the marginal distributions and follow up with sensitivity analyses for nonzero parameter covariances.

For large simulation models, we may have estimates for some of the parameters, but not others. Often we lack parameter estimates, because data are unavailable at appropriate scales. Here it may be helpful to return to the concept of a prior-likelihood-update cycle, where information is continually assimilated in the form of a new posterior or predictive distribution. If there is no posterior, because data are not yet available, we are at the beginning of this cycle. We have no choice other than to substitute a prior for the posterior. If this sounds arbitrary, consider the more common situation in ecological modeling, where parameters are typically assumed to be constants. All we have done is allow for the uncertainty in parameter values, which is typically ignored. These priors may come from expert opinion, which may be formally integrated into a proper distribution (Winkler 1981; Reckhow 1994a).

Expert opinion is most commonly practiced in the context of decision analysis, but it can be used for prior elicitation. Here is a simple example. A manager needs to make a decision that depends on her confidence in estimates of θ. Available are the opinions of n experts. Experts have on hand different amounts of information. We wish to determine the manager's posterior confidence in θ based on the combined knowledge of experts. This example is modified from Winkler (1981).

For simplicity, suppose that an expert's belief about θ is described by a Gaussian distribution. Each expert provides a prediction of θ, call it μ_i. The variance of the distribution reflects an expert's level of information, skill, and so forth; a large variance σ_i^2 indicates limited information, and vice versa. The covariances among experts come from past experience. We expect (but do not require) positive covariances among experts, because at least some of the information upon which a prediction is made will have been available to multiple experts. Based on past performance, we might know something about an expert's forecasting skill, including any systematic bias.

The covariance between two experts i and j is given by $\rho_{ij}\sigma_i\sigma_j$, where the correlation coefficient ρ_{ij} is positive, if experts tend to provide similar predictions. This is the expected situation; we expect strong, positive correlations between weather forecasts from different meteorologists. The covariance matrix for expert opinions is $\mathbf{\Sigma}$. To obtain the required posterior, we will treat this joint distribution as a likelihood. The decision maker has a prior mean μ_0 and variance τ^2. The model then has a Gaussian likelihood, a flat prior on θ, and no prior for $\mathbf{\Sigma}$, which is defined in advance. The model is

$$p(\theta|\mathbf{\mu},\mathbf{\Sigma}) \propto N(\mathbf{\mu}|\mathbf{1}\theta,\mathbf{\Sigma})N(\theta|\mu_0,\tau^2)$$

where $\mathbf{1}$ is the length-n column vector of ones. The posterior estimate of θ is $N(\theta|V v,V)$, with $\mathbf{V}^{-1} = \mathbf{1}^T\mathbf{\Sigma}^{-1}\mathbf{1} + \tau^{-2}$, and $\mathbf{v} = \mathbf{1}^T\mathbf{\Sigma}^{-1}\mathbf{\mu} + \tau^{-2}\mu_0$. This is simply the weighted average of experts and prior, with weights being determined by elements of $\mathbf{\Sigma}$. Clark et al. (2003b) provide an example where a Dirichlet distribution is used to summarize expert opinion on dispersal in terms of categorical distance classes. Martin et al. (2005) used Gaussian priors to incorporate expert opinion on bird responses to grazing.

The use of priors or expert opinion, in lieu of fitted parameter values, can be an initial step to be followed by adaptive management strategies, where learning is among the primary goals. Shea and Possingham (2000) examined simulation models of release strategies for biological control agents. Among their recommendations were the application of mixed releases early in a control program, primarily to accelerate understanding of what works. Conversely, model analysis may show that parameters deemed important at the time of model construction have limited impact and so do not warrant data collection efforts. The use of model output as basis for management can be done formally, using decision theory (Section 11.3.7).

11.3.6. An Example

Wikle and Royle (2005) constructed predictive intervals for migratory bird abundances using techniques similar to those discussed in Section 10.10. Data consist of counts for mallard lone drakes and pairs from the U.S. Fish and Wildlife Service's Breeding Population Survey. A Bayesian hierarchical spatio-temporal model was constructed whereby observations from 1956 through 1998 were modeled with a Poisson sampling distribution, with log intensity modeled as a dynamic process at the next stage, including process error. The spatio-temporal dynamical process depends on an index of drought over the U.S. prairie pothole region.

The maps presented in Figure 11.3 demonstrate a clear capacity to forecast one year ahead. The forecast standard deviations incorporate parameter and process uncertainty. The stochasticity due to sampling might be included for the comparison with data, in which case additional Poisson variability will be mixed. Even without the stochasticity associated with sampling, forecast and observations show strong agreement. Such predictions have obvious utility for decisions concerning management in the context of interannual climate variability.

11.3.7 Predictions in a Decision Framework

Carpenter (2002) suggests three ways in which science can contribute to decision making, including (1) improved understanding of ecological thresholds and dynamics; (2) assessing uncertainty more rigorously; and (3) using scenarios as tools for thinking through the possible consequence of decisions and the ways in which future unexpected events may influence their outcomes. Model uncertainty is an important component to include, but it may not be subject to formal analysis (Clark et al. 2001a). Decision theory has a vast literature that I do not review here. I do provide a brief summary of the components of a decision analysis, because it can be viewed as a simple extension of inferential and predictive tools used in this book.

Decision analysis provides a framework for management in the face of uncertainty (Reckhow 1994a). It brings an additional step in the sense that, conditional on uncertainty in estimates or predictions, we optimize over choices (Berger 1985). For a given decision d, there is an outcome q. Outcomes may be tangible, such as cost or profit, information, life expectancy, and so on. They may be intangible, such as reputation or quality of life. They may be multivariate, such as water quality (Reckhow 1994b) or ecosystem health, which could entail a vector of outcomes $\mathbf{q}$, including such things as biodiversity, carbon storage, and water quality (Costanza et al. 1997). Outcomes may be viewed as parameters, in the form of a posterior, or state variables, in the

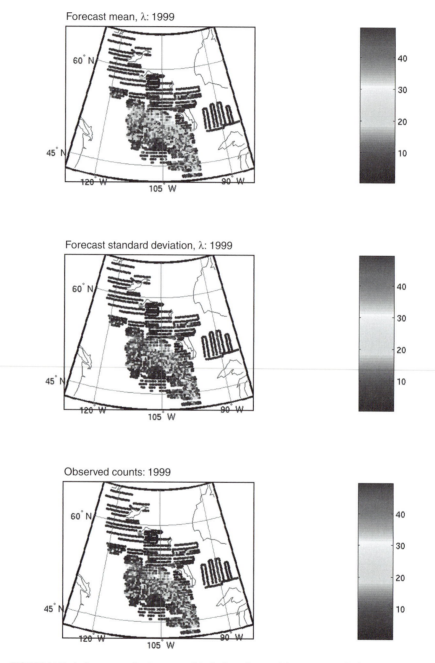

FIGURE 11.3. A forecast of migratory bird abundance. The top panel shows the forecast mean of the posterior predictive distribution for the Poisson sampling process for 1999 given observed pair counts from 1956 to 1998 and the model. The second panel shows the posterior predictive standard deviation for the Poisson intensity process for 1999. The bottom panel shows the observations for 1999. From Wikle and Royle (2005).

form of a predictive distribution. The density of outcomes is conditioned on the decision $p(q|d)$. The utility function $u(q)$ translates an outcome to value. For example, an ecologist might place more value on ecosystem health and less on profit margin than might an investor. In principle, one calculates the expected utility (over the set of decisions under consideration), and chooses the one having the highest expected utility.

There are a number of difficulties with formal decision analysis, not the least of which are the definitions of $p(q|d)$ and of $u(q)$. The density $p(q|d)$ describes subjective probability. It may be difficult to quantify. Consider the example of climate change, having effects that are associated with the future (Layton and Levine 2003). We must weigh costs of current mitigation against those of future impacts. How strongly do experts, policy makers, and the public at large believe that an action d taken today will affect costs in the future? We must answer this question in order to define $p(q|d)$.

How do we define utility in the future? Utility itself cannot be observed, but, sometimes, decisions that reflect utility can. Given relative differences in unobserved $u(d)$, we may have opportunity to collect data on choices. (Of course, this situation pertains to a group of individuals and is generally not relevant for problems involving individual choice.) If choices cannot be observed, one may attempt to collect information on relative preferences for the feasible alternatives. There are still substantial challenges. Different parts of society might not all agree on the trade-offs between ecosystem health and economic growth, particularly when costs are dispersed in time and space. The willingness to mitigate depends on how far into the future the costs will be incurred. Standard practice involves exponential discounting $e^{-\rho t}$, where t is lead time (typically years) and ρ is the discount rate. For some effects of climate change that will be realized decades from now, small changes in ρ have substantial impact on the assessment. Economists may disagree on the value of ρ (Weitzman 2001). For climate mitigation, the trade-offs are unobservable: we cannot collect data on choices for past climate changes that are relevant to the current situation. Moreover, survey data can be misleading. The willingness to pay for mitigation claimed on a survey may not represent an individual's observed behavior.

The Bayesian machinery can be applied to formal decision analysis. An overview of formal decision analysis is given by Berger (1985). Gelman et al. (2006) provide a recent perspective with several examples. Dorazio and Johnson (2003) discuss an application within an adaptive management framework. For a sample of survey respondents on climate change in Colorado, Layton and Levine (2003) developed a mixed model for utility, with effects including variables related to price, forest loss, several abatement options, and no action. Time horizons were 60 and 150 years. Utility itself was modeled as a latent variable. Modeling was hierarchical Bayes, and it was analyzed with MCMC methods discussed in this book.

As a second example, Peterson et al. (2003) describe a case where lake management is based on the perception of lake status, which is incompletely known and subject to time lags and threshold behavior. They consider the possibility that belief in lake status can build over time, solidifying management policies that are ill suited to the rapid shifts in status that can occur. Belief in an incorrect model, coupled with human behavior, can result in ecosystem collapse, despite apparently rational behavior. The Peterson et al. (2003) example emphasizes the importance of considering multiple models, not just those that best explain recent observations.

In general terms, formal decision analysis can be viewed as an extension of inference, to include aspects of a problem that affect choices. The analytical tools are aimed at facilitating decision, based on uncertainty and risk. The multiple outcomes of a decision may each be described by a probability distribution. Costs, benefits, and risks can all depend on these outcome(s). There are often competing decisions, with benefits and costs inequitably spread among interested parties. For example, lake management in agricultural areas may entail competing interests of water supply, fishing and other recreational activities, and disposal of phosphorus pollution from surrounding farms (Peterson et al. 2003). Each party may perceive a different set of benefits and risks from a similar decision.

Application of formal methods may be limited by information and problems quantifying components related to what we can know, how well we know it, and consequences of alternative decisions. Surprises, such as sudden disease outbreaks, may be better anticipated through a process that seeks to assess vulnerabilities than through a formal decision analysis based on the current setting. Whether or not a formal statistical approach is adopted, efforts to translate inference and prediction into a decision framework may guide future research in the direction of reducing uncertainty (Reckhow 1994b).

11.4 Some Remarks on Software

Obviously, the accompanying lab manual, based on R, is not the only software option. Others range from low-level languages, such as FORTRAN and C, to high-level languages, such as R and MATLAB, to the Gibbs sampling program BUGs (www.mrc-bsu.cam.ac.uk/bugs/winbugs). The latter will be widely used by ecologists who have limited understanding of the underlying assumptions. Experience tells me that turning nonstatisticians loose on BUGs is like giving guns to children. The typical student in environmental sciences, even with several statistics courses, does not possess the knowledge of distribution theory needed to execute and interpret a sophisticated analysis. An analysis by the uninitiated requires collaboration with one who is familiar with the modeling and computational issues. In short, I do not view BUGs as an alternative to learning the basics covered in this book. Having said that, BUGs is an extraordinarily powerful tool for those who do grasp the basics of modeling and computation. And BUGs can facilitate education. Students still need not only the fundamental understanding, but also the experience with programming their own MCMC code—a motivation for this text.

Appendix A

Taylor Series

Nonlinear equations often cannot be solved. It may be possible to approximate nonlinear equations with a linear model that is close enough. First, some notation. Think of the simple linear equation

$$y = b_0 + b_1 x \qquad \text{A.1}$$

The intercept b_0 and slope b_1 translate a variable x into another variable $y = f(x)$. The intercept b_0 is the value $f(0)$ (to see this, set $x = 0$) and the slope b_1 is change in $f(x)$ for a unit change in x,

$$b_1 = df(x)/dx$$

Thus, another way of writing Equation A.1 is

$$f(x) = f(0) + x \frac{df}{dx}\bigg|_{x=0} \qquad \text{A.2}$$

The notation for the second term indicates that the derivative is evaluated at $x = 0$. So far I have only traded one notation for another—Equation A.2 is a simple linear model, with the derivative and the constant evaluated at the same value of x (in this case, zero).

Now suppose that $f(x)$ is nonlinear in x. Because it is nonlinear, it may be difficult to solve or to analyze. A Taylor series allows us to approximate $f(x)$ with a polynomial, each successive term containing a successively higher derivative of $f(x)$. Calculus tends to be simple with polynomials, because they can be treated term by term. Here I describe a Taylor series to approximate $f(x)$ with a polynomial.

A Taylor series exploits the fact that two functions having all of the same derivatives are equivalent. A Taylor expansion makes use of this fact to express $f(x)$ as a series of terms, each of which contains a higher-order derivative of $f(x)$ and a power of x. Equation A.2 contains the first two terms of the Taylor expansion, which includes the intercept and the linear term (first derivative).

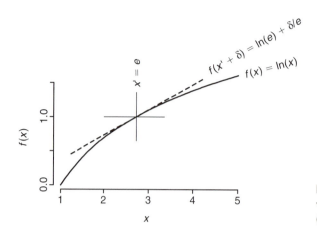

FIGURE A.1. The function $\ln(x)$ with a first order approximation (dashed line).

If the function $f(x)$ is nonlinear, we could draw a tangent line at a particular x value, call this value x', and follow that line to the right or left to obtain approximate values of $f(x)$ at values of x that are near x' (Figure A.1). In other words, the tangent line acts as a *first-order approximation* to $f(x)$ near the value x'. Suppose we wish to approximate the function at the location $x = x' + \delta$. In notation like that used for Equation A.2, the approximation for $f(x)$ looks this

$$f(x' + \delta) = f(x') + \delta \frac{df}{dx}\bigg|_{x'} \qquad \text{A.3}$$

Figure A.1 shows this approximation for the natural log function, $f(x) = \ln(x)$. The solution is easy for the case of $x' = e$, the base of natural logarithms, $e = 2.718282\ldots$ That answer is $f(e) = 1$. We can use this knowledge to obtain approximate answers for other values of the natural log function. For $x = 3$, we have $x' = e$, and $\delta = 0.282$, that is, the difference between 3 and e. Equation A.3 becomes

$$f(3) = f(e + 0.282) \approx f(e) + 0.282 \times \frac{df}{dx}\bigg|_{e}$$

$$= 1 + 0.282 \times \frac{1}{x}\bigg|_{e}$$

$$= 1 + 0.282 \times \frac{1}{e}$$

$$= 1.104$$

This approximate answer is only slightly larger than the direct calculation $\ln(3) = 1.099$. Note from Figure A.1 that this approximation would become unacceptable for values farther from e.

Example A.1. Linear approximation to the exponential function
Let $f(x)$ be an exponential function

$$f(x) = e^{-ax}$$

We wish to approximate this function with a linear equation, the first two terms of the Taylor series. Using Equation A.2, the first-order (linear) approximation from $x = 0$ is

$$f(x) \approx e^{-a \times 0} + x \cdot (-ae^{-a \times 0}) = 1 - ax$$

Figure A.2a shows the function (solid line) and the linear approximation (dashed line).

We can extend this method to approximate values of x farther from $x' = e$. This extension involves not only an intercept and slope, but also the curvature in $f(x)$. This curvature is described by higher-order derivatives. It can be useful to think of higher-order derivatives as providing information about x at values that are increasingly distant from x'. The farther we are from the reference point x', the more terms we need. This method is termed a *Taylor expansion*, and it can be used for an arbitrary number of terms with the formula,

$$f(x' + \delta) = \sum_{k=0}^{\infty} \frac{\delta^k}{k!} \frac{d^k f}{dx^k}\bigg|_{x'}$$

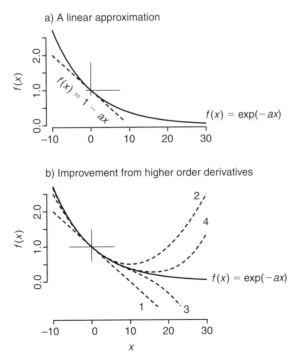

a) A linear approximation

b) Improvement from higher order derivatives

FIGURE A.2. (*a*) A graph of e^{-ax} compared with the linear approximation $1 - ax$ that comes from the first two terms of the Taylor series. Both functions have the same value and the same first derivative at the point of evaluation, $x' = 0$. In this example $a = 0.1$. (*b*) Addition of successively higher terms in the Taylor series allow for closer approximation to the true function. These additional terms are needed as we move farther from the point of expansion, $x' = 0$.

or, equivalently,

$$f(x) = \sum_{k=0}^{\infty} \frac{(x - x')^k}{k!} \frac{d^k f}{dx^k}\bigg|_{x'}$$

The Taylor series is best illustrated with an example. Return to the exponential function of Example A.1. The first term is the intercept $f(0)$, and the second term includes the first derivative evaluated at $x = 0$. So long as x is not too far from zero, this linear approximation is not too bad. For example, suppose $a = 0.1$. Then the linear approximation

$$f(1) = 1 - 0.1 = 0.9$$

is close to the actual value of 0.905. In fact, as x approaches zero the linear approximation converges with the actual value.

We can improve on the estimate of $f(x)$ by including more derivatives, that is, information about $f(x)$ farther from $x = 0$. Consider the Taylor series where we expand about $x = 0$. Add more terms of the series:

$$f(x) = f(x)_{x=0} + x\frac{df}{dx}\bigg|_{x=0} + \frac{x^2}{2}\frac{d^2 f}{dx^2}\bigg|_{x=0} + \frac{x^3}{6}\frac{d^3 f}{dx^3}\bigg|_{x=0} + \frac{x^4}{24}\frac{d^4 f}{dx^4}\bigg|_{x=0}$$

The kth derivative of $f(x)$ is

$$\frac{d^k}{dx^k}(e^{-ax}) = (-1)^k a^k e^{-ax}$$

Find this result by differentiating once, then again, and so on. When evaluated at $x = 0$, the factors involving e disappear, and we have

$$f(x) = 1 + (-ax) + \frac{(ax)^2}{2} + \frac{-(ax)^3}{6} + \frac{(ax)^4}{24}$$

Now, for $a = 0.1$, $f(1) = 0.9048$. Using the Taylor series we get this answer with only four terms,

Order:	0	1	2	3
Successive Estimate:	1	0.90	0.905	0.9048

A graph of e^{-ax} is compared with the linear approximation $(1 - ax)$ that comes from the first two terms of the Taylor series in Figure A.1. Both functions have the same value and the same first derivative at the point of evaluation, $x = 0$.

For $x = 10$, $f(10) = 0.3679$. We need more terms to obtain a useful approximation, because 10 is farther from $x = 0$:

Order:	0	1	2	3	4	5
Successive Estimates:	1	0	0.50	0.333	0.375	0.367

Thus the Taylor series for this function is

$$e^{-ax} = \sum_{k=0}^{\infty} \frac{(-1)^k (ax)^k}{k!}$$

Figure A.2b shows the approximations using terms 1 through 4.

Example A.2. Proportionate changes in population density

In Chapter 2 I used change in log density to represent per capita rates of population growth. Here I use the Taylor series to show why proportionate rates are represented by change in log values.

Consider a population with density changing at rate dn/dt. Proportionate change in density can be written as

$$\frac{1}{n_t} \frac{dn}{dt}$$

Change in log density is

$$\frac{\ln(n_{t+dt}) - \ln(n_t)}{dt}$$

We wish to show that these two are nearly equivalent,

$$\frac{1}{n_t} \frac{dn}{dt} \approx \frac{\ln(n_{t+dt}) - \ln(n_t)}{dt} \qquad \text{A.4}$$

The first term in the numerator on the right-hand side contains the arbitrary quantity dt: On the left-hand side, dt is the differential. To equate these two sides, we should understand how n changes on the right-hand side as dt gets small. First redefine n at time $t + dt$ as n at t plus the change in n, or dn. Then I can write $\ln(n_{t+dt}) = \ln(n_t + dn)$. Here I am saying that density changes by a small increment dn in the elapsed time dt—the density at the next time step, $t + dt$, will be the current density plus the change in density, $n + dn$. I now have a log function of a sum. I cannot simplify this log function, so I cannot get any closer to the goal of demonstrating that these relationships are nearly equivalent.

The Taylor series can help. Using Equation A.2, I can write the expansion,

$$f(n_t + dn) \equiv \ln(n_t + dn) \approx f(n_t) + dn \left. \frac{df}{dn} \right|_{n_t}$$

$$= \ln(n_t) + dn \times \frac{1}{n_t}$$

Substituting back into the right-hand side of A.3,

$$\frac{1}{n_t}\frac{dn}{dt} \approx \frac{\ln(n_t) + dn/n_t - \ln(n_t)}{dt}$$

$$= \frac{dn/n_t}{dt}$$

To arrive at this near-equality I used an approximation, represented by the first two terms of the Taylor series. This approximation will be good, provided dt is not too large. As dt approaches zero, the two converge.

Example A.3. A second view of discrete and continuous models
To make the transition from a discrete time model to a continuous one, we use limits. We can view the relationship between them using the Taylor series in a different way. Begin with the linear difference equation model for population growth

$$n_{t+1} = n_t(1 + b - \delta)$$

This model describes change that occurs at discrete intervals. It applies to a population having fixed gains b and losses δ at fixed time increments. It can also be useful for census data, where the data are collected at intervals. For populations that reproduce and experience mortality more or less continuously, we can translate this model into a differential equation. I first consider how a discrete model converges to the differential equation as time intervals become small.

Suppose there is some arbitrary time increment between reproductive bouts and mortality events dt. Consider a new parameter that summarizes births and deaths per unit time, $r = b - \delta$. Note that parameter r is positive so long as the birth rate exceeds the mortality rate. Now for the discrete time interval dt, write the same model like this:

$$n_{t+dt} = n_t[1 + rdt]$$

where the rate parameter r is scaled by the elapsed time between censuses dt.

The interval dt could be arbitrarily small, and we would like to express population growth rate in terms of some standard time units, such as $t = $ days or $t = $ years. Thus, rescale the arbitrary time interval dt so as to express growth in terms of some standard unit for t. For elapsed time t, the rescaling in terms of dt units is

$$n_t = n_0[1 + rdt]^{t/dt}$$

The continuous model is the limiting result as dt becomes small. From Example A.1, recognize the bracketed expression as the first two terms of the expansion for e^{rdt}. Taken in the limit as dt tends to zero this linear approximation is exact, so we write

$$\lim_{dt \to 0} [1 + r \cdot dt]^{t/dt} \to (e^{r\,dt})^{t/dt} = e^{rt}$$

Then substitution gives

$$n_t = n_0 e^{rt}$$

This result is equivalent to the linear difference of Section 2.3,

$$n_t = n_0 \lambda^t$$

Appendix B

Some Notes on Differential and Difference Equations

This appendix provides background for difference and differential equation models used in this book. A full overview is the subject of whole texts. This very brief introduction mentions some the basic principles that arise here and is intended for reference only.

B.1 Guessing a Solution for a Differential Equation

Consider some alternative ways to solve a linear differential equation. In introductory calculus we learn to solve equations of the form

$$\frac{dx}{dt} = f(t)$$

by antidifferentiation. The right-hand side is a function of t. Because the derivative is written with respect to t, we simply antidifferentiate with respect to t.

Equations for population growth often have the form

$$\frac{dn}{dt} = f(n)$$

This equation cannot be solved by antidifferentiation, because we do not know what function of time is involved; the derivative is defined with respect to time, but the right-hand side is expressed in terms of state n. We know that n changes over time—if it did not, we would not bother writing a rate equation for it; so assume this derivative is not everywhere zero.

Begin by looking for *constant solutions*, that is, solutions that apply when n does not change. Constant solutions satisfy $dn/dt = 0$. The linear growth rate equation

$$\frac{dn}{dt} = rn \qquad\qquad\qquad \text{B.1}$$

has the constant solution $n = 0$, which is obtained by setting the derivative equal to zero. Another common population growth model, the logistic model,

$$\frac{dn}{dt} = rn\left(1 - \frac{n}{K}\right)$$

is nonlinear and has constant solutions at $n = 0$, and $n = K$. I consider this model shortly.

For a more general solution, the best starting point is often a guess. The exponential function is valuable, because it figures prominently in the solutions to many differential equations. Consider that the exponential function does not change upon differentiation:

$$n(t) = e^t$$
$$\frac{dn}{dt} = e^t \qquad\qquad \text{B.2}$$
$$\frac{d^2n}{dt^2} = e^t$$

With a negative exponent, higher-order derivatives alternate in sign:

$$n(t) = e^{-t}$$
$$\frac{dn}{dt} = -e^{-t}$$
$$\frac{d^2n}{dt^2} = e^{-t}$$

These relationships help us guess solutions to differential equations. Further use of the exponential function comes with application of the chain rule. Thus, if

$$n = e^{at}$$

then

$$\frac{dn}{dt} = ae^{at} = an$$

The last step simply substitutes n for e^{at}.

B.2 Solving the Linear Differential Equation

There are a few more tools we can bring to bear on linear models. They benefit from some definitions. (First-order) linear differential equations have the form

$$f(t)\frac{dn}{dt} - g(t)n = h(t) \qquad\qquad \text{B.3}$$

Either function of time can be equal to zero or to a constant. For instance, Equation B.1,

$$\frac{dn}{dt} - rn = 0$$

is a linear differential equation having $f(t)$ equal to one, $g(t)$ equal to r, and $h(t)$ equal to zero. This example is termed a *homogeneous linear differential equation*, representing the special case where $h(t) = 0$.

B.2.1 Superposition

There can be many solutions to linear differential equations that are described by the *superposition theorem*. For homogeneous linear (first-order) differential equations,

$$f(t)\frac{dn}{dt} - g(t)n = 0$$

the theorem says that, if $n = m(t)$ is *a* solution, then *all* solutions are given by

$$n = cm(t)$$

where c is a constant. This simply says that all functions that differ only by a scalar value c have the same derivative. Thus, the linear homogeneous equation

$$\frac{dn}{dt} = rn$$

has solutions

$$n(t) = ce^{rt}$$

Note that we could know the particular solution simply by knowing the density n at time $t = 0$, that is, n_0. We could then replace c by n_0. When I say that there are many solutions to this differential equation, I mean that n_0 could take on any value without changing the differential equation.

Inhomogeneous linear differential equations do not have $h(t) = 0$, that is,

$$f(t)\frac{dn}{dt} - g(t)n = h(t)$$

The example for a population with immigration rate I,

$$\frac{dn}{dt} - rn = I$$

is inhomogeneous, because $h(t)$ does not equal zero (here $h(t) = I$). The super-position theorem for inhomogeneous equations says that if $n = m_p(t)$ is a particular solution, and $n = m(t)$ is *any* solution, then all solutions are given by

$$n = m_p(t) + cm(t)$$

Note that I have simply taken the result for the homogenized version, $cm(t)$, and added to it the particular solution $m_p(t)$.

Example B.1. A sink population using superposition
Linear population growth with a constant rate of immigration says that I new individuals enter the population from elsewhere per unit time,

$$\frac{dn}{dt} = rn + I$$

This is a linear differential equation, because the second derviative with respect to n is zero. It is inhomogeneous. A particular (constant) solution is found by setting the derivative equal to zero,

$$n = \frac{-I}{r}$$

Because I is positive (by definition), this model can have a constant solution only if r is negative. This is the definition of a sink population (Pulliam 1988). To check this particular solution, substitute back into the original equation to obtain

$$0 = -I + I$$
$$= 0$$

Now a homogenized version of the equation is

$$\frac{dn}{dt} = rn$$

(it is homogenized by setting I equal to zero). From previously, we know this to have the solution

$$n(t) = ce^{rt}$$

Thus, the sum of these solutions gives all of the solutions,

$$n(t) = \frac{-I}{r} + ce^{rt}$$

Solving for the constant c gives the solution

$$n(t) = \frac{-I}{r} + \left(n_0 + \frac{I}{r}\right)e^{rt}$$

B.2.2 Separable Equations

The *general* solution must satisfy the rate equation, and it must express density as a function of time,

$$n(t) = f(t)$$

The specific solution must also satisfy additional constraints, including initial and boundary conditions. In this case, we have an initial condition $n(0) = n_0$. The notation indicates that density $n(t)$ at time $= 0$ is equal to the constant n_0.

Separable equations have the form

$$\frac{dn}{dt} = f(n)g(t)$$

$g(t)$ can equal one. These equations can be solved by separation, by collecting all of the n's on one side and all of the t's on the other:

$$\frac{dn}{f(n)} = g(t)dt$$

Now integrate,

$$\int_{n(0)}^{n(t)} \frac{dn}{f(n)} = \int_0^t g(t')dt'$$

Example B.2. Solve the linear growth equation using separation of variables. Collect the n's on one side and the t's on the other,

$$\frac{dn}{n} = rdt$$

Now integrate

$$\int \frac{dn}{n} = \int rdt$$

or $\ln n = rt + c_1$, where c_1 is an integration constant. Exponentiate to obtain

$$n(t) = c_2 e^{rt}$$

where the constant $c_2 = e^{c_1}$, and solve for c_2

$$n(0) = c_2 e^{r \times 0} = c_2$$

to yield the solution

$$n(t) = n(0)e^{rt}$$

We could also have solved this as a definite integral

$$\int_{n(0)}^{n(t)} \frac{dn}{n} = \int_0^t r\, dt'$$

Take antiderivatives of both sides and substitute for limits

$$\ln n \big|_{n(0)}^{n(t)} = rt' \big|_0^t$$

$$\ln n(t) - \ln n(0) = rt$$

Example B.3. The immigration model of Example B.1 is separable. I solve it here to demonstrate that either method works. Upon separating the differential equation obtain

$$\frac{dn}{rn + I} = dt$$

Note that all n's are on the left. To perform the integration

$$\int \frac{dn}{rn + I} = \int dt$$

make the substitution $u = rn + I$. Because we will integrate over a new variable u, rather than n, we must replace dn on the left-hand side with du. This is accomplished by differentiating u with respect to n,

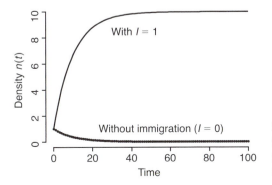

FIGURE B.1. A plot of the immigration model for a "sink" population (Pulliam 1988), with parameter values $r = -0.1$, $n(0)=1$.

$$\frac{du}{dn} = \frac{d}{dn}(rn + I) = r$$

or, equivalently, $dn = du/r$. Now substituting for dn in the integral we have

$$\int \frac{du}{ru} = \int dt$$

$$\ln u = rt + c$$

where c is an integration constant. Exponentiating and substituting for u gives

$$rn + I = ce^{rt}$$

The constant c is still undetermined. Solving for n yields the same answer we obtained in the previous example,

$$n(t) = \frac{-I}{r} + ce^{rt}$$

For the initial conditions $t = 0$ and $n(0)$ we can determine the value of the constant. This gives the result for the constant,

$$c = n(0) + \frac{I}{r}$$

B.3 Constant Solutions for Nonlinear Differential Equations

To what values would the state variables (e.g., population density) tend if we allowed a population to grow to the point where density dependence became important? Constant solutions can be useful benchmarks, especially for non-linear models. For a rate equation, a necessary condition is that the rate of change tends to zero. Let $f(n)$ be the growth rate of a population of density n,

$$f(n) \equiv \frac{dn}{dt}$$

The equilibrium (steady state) is a constant solution and, thus, satisfies

$$f(n) = 0$$

Example B.4. Continuous logistic model
For the continuous logistic,

$$f(n) = rn\left(1 - \frac{n}{K}\right)$$

setting $f(n) = 0$ and solving for density gives constant solutions $n^* = 0$ and $n^* = K$. These constant solutions represent the places where the plot of population growth rate against density equals zero (right-hand side of Figure 2.4b).

The logistic equation is nonlinear, but can still be solved by separation of variables. Write the logistic growth equation as

$$\frac{dn}{dt} = \frac{rn(K - n)}{K}$$

Separate the equation by collecting the n's on the left-hand side and t on the right-hand side

$$\int_{n_0}^{n(t)} \frac{K}{n(K - n)} \, dn = \int_{0}^{t} r \, dt'$$

Use partial fractions on the left-hand side to obtain

$$\int_{n_0}^{n(t)} \frac{K}{n(K - n)} dn = \int_{n_0}^{n(t)} \left(\frac{1}{n} + \frac{1}{K - n} \right) dn = [\ln n - \ln(K - n)]_{n_0}^{n(t)}$$

The method of partial fractions is like the inverse of finding a least common denominator. The last step made use of the substitution $u = K - n$. After substituting the limits for n, integrate the right - hand side to obtain the equality

$$\ln\left(\frac{n(t)(K - n_0)}{n_0(K - n(t))} \right) = rt$$

Now exponentiate and solve in terms of $n(t)$.

B.4 Graphs for Difference Equations

Graphical methods can provide insight into the behavior of difference equations. Consider a linear difference equation, written in terms of equal time increments 1, 2, . . . , and n_0 is initial density. If

$$n_1 = n_0\lambda$$

then density at the next time step is

$$n_2 = n_1\lambda$$

Now substitute for n_1

$$n_2 = (n_0\lambda)\lambda = n_0\lambda^2$$

This iterative (inductive) approach suggests the solution

$$n_t = n_0 \lambda^t$$

Few difference equation models have solutions. Because a general solution may not be available, we look for specific solutions that may provide insight. For example, simply knowing the parameter values at which the population is expected to increase, decrease, or remain constant is better than no information at all.

Recall that a constant solution is defined by the value(s) where there is no change in the state variable(s), that is, $n_t = n_{t+1} = n_{t+2} \ldots$ Refer to this constant solution with the notation n^*. To find this constant solution replace all n's with n^* and solve. For the previous linear model, this would be

$$n^* = n^* \lambda$$

Clearly, a constant solution obtains when $\lambda = 1$. The population grows if $\lambda > 1$, and it declines if $\lambda < 1$.

Linear population models predict exponential growth or decline. There is no *population regulation*, in the sense that populations may increase indefinitely (Figure B.2). This is easy to see from the solution $n_t = n_0 \lambda^t$ If $\lambda > 1$, then this model blows up. Because populations cannot grow linearly for long, linear difference equation models are typically used to describe short-term population growth.

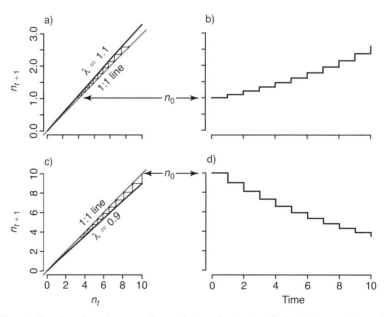

FIGURE B.2. Graphical projection of population densities off the 1:1 line. Cobweb diagram (*left*) and population growth for values for two values of λ.

Nonlinear difference equation models of Section 2.2.3 and Chapter 9 usually cannot be solved. A plot of n_{t+1} against n_t helps us visualize the behavior of difference equation models (May and Oster 1976). Consider again the linear equation $n_{t+1} = n_t\lambda$, a straight line with zero intercept and slope λ. A constant solution is $n_{t+1} = n_t$, and $\lambda = 1$. Note on Figure B.2a that projection off the 1:1 line puts us back where we began. By "projection off the 1:1 line" I mean the following: draw a vertical line from a starting value of n_t (a point along the horizontal axis) to the solid line having slope λ. In the case of Figure B.2a, $\lambda = 1.1$. This line is the plot of n_{t+1} vs n_t. From this point draw a horizontal line to the 1:1 line (dashed line in Figure B.2). From this point, draw a vertical line back to the solid line. Continuing in this fashion we obtain the projection of population growth. The plots on the right-hand side of Figure B.2 show the projection plotted against time. Figure B.2 is sometimes termed a *cobweb diagram* for reasons that will become clear when we consider nonlinear models. The projection method allows us to project population density forward, and it will help us understand more complicated models. Example B.3 shows that this type of projection can even be useful for some linear models.

Example B.5. A constant solution with emigration or harvesting
Suppose that a constant density E is removed at each time increment due to harvesting or emigration. Obviously, this density E can only be removed so long as $n_t > E$. Is there a constant solution for the population?

The linear model with emigration could be written as

$$n_t = n_{t-1}\lambda - E$$

To find a constant solution, substitute for the n's

$$n^* = n^* \lambda - E$$

and solve

$$n^* = \frac{E}{\lambda - 1}$$

Because E is positive (by definition of the problem), there can only be a constant solution if $\lambda > 1$. Otherwise, the constant solution is negative, and the population declines. Provided that $\lambda > 1$, can we conclude that there is a density n^* that represents a point of transition between population growth and decline due to harvesting or emigration pressure? To say this for sure, we need an additional tool, termed *stability analysis* (Section B.5).

For the harvesting Example B.5, the straight line represented by the model has a negative intercept and crosses the 1:1 line at the constant solution (Figure B.3a). For an initial density above the constant solution, projection off the 1:1 line results in exponential growth. For an initial density below the constant solution, the density declines to extinction. In other words, the fate of the population depends entirely on the initial density. So the term *constant*

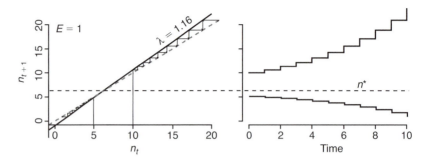

FIGURE B.3. The harvest model from Example B.5 having constant solution $n^* = 6.2$. At *left* is a cobweb diagram, which can be used to determine if the constant solution is stable. This diagram tracks the path of change in density from n_t to n_{t+1}. It is constructed by starting from an initial density on the horizontal axis and then projecting off the dashed 1:1 line to obtain the next density. So long as density exceeds n^* (e.g., $n_t = 10$), the population size can increase (right-hand side). If density is less than n^* (e.g., $n_t = 5$), the population declines.

solution does not tell us about behavior of the model. Rather, it represents a threshold value between exponential growth and extinction.

Graphical methods can be especially valuable for nonlinear difference equations. We can learn something of their behavior by finding constant solutions and by exploring their behavior near such solutions and by graphical analysis. A graph of the Ricker model in Figure B.4 shows a population that begins at a density above the carrying capacity, which, here, is set to $K = 1$. Strong density dependence results in such a sharp decline that the next density is well below the carrying capacity. Thereafter, density tends to oscillate around K, due to the feedback of density on growth rate and the inherent lag represented by the time step of one unit. The constant solution for the Ricker model is the point of intersection of the model (the curve in Figure B.4a) and the 1:1 line on the cobweb diagram. If the parameter r is not too large, then projection off the 1:1 line brings us to the constant solution K, that is, the point of intersection. Stability analysis is typically used to analyze model behavior near constant solutions.

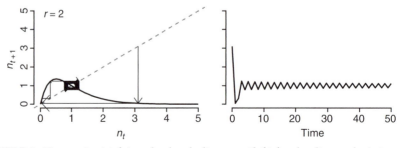

FIGURE B.4. Time series (*right*) and cobweb diagrams (*left*) for the discrete logistic equation for $r = 2$. Left, the dashed line is 1:1.

The Ricker model is but one of a number of nonlinear difference equation models commonly used to model population growth. Other examples include the discrete logistic model (May 1973) and the Beverton-Holt stock recruitment model, which arises in Section 9.2.4. Models differ in how they embody the feedback effect on population growth rate. They share the feature that feedback effects determine the equilibrium properties of models, and those effects depend on the time lag. This lag can be implicit, in terms of the time step of the model, or they can be incorporated explicitly (Section 9.12).

Example B.6. The Ricker model
Find the constant solution for the Ricker model,

$$n_{t+1} = n_t \exp\left[r\left(1 - \frac{n_t}{K} \right) \right]$$

The constant solution obtains when the density remains the same, or $n_{t+1} = n_t$. Let n^* be the value of this constant solution, that is, $n^* = n_{t+1} = n_t$. Substitution gives

$$n^* = n^* \exp\left[r\left(1 - \frac{n^*}{K} \right) \right]$$

Now solve for n^* to obtain the constant solutions $n^* = (0, K)$. This result says that density will not change only if density is precisely zero or K. From any nonzero density, population growth will cause density to converge to K.

Examples B.5 and B.6 seem to say conflicting things about the constant solution. In the emigration or harvesting model of Example B.5, the constant solution was the density from which growth causes divergence. For the logistic growth model of Example B.6 the constant solution K is an attractor; density increases or decreases to this value. So finding the constant solution is not enough. We also need to determine whether that solution represents a value to which densities might tend to converge or diverge. Stability analysis can help us distinguish them. The Taylor series in Appendix A is a basic tool for stability analysis.

B.5 Equilibria and Stability

We can sometimes solve equations or use graphs (previous section) to identify equilibria, that is, constant solutions. Knowing the value of a constant solution is not enough, because models can have constant solutions that are *neutrally stable* or *unstable*. Stability analysis helps us understand whether a system is likely to exist at all—unstable systems may not survive to be observed. Stability analysis can be used to determine properties of constant solutions.

I include discussion of stability here primarily to emphasize its relevance for inference. There is a substantial literature on stability of food webs and ecosystems, including basic discussions in a number of excellent texts. The tendency is to focus on the potential for complex dynamics in the context of parameter values. After reviewing some basics, I bring up some considerations for the context of inference, where difference equations are often substituted for differential equations.

B.5.1 Types of Stability

An equilibrium value is one for which a state variable has first derivative(s) equal to zero. If we view the state variable as the location of a ball on a rough landscape, all of the examples in Figure B.5 satisfy the necessary conditions for an equilibrium value.

These equilibrium values differ in their stability characteristics. Local stability is examined using linear (neighborhood) stability analysis, which tells us what happens to the state variable if we move it a small distance from the equilibrium. For a population model, add a small amount of density to the equilibrium value and observe what happens to the growth rate. If the resultant growth rate is negative, then density declines, thus returning to the equilibrium value. If positive, the population grows away from the equilibrium value. In the analogy of Figure B.5, stability analysis is equivalent to moving the ball a small distance to the right or left and then observing whether it returns to the initial location. Stability can be local or global, because the surface may be complex. Linear stability analysis makes use of a Taylor expansion of the growth rate function (Appendix A).

B.5.2 Linear Stability Analysis for Continuous Models

A model of continuous population growth could be represented by a function of density $f(n)$,

$$f(n) \equiv \frac{dn}{dt}$$

A plot of this function for the logistic model is shown in Figure 2.4b. There are two constant solutions, represented by points where the function is equal to zero. For the logistic model these constant solutions occur at $n = 0$ and $n = K$. The growth rate is zero at these points, but will it remain so? A local stability analysis will tell us if a small change in density away from the constant solution would be followed by return to that constant solution.

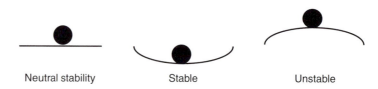

Neutral stability   Stable   Unstable

FIGURE B.5. A standard stability cartoon for a ball on a landscape.

To analyze stability add a small amount of density to the constant solution n^*, say n', and determine the subsequent growth rate of the population. At time t, the population is at density

$$n(t) = n^* + n'$$

Now follow what happens to this perturbation by determining the growth rate of the perturbation $n'(t)$ (Figure B.6). The next few steps illustrate that the perturbation can be approximately represented by the homogeneous linear differential equation

$$\frac{dn'}{dt} = an'$$

To arrive at this result use a Taylor expansion of the growth rate and ignore all but the linear term. Consider the first two terms of the Taylor series for $f(n)$ near the value n^*,

$$f(n) \approx f(n^*) + n'\frac{df}{dn}\bigg|_{n^*}$$

Because n^* is a constant solution, $f(n^*) = 0$, we can ignore the first term on the right-hand side. We now have

$$f(n) = n'\frac{df}{dn}\bigg|_{n^*}$$

Write the left-hand side as

$$f(n) \equiv \frac{dn}{dt} = \frac{d}{dt}[n^* + n'(t)] = 0 + \frac{dn'}{dt}$$

Again, the zero comes from the fact that n^* is a constant solution. With this substitution we have

$$\frac{dn'}{dt} = n'\frac{df}{dn}\bigg|_{n^*} = an'$$

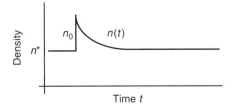

FIGURE B.6. A perturbation to the equilibrium value n^*.

The equation contains a rate parameter a, which is a linear approximation of the effect of density on population growth rate. This first derivative is a slope parameter that determines the stability characteristics. To see this, note that the solution to the rate equation for dn'/dt is

$$n'(t + \Delta t) = n'(t)e^{a\,\Delta t}$$

If a is positive this perturbation grows away from the initial value near n^*, and, thus, n^* is unstable and vice versa. For the logistic model these tangent lines are shown for the two constant solutions in Figure B.7. In summary, a simple recipe is:

Step 1. Determine how density n affects population growth rate $f(n) \equiv dn/dt$,

$$\frac{df(n)}{dn} \equiv \frac{d}{dn}\left(\frac{dn}{dt}\right)$$

Step 2. Evaluate this derivative at the equilibrium density n^*. Let a represent this derivative,

$$a \equiv \frac{df(n)}{dn}\bigg|_{n=n^*}$$

If a is negative, then adding a small amount of density results in a negative growth rate, and density declines back toward the equilibrium; the equilibrium is therefore *stable*. If positive, density increases away from the equilibrium, and the equilibrium is unstable. If zero, density has no effect on growth rate, and the equilibrium is neutrally stable.

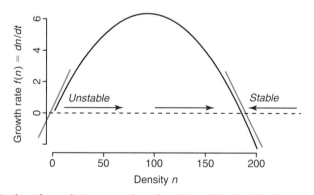

FIGURE B.7. A plot of population growth as function of density for the continuous logistic model. The constant solution at $n = 0$ is unstable, because increasing density results in a positive growth rate. The derivative a is represented by the tangent at $n = 0$, represented here by a solid line. The constant solution at $n = K$ is stable, because increasing density results in a negative population growth rate, and density declines. In other words, at $n = K$, $a < 0$.

Example B.7. Exponential growth
Consider a population that grows exponentially,

$$\frac{dn}{dt} = n(b - \delta)$$

A constant solution is $n^* = 0$ (the density satisfying $dn/dt = 0$). The rate at which density changes following the perturbation to n^* is

$$a \equiv \left.\frac{df(n)}{dn}\right|_{n=0} = b - \delta = r$$

Because n^* is stable when $a < 0$, stability requires that $\delta > b$. Thus, the model is stable only for an extinct population ($n = 0$) having a (potentially) higher mortality rate than birth rate. If birth rate could be higher than mortality rate, a would be positive, and, following any small introduction, the population would begin to increase. Because there is no other constant solution, this model population tends to infinity with time. Thus, the constant solution $n = 0$ is unstable for any population capable of positive growth rate.

Example B.8. Continuous logistic growth
The continuous logistic

$$f(n) = rn\left(1 - \frac{n}{K}\right)$$

has constant solutions at $n^* = 0$ and $n^* = K$. The derivative

$$\frac{df(n)}{dn} = r - \frac{2rn}{K}$$

is evaluated at the equilibria. For the first constant solution, $n^* = 0$,

$$\left.\frac{df(n)}{dn}\right|_{n=0} = r$$

is positive and therefore unstable. The population will increase away from $n = 0$ provided there is at least a small amount of density to get it started. The second constant solution at $n^* = K$

$$\left.\frac{df(n)}{dn}\right|_{n=K} = -r$$

is negative, and the critical point is stable. In fact, it is globally stable (Figure B.7).

Example B.9. Equilibrium and stability properties for a continuous model with immigration and emigration

Suppose the birth rate b consists of a fraction d disperser progeny and $(1 - d)$ progeny that remain within the local population of size n. Then dispersers are produced at a rate

$$n \, bd$$

and non dispersers at a rate

$$nb \, (1 - d)$$

Immigrants I are not produced by the local population. Together these processes imply that local population density will change at a rate given by

$$\frac{dn}{dt} = nb(1 - d) - \delta n + I$$
$$= \text{progeny that do not disperse} - \text{deaths}$$
$$+ \text{ immigrants}$$

Note again that births and deaths depend on local density, whereas immigration (in this case) does not. An equilibrium exists at

$$n^* = \frac{I}{\delta - b(1 - d)}$$

Without immigration, there can only be an equilibrium at $n^* = 0$. With immigration, there can only be a positive equilibrium if immigrants are offset by net mortality. Analysis of the effect of density on growth rate yields the slope parameter

$$a = b \, (1 - d) - \delta$$

The population is stable at a positive density n^* if mortality exceeds the rate of new births that do not disperse. If $\delta < b(1 - d)$, the population grows indefinitely. Note that the rate of immigration determines the population size, but it has no effect on whether or not the population is stable—stability only requires mortality high enough to offset births. Immigration increases the equilibrium population size, but it cannot destabilize the population. Thus, immigration allows the maintenance of stable population density in an environment where it would otherwise go extinct. This is a sink population, which is maintained by some other source population responsible for immigration rate I.

B.5.3 Linear Stability Analysis for Difference Equations

Stability analysis of a discrete model uses a similar approach. A small perturbation from the equilibrium n^*, call it n', may increase or decrease back to n^*. Consider the difference equation

$$n_{t+1} = f(n_t)$$

Recall that a constant solution for this difference equation is

$$n_{t+1} = n_t \equiv n^*$$

or, alternatively,

$$n^* = f(n^*)$$

For a difference equation, the constant solution is unstable if the effect of density on the discrete growth rate is greater than one, not zero. To be precise, the absolute value of the derivative of the model must be less than one. In the discrete case we ask how increasing density would affect this discrete growth rate at the constant solution,

$$n_{t+1} \approx n_t \frac{df}{dn}\bigg|_{n^*}$$

The population will increase away from the constant solution if the derivative is greater than one. If the derivative is between zero and one, the solution will be stable.

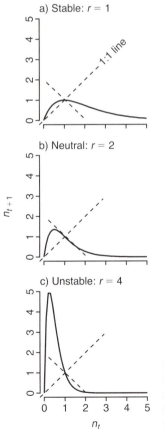

FIGURE B.8. Plots of the logistic model showing the slope of the model relative to a slope of one. If steeper than 1, the constant solution is unstable.

Example B.10. A saturating model
Consider the nonlinear difference equation

$$n_{t+1} = \frac{cn_t}{b + n_t}$$

This equation is at steady state when $n_{t+1} = n_t = n^*$. Substitution gives

$$n^* = \frac{cn^*}{b + n^*}$$

Solving for density gives the critical points $n^* = c - b$, $n^* = 0$. Again, the steady state may not be stable; if it is unstable the slightest perturbation will send the population off in another direction. Let

$$f(n_t) = \frac{cn_t}{b + n_t}$$

Solve for the slope parameter,

$$a \equiv \frac{df}{dn}\bigg|_{n^*} = \frac{bc}{(b + n^*)^2}$$

Now a is evalutated at $n^* = c - b$, so

$$a = b/c$$

For stability $|a| < 1$ or

$$\left|\frac{b}{c}\right| < 1$$

Both b and c are positive in the model, and $c > b$. The equilibrium exists only if $c > b$, so it must be stable.

B.5.4 Stability for a Nonlinear Difference Equation Using Graphs

On the phase plot of n_{t+1}/n_t stability is determined by a, the slope of the tangent at the constant solution (Figure B.8). At equilibrium that tangent is the parameter I have called a. I now rename it λ to call attention to its relationship to eigenvalues, which come up in Appendix C,

$$\frac{df(n)}{dn}\bigg|_{n=n^*} = \lambda$$

This parameter determines the strength of the feedback around the equilibrium point. If λ is small, the function $f(n)$ is rather flat and the equilibrium point is an attractor. If steep, it is a repellor, and the population behaves chaotically. Consider the example of May and Oster (1976). The Ricker model

$$n_{t+1} = f(n_t)$$

$$f(n_t) = n_t \exp[r(1 - n_t)]$$

has the equilibrium point $n^* = 1$, because I have set $K = 1$. The density dependence

$$\frac{df}{dn} = e^{r(1-n_t)} + n \cdot (-r)e^{r(1-n_t)}$$

evaluated at $n^* = 1$ is $1 - r$. The slope parameter $\lambda = 1 - r$ is affected by r in the following manner:

$1 > \lambda > 0$ monotonic damping
$0 > \lambda > -1$ damped oscillations
$-1 > \lambda$ diverging oscillations
$\lambda = 1$ neutrally stable oscillations
$\lambda > 1$ monotonic divergence

The examples in Figure B.9 show how the steepness of the function increases at the critical point (i.e., the intersection with the dashed line) as r increases.

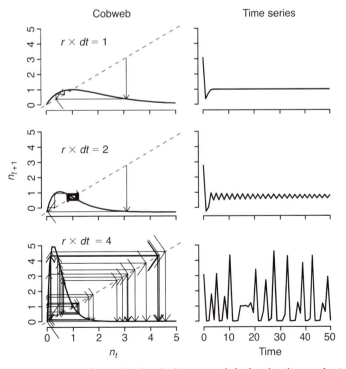

FIGURE B.9. Time series (*right*) and cobweb diagrams (*left*) for the discrete logistic equation with three different values for r.

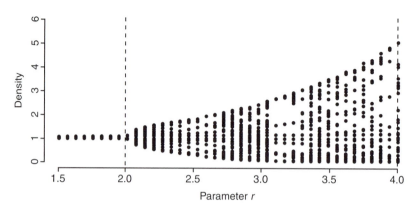

FIGURE B.10. Bifurcation diagram for the discrete logistic growth model.

These increasing values are associated with monotonic convergence to the critical point ($r = 1$ or $\lambda = 0$) to divergence away from the critical point as r becomes large. Regardless of where we start, the stable case always tends monotonically to the equilibrium at the intersection of the two lines. The neutral equilibrium at $r = 2$ ($\lambda = 1$) shows the oscillations. The chaotic behavior for $r = 4$ results from the steep angle at the repellor n^*.

How do we interpret these results? The stability analysis tells us that if the per capita growth rate gets too large, then the attractor at n^* is no longer a point equilibrium. The population begins to fluctuate, first in a cyclic fashion. If r gets too large we get chaos. These *complex dynamics* result from the combination of strong feedback and the time lag. Together, these qualities allow density to overshoot the equilibrium value. The bifurcation diagram (Figure B.10) indicates how density changes with increasing values of r.

There is a large literature on chaos in population and community ecology that is readily accessible (e.g., Hastings et al. 1993). For our purposes it is important to know that nonlinear difference equations can exhibit deterministic dynamics that are difficult to identify as such. Sensitivity to initial conditions can be large. For nonlinear models, one must be aware that changes in the step size of the model can have profound implications.

B.5.5 Comparison of Discrete and Continuous Logistic Models

Constant values of a model provide some insight into its behavior. Because we often cannot solve nonlinear models (i.e., write density as some explicit function of time), the constant solution may provide guidance concerning where density tends with time. When derivatives are used as the basis for identifying constant solutions, stability analysis is needed to provide additional insight. If a constant solution is unstable, we might suspect particular types of complex behavior, or we might question whether such a population could persist. The techniques illustrated here are readily extended to systems of nonlinear difference equations. May (1973) used such methods to ask how the complexity of a food web might affect stability.

The combination of three considerations makes these stability properties pertinent to inference, the topic of this book. First, ecologists have long been preoccupied with complex dynamics, and have devoted substantial effort to identification of parameter values that might be needed to cause them (e.g., Example B.10). Second, fitting models to data often involves discretization of what were initially formulated as differential equations. Finally, the time step employed for such models often has little to do with known time lags, depending instead on availability of data or convenience. There is an implicit time lag of one in the difference equations of the foregoing section. Unless a time lag is explicitly included in a differential equation (e.g., Nisbet and Gurney 1982), a simple differential equation will not produce complex dynamics, because there is no delay in feedbacks. Without the time lag there can be no overshoot. For difference equations, the strength of the feedback operates together with the time scale of the process.

In Section 2.2.3, I discuss a standard discretization for a differential equation, the example being the logistic,

$$n_{t+dt} = n_t + rn_t\left(1 - \frac{n_t}{K}\right)dt$$

This is not the only way to discretize the logistic model. For example, I could have discretized this way,

$$n_{t+dt} = n_t\left[1 + rn_t\left(1 - \frac{n_t}{K}\right)\tau\right]^{dt/\tau}$$

Or I could have written other functional forms having similar behavior (e.g., Chapter 9). Regardless, if dt is small relative to the values of other rate parameters in the model (in this case, r has units t^{-1}), these discretized versions will behave like the differential equation from which they were derived. This is not the case if dt is large relative to r. Taking $dt = 1$ for the first of these two equations, stability analysis indicates diverging oscillations if $r > 2$. This is actually a special case for $dt = 1$. For explicit dt, we find diverging oscillations if $r \cdot dt > 2$. This has obvious implications for the interpretation of the relationship between r and potential for complex dynamics. If the time step used in the model is not an actual lag in nature (e.g., one generation per time step), then there is no meaningful interpretation of complex dynamics.

Moreover, for purposes of inference, we are typically concerned with approximating a rate equation with a difference equation. Thus, the goal is to select a time step for the model that is not so large as to result in behaviors that are substantially different from the differential equation that served as motivation. I point out in Chapters 2 and 9 that discretization is less of a concern with inference than it is for interpretation of complex dynamics. Still, it is important to recognize the potential for instability, particularly where nonlinear models may be applied to data at unequal time steps (Section 9.8.5). As a specific example, calculation of predictive loss may involve projecting the model forward for stochastic parameter values, bringing in complex behavior (lab manual).

Appendix C

Basic Matrix Algebra

Some fundamentals of matrix algebra are needed for linear models. Here are just the basics.

C.1 Notation

Matrixes are used to summarize systems of equations. Consider the system

$$
\begin{array}{rrcr}
4x_1 & -6x_2 & = & 10 \\
-x_1 & +2x_2 & = & -2
\end{array}
$$

To find values of the variables that satisfy both equations, multiply the second equation by 4 and add the two equations. The solution (x_1, x_2) is (4, 1). But larger systems are harder to solve. Some basic notation will help.

Write this system of two equations more compactly in terms of a matrix of coefficients

$$
\mathbf{A} = \begin{bmatrix} 4 & -6 \\ -1 & 2 \end{bmatrix}
$$

a vector of x's

$$
\mathbf{x} = \begin{bmatrix} x_1 \\ x_2 \end{bmatrix}
$$

and a vector of coefficients for the right-hand side y,

$$
\mathbf{y} = \begin{bmatrix} y_1 \\ y_2 \end{bmatrix}
$$

In this new notation, the system becomes

$$\mathbf{Ax} = \mathbf{y}$$

The dimensions of these arrays are given by this notation:

$$\underset{(2\times2)}{\mathbf{A}} \quad \underset{(2\times1)}{\mathbf{x}} \quad = \quad \underset{(2\times2)}{\mathbf{y}} \qquad \text{C.1}$$

Below each array is a number (n by m), where n is the number of rows and m is the number of columns. The notation a_{ij} designates a particular element of the matrix $\mathbf{A}$, and the notation x_i refers to a particular element of the vector $\mathbf{x}$. Thus, the element in the second row and first column of $\mathbf{A}$ is

$$a_{21} = -1$$

In terms of these indexes, we could write the matrix as

$$\mathbf{A} = \begin{bmatrix} a_{11} & a_{12} \\ a_{21} & a_{22} \end{bmatrix}$$

The notation suggests that we can multiply the matrix $\mathbf{A}$ by the vector $\mathbf{x}$ to obtain the vector $\mathbf{y}$. This notation provides a method for solving this system of equations that involves eigenanalysis. First consider some standard operations in matrix algebra.

Vectors and matrixes often have subscripts. Variables in italics are scalar quantities and may have subscripts indicating any stratification, such as $y_{i,t}$ for a state variable y observed for individual i at time t. A set of such observations taken over a particular stratification is assembled in a vector, indicated by lowercase bold font, omitting the subscripts for that stratification. Thus $\mathbf{y}_t = [y_{1,t}, \ldots, y_{n,t}]^T$ is the set of observations on n individuals at time t, and $\mathbf{y}_i = [y_{i,1}, \ldots, y_{i,T}]^T$ is the sequence of observations on individual i from time 1 to T. The superscript T means *transpose* (Section C.4). Matrixes are bold and capitalized. The set of all observation on y can be assembled in an n by T matrix $\mathbf{Y}$. If there is a matrix of such observations at each location j, we have $\mathbf{Y}_j$, with rows $\mathbf{y}_{ij}$, and elements $y_{ij,t}$.

For parameters, I have not uniformly used bold and capital letters in this manner, because I suspected that many readers would not match the lower- and uppercase Greek letters, and some Greek letters are reserved (e.g., Γ for the gamma function). I have maintained the convention on subscripting. Thus, if there is a matrix of parameter values θ_{ij}, then the vector of rows is θ_i and the complete matrix is θ. Context should make this notation apparent.

There is an exception to this notation for subscripting symmetric matrixes. For symmetric matrixes, I sometimes include a subscript to indicate the dimension of the matrix. Thus, the n by n identity matrix (Section C.5) is sometimes indicated as $\mathbf{I}_n$ and an n by n covariance matrix by Σ_n. For a parameter covariance matrix, I sometimes include a subscript to indicate that parameter vector. Thus, for the length-p parameter vector $\beta = [\beta_1, \ldots, \beta_p]^T$ has a covariance matrix $\mathbf{V}_\beta$. This notation will be clear from context.

C.2 Matrix Addition

Matrix addition is accomplished by adding elements of the two matrixes having the same locations. Addition of two matrixes A and B is done like this:

$$\underset{(n \times m)}{\mathbf{A}} + \underset{(n \times m)}{\mathbf{B}} = \begin{bmatrix} a_{11} + b_{11} & a_{12} + b_{12} & \cdots & a_{1m} + b_{1m} \\ a_{21} + b_{21} & a_{22} + b_{22} & & \\ \vdots & \vdots & \ddots & \\ a_{n1} + b_{n1} & a_{n2} + b_{n2} & & a_{nm} + b_{nm} \end{bmatrix} \qquad \text{C.2}$$

The matrixes must have the same dimensions; otherwise, there will not be matching elements from both matrixes.

C.3 Multiplication

Matrixes can be multiplied by a scalar s, in which case the product is a new matrix containing elements sa_{ij},

$$s\mathbf{A} = \begin{bmatrix} sa_{11} & sa_{12} \\ sa_{21} & sa_{22} \end{bmatrix} \qquad \text{C.3}$$

Note that $s\mathbf{A} = \mathbf{A}s$.

Multiplication of a matrix $\mathbf{A}$ by a vector $\mathbf{x}$ produces a new vector $\mathbf{y}$, where the ith element of $\mathbf{y}$ is the sum of the product of each element of $\mathbf{x}$ and each element of the ith row of $\mathbf{A}$,

$$\underset{(2 \times 2)(2 \times 1)}{\mathbf{A} \quad \mathbf{x}} = \begin{bmatrix} a_{11} & a_{12} \\ a_{21} & a_{22} \end{bmatrix} \begin{bmatrix} x_1 \\ x_2 \end{bmatrix} = \begin{bmatrix} a_{11}x_1 + a_{12}x_2 \\ a_{21}x_1 + a_{22}x_2 \end{bmatrix} = \begin{bmatrix} \sum_{j=1}^{m} a_{1j}x_j \\ \sum_{j=1}^{m} a_{2j}x_j \end{bmatrix} \qquad \text{C.4}$$

In other words, the ith element of $\mathbf{y}$ is

$$y_i = \sum_{j=1}^{m} a_{ij}x_j \qquad \text{C.5}$$

Note that multiplication of a matrix by a vector requires that the number of columns in the matrix equal the number of rows in the vector (Equation C.1). For example,

$$\begin{bmatrix} 4 & -6 \\ -1 & 2 \end{bmatrix} \begin{bmatrix} 4 \\ 1 \end{bmatrix} = \begin{bmatrix} 4.4 - 6.1 \\ -1.4 + 2.1 \end{bmatrix} = \begin{bmatrix} 10 \\ -2 \end{bmatrix}$$

Multiplication of a matrix by a vector is a special case of matrix multiplication. For two matrixes, we have

$$
\underset{(n \times m)(m \times p)}{\mathbf{A} \quad \mathbf{B}} =
\begin{bmatrix}
\sum_{j=1}^{m} a_{1j}b_{j1} & \sum_{j=1}^{m} a_{1j}b_{j2} & \cdots & \sum_{j=1}^{m} a_{1j}b_{jp} \\[2ex]
\sum_{j=1}^{m} a_{2j}b_{j1} & \sum_{j=1}^{m} a_{2j}b_{j2} & \cdots & \sum_{j=1}^{m} a_{2j}b_{jp} \\[2ex]
\vdots & \vdots & \ddots & \\[2ex]
\sum_{j=1}^{m} a_{nj}b_{j1} & \sum_{j=1}^{m} a_{nj}b_{j2} & & \sum_{j=1}^{m} a_{nj}b_{jp}
\end{bmatrix}
\qquad \text{C.6}
$$

For the ik^{th} element, sequentially multiply each element of the ith row of $\mathbf{A}$ by the kth column of $\mathbf{B}$. Then sum,

$$
y_{ik} = \sum_{j=1}^{m} a_{ij}b_{jk} \qquad \text{C.7}
$$

For example,

$$
\begin{bmatrix} 4 & -6 \\ -1 & 2 \end{bmatrix}
\begin{bmatrix} 1 & 0 \\ 0 & 1 \end{bmatrix} =
\begin{bmatrix} 4+0 & 0-6 \\ -1+0 & 0+2 \end{bmatrix} =
\begin{bmatrix} 4 & -6 \\ -1 & 2 \end{bmatrix}
$$

In this case, matrix multiplication produced the same matrix I started with. The *identity matrix*, consisting of a diagonal of ones and zeros everywhere else, acts like a "1" in standard multiplication. I use the notation $\mathbf{I}$ or $\mathbf{I}_n$, the latter indicating the dimension of the identity matrix.

Note that matrix multiplication requires that the number of columns in the first matrix (say, m) must equal the number of rows in the second matrix. The matrix that results from matrix multiplication takes its first dimension from the first matrix and its second dimension from the second matrix. In other words,

$$
\underset{(n \times m)(m \times p)}{\mathbf{A} \quad \mathbf{B}} = \underset{(n \times p)}{\mathbf{Z}}
$$

Thus, unlike standard multiplication, where $a \times b = b \times a$, for matrix multiplication $\mathbf{AB} \neq \mathbf{BA}$. Note the following additional relationships:

$$
\mathbf{A(BC)} = \mathbf{(AB)C}
$$

$$
\mathbf{A(B+C)} = \mathbf{AB} + \mathbf{AC}
$$

$$
\mathbf{(A+B)C} = \mathbf{AC} + \mathbf{BC}
$$

C.4 Transpose

To transpose a vector or matrix, switch the rows and columns. If

$$\mathbf{y} = \begin{bmatrix} y_1 \\ y_2 \\ y_3 \end{bmatrix}$$

then

$$\mathbf{y}^T = \begin{bmatrix} y_1 & y_2 & y_3 \end{bmatrix}$$

or, equivalently,

$$\mathbf{y} = \begin{bmatrix} y_1 & y_2 & y_3 \end{bmatrix}^T$$

For the matrix

$$\mathbf{A} = \begin{bmatrix} a_{11} & a_{12} \\ a_{21} & a_{22} \end{bmatrix}$$

the transpose is

$$\mathbf{A}^T = \begin{bmatrix} a_{11} & a_{21} \\ a_{12} & a_{22} \end{bmatrix}$$

For example, the matrix

$$\mathbf{A} = \begin{bmatrix} 4 & -6 \\ -1 & 2 \end{bmatrix}$$

has transpose

$$\mathbf{A}^T = \begin{bmatrix} 4 & -1 \\ -6 & 2 \end{bmatrix}$$

Note the following relationships:

$$(\mathbf{A}^T)^T = \mathbf{A}$$

$$(\mathbf{A} + \mathbf{B})^T = \mathbf{A}^T + \mathbf{B}^T$$

$$(\mathbf{AB})^T = \mathbf{B}^T\mathbf{A}^T$$

C.5 Identity Matrix

The identity matrix is a diagonal matrix of ones,

$$\mathbf{I} = \begin{bmatrix} 1 & 0 & \cdots & 0 \\ 0 & 1 & \cdots & 0 \\ \vdots & \vdots & \ddots & \\ 0 & 0 & & 1 \end{bmatrix}$$

It has the important property that

$$\underset{(n \times m)}{\mathbf{A}} \underset{(m \times m)}{\mathbf{I}} = \mathbf{A}$$

and

$$\underset{(n \times n)}{\mathbf{I}} \underset{(n \times m)}{\mathbf{A}} = \mathbf{A}$$

Diagonal matrixes are sometimes represented with the notation $Diag(\mathbf{x})$, where $\mathbf{x}$ is a vector that represents the diagonal. For example, $Diag(\mathbf{x}) = \mathbf{I}_n$ implies that $\mathbf{x}$ is a length n vector of ones.

C.6 Hadamard Product

The Hadamard product of two matrixes is a new matrix having elements that are the products of elements having the same positions,

$$\underset{(n \times m)}{\mathbf{A}} \circ \underset{(n \times m)}{\mathbf{B}} = \begin{bmatrix} a_{11}b_{11} & a_{12}b_{12} & \cdots & a_{1m}b_{1m} \\ a_{21}b_{21} & a_{22}b_{22} & \cdots & a_{2m}b_{2m} \\ \vdots & \vdots & \ddots & \\ a_{n1}b_{n1} & a_{n2}b_{n2} & & a_{nm}b_{nm} \end{bmatrix}$$

Of course, matrixes must have the same dimensions.

C.7 Determinant of a Matrix

The determinant of a square matrix is a quantity that tells us if a matrix is *singular*: If it is singular, the determinant is zero. For a 2×2 matrix the determinant is

$$|\mathrm{A}| = a_{11}a_{22} - a_{21}a_{12} \qquad \text{C.8}$$

For example,

$$\begin{vmatrix} 4 & -6 \\ -1 & 2 \end{vmatrix} = 8 - 6 = 2$$

The equation

$$\mathbf{Ax} = 0 \tag{C.9}$$

has a nontrivial solution if the determinant of $\mathbf{A}$ is zero, that is, if it is singular.

C.8 Matrix Inversion

Let us solve for $\mathbf{x}$ in the equation

$$\mathbf{Ax} = \mathbf{y}$$

where $\mathbf{A}$ is a nonsingular $n \times n$ matrix, and $\mathbf{x}$ and $\mathbf{y}$ are length-n vectors. Unlike standard algebra, we cannot simply divide both sides by $\mathbf{A}$. The solution requires a new $n \times n$ matrix $\mathbf{B}$ such that

$$\mathbf{BA} = \mathbf{I}$$

Then multiplying both sides by $\mathbf{B}$ yields

$$\mathbf{BAx} = \mathbf{By}$$

If $\mathbf{BA} = \mathbf{I}$, then this reduces to

$$\mathbf{x} = \mathbf{By}$$

We now have $\mathbf{x}$ on the left-hand side, so we need to determine $\mathbf{B}$.

For the matrix $\mathbf{B}$ to exist, then both $\mathbf{A}$ and $\mathbf{B}$ must be nonsingular. Matrix $\mathbf{B}$ is called the *inverse* of matrix $\mathbf{A}$, indicated by the notation $\mathbf{A}^{-1}$. It has the property

$$\mathbf{AA}^{-1} = \mathbf{A}^{-1}\mathbf{A} = \mathbf{I} \tag{C.10}$$

Use a computer to find a matrix inverse. But for a 2×2 matrix it is useful to know that the inverse is

$$\mathbf{A}^{-1} = \frac{1}{|\mathbf{A}|} \begin{bmatrix} a_{22} & -a_{12} \\ -a_{21} & a_{11} \end{bmatrix} \tag{C.11}$$

For the matrix

$$\mathbf{A} = \begin{bmatrix} 4 & -6 \\ -1 & 2 \end{bmatrix}$$

the inverse is

$$\mathbf{A}^{-1} = \frac{1}{4 \cdot 2 - 6 \cdot 1} \begin{bmatrix} 2 & 6 \\ 1 & 4 \end{bmatrix} = \begin{bmatrix} 1 & 3 \\ 1/2 & 2 \end{bmatrix}$$

To check this, see that

$$\mathbf{A}\mathbf{A}^{-1} = \begin{bmatrix} 4 - 3 & 12 - 12 \\ -1 + 1 & -3 + 4 \end{bmatrix} = \begin{bmatrix} 1 & 0 \\ 0 & 1 \end{bmatrix}$$

Now the solution to the equation is

$$\mathbf{x} = \mathbf{A}^{-1}\mathbf{y}$$

$$= \begin{bmatrix} 1 & 3 \\ 1/2 & 2 \end{bmatrix} \begin{bmatrix} 10 \\ -2 \end{bmatrix} = \begin{bmatrix} 10 - 6 \\ 5 - 4 \end{bmatrix} = \begin{bmatrix} 4 \\ 1 \end{bmatrix}$$

Example C.1. Regression provides an opportunity to practice several of these operations, involving matrix multiplication, transposition, and inversion (and, thus, the determinant). The two-observation case I use as illustration is not entirely trivial, because multiple observations are handled in the same way.

The standard linear regression model for two observations can be written as

$$y_1 = b_0 + b_1 x_{11} + \varepsilon_1$$

$$y_2 = b_0 + b_1 x_{12} + \varepsilon_2 \qquad \text{C.12}$$

In matrix notation this becomes

$$\mathbf{y} = \mathbf{X}\mathbf{b} + \varepsilon, \qquad \text{C.13}$$

where observations are contained in the $n \times 1$ vector

$$\mathbf{y} = \begin{bmatrix} y_1 \\ y_2 \end{bmatrix}$$

parameters (to be estimated) are the $(p \times 1)$ vector

$$\mathbf{b} = \begin{bmatrix} b_0 \\ b_1 \end{bmatrix}$$

variables are the $(n \times p)$ design matrix

$$\mathbf{X} = \begin{bmatrix} 1 & x_1 \\ 1 & x_2 \end{bmatrix}$$

and random errors are the vector

$$\varepsilon = \begin{bmatrix} \varepsilon_1 \\ \varepsilon_2 \end{bmatrix}$$

If errors are normally distributed, then the maximum likelihood parameter estimates are given by

$$\hat{\mathbf{b}} = (\mathbf{X}^T\mathbf{X})^{-1}\mathbf{X}^T\mathbf{y} \qquad\qquad \text{C.14}$$

If the measured variables have values $\mathbf{x}^T = [2\ 3]$ and the observations are $\mathbf{y}^T = [1\ 2]$, what are the parameter estimates?

Consider just the mean structure of the model $\mathbf{y} = \mathbf{Xb}$. Because stochasticity is Gaussian with mean zero, it is reasonable to expect that the MLE for $\mathbf{b}$ would be the solution to the deterministic equation $\mathbf{y} = \mathbf{Xb}$. To solve this, we cannot premultiply both sides by the inverse of $\mathbf{X}$, because $\mathbf{X}$ is not a square symmetric matrix—it cannot have an inverse. The solution for $\mathbf{b}$ can be obtained by a different route, after introducing the concept of *rank*.

If the matrix $\mathbf{X}$ has dimensions n by p, then the rank of $\mathbf{X}$ is

$$rank(\mathbf{X}) \le \min(n,p)$$

The inequality depends on collinearity in the columns of $\mathbf{X}$. If columns are not collinear, then the matrix is *full rank*, and $rank(\mathbf{X}) = \min(n,p)$. In regression applications we expect $rank(\mathbf{X})=p$. If $\mathbf{X}$ is full rank, it is possible to invert the *cross product* of $\mathbf{X}$, which is the symmetric p by p matrix $\mathbf{X}^T\mathbf{X}$. Returning to the equation $\mathbf{y} = \mathbf{Xb}$, note that we can premultiply both sides by $\mathbf{X}^T$,

$$\mathbf{X}^T\mathbf{y} = \mathbf{X}^T\mathbf{Xb}$$

The cross product of $\mathbf{X}$ on the right-hand side symmetric; provided $\mathbf{X}$ is full rank, we can invert $\mathbf{X}^T\mathbf{X}$. Premultiplication by its inverse will leave us with $\mathbf{b}$ on the right-hand side, because $(\mathbf{X}^T\mathbf{X})^{-1}\mathbf{X}^T\mathbf{X} = \mathbf{I}$. So

$$(\mathbf{X}^T\mathbf{X})^{-1}\mathbf{X}^T\mathbf{y} = (\mathbf{X}^T\mathbf{X})^{-1}\mathbf{X}^T\mathbf{Xb}$$

$$(\mathbf{X}^T\mathbf{X})^{-1}\mathbf{X}^T\mathbf{y} = \mathbf{b}$$

This is the MLE of $\mathbf{b}$.

Intuition tells us that the parameters should be $\mathbf{b} = [-1\ 1]$. Here are the basic matrix manipulations used to arrive at the regression result. The design matrix is

$$\mathbf{X} = \begin{bmatrix} 1 & 2 \\ 1 & 3 \end{bmatrix}$$

Begin with the *cross product*, a term used to describe a specific type of matrix multiplication

$$\mathbf{X}^T\mathbf{X} = \begin{bmatrix} 1 & 1 \\ 2 & 3 \end{bmatrix}\begin{bmatrix} 1 & 2 \\ 1 & 3 \end{bmatrix} = \begin{bmatrix} 2 & 5 \\ 5 & 13 \end{bmatrix}$$

and then invert,

$$(\mathbf{X}^T\mathbf{X})^{-1} = \begin{bmatrix} 2 & 5 \\ 5 & 13 \end{bmatrix}^{-1} = \frac{1}{26-25}\begin{bmatrix} 13 & -5 \\ -5 & 2 \end{bmatrix} = \begin{bmatrix} 13 & -5 \\ -5 & 2 \end{bmatrix}$$

Multiply by the transpose of $\mathbf{X}$

$$(\mathbf{X}^T\mathbf{X})^{-1}\mathbf{X}^T = \begin{bmatrix} 13 & -5 \\ -5 & 2 \end{bmatrix}\begin{bmatrix} 1 & 1 \\ 2 & 3 \end{bmatrix} = \begin{bmatrix} 3 & -2 \\ -1 & 1 \end{bmatrix}$$

and then by $\mathbf{y}$

$$(\mathbf{X}^T\mathbf{X})^{-1}\mathbf{X}^T\mathbf{y} = \begin{bmatrix} 3 & -2 \\ -1 & 1 \end{bmatrix}\begin{bmatrix} 1 \\ 2 \end{bmatrix} = \begin{bmatrix} -1 \\ 1 \end{bmatrix}$$

The regression model is now

$$y_i = -1 + x_i + \varepsilon_i.$$

that is, the same answer we arrived at by intuition and by a direct solution of $\mathbf{X}^{-1}\mathbf{y}$.

C.9 Positive Definite Matrixes

Covariance matrixes are square and symmetric about the diagonal, which holds the variances (Chapter 5). The normal distribution presents us with an exponent containing the Mahalanobis distance, a scalar quantity that results from the matrix multiplication

$$\mathbf{x}^T\mathbf{C}\mathbf{x} = \sum_{i=1}^{n}\sum_{j=1}^{n} c_{ij}x_i x_j$$

where $\mathbf{C}$ is the inverse of the n by n covariance matrix. This scalar quantity must be positive, which can only occur if the covariance matrix is *positive definite*. Positive definite matrixes are nonsingular. All eigenvalues (next section) are positive. Recall that the normal distribution has the determinant of the covariance matrix in the denominator. We cannot divide by zero. Furthermore, a negative variance is not permitted.

C.10 Eigenanalysis

Eigenanalysis provides a way to summarize the information contained in a matrix, such as a transition matrix used in demographic analysis and patch occupancy models, and in correlation and covariance matrixes. The multiplication of a square matrix by a vector can be equivalent to multiplying that vector by a scalar,

$$\mathbf{Ax} = \lambda\mathbf{x}, \qquad \text{C.15}$$

where $\mathbf{A}$ is a n by n matrix, $\mathbf{x}$ is a length-n vector, and λ is a scalar. There can be as many as n different values of λ that satisfy this equation, and there will be a different vector $\mathbf{x}$ associated with each λ. The scalar λ is called an *eigenvalue*, and $\mathbf{x}$ is a (right) *eigenvector*. It is called a *right* eigenvector, because it lies to the right of $\mathbf{A}$ and of λ.

To find the roots λ we solve as follows. First subtract $\lambda\mathbf{x}$ from both sides,

$$\mathbf{Ax} - \lambda\mathbf{x} = 0$$

or

$$(\mathbf{A} - \lambda\mathbf{I})\mathbf{x} = 0 \qquad \text{C.16}$$

where $\mathbf{I}$ is the n by n identity matrix. From Equation C.9 we know that the roots to this equation satisfy

$$|\mathbf{A} - \lambda\mathbf{I}| = 0 \qquad \text{C.17}$$

This is termed the *characteristic equation* for matrix $\mathbf{A}$. The roots λ obtained from C.17 can be used to determine the *characteristic* (eigen-) vectors from Equation C.16.

Example C. 2. The stage-structured population growth model can be written as

$$\mathbf{n}_{t+1} = \mathbf{An}_t$$

where the vector $\mathbf{n}_t$ contains the distribution of stage classes at time t. If there is a *stable stage distribution*, there must be a scalar quantity λ that proportionately scales the density in each age class from one time step to the next,

$$\mathbf{n}_{t+1} = \lambda \mathbf{n}_t$$

Equating these two equations gives

$$\mathbf{A}\mathbf{n}_t = \lambda \mathbf{n}_t \qquad\qquad \text{C.18}$$

From Equation C.15 we see that λ is an eigenvalue of $\mathbf{A}$, and (once a stable stage distribution is attained) the vector of densities is the corresponding right eigenvector of the transition matrix $\mathbf{A}$. It follows that the right eigenvector $\mathbf{n}_t$ is proportional to the stable stage distribution.

Eigenanalysis is done with a computer. An exception is made for a 2×2 matrix, because this special case is easy and provides insight concerning the origin of complex eigenvalues. The determinant in the characteristic Equation C.17 is written as

$$|\mathbf{A} - \lambda\mathbf{I}| = \left| \begin{bmatrix} a_{11} & a_{12} \\ a_{21} & a_{22} \end{bmatrix} - \begin{bmatrix} \lambda & 0 \\ 0 & \lambda \end{bmatrix} \right| = \begin{vmatrix} a_{11} - \lambda & a_{12} \\ a_{21} & a_{22} - \lambda \end{vmatrix}$$

Using Equation C.8 obtain

$$\begin{vmatrix} a_{11} - \lambda & a_{12} \\ a_{21} & a_{22} - \lambda \end{vmatrix} = (a_{11} - \lambda)(a_{22} - \lambda) - a_{12}a_{21}$$

Some manipulation achieves the standard form for a quadratic equation,

$$\lambda^2 - (a_{11} + a_{22})\lambda + a_{11}a_{22} - a_{12}a_{21} = 0$$

The coefficient of λ is the sum of the diagonal elements of matrix $\mathbf{A}$, which is known as the *trace*. Together, the last two terms comprise the determinant. A simple way to remember the characteristic equation for a 2×2 transition matrix is

$$\lambda^2 - \text{Trace}\mathbf{A}\, \lambda + |\mathbf{A}| = 0 \qquad\qquad \text{C.19}$$

As with any quadratic equation there are two roots, which I designate as λ_1 and λ_2. One or more may be *complex*, because the solution will involve a square root (because the equation contains λ^2). Square roots of negative numbers are complex. To determine the corresponding eigenvectors we substitute each eigenvalue into C.15 and solve for $\mathbf{x}$.

Just to be clear, let us plug in some demographic parameters. The matrix equation for a population with two age classes that reproduces only in the second class looks like this:

$$\mathbf{A} - \lambda\mathbf{I} = \begin{bmatrix} 0 & m \\ s & 0 \end{bmatrix} - \lambda \begin{bmatrix} 1 & 0 \\ 0 & 1 \end{bmatrix} = \begin{bmatrix} -\lambda & m \\ s & -\lambda \end{bmatrix}$$

From Equation C.19, this has the roots

$$\lambda_{1,2} = \pm\sqrt{ms}$$

The subscript 1,2 indicates that there are two eigenvalues. There will be as many eigenvalues as there are stages in the model. There is an eigenvector corresponding to each eigenvalue. The right eigenvector for λ_1 is

$$\mathbf{Aw} = \lambda_1 \mathbf{w}$$

This equation says

$$\begin{bmatrix} 0 & m \\ s & 0 \end{bmatrix}\begin{bmatrix} w_1 \\ w_2 \end{bmatrix} = \lambda_1\begin{bmatrix} w_1 \\ w_2 \end{bmatrix}$$

and represents a pair of equations

$$mw_2 = \lambda_1 w_1$$
$$sw_1 = \lambda_1 w_2$$

The right eigenvector is proportional to the age structure

$$w_1 = w_2\sqrt{\frac{m}{s}}$$

which is proportional to $\mathbf{n}_t$ in Equation C.18.

Example C.3. Consider the Leslie matrix for a population with a biennial life history that has a probability of $s = 0.5$ of surviving to the second year, at which time it produces two offspring and then dies. Determine the growth rate and age structure using life table methods and eigenanalysis.

If an individual survives the first year, then produces two offspring, the Leslie matrix is

$$\mathbf{A} = \begin{bmatrix} 0 & 2 \\ 0.5 & 0 \end{bmatrix}$$

The survival function (Chapter 2) is

$$l_x = \prod_{i=1}^{x-1} s_i$$

and the life table looks like this:

Age x	m_x	l_x	$l_x m_x$
1	0	1	0
2	2	0.5	1
sum		1.5	1

Note that the net reproductive rate is 1. The equilibrium age structure from the survivor function (Appendix E) is

$$w_x = \begin{bmatrix} 1.0/1.5 \\ 0.5/1.5 \end{bmatrix}$$

Using eigenanalysis to determine the growth rate and age distribution of the population described by this transition matrix, calculate the eigenvalues of $\mathbf{A}$,

$$\mathbf{A} - \lambda\mathbf{I} = \begin{bmatrix} 0 & 2 \\ 0.5 & 0 \end{bmatrix} - \lambda\begin{bmatrix} 1 & 0 \\ 0 & 1 \end{bmatrix} = \begin{bmatrix} -\lambda & 2 \\ 0.5 & -\lambda \end{bmatrix}$$

Set the determinant equal to zero to obtain

$$\lambda \pm <1$$

The right eigenvector for $\lambda_1 = 1$ is

$$\begin{bmatrix} 0 & 2 \\ 0.5 & 0 \end{bmatrix}\begin{bmatrix} w_1 \\ w_2 \end{bmatrix} = 1\begin{bmatrix} w_1 \\ w_2 \end{bmatrix}$$

yielding the pair of equations

$$2w_2 = w_1$$

$$0.5w_1 = w_2$$

which gives the age structure

$$w = \begin{bmatrix} 0.67 \\ 0.33 \end{bmatrix}$$

C.10.1 Complex Numbers

Roots of negative numbers often arise in the context of solving equations. For instance, the quadratic equation for eigenvalues of a 2×2 matrix has the square root of the difference of two terms. If the result is negative, we have the conjugate pair of roots,

$$\lambda_{1,2} = a \pm bi$$

where

$$i \equiv \sqrt{-1}$$

The *real* part of each root is represented by *a*. The *imaginary* part is represented by *b*. The imaginary part arises when we solve for roots. For a square root of a negative number, write the following:

$$\sqrt{-x} = \sqrt{x} \cdot \sqrt{-1} \equiv \sqrt{x} \cdot i$$

In other words, *i* represents the square root of -1. The first term of each eigenvalue is the *real part*, because it is a real number. The second term is the imaginary part and is the coefficient of *i*. Such complex numbers are expressed by coordinates (a, b) or in polar coordinates

$$a = r\cos\varphi$$

$$b = r\sin\varphi$$

for the vector of length

$$r = \sqrt{a^2 + b^2}$$

and angle

$$\varphi = \arctan(b/a)$$

to the positive real axis.

C.10.2 Application: Eigenanalysis of Stage Structure

The northern spotted owl model is a stage-structured model, because adults are lumped into a single stage (there is a nonzero element in lower right corner of the projection matrix). We analyze the matrix model to determine growth rate, equilibrium age structure, and elasticities. Eigenanalysis of matrixes from Section 2.2.4 yields the eigenvalues $-0.0305 + 0.236i$, $-0.0305 - 0.236i$, and $0.929 + 0i$. These conjugate pairs have real and imaginary parts. The dominant eigenvalue is that having the largest real part. In this case, it is 0.93, which agrees with the calculation from Lotka's equation (Appendix E). The equilibrium stage structure comes from the right eigenvector associated with this eigenvalue,

$$\frac{\mathbf{w}}{\sum_i w_j} = [0.278 \quad 0.0475 \quad 0.675]^T \qquad \text{C.20}$$

Note that the eigenvector is normalized to sum to one. This structure has most individuals in the adult class, because this class includes all individuals greater than two years old. Next most abundant are juveniles, representing the combined reproduction from adults. Subadults are last, because juvenile survival is low (few enter at each time step), and all individuals exit this class after a single year.

The reproductive value (Appendix E) comes from the left eigenvector associated with the dominant eigenvalue,

$$\frac{\mathbf{v}}{v_1} = \begin{bmatrix} 1 & 5.84 & 6.25 \end{bmatrix}^{\mathrm{T}} \qquad\qquad \text{C.21}$$

The left eigenvector is often divided by the first element of the vector. Reproductive value is high for adults, because they have survived. Juveniles have low reproductive value, because many will die before reaching reproductive age.

For more background on matrix algebra, the reader is referred to Caswell (2001) for applications to population biology and to any standard regression text for applications to statistical inference.

Appendix D

Probability Models

Using models and data to gain understanding requires some distribution theory. In this appendix I summarize some of the underlying elements of probability theory and the concept of moments. I use several examples to illustrate concepts, the first being a simple checkerboard distribution of fruit pigeons.

Diamond (1975) considered the role of competition in structuring communities in a paper that reported the tendency for similar bird species on archipelagos to occupy different islands. He pointed out that it would be hard to explain this checkerboard pattern on the basis of chance alone. Several factors could contribute to it, one of which is competition. Debate followed, with many ecologists contributing their interpretations of such patterns. I use this example to introduce some basic concepts in probability.

D.1 Conditional Probability and Bayes' Theorem

Think of R and S as the events that the fruit pigeons *Ptilinopus rivoli* and *P. solomonensis* occupy a given island, respectively (Diamond 1975). The complementary events are absence of *P. rivoli*, designated as R^c, and absence of *P. solomonensis*, designated as S^c. Further, let (r, s) represent an island's status, where $r = R$ for occupied islands and $r = R^c$ for unoccupied islands. Thirty-two islands are shown as having one species (R, S^c) or (R^c, S), both species (R, S), or neither species (R^c, S^c) (Figure D.1). The area within the outer box in Figure D.2 encompasses the total probability of all combinations of observing the two species. This total probability equals one.

We wish to estimate probabilities of observing each of the four outcomes. These probabilities together with estimates based on thirty two islands are listed in Table D.1.

The probability of observing any combination of the two species is termed the *joint probability*. The probability of observing a species is the sum of the probability of observing it alone and in combination with the other. The two probabilities, Pr{R} and Pr{S}, together with their complements, comprise the *marginal probabilities* of the two events. To see why these are termed *marginal*, consider that we obtain them by summing rows or columns. They are entered along the

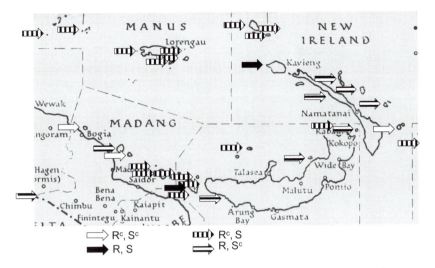

FIGURE D.1 Islands occupied by two species of fruit pigeon. From Diamond (1975).

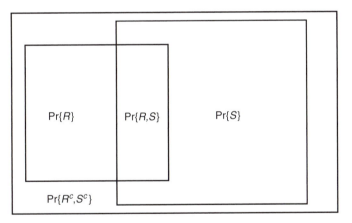

FIGURE D.2. A probability space for r and s.

TABLE D.1.
Probability Estimates Based on the Numbers of Islands Where R and S Occur

Event	Probability	Estimate from Data
r but not s	$\Pr\{R, S^c\}$	$9/32 = 0.281$
s but not r	$\Pr\{R^c, S\}$	$18/32 = 0.563$
Both	$\Pr\{R, S\}$	$2/32 = 0.0625$
Neither	$\Pr\{R^c, S^c\}$	$3/32 = 0.0938$
Total		1

margins of Table D.2, and represent the fraction of all islands occupied by the species in question. They are not the actual probabilities themselves, which are unknown, but rather represent estimates thereof, based on a set of observations.

Interpreting the pattern of occupancy involves several of these quantities. For example, if the probability of observing both species together is viewed as lower than expected by chance, then we might compare the observation of $R \cap S$, taken to be an estimate of $\Pr\{R, S\}$, or 0.0625, with the value predicted by some null hypothesis. If the null hypothesis says that events R and S are *independent*, then the expected value is

$$\Pr\{R,S\} = \Pr\{R\}\ \Pr\{S\} \qquad \qquad \text{D.1}$$

For this example, the estimate for the right-hand side is $\Pr\{R\} \times \Pr\{S\} = 0.344 \times 0.625 = 0.215$. This estimate is much greater than the observed number of islands having both species, which is $\frac{2}{32}$, for value of $\Pr\{R, S\} = 0.0625$. This tendency for the two species to occupy different islands is described with a modification of Equation D.1. If the two events are not independent, then the joint probability is calculated as

$$\Pr\{R,S\} = \Pr\{R|S\}\ \Pr\{S\} = \Pr\{S|R\}\ \Pr\{R\} \qquad \text{D.2}$$

Equation D.2 contains *conditional probabilities*, the probability of observing an event given that the second event has occurred. This notation means that the event to the left of the vertical bar is conditioned on the event to the right of the vertical bar. If the two events are independent, then $\Pr\{R\} = \Pr\{R|S\}$, $\Pr\{S\} = \Pr\{R\}\Pr\{S\}$, and Equations D.1 and D.2 yield the same answers. For this example, the estimates for these conditional probabilities are given in Table D.3. From Equation D.2 we obtain for our example

$$\Pr\{R,S\} = 0.1 \times 0.625 = 0.0625$$

or

$$\Pr\{R,S\} = 0.182 \times 0.344 = 0.0625$$

Note that this matches the value obtained as $\frac{2}{32}$, the fraction of islands containing both species.

A handy rule, termed *Bayes' theorem*, results from some manipulation of Equation D.2. Rearrange,

$$\Pr\{R|S\} = \frac{\Pr\{R,S\}}{\Pr\{S\}}$$

TABLE D.2.
Estimates of Marginal Probabilities in the Table Margins

Events	S	S^c	Marginal
R	2	9	$\Pr\{R\} = 11/32$
R^c	18	3	$\Pr\{R^c\} = 21/32$
Marginal	$\Pr\{S\} = 20/32$	$\Pr\{S^c\} = 12/32$	32

TABLE D.3.
Conditional Probability Estimates for R and S

Event	Probability	Estimate from Data
R given S	$\Pr\{R\|S\}$	$2/20 = 0.100$
S given R	$\Pr\{S\|R\}$	$2/11 = 0.182$

and substitute for the numerator to obtain

$$\Pr\{R\,|\,S\} = \frac{\Pr\{S|R\}\,\Pr\{R\}}{\Pr\{S\}} \qquad \text{D.3}$$

This theorem is used in many contexts.

The *joint, marginal,* and *conditional* probabilities provide a basis for evaluating a simple hypothesis concerning whether or not species tend to occupy different islands. In the case of checkerboard distributions on archipelagos, manipulation of these quantities, sometimes in sophisticated ways, has been the basis for lively debate that has spanned several decades. These concepts will be important as the foundation for stochastic processes and data modeling.

D.2 Density and Distribution Functions

Evaluation of ecological models is based on properties of samples. Sample properties depend, in turn, on the population distributions from which they are drawn. Here I consider discrete and continuous random variables and their moments.

D.2.1 Empirical Distribution Functions

An empirical distribution describes the frequency of an observation in a sample. For a sample of size n, simply assign a value of $\frac{1}{n}$ to each observation

$$p(y_i) = \frac{1}{n} \qquad \text{D.4}$$

where y_i is the ith observation. This distribution is empirical, because it is based on the sample itself and makes no assumptions regarding distribution shape. Below is a sample of ten observations, which I drew from a binomial distribution having parameter values $n = 32$ and probability parameter $\theta = 0.344$.

i:	1	2	3	4	5	6	7	8	9	10
y_i:	14	11	11	10	17	13	9	9	11	15

The empirical distribution says that each observation has a frequency of $1/n = 0.1$. A histogram represents these data graphically (vertical bars in Figure D.3). The total frequency of the value 9 is obtained by summing frequencies of the seventh and eighth observations, that is, $\frac{2}{10}$. Note that

$$\sum_{i=1}^{n} p(y_i) = \sum_{i=1}^{n} \frac{1}{n} = n \times \frac{1}{n} = 1$$

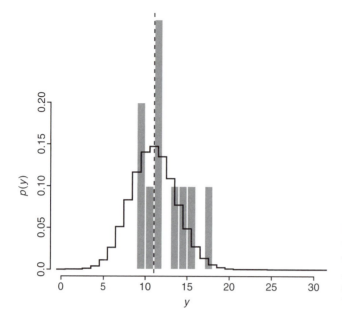

FIGURE D.3. A random sample of ten draws (shaded histogram) from the binomial with parameters $n = 32$ and $\theta = 0.344$ (step curve). The mean of the binomial is indicated by the dashed line.

The empirical distribution function is not a collection of probabilities. It is the frequency of each observed event. We might ponder the probability of an asteroid striking the earth, but it is not sensible to speak of it in terms of frequencies. Likewise, the frequencies in Figure D.3 provide insights into the underlying process (in this case, the binomial step curve in Figure D.3), but they are not probabilities. The empirical distribution function may help us estimate probabilities, and the mathematics of frequencies and probabilities are often the same. I purposefully draw the parallel here, because it will help us see the relationship between sample moments and parametric moments.

D.2.2 Parametric Distribution Functions

Parametric distributions have location and shape determined by one or more parameters. Parametric distributions are powerful inference tools, because we can often estimate the parameters that describe this shape with relatively few observations. A large number of observations might be required to learn about a random process based on an empirical distribution (e.g., the histogram in Figure D.3), whereas we can often estimate parameters with relatively few observations. Parametric distributions are most powerful when there is theoretical justification for them (e.g., the Central Limit Theorem for the normal distribution or the binomial for a Bernoulli [e.g., coin-flipping] process).

Discrete random variables assume integer values. The probability distribution of the discrete random variable is the collection of probabilities assigned to the possible outcomes. A particular outcome might be that y of the total islands in Figure D.1 are occupied by species r. The distribution $p(y)$ contains non-negative values for all possible outcomes y, which are mutually exclusive (nonoverlapping) and all-inclusive. Thus, the sum of all possible outcomes must equal unity,

$$\sum_y p(y) = 1$$

(Figure D.4a). Because the $p(y)$ are additive, the probability of any one event has the same dimensions as the sum of the individual probabilities (i.e., none). This seems obvious, but it does not apply to continuous random variables.

Continuous random variables vary continuously, so the probability of observing any given value is infinitely small. Consider a continuous random variable y, such as an individual's height. Let y_i represent the value of y assigned to an individual selected at random. Because y varies continuously, there is no chance that we can obtain a value of precisely y, that is,

$$\Pr\{y_i = y\} = 0$$

where y_i is a continuous random variable. If there is no chance of observing a specific y, then how can we talk about probabilities of continuous variables?

We can only speak of the probability of a range of values. Although there is no chance that a random variate y_i will precisely equal y, there is a chance that y_i may fall within an interval of width, say, dy, that is, within the interval $(y, y + dy)$. As with discrete variables, we represent continuous random variables with probability models. Using such a model $P(y)$, termed a *distribution function*, we could state that the probability that y_i is less than y (it falls within the interval $(-\infty, y)$) is

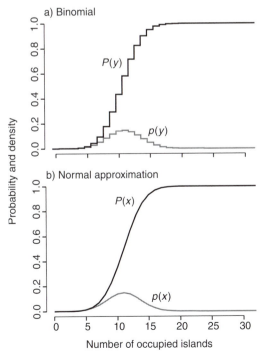

FIGURE D.4. Discrete and continuous distributions. (*a*) Binomial distribution and cumulative functions for parameters $n = 32$ and $\theta = 0.344$. (*b*) Normal approximation by matching moments, with mean = $n\theta$ and variance $n\theta(1 - \theta)$.

$$P(y) = \Pr\{y_i \leq y\}$$

The distribution function describes the total probability that y_i lies to the left of y. Thus, $P(y)$ approaches zero for small values of y, and it approaches one for large values of y (Figure D.4b). The probability that y_i is contained by the interval $(y, y + dy)$ is

$$\Pr\{y < y_i < y + dy\} = P(y + dy) - P(y) \qquad \text{D.5}$$

Like all probabilities, $P(y)$ is dimensionless.

This is fine for an interval, but what if we desire a description of probability near some value of y—something like the distribution for a discrete random variable? We could always choose some arbitrary interval of width dy and use Equation D.5 to arrive at a probability. But any interval we choose will produce a different answer (the larger the interval dy, the greater the probability). We need a standard.

The standard for continuous random variables is a probability *density function*, which is obtained by differentiating the distribution function

$$p(y) = \frac{d}{dy}P(y) \qquad \text{D.6}$$

(Figure D.4b). Conversely, the distribution function is the integral of the density function over everything $< y_i$

$$P(y) = \int_{-\infty}^{y} p(z)dz$$

It might be helpful to put these two together and think of the density as the rate of change in the distribution function at the integration limit y,

$$p(y) = \frac{d}{dy}P(y) = \frac{d}{dy}\int_{-\infty}^{y} p(z)dz$$

The probability that y_i lies in interval $(y, y + dy)$ can be determined by integrating the density function

$$\Pr\{y < y_i < (y + dy)\} = \int_{y}^{y+dy} p(z)dz = P(y + dy) - P(y) \qquad \text{D.7}$$

$P(y)$ is the *cumulative distribution function*.

These definitions illustrate an important difference between discrete distributions and continuous density functions: whereas the discrete distribution is dimensionless (it is a probability), the density function $p(y)$ has units of $\frac{1}{y}$. The density function is not a probability. For example, $p(y)$ can assume values greater then one. Remember rules of integration to convince yourself that $P(y)$ has units of a probability. $P(y)$ can never exceed one.

D.2.3 Application: A Distribution for Archipelagos

Let us use a few of these relationships to reconsider birds on archipelagos. The probability that *Ptilinopus rivoli* occupies an island can be represented by a *parameter* θ that is estimated from data. This parameter represents the probability $\Pr\{R\}$ and is estimated as eleven occupied islands divided by the total thirty-two,

$$\hat{\theta} = 11/32 = 0.344$$

The "hat" notation indicates that this is an estimate of the probability, which is, in fact, unknown.

If the estimate $\hat{\theta}$ were the true probability of occupancy, and it applied identically to all islands, what would be the probability of observing a number of occupied islands different from eleven? We might observe as many as thirty two or as few as zero occupied islands. If all islands have the same probability of occupancy θ, then the distribution is *binomial*,

$$p(y) = \binom{n}{y}\theta^y(1 - \theta)^{n-y} \qquad \text{D.8}$$

(Appendix F), where $n = 32$ is the total number of islands and y is the number of islands occupied. The binomial coefficient is a combinatorial for the number of ways in which we could obtain y occurrences out of n total (Box D.1). There is only one way to obtain $y = 0$ (all islands unoccupied) and $y = 32$ (all islands occupied), but there are many combinations that could yield intermediate levels of occupancy. The notation

$$y \sim Bin(n,\theta)$$

means that y is distributed as a binomial with parameters θ and n. I sometimes write the distribution or density of y as

$$p(y) = Bin(y|n,\theta)$$

This notation says that the distribution of y (placed to the left of the vertical bar) is binomial with parameters placed to the right of the bar.

The location and shape of the distribution in Figure D.4a is fully determined by n and θ. To interpret Figure D.4 remember that we assumed that all islands in the entire archipelago are subject to a single value of θ. For species r, the most probable number of occupied islands (the mode) is $n\hat{\theta} = 11$. This is also the mean of the binomial distribution. We believe this to be the most probable value, because it explains the data better than does any other value. If we could observe a large number of archipelagos, each including thirty two islands and having an independent probability of event R equal to θ, then the binomial distribution represents the proportion of archipelagos that would have y islands occupied by species r. A sample of ten archipelagos might look like the histogram in Figure D.5a. A sample of hundred archipelagos looks more like the underlying (assumed) binomial (Figure D.5b), a sample of thousand archipelagos is a close fit (Figure D.5c). This frequency view of probability only applies if we have many identical archipelagos.

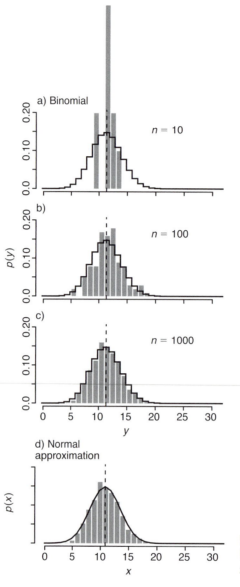

FIGURE D.5. Random samples from binomial with a normal approximation in *d*.

D.3 Expectation and Moments

Knowing the parameter values for a probability model can often allow us to visualize the shape of a distribution. For example, the normal distribution has a location parameter (the mean) and a scale parameter (the standard deviation or variance). The normal density in Figure D.5d has a mean of 11 and a standard deviation of 2.69. If I tell you the mean and standard deviation of a sample, you can picture the full distribution.

Not all distributions possess such evocative parameters. If I tell you the values for the scale and location parameters of a gamma distribution, call them α and

β, many of you would not accurately visualize its shape. I could help by saying that the mean of the distribution is $\frac{\alpha}{\beta}$ and the variance is $\frac{\alpha}{\beta^2}$. The mean and variance are moments. Moments provide a common basis for communicating information about distributions. *Moments* are the expected values of powers of a random variable. They exist for all empirical samples and for many parametric distributions. Moments are valuable as summaries that give insight into the nature of distributions. The mth moment for a random variable x is assigned the notation $E[x^m]$, read as the *expected value* of the thing in brackets (in this case, x^m). The 0th moment ($m = 0$) is one, because any number raised to the power 0 is one. The first moment ($m = 1$), or the expected value of x, is the *mean*, written as $E[x]$. The expected value of any constant quantity is that quantity itself. Thus, constants can be pulled out of the brackets. If c is a constant, then

$$E[cx] = cE[x]$$

We will need this rule when we take expectations of functions that contain both random variables and constants. Also note that we can take an expectation term by term,

$$E[x + y] = E[x] + E[y]$$

Central moments differ only in that they are measured as departures from the mean,

$$E[(x - E[x])^m] \qquad\qquad \text{D.9}$$

The first central moment is zero (the average deviation from the mean). The second central moment ($m = 2$) is the expected value of the squared deviations from the mean, that is, the *variance*. Note that the variance can be obtained from either central or noncentral moments:

$$\begin{aligned}
\text{var}[x] &= E[(x - E[x])^2] &&\text{D.10}\\
&= E[x^2 - 2xE[x] + (E[x])^2] &&\textit{expand the quadratic}\\
&= E[x^2] - 2E[x]E[x] + (E[x])^2 &&\textit{expectation term by term}\\
&= E[x^2] - (E[x])^2 &&\textit{collect terms}
\end{aligned}$$

The first line says that the variance is the expected squared departure from the mean. In the third step, note that the expectation of a constant (the mean is a constant) is the constant itself. The last line shows the variance to be the difference between the second moment and the first moment squared.

D.3.1 Empirical Moments

Consider a random variable x_i having a distribution function given by Equation D.4. An empirical moment for a sample of size n can be viewed as a weighted average of observations raised to a power m,

$$E[x^m] = \sum_{i=1}^{n} x_i^m \, p(x_i) \qquad\qquad \text{D.11}$$

From Equation D.4 this is equivalent to

$$E[x^m] = \frac{1}{n} \sum_{i=1}^{n} x_i^m \qquad\qquad \text{D.12}$$

This is the definition of the mth *sample moment* (or empirical moment). The first moment is the sample mean. I use the notation $\bar{x}$ to represent the first empirical moment, the mean. From Equations D.8 and D.12 the mth central moment is

$$E[(x - \bar{x})^m] = \sum_{i=1}^{n} (x_i - \bar{x})^m p(x_i) = \frac{1}{n} \sum_{i=1}^{n} (x_i - \bar{x})^m \qquad \text{D.13}$$

To obtain the *sample variance* from noncentral moments, expand the quadratic that results when $m = 2$ in Equation D.13:

$$\text{var}[x] = \frac{1}{n} \sum_{i=1}^{n} (x_i - \bar{x})^2 = \frac{1}{n} \sum_{i=1}^{n} (x_i - 2x_i\bar{x} + \bar{x}^2) \qquad \textit{expand quadratic}$$

$$= \frac{1}{n} \sum_{i=1}^{n} x_i^2 - 2\bar{x}\frac{1}{n} \sum_{i=1}^{n} x_i + \frac{1}{n} \times n\bar{x}^2 \qquad \textit{sum term by term}$$

$$= \overline{x^2} - 2\bar{x}^2 + \bar{x}^2 \qquad \textit{collect terms}$$

$$= \overline{x^2} - \bar{x}^2 \qquad\qquad \text{D.14}$$

$\overline{x^2}$ is the average of the squared observations (the *second sample moment*). There are times when both central and noncentral moments will be useful.

These few moments provide useful summaries of samples and are termed *descriptive statistics*. In statistical applications, designate the sample variance to be

$$\text{var}[x] = \frac{1}{n} \sum_{i=1}^{n} (x_i - \bar{x})^2$$

Because these results underestimate the true variance when n is small, I sometimes divide by $n - 1$, rather than by n. Context will indicate which version is used.

D.3.2 Parametric Moments for Discrete Variables

Although parametric distributions are defined by parameters, rather than individual observations, they have moments that are calculated in an analogous way. The mth moment of a discrete distribution $p(x)$ is

$$E[x^m] \equiv \mu_m = \sum_{x=0}^{\infty} x^m p(x) \qquad\qquad \text{D.15}$$

Note the relationship between Equation D.11 and Equation D.15. Parametric moments share many features of empirical moments. The 0th moment $(m = 0)$

is equal to one, being the area under the distribution. This is true for both the empirical and the parametric distributions in Figure. D.7. The first moment ($m = 1$) is the mean. To distinguish parametric moments from sample moments, I use μ_1 to represent the parametric mean.

Parametric distributions also have central moments taken about the mean:

$$E[(x - \mu_1)^m] \equiv \widetilde{\mu_m} = \sum_{x=0}^{\infty} (x - \mu_1)^m \, p(x) \qquad \text{D.16}$$

(compare with Equation D.13). The variance is obtained as the second central moment or as the difference between the second moment and the squared first moment.

$$\text{var}[x] \equiv \overline{\mu}_2 = E[(x - \mu_1)^2] = \mu_2 - \mu_1^2 \qquad \text{D.17}$$

(compare Equation D.14).

Other indexes of shape are based on central moments,

$$\text{skewness} \equiv \frac{\overline{\mu}_3}{\overline{\mu}_2^{3/2}} \qquad \text{D.18}$$

$$\text{kurtosis} \equiv \frac{\overline{\mu}_4}{\overline{\mu}_2^2} \qquad \text{D.19}$$

and they can also be estimated empirically. Unlike the mean and variance, these quantities are dimensionless (standarized by the second moment) so as to be unaffected by scale.

How do these parametric moments compare with empirical ones? The sample in Figure D.3 was drawn at random from the binomial distribution plotted with it. The sample mean is 12 with a variance of 7.111. The parametric mean of the distribution from which it is drawn is 11 with a variance 7.22. So even with only ten observations the moments are close to those used to generate the data, despite big differences in the distributions themselves (Figure D.3). We could use the sample moments to draw a sample distribution that looks much like the binomial one from which this sample was obtained.

D.3.3 Parametric Moments for Continuous Random Variables

Moments for *continuous variables* are determined by integration rather than by summation. Let x_i be a continuous random variable having probability density function $p(x)$. The mth moment is given by

$$E[x^m] \equiv \mu_m = \int_{-\infty}^{\infty} x^m \, p(x)dx \qquad \text{D.20}$$

and the mth central moment is

$$E[(x - \mu_1)^m] = \int_{-\infty}^{\infty} (x - \mu_1)^m \, p(x)dx \qquad \text{D.21}$$

As before, the 0th moment is one (the area under the density), the first moment $\mu_1 \equiv E[x]$ is the mean, the first central moment, $\widetilde{\mu}_1 = E[(x - \mu_1)]$, is equal to zero, and the second central moment, $\widetilde{\mu}_2 = E[(x - \mu_1)^2]$, is the variance. As for discrete variables, the variance can also be calculated from central or noncentral moments

$$
\begin{aligned}
\text{var}[x] &= \int_{-\infty}^{\infty} (x - \mu_1)^2 p(x) dx \\
&= \int_{-\infty}^{\infty} (x^2 - 2\mu_1 x + \mu_1^2) p(x) dx && \textit{expand quadratic} \\
&= \int_{-\infty}^{\infty} x^2 p(x) dx - 2\mu_1 \int_{-\infty}^{\infty} x p(x) dx \\
&\quad + \mu_1^2 \int_{-\infty}^{\infty} p(x) dx && \textit{integrate term by term} \\
&= \mu_2 - 2\mu_1 \cdot \mu_1 + \mu_1^2 \cdot 1 && \textit{integrate to obtain noncentral} \\
& && \textit{moments (use Equation D.20)} \\
&= \mu_2 - \mu_1^2 && \textit{collect terms} \qquad \text{D.22}
\end{aligned}
$$

(compare with the discrete version in Equation D.14).

D.3.4 Application: Normal Approximation to the Binomial

Moments have many applications beyond simple description. For example, with a large sample size, the binomial tends to a normal distribution. We could draw this normal approximation using the moments that describe the binomial. The normal distribution has two parameters, the mean μ_1 (first moment) and variance $= \sigma^2 = \widetilde{\mu}_2$ (second central moment). To approximate the binomial, simply equate the moments. For the binomial,

$$\mu_1 = n\theta,$$

and

$$\widetilde{\mu}_2 = n\theta(1 - \theta)$$

(Appendix E). For a random variable x_i, the normal distribution having these moments is designated as

$$x_i \sim N(\mu_1, \sigma^2) = N(n\theta, n\theta(1 - \theta))$$

where the first parameter is the mean and the second is the variance (Figure D.5d). In Section 3.11 I use this moment matching as one method for fitting a distribution to data.

D.4 Jointly Distributed Random Variables

Data confront us with more than one random element. After all, we typically associate some stochasticity with each observation. Stochasticity is used to represent uncertainty in models. In this section I consider ways to use joint distributions for understanding probabilities of multiple events. At the beginning of this appendix I considered several relationships of jointly distributed random variables that merit restatement in the form of some simple rules.

1. *Joint distributions can be factored into univariate distributions*—Joint distributions can be difficult to use. We can simplify problems by factoring a joint distribution into marginal and conditional distributions:

$$P(A,B) = p(A|B)p(B)$$

$$= p(B|A)p(A)$$

   Note that distributions on the right-hand side are univariate, both being the product of a conditional and a marginal. Thus, if we can write down these two univariate distributions, which may be simple, we obtain the more complex, joint distribution, by multiplication. Of course, if the joint distribution is complex, there will be many factors.

2. *Bayes' theorem*—Combining the factored distributions above we arrive at a relationship between unvariate distributions, termed *Bayes' theorem*,

$$p(B|A) = \frac{p(A|B)p(B)}{p(A)}$$

   This is the basis for Bayesian analysis. It is important, because, among other things, it allows us to express parameters B as a probability distribution that is conditioned on data A.

3. *The conditional distribution is proportional to the joint distribution*—We often require the conditional distribution of a single variable. From the first rule, note that a conditional distribution differs from the joint distribution only by a constant. In other words $p(A|B)$ is proportional to $p(A,B)$, because, given B, its shape is not affected by the factor $p(B)$. In other words,

$$p(A|B) \propto p(A,B)$$

We can combine some of these rules to illustrate their utility. Consider the joint distribution $p(A,B,C)$. In most cases a three-dimensional distribution will be intractable. By factoring we can express it hierarchically as a product of univariate distributions. Here is a first factorization,

$$p(A,B,C) = p(A,B|C)p(C)$$

Now factor again,

$$p(A,B,C) = p(A|B,C)p(B|C)p(C)$$

We have now simplified the problem to one of a product of univariate distributions.

Example D.1. The kidney stone example from Figure. 1.1 involves three events having joint distribution,

$$p(I,D,C) = p(C|D,I)p(D|I)p(I)$$

Because there is no arrow connecting C and I, the first probability on the right-hand side can be simplified to

$$p(C|D,I) = p(C|D)$$

In other words, given D, we learn nothing more about C from knowing I; everything I tells us about C is already provided by D. For this example, if we know whether the stone is disintegrated (D), the quality of the image (I) adds no new information. The joint distribution for this chain of events can thus be written as

$$p(I,D,C) = p(C|D)p(D|I)p(I)$$

Example D.2. In Section 2.12.2 I consider probability functions associated with demography. Age-specific mortality rate was expressed as a conditional probability

$$h(a)da \approx \Pr \{a_i \in (a,a + da) \mid a_i \geq a\}$$

where a_i is age of death. Recall that the survivor function is the probability of survival to age a,

$$l(a) = \Pr\{a_i \geq a\}$$

and the probability of dying in the interval $[a,a + da]$ is

$$f(a)da = h(a)da \times l(a)$$

Using the first rule listed above, recognize the right-hand side to be the joint distribution of surviving until and dying within the specified interval. In other words, we have

$$\Pr\{\text{death in } (a,a + da)\} = \Pr\{\text{death in } (a,a + da)|\text{survive to } a\} \Pr \{\text{survive to } a\}$$

In the notation of Section 2.12.2, we have

$$f(a)da = \Pr\{a_i \in (a,a + da)|a_i \geq a\} \times \Pr\{a_i \geq a\} \qquad \text{this is } h(a)da \times l(a)$$

This is now a joint probability of surviving until and dying at age a. Likewise, from the second rule, we have

$$h(a)da = \frac{f(a)da}{l(a)}$$
$$= \frac{\Pr\{a_i \epsilon (a, a + da)\}}{\Pr\{a_i \geq a\}}$$

You can think of the various functions describing mortality in the context of Bayes' theorem.

D.4.1 Covariance and Correlation

Francis Galton (1822–1911) sought to understand patterns of inheritance, in part, through examination of how traits vary among family members. As part of this effort he collected data on heights of men and their brothers (Galton 1889) (Table D.4). I use these data to demonstrate the concept of correlation introduced by Galton, although the definition we use today was derived later (Pearson and Filon 1898). You could think of these data as consisting of $n = 1348$ pairs of observations on $p = 2$ variables x ("a man") and y ("a man's brother").

Before considering Galton's data set, define the *covariance*, which comes from the *mixed moment*, or the expected product of deviations from the mean values

$$cov[x,y] = E[(x - E[x])(y - E[y])] \qquad \text{D.23}$$

TABLE D.4.
Galton's Data on Height of Brothers

Men's Height	< 63	63.5	64.5	65.5	66.5	67.5	68.5	69.5	7.5	71.5	72.5	73.5	> 74
< 63	5	5	3	3	4	2							1
63.5	5	2	8	3	3	4	1	1		1			1
64.5	3	8	12	15	1	8	5	2	1				
65.5	3	3	15	18	33	36	8	2	1	1			
66.5	4	3	1	33	28	35	2	12	7	2	1		
67.5	2	4	8	26	35	38	38	2	18	8	1	1	
68.5		1	5	9	18	38	46	36	3	11	6	3	
69.5		1	2	1	11	2	36	55	44	17	5	4	2
7.5			1	1	7	19	3	45	36	14	9	8	1
71.5		1		1	2	8	11	18	14	2	9	4	
72.5					1	1	6	5	9	9	8	3	5
73.5						1	3	4	8	3	3	2	3
> 4	1	1						1	1		5	3	12

Like other second central moments, this can be expressed in terms of the cross product and the product of the means,

$$= E[xy] - 2E[x]E[y] + E[x]E[y]$$

$$= E[xy] - E[x]E[y] \qquad \text{D.24}$$

If x and y are the same variable, this is the formula for the variance (Equation D.9).

Now consider the distribution of brothers' heights. The joint empirical distribution of x and y is

$$p(x_i, y_i) = \frac{1}{n} \qquad \text{D.25}$$

In other words, each observation contributes $\frac{1}{n}$ to the full data set. To obtain $p(x_i, y_i)$ simply divide each value of Table D.4 by $n = 1348$. This distribution is represented as a contour plot in Figure D.6a. The mean values for x and y are 68.42 and 68.40 inches, respectively.

The m,n^{th} central mixed moment of x,y is defined analogous to that for a single variable (Equation D.12),

$$E[(x - \bar{x})^m (y - \bar{y})^n] = \sum_{i=1}^{n} (x_i - \bar{x})^m (y_i - \bar{y})^n p(x_i, y_i) \qquad \text{D.26}$$

Substituting from Equation D.25 yields

$$= \frac{1}{n} \sum_{i=1}^{n} (x_i - \bar{x})^m (y_i - \bar{y})^n$$

The *sample covariance* is an *empirical moment*. By direct analogy with Equation D.24,

$$\begin{aligned}
\text{cov}[x,y] &= E[(x - \bar{x})(y - \bar{y})] \\
&= \frac{1}{n}\left[\sum_{i=1}^{n} x_i y_i - \bar{x}\sum_{i=1}^{n} y_i - \bar{y}\sum_{i=1}^{n} x_i + \bar{x}\cdot\bar{y} \right] \\
&= \overline{xy} - 2\bar{x}\cdot\bar{y} + \bar{x}\cdot\bar{y} \\
&= \overline{xy} - \bar{x}\cdot\bar{y} \qquad \text{D.27}
\end{aligned}$$

The *covariance matrix* for these data

$$\mathbf{C} = \begin{bmatrix} \text{var}[x] & \text{cov}[x,y] \\ \text{cov}[x,y] & \text{var}[y] \end{bmatrix} \equiv \begin{bmatrix} c_1 & c_{12} \\ c_{12} & c_2 \end{bmatrix} = \begin{bmatrix} 6.29 & 3.99 \\ 3.99 & 6.22 \end{bmatrix}$$

is symmetric about the diagonal, with diagonal elements being the variances. The *correlation* between x and y is given by

$$cor[x,y] = \frac{\text{cov}[x,y]}{\sqrt{\text{var}[x]\text{var}[y]}} \equiv \frac{c_{12}}{\sqrt{c_1 c_2}} \qquad \text{D.28}$$

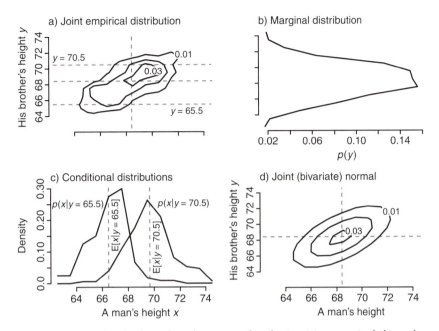

FIGURE D.6. Galton's height data plotted as a joint distribution (*a*), a marginal (*b*), and two conditionals (*c*). The conditional distributions for *x* in *c* are taken at values of *y* shown as dashed lines in *a*. Vertical dashed lines in *c* are conditional means. In *d* is shown a normal distribution parameterized with empirical moments taken from the data.

The *correlation matrix*

$$\mathbf{R} = \begin{bmatrix} \dfrac{c_1}{\sqrt{c_1 c_1}} & \dfrac{c_{12}}{\sqrt{c_1 c_2}} \\ \dfrac{c_{12}}{\sqrt{c_1 c_2}} & \dfrac{c_2}{\sqrt{c_2 c_2}} \end{bmatrix} = \begin{bmatrix} 1 & 0.638 \\ 0.638 & 1 \end{bmatrix} \qquad \text{D.29}$$

is also symmetric about the diagonal. The ones along the diagonal indicate the correlation of a variable with itself. The high correlation between brothers is inflated because of the way Galton counted combinations of brothers within the same family (Stigler 1986).

If x and y are independent, then $\text{cov}[x,y]$ is zero and $E[xy] = E[x]E[y]$. The correlation coefficient is always ≤ 1, because

$$\text{cov}[x,y] \leq \sqrt{\text{var}[x]\text{var}[y]}$$

This follows from the Cauchy-Schwartz inequality (the product moment cannot exceed the product of the first moments).

Covariances provide a simple summary of the linear relationship between two random variables, and thus represent a useful descriptive statistic. The positive covariance in this example describes the tendency for tall men to have tall brothers (Figure D.6a). Together with the sample means, which locate the center of the distribution, the covariance matrix helps us visualize (and draw) the distribution shape. Using just the means and covariance matrix, we can construct a bivariate normal distribution (Figure D.6d) that captures the main features of the data (Figure D.6a).

D.4.2 A Joint Distribution of θ and y

The island occupancy example provides opportunity to examine joint distributions in a different context. Figure D.5 assigns a probability to all possible levels of occupancy (0 to 32) based on a parameter estimate that comes from observation of a single archipelago. This might seem optimistic (or even pretentious), but this is an implication of Figure D.5. We might object to this analysis if we are unwilling to accept that there is a fixed value θ that applies to all islands alike, that one archipelago is sufficient to estimate θ, or both. A glance at the map (Figure. D.1) does not inspire the notion of coin flipping. We can think of many reasons why occupancy is far more variable than the flip of a coin.

How does variability in θ affect the interpretation? There are several ways forward. We might try to determine how the value of θ could differ among archipelagos and/or islands and the variables that affect it. Then θ becomes a function of these variables, such as the generalized linear models of Section 8.2. Alternatively, we might not know much about the factors that affect θ. In this case, we can allow for variability, even when we do not know the cause. Examples of this approach are found in Section 8.3. Here I simply point out how rules for jointly distributed random variables can help determine the effects of variability in θ on the distribution of y. The approach involves assigning a distribution to θ and analyzing its impact on y.

The concept of a distribution of θ stems from the idea that there may be some average probability that a fruit pigeon population occupies an island or islands within an archipelago, but there is variability about that average value that might depend on the island and on many factors that we cannot measure. I continue to assume that, given a particular value of θ, occupancy is binomial, but just accept that θ is not fixed. The binomial probability is subject to random effects.

To incorporate variability of θ into the analysis of hypothetical archipelagos, I use properties of jointly distributed random variables to derive the marginal distribution of y that accounts for variability in θ. For this example I use a standard density for a quantity that varies between zero and one, the beta density,

$$\theta \sim Beta(\alpha,\beta)$$

or

$$p(\theta) = Beta(\theta|\alpha,\beta) = \frac{\theta^{\alpha-1}(1-\theta)^{\beta-1}}{B(\alpha,\beta)} \qquad \text{D.30}$$

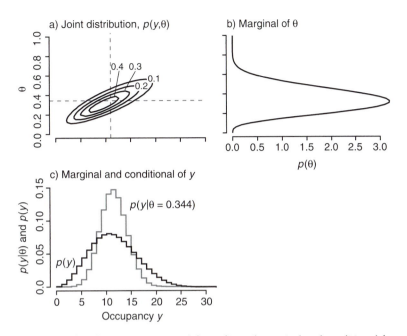

Figure. D.7. Joint distribution (*a*), marginal for θ (*b*), and marginal and conditional for y (*c*). Note that the distribution of y is discrete, whereas θ has a continuous density. The marginal distribution of y is broader than the conditional, because it integrates the variability in θ.

(Appendix F). The beta function $B(\alpha, \beta)$ is the normalization constant for this density

$$B(\alpha,\beta) = \int_0^1 \theta^{\alpha-1}(1 - \theta)^{\beta-1}d\theta$$

The relationships among beta functions, gamma functions, and combinatorials are outlined in Box D.1. There is no theoretical basis for use of the beta density here.

Box D.1 Combinatorials, beta, and gamma functions

This box contains technical material relating functions that arise in binomial, beta, and gamma distributions.

The binomial coefficient

$$\binom{n}{y} = \frac{n!}{y!(n - y)!}$$

involves factorials, which are defined only for integers. It is the number of ways in which y objects can be drawn from n total. Gamma functions provide the factorial

---

(Continued on following page)

for integers, but they are defined for nonintegers as well. The gamma function is like a continuous version of the factorial. It has the form of an integral equation

$$\Gamma(a) = \int_0^\infty x^{a-1} e^{-x} dx$$

for positive values of the argument a. Note that x is a dummy variable that we integrate out of existence. This function blows up with large values of a (like the factorial). For integer values of a the relation to the factorial is

$$a! = \Gamma(a + 1) = a\Gamma(a)$$

Gamma integrals arise in many contexts. When faced with an integration problem involving θ check for substitutions that will yield a gamma function. The gamma density (see below) is the most obvious example; the gamma function is the integration constant.

When evaluating factorials numerically, it can be convenient to do so using gamma functions. Written in terms of gamma functions, the binomial coefficient is

$$\binom{n}{y} = \frac{n!}{y!(n-y)!} = \frac{\Gamma(n+1)}{\Gamma(y+1)\Gamma(n-y+1)}$$
$$= \frac{n}{y(n-y)} \times \frac{\Gamma(n)}{\Gamma(y)\Gamma(n-y)}$$

The *beta function* is an integral that may appear in problems involving the binomial or Student's t. The beta function is the integration constant for the beta density. Beta

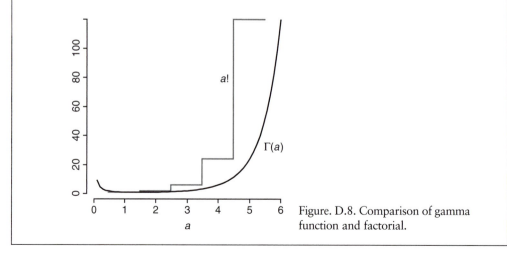

Figure. D.8. Comparison of gamma function and factorial.

integrals can be written in several ways. Here is one of the more common ways, together with its relationship to the gamma function,

$$B(a,b) = \int_0^1 z^{a-1}(1 - z)^{b-1}dz = \frac{\Gamma(a)\Gamma(b)}{\Gamma(a + b)}$$

Just as you might think of a gamma function as a continuous version of a factorial, a beta function acts like a continuous version of a combinatorial,

$$\frac{n!}{y!(n - y)!} = \frac{1}{B(y + 1, n - y + 1)}$$

$$= \frac{n}{y(n - y)} \times \frac{1}{B(y, n - y)}$$

Box D.2 shows an example where these relationships can be valuable.

The two quantities y and θ have a *joint distribution* $p(y,\theta)$ that describes the probability of all combinations of y and θ (Figure. D.7a). From Equation D.2, write the joint distribution

$$p(y,\theta) = p(y|\theta)p(\theta) = p(\theta|y)p(y) \qquad \text{D.31}$$

as products involving *conditionals*, $p(y|\theta)$ and $p(\theta|y)$, and *marginals*, $p(y)$ and $p(\theta)$. In this instance, one of the quantities is continuous (θ) and one is discrete (y). For simplicity, I refer to both as distributions.

The conditionals describe the distribution of one variable corresponding to a specific value of the other variable. A conditional distribution is proportional to a horizontal or vertical transect through the joint distribution (Figure. D.7a). Figure D.7c includes the conditional density of y for the value of θ estimated from the single archipelago ($\theta = 0.344$),

$$\Pr\{y \text{ given } \theta\} = p(y|\theta)$$

This distribution appeared in Equation D.7, but I did not then recognize it as conditional. I referred to it then as $p(y)$, rather than $p(y|\theta)$, because we might not bother explicitly to condition on things that are always fixed. In fact, all densities are conditioned on many things that are unspecified. The conditional density describes the density we would obtain if θ was a fixed value. It is now specified explicitly, because we recognize that it need not be fixed. The notation $p(y|\theta)$ means that the density applies at this fixed value, but we acknowledge that θ might assume other values.

We are now prepared to address the question of how y is affected by this new source of variability. The answer requires the marginal distribution of y. It can be found by integrating away the variability in θ,

$$p(y) = \int_0^1 p(y,\theta)d\theta \qquad \text{D.32}$$

The integration limits reflect the fact that θ represents a probability, so its density is defined only on the interval $(0, 1)$. The marginal density can be viewed as a weighted average of all possible conditional densities, each weighted by its relative contribution $p(\theta)$. We factor it into two univariate distributions,

$$p(y) = \int_0^1 p(y|\theta)p(\theta)d\theta \qquad \text{D.33}$$

Substituting the binomial conditional and the beta marginal, we obtain

$$p(y) = \int_0^1 p(y|\theta)p(\theta)d\theta = \int_0^1 \binom{n}{y}\theta^y(1-\theta)^{n-y} \times \frac{\theta^{\alpha-1}(1-\theta)^{\beta-1}}{B(\alpha,\beta)}d\theta$$

Now collect the exponents, and write the resulting integral as a beta function,

$$= \frac{\binom{n}{y}\int_0^1 \theta^{y+\alpha-1}(1-\theta)^{n-y+\beta-1}d\theta}{B(\alpha,\beta)} = \binom{n}{y}\frac{B(y+\alpha, n-y+\beta)}{B(\alpha,\beta)}$$

This marginal distribution of y no longer contains θ. Instead, it is a distribution defined by the parameters that control the distribution of θ. To make this relationship explicit, we could write Equation D.26 as

$$p(y|\alpha,\beta) = \int_0^1 p(y|\theta)p(\theta|\alpha,\beta)d\theta$$

In Figure D.7 I use values for parameters α and β such that θ has a mean value equal to the observed mean of $\frac{11}{32}$. A plot of the density shows that values of 0.2 and 0.4 are about half as probable as the mean, and values less than 0.05 and greater than 0.7 are very unlikely (Figure D.7b). To select parameters with specific means and variances, I could use moment matching (Chapter 3).

The marginal distribution of y has larger variance than does the conditional distribution (Figure D.7c). This relationship is expected, because we have added an additional source of variability. Indeed, if all values of θ are equally likely (obtained here if $p(\theta) = 1$ for all values of θ), then the distribution of y is uniform (Box D.2).

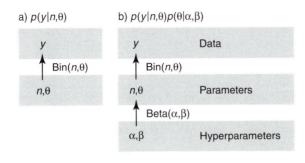

a) $p(y|n,\theta)$

b) $p(y|n,\theta)p(\theta|\alpha,\beta)$

FIGURE D.9. An added level of stochasticity in the binomial model
results when a parameter varies.

Box D.2 The binomial/negative binomial mixture

The negative binomial mixture can be used for distributions of survivors, where
the initial number at risk is variable and, potentially, unknown. In the seedling
Example D.4, I considered a mixture of three stochastic processes: variable seed
dispersal, Poisson distribution of seed, and binomial survival. The distribution of
seedlings was solved as

$$p(y|\alpha,\beta,\theta) = \int_0^\infty \sum_{n=0}^\infty Bin(y|n,\theta)Pois(n|\lambda)Gam(\lambda|\alpha,\beta)d\lambda$$

I first mixed the binomial (survival to seedling stage) and Poisson process (distribution of seed), and then integrated the variability in λ (variation in dispersal process)
to arrive at a marginal negative binomial distribution,

$$NB(y|\alpha,\beta/\theta) = \int_0^\infty Pois(n|\theta\lambda)Gam(\lambda|\alpha\beta)d\lambda$$

Figure D.10b depicts the same mixture, but starts with the variability in λ and
progresses through the distribution of seeds and then seedlings. We should arrive
at the same answer regardless of starting point. This alternative starting point
means that we begin by integrating over λ followed by summing over n,

$$p(y|\alpha,\beta,\theta) = \sum_{n=0}^\infty Bin(y|n,\theta)\int_0^\infty Pois(n|\lambda)Gam(\lambda|\alpha,\beta)d\lambda$$

The first integration yields the negative binomial distribution in parameters α and
β, leaving

$$= \sum_n Bin(y|n,\theta)NB(n|\alpha,\beta)$$

(Continued on following page)

We now have a mixture involving binomial survival, where the number of total individuals at risk varies as a negative binomial. To solve this, first write out the distributions

$$= \sum_n \frac{n!}{y!(n-y)!} \theta^y (1-\theta)^{n-y} \frac{(n+\alpha-1)!}{n!(\alpha-1)!} \left(\frac{\beta}{\beta+1}\right)^\alpha \left(\frac{1}{\beta+1}\right)^n$$

and pull out of the summation coefficients that do not depend on n

$$= \frac{\theta^y}{y!(\alpha-1)!} \left(\frac{\beta}{\beta+1}\right)^\alpha \sum_{n=y}^\infty \frac{(n+\alpha-1)!(1-\theta)^{n-y}}{(n-y)!} \left(\frac{1}{\beta+1}\right)^n$$

The goal is to eliminate n and to express the result in terms of a standard distribution. The series has the key ingredients of a negative binomial distribution in n. If we can complete this distribution, we might then sum it out of existence. I make several substitutions to achieve this end. Let $z = n - y$, and multiply by $(y + \alpha - 1)!/(y + \alpha - 1)!$

$$= \frac{(y+\alpha-1)!}{y!(\alpha-1)!} \left(\frac{\beta}{\beta+1}\right)^\alpha \left(\frac{\theta}{\beta+1}\right)^y \sum_{z=0}^\infty \frac{(z+y+\alpha-1)!}{z!(y+\alpha-1)!} \left(\frac{1-\theta}{\beta+1}\right)^z$$

and

$$q = \frac{\beta+\theta}{1-\theta}$$

Then

$$= \binom{y+\alpha-1}{\alpha-1} \left(\frac{\beta}{\beta+1}\right)^\alpha \left(\frac{\theta}{\beta+1}\right)^y \sum_{z=0}^\infty \binom{z+y+\alpha-1}{y+\alpha-1} \left(\frac{1}{q+1}\right)^z$$

Now complete the negative binomial with the inclusion of the constant factor

$$\left(\frac{q}{q+1}\right)^{y+\alpha}$$

Dividing through by this coefficient to obtain

$$= \binom{y+\alpha-1}{\alpha-1} \left(\frac{\beta}{\beta+1}\right)^\alpha \left(\frac{\theta}{\beta+1}\right)^y \left(\frac{q+1}{1}\right)^{y+\alpha} \sum_{z=0}^\infty NB(z|y+\alpha,q)$$

The series now sums to one. Substitute back for q to obtain

$$= \binom{y+\alpha-1}{\alpha-1} \left(\frac{\beta}{\beta+1}\right)^\alpha \left(\frac{\theta}{\beta+1}\right)^y \left(\frac{\beta+1}{\beta+\theta}\right)^{y+\alpha}$$

$$= \binom{y + \alpha - 1}{\alpha - 1} \left(\frac{\beta/\theta}{\beta/\theta + 1}\right)^\alpha \left(\frac{1}{\beta/\theta + 1}\right)^y$$

$$= NB(y|\alpha, \beta/\theta)$$

Thus, the result is the same, regardless of where we begin. Figure D.10b shows how the variability in Poisson parameter makes for a seedling distribution with larger variance than the Poisson (compare D.10a). This broad variance, in turn, produces a broader distribution of seedlings.

How does this analysis change our view of the problem? The initial analysis (Figure D.7a) assumed that there is a parameter θ that could be estimated from data like Figure D.1. Stochasticity comes from the binomial process that generates the data y. We have now admitted that θ can vary (Figure D.7b). This adds an additional stage to the model. We have a process that generates stochastic θ with a superimposed process that generates stochastic y. The resultant mixture of processes results in increased variability (Figure D.7c). By changing our view that θ represents a process rather than a parameter (Figure D.7a), we estimate more uncertainty in the probability of occupancy than is reported in standard analyses.

This example demonstrates relationships involving joint distributions that we might view in the context of how variability at one level affects variability at the next. These relationships are the basis for hierarchical models (Chapter 8).

D.4.3 Jointly Distributed Variables as Mixtures

The foregoing examples represent different ways of thinking about jointly distributed random variables. In Galton's height data, I discussed both x and y as data that result from some partially known data-generating mechanism (Figure D.6a). In the case of island occupancy, I began with a conditional distribution of a variable y given a fixed value of a parameter θ (Figure D.8a). I then asked how the distribution of y changes after introducing variability in θ (Figure D.8b). The marginal distribution of y integrates over variability in θ and, thus, is much broader than the conditional distribution (Figure D.7c). This marginal distribution is termed a *mixture*, because it arises from two sources of variability. Such hierarchical models have wide application. Here I consider some specific examples of mixtures and mention a few more of their properties. I begin with an example demonstrating how to marginalize over the variability at one stage when both stages are discrete.

Example D.3. To fit mortality rates to seedling censuses, Lavine et al. (2002) modeled the distribution of seedlings as the result of binomial survival of Poisson-distributed seeds. The model includes two random variables, y, the number of seedlings, and n, the number of seeds. The data-generating process is binomial, but one of the binomial parameters is a variable

$$y|n,\theta \sim Bin(n,\Theta)$$

$$n|\lambda \sim Pois(\lambda)$$

(Figure. D.9a). This model can be viewed as a mixture of two discrete processes, the binomial survival that follows the initial Poisson distribution. To obtain the marginal distribution of seedling densities we must sum over the variability introduced by the Poisson seed distribution,

$$p(y) = \sum_n p(y,n)$$

As previously, factor the joint distribution into two univariate ones,

$$p(y) = \sum_n p(y|n)p(n)$$

Substituting for the appropriate distributions we have

$$p(y|\theta,\lambda) = \sum_{n=0}^{\infty} Bin(y|n,\theta)Pois(n|\lambda)$$

$$= \sum_{n=y}^{\infty} \binom{n}{y}\theta^y(1 - \theta)^{n-y}\frac{e^{-\lambda}\lambda^n}{n!}$$

This summation begins at $n = y$, because the binomial is zero for any $n < y$ (there is zero probability of more survivors than initial seeds).

The goal is now to simplify this expression in terms of a standard distribution. To this end, write out the combinatorial (Box D.1) and pull out of the summation coefficients that do not depend on n,

$$= \frac{\theta^y e^{-\lambda}}{y!} \sum_{n=y}^{\infty} \frac{(1 - \theta)^{n-y}\lambda^n}{(n - y)!}$$

The summation is beginning to look like a Taylor series for $\exp[\bullet]$ (Appendix A). If we multiply by λ^y/λ^y and then substitute $z = n - y$, we have a Taylor series

$$= \frac{\lambda^y\theta^y e^{-\lambda}}{y!} \sum_{z=0}^{\infty} \frac{[(1 - \theta)\lambda]^z}{z!}$$

with the summation being equal to $e^{(1-\theta)\lambda}$. This simplifies to

$$= \frac{(\lambda\theta)^y e^{-\lambda\theta}}{y!}$$

$$= Pois(y|\theta\lambda)$$

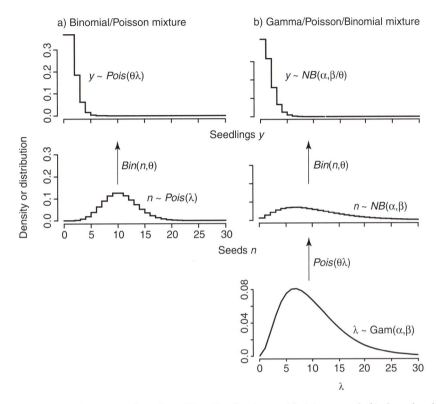

Figure. D.10. Two examples of seedling distributions with (*a*) two and (*b*) three levels of variability. Parameter values are $\theta = 0.1$, $\alpha = 3$, $\beta = 0.3$, $\lambda = 10$.

(Figure D.10). Thus, the marginal (mixture) distribution is Poisson with a parameter (mean value) equal to the mean initial seed density times the probability of survival. This result is highly intuitive.

Example D.4. The distribution of initial seeds might be more variable than a Poisson. Suppose seed arrival is essentially Poisson, but that the Poisson process (represented by the Poisson parameter) varies from place to place. Clark et al. (1998) assumed that the Poisson parameter has a gamma distribution. The mixture then is

$$y|\theta\lambda \sim Pois(\theta\lambda)$$

and

$$\lambda \sim Gam(\alpha, \beta)$$

The marginal distribution of seedlings now involves integration, because λ is a continuous parameter,

$$p(y) = \int_0^\infty p(y,\lambda)d\lambda = \int_0^\infty p(y|\lambda)p(\lambda)d\lambda$$

$$= \int_0^\infty \frac{(\theta\lambda)^y e^{-\theta\lambda}}{y!} \times \frac{\beta^\alpha \lambda^{\alpha-1} e^{-\beta\lambda}}{\Gamma(\alpha)} d\lambda$$

Pull out of the integrand all coefficients that do not depend on λ,

$$= \frac{\theta^y \beta^\alpha}{y!\Gamma(\alpha)} \int_0^\infty \lambda^{y+\alpha-1} e^{-\lambda(\theta+\beta)} d\lambda$$

Now substitute a new variable $u = \lambda(\theta + \beta)$ and write the integral in terms of a gamma function. The λ's become $u/(\theta + \beta)$. We must also substitute for $d\lambda$. Note that $du/d\lambda = \theta + \beta$. Thus, in terms of u,

$$= \frac{\theta^y \beta^\alpha}{y!\Gamma(\alpha)(\theta + \beta)^{y+\alpha}} \int_0^\infty u^{y+\alpha-1} e^{-u} du$$

$$= \frac{\theta^y \beta^\alpha \Gamma(y + \alpha)}{y!\Gamma(\alpha)(\theta + \beta)^{y+\alpha}}$$

Recalling the relationship among gamma functions, factorials, and combinatorials from Box D.1, we can write this as

$$= \frac{(y + \alpha - 1)!}{y!(\alpha - 1)!} \frac{\theta^y \beta^\alpha}{(\theta + \beta)^{y+\alpha}} = \binom{y + \alpha - 1}{\alpha - 1}\left(\frac{\beta/\theta}{\beta/\theta + 1}\right)^\alpha \left(\frac{1}{\beta/\theta + 1}\right)^y$$

This mixture of a Poisson and a gamma process is known as a *negative binomial distribution*,

$$y \sim NB(\alpha, \beta/\theta)$$

(Figure D.11b). The distribution of seedlings is now expressed solely in terms of the parameters that describe the variability in n and λ. The mean of this distribution is $\alpha\theta/\beta$.

D.4.4 Conditional Expectation

The expectation of x for a given value of y is termed the *conditional expectation*

$$\mu_{x|y} \equiv E[x|y] = \int_{-\infty}^\infty xp(x|y)dx \qquad \text{D.34a}$$

or, for a discrete variable,

$$\mu_{x|y} = \sum_{x=0}^\infty xp(x|y) \qquad \text{D.34b}$$

These expressions are identical to Equation D.19 and D.14, respectively, if we simply drop the conditioning on y. The (unconditional) expectation of x requires a second integration,

$$\mu_x \equiv E[x] = E[E[x|y]] = \int_{-\infty}^{\infty} \left[\int_{-\infty}^{\infty} xp(x|y)dx \right] p(y)dy$$

$$= \int_{-\infty}^{\infty} E[x|y]p(y)dy \qquad \text{D.35}$$

D.4.5 Variance of x That Depends on y

When variables x and y covary, the variance of x depends on y. Note that

$$\text{var}[x] = E[(x - E[x])^2]$$

By adding and subtracting $E[x|y]$, followed by some tedious algebra (e.g., Casella and Berger 1990), we arrive at a useful formula for the variance of x,

$$\text{var}[x] = E[\text{var}[x|y]] + \text{var}[E[x|y]] \qquad \text{D.36}$$

Example D.5. Population abundances are typically estimated with error. If the true abundance is n, and y individuals are counted, then y and n are related by

$$E[y|n] = \theta n$$

where θ is the probability of being counted. This is the *conditional expectation* of y. Both y and n are random variables. Suppose that n is sufficiently large that we can treat it as normally distributed with mean μ and variance σ^2. The distribution of y is binomial,

$$p(y|n,\theta) = Bin(y|n,\theta)$$

having conditional variance

$$\text{var}[y|n] = n\theta(1 - \theta)$$

The expectation of this variance implies that n assumes its mean value $n = \mu$, or

$$E[\text{var}[y|n]] = \mu\theta(1 - \theta)$$

The variance of a constant θ times n is simply the constant squared times the variance of n, that is,

$$\text{var}[E[y|n]] = \text{var}[\theta n] = E[(\theta n - E[\theta n])^2] = E[(\theta n)^2 - 2\theta nE[\theta n] + (E[\theta n])^2]$$

$$= \theta^2[n^2] - 2\theta^2(E[n])^2 + \theta^2(E[n])^2$$

$$= \theta^2\{E[n^2] - (E[n])^2\}$$

$$= \theta^2\text{var}[n] \qquad \text{D.37}$$

Thus,

$$\text{var}[E[y|n]] = \theta^2 \text{var}[n] = \theta^2 \sigma^2$$

and the variance in y is

$$\text{var}[y] = \mu\theta(1 - \theta) + \theta^2\sigma^2$$

Note that the variance in the sample densities is like the variance of the binomial sampling distribution with a second term. This second term means that the sample variance is greater than a binomial.

The foregoing relationships can be written in matrix notation. Consider a matrix of constants $\mathbf{X}$ and a random vector $\mathbf{y}$ with covariance matrix $\mathbf{V}_y$. We have the means and covariances

$$E[\mathbf{Xy}] = \mathbf{X}E[\mathbf{y}] \qquad\qquad \text{D.38a}$$

$$var[\mathbf{Xy}] = \mathbf{X}\mathbf{V}_y\mathbf{X}^T \qquad\qquad \text{D.38b}$$

D.4.6 Covariance of Continuous Variables

The *covariance* between two continuous random variables, x and y, having means μ_x and μ_y and joint density $p(x,y)$, is given by

$$E[(x - \mu_x)(y - \mu_y)] = \int_{-\infty}^{\infty}\int_{-\infty}^{\infty} (x - \mu_{xy})(y + \mu_y)p(x,y)dxdy \quad \text{D.39}$$

Begin by expanding the cross product

$$= \int_{-\infty}^{\infty}\int_{-\infty}^{\infty} (xy - \mu_y x - \mu_x y + \mu_x\mu_y)p(x,y)dxdy$$

and integrate term by term,

$$= \int_{-\infty}^{\infty} y \int_{-\infty}^{\infty} xp(x,y)dxdy - \mu_y \int_{-\infty}^{\infty}\int_{-\infty}^{\infty} xp(x,y)dxdy$$

$$- \mu_x \int_{-\infty}^{\infty} y \int_{-\infty}^{\infty} p(x,y)dxdy + \mu_x\mu_y \int_{-\infty}^{\infty}\int_{-\infty}^{\infty} p(x,y)dxdy$$

For each term, the first integration over x results in integrands that contain marginal densities $p(y)$ and conditional means $\mu_{x|y}$ (Equation D.34). The last term contains an integral over the entire joint density, and is equal to one. We now have

$$= \int_{-\infty}^{\infty} y \cdot \mu_{x|y}p(y)dy - \mu_y \int_{-\infty}^{\infty} \mu_{x|y}p(y)dy - \mu_x \int_{-\infty}^{\infty} yp(y)dy + \mu_x\mu_y$$

Integrating the first term now gives the expected product xy. The integrand is the product of y and the mean of x given y weighted by the probability of y. The middle two terms both yield the product of the means. Together we now have

$$= \mu_{xy} - \mu_y\mu_x - \mu_x\mu_y + \mu_x\mu_y$$

or, simply,

$$= \mu_{xy} - \mu_y\mu_x \qquad \text{D.40}$$

D.5 Functions of Variables and Changing Variables

D.5.1 Expectation of a Function

We are often interested in predicting the behavior of a model that is parameterized with data. Data may be represented by a distribution or by moments. Here I extend the discussion of random variables to include their implications for model predictions. The expectation of a function of x can be approximated by the approach used for moments, provided we have some information regarding variability in x. Let $\phi(x)$ be a function of a random variable x, and x has the density $p(x)$. Recall that the expectation of x is obtained by

$$E[x] = \int_{-\infty}^{\infty} xp(x)dx$$

We can replace x with some function of x to obtain the expectation of that function,

$$E[\phi(x)] = \int_{-\infty}^{\infty} \phi(x)p(x)dx \qquad \text{D.41}$$

The calculation of the variance was an example of this method. To see this, let $\phi(x)$ be the squared difference between x and its mean,

$$\phi(x) = (x - \mu_1)^2$$

By definition, the variance is the expectation of $\phi(x)$, that is,

$$\text{var}[x] = E[\phi(x)] = E[(x - \mu_1)^2] = \int_{-\infty}^{\infty} (x - \mu_1)^2 \, p(x)dx$$

Unfortunately, we may not know the density $p(x)$. An approximation of $E[\phi]$ can be had using a Taylor series. Here is how to obtain it.

We expect $E[\phi(x)]$ to lie somewhere in the neighborhood of $\phi(E[x])$. That is, the mean of the function is probably close to the function of the mean. We can therefore approximate $E[\phi]$ by expanding $\phi(x)$ about $E[x]$ and integrating the resultant polynomial. I use the sample mean $\bar{x}$ as an estimate of $E[x]$. The expansion about the mean value of x looks like this:

$$\phi(x) = \sum_{m=0}^{\infty} \frac{(x - \bar{x})^m}{m!} \frac{d^m \phi}{dx^m}\bigg|_{x=\bar{x}} \qquad \text{D.42}$$

(Appendix A). Now substitute this series into the integral Equation D.41 and switch the order of integration and summation. This switch is permitted, because ϕ is being evaluated at a constant $x = \bar{x}$. The result is a series

$$E[\phi(x)] = \sum_{m=0}^{\infty} \phi^{(m)}(\bar{x}) \frac{\tilde{\mu}_m}{m!}$$

where

$$\phi^{(m)}(\bar{x}) = \frac{d^m \phi}{dx^m}\bigg|_{\bar{x}}$$

is the mth derivative of $\phi(x)$ evaluated at the expected value of x, and

$$\tilde{\mu}_m = \int_{-\infty}^{\infty} (x - \mu_1)^m \, p(x) dx$$

is the mth central moment of x (see Equation D.20). Writing out the first few terms of the series,

$$E[\phi(x)] = [\phi(\bar{x}) \times 1] + [\phi'(\bar{x}) \times 0] + \left[\phi''(\bar{x}) \times \frac{\mu_2}{2}\right] + \cdots$$

The terms on the right are:

Term	Order	Explanation
first	0	the function of the mean times the zero order, (always equal to one for a probability density function)
second	1	the first *central* moment is always zero, because we are expanding about the mean → this term disappears
third	2	the product of the second derivative and the second central moment, which is, of course, the variance

The approximation is then, to second order,

$$E[\phi(x)] \approx \phi(\bar{x}) + \phi''(\bar{x}) \frac{\text{var}[x]}{2} \qquad \text{D.43}$$

We might view this expression as the function of the mean corrected by the contribution of the variance. If the second derivative is negative, this contribution lowers the estimate of the mean, and vice versa. Moreover, if the function ϕ is linear, then the second derivative is zero, and variability in x does not affect the expectation of ϕ.

D.5.2 Variance of a Function of a Random Variable

A function of a random variable has a variance that can be approximated by similar methods. The variance in $\phi(x)$ can be written as

$$\text{var}[\phi(x)] = E[(\phi(x) - E[\phi(x)])^2]$$

To first order, estimate ϕ to be

$$\phi(x) \approx \phi(\mu_1) + \phi'(\mu_1)(x - \mu_1)$$

Taking expectations of both sides we obtain

$$E[\phi(x)] \approx E[\phi(\mu_1) + \phi'(\mu_1)(x - \mu_1)]$$

$$= \phi(\mu_1) + \phi'(\mu_1)E[x - \mu_1] \quad \textit{expectations of constants are constants}$$

the last step following from the fact that the functions are evaluated at the constant μ_1. The second term disappears, because the expected deviation from the mean (the first central moment) is zero. Thus,

$$E[\phi(x)] \approx \phi(\mu_1)$$

Using the fact that

$$\phi(x) - \phi(\mu_1) \approx \phi'(\mu_1)(x - \mu_1)$$

we substitute into the equation for the variance

$$\text{var}[\phi] = E[(\phi'(\mu_1)(x - \mu_1))^2] = \phi'(\mu_1)^2 E[(x - \mu_1)^2]$$

Recognizing $E[(x - \mu_1)]^2$ as the variance in x, we have

$$\text{var}[\phi] \approx (\phi'(\mu_1))^2 \text{var}[x] \qquad \text{D.44}$$

Example D.6. Real and Ellner (1992) considered a trait x that affects geometric mean fitness (number of offspring) as

$$\phi(x) = \ln x$$

Variability in the trait was described as

$$x_i = A(b + \varepsilon_i)$$

Parameter b might be the maximum value of a trait, in which case A is an allocation fraction that assumes values between zero and one. The authors did not assume a distribution for the random variates ε_i, only that the mean is zero, and the variance might be estimated as σ_ε^2. They asked how this variability affects fitness.

The mean of x is simply Ab, because the equation is linear in ε. Thus, from Equation D.43, we see that the expectation of x is unaffected by variability in ε. The variance in x is obtained using Equation D.44,

$$\mathrm{var}[x] = (x'(\bar{\varepsilon}))^2 \, \sigma_\varepsilon^2 = A^2 \, \sigma_\varepsilon^2$$

where the derivative is

$$x'(\bar{\varepsilon}) \equiv \left. \frac{dx}{d\varepsilon} \right|_{\bar{\varepsilon}} = A$$

Using Equation D.43 the expected fitness is

$$E[\phi(x)] = \ln(Ab) - \frac{\mathrm{var}[x]}{2(Ab)^2}$$

$$= \ln(Ab) - \frac{\sigma_\varepsilon^2}{2b^2}$$

In this case, the expected fitness increases with the expected allocation to trait x. Because, there are diminishing returns (fitness increases with the log of A), variability decreases the expected fitness. The variance in the function is

$$\mathrm{var}[\phi] = (\phi'(\mu_1))^2 \mathrm{var}[x] = (Ab)^{-2} A^2 \sigma^2 = \sigma^2/b^2$$

D.5.3 Variance of a Function of Several Variables

Suppose that ϕ is an arbitrary function of random variables $x_1, x_2, \ldots x_k$ having mean values $\mu_1, \ldots, \mu_k$ represented by vectors x and μ, respectively. The variance of ϕ can be determined in a manner like that used for a function of a single variable. The first order of the Taylor expansion of ϕ is

$$\phi(\mathbf{x}) \approx \phi(\mu) + \sum_{i=1}^{k} (x_i - \mu_i)\phi_i'(\mu)$$

Note that we are summing over the contribution of each of k linear terms. ϕ_i' is the first derivative with respect to the ith variable x_i evaluated at the mean for all variables. now take expectations of both sides,

$$E[\phi(\mathbf{x})] \approx E\left[\phi(\mu) + \sum_{i=1}^{k}(x_i - \mu_i)\phi_i'\mu\right]$$

Taking account of the constant terms on the right-hand side,

$$E[\phi(x)] \approx \phi(\mu) + \sum_{i=1}^{k} E[x_i - \mu_i]\phi_i'(\mu)\mu$$

Because the expectation of $x_i - \mu_i$ is zero, the series disappears, and we have

$$E[\phi(\mathbf{x})] \approx \phi(\mu)$$

Now, recall that the variance is the expected squared deviation from the mean

$$\text{var}[\phi] \approx E[(\phi(\mathbf{x}) - \phi(\mu))^2]$$

Using the fact that $\phi(\mathbf{x}) - \phi(\mu) \approx \sum_{i=1}^{k}(x_i - \mu_i)\phi_i'(\mu)$, substitute

$$\text{var}[\phi] \approx E\left[\left(\sum_{i=1}^{k}(x_i - \mu_i)\phi_i'(\mu)\right)^2\right]$$

This can be shown to be approximately equal to

$$\text{var}[\phi] \approx \sum_{i=1}^{k}(\phi_i'(\mu))^2\,\text{var}[x_i] + \sum_{i,j|i\neq j}^{k}\phi_i'(\mu)\phi_j'(\mu)\,\text{cov}[x_i x_j] \quad \text{D.45}$$

This is simply the sum of the all elements of the covariance matrix, each scaled by the square of the first derivatives. This approximation only uses first derivatives.

Example D.7. Variance of a sum
The variance of the sum of two variables x and y depends not only on their individual variances, but also on their covariance. Before applying the method outlined above, consider a direct calculation. The variance can be written as

$$\text{var}[x + y] = E[((x + y) - E[x + y])^2]$$

In keeping with the definition of a variance, this is the expected squared deviation from the mean $E[x + y]$. To see this more clearly represent, this sum as the function $\phi(x,y) = x + y$. Multiplying through, we have

$$E[(\phi - E[\phi])^2] = E[(\phi - E[\phi])(\phi - E[\phi])]$$

We can apply the same algebra used to obtain Equation D.9 to demonstrate that this is the difference between the second moment and the first moment squared,

$$= E[\phi^2] - E[\phi]^2$$

Now substitute for $\phi = x + y$,

$$= E[(x + y)^2] - (E[x] + E[y])^2$$

Expanding both terms,

$$= E[x^2 + 2xy + y^2] - (E[x]^2 + 2E[x]E[y] + E[y]^2)$$
$$= E[x^2] + 2E[xy] + E[y^2] - E[x]^2 - 2E[x]E[y] - E[y]^2$$

Now recall that

$$var[x] = E[x^2] - E[x]^2$$
$$var[y] = E[y^2] - E[y]^2$$

and

$$cov[x,y] = E[xy] - E[x]E[y]$$

Substituting for the variances and covariance we have the variance of the sum,

$$var[x + y] = var[x] + var[y] + 2cov[x,y] \qquad \text{D.46}$$

If the covariance is positive, the variance of the sum is greater than the sum of the variances, and vice versa.

To apply the approximation D.45, note the following:

$$\phi(\mu) = \mu_x + \mu_y$$

$$\phi_1'(\mu) = \phi_2'(\mu) = 1$$

Then, by Equation D.45 we have

$$var[\phi] \approx \sum_{i=1}^{k} (\phi_i'(\mu))^2 \, var[x_i] + \sum_{i,j,|i \neq j} \phi_i'(\mu)\phi_j'(\mu) \, cov[x_i, x_j]$$

$$\approx 1 \cdot var[x] + 1 \cdot var[y] + 2 \, cov[x, y]$$

Thus, we see that the approximation, in this instance, is exact.

The foregoing example is readily extended to the sum of n random variables, in which case

$$\phi(\mu) = \sum_{i=1}^{n} \mu_i$$

$$\phi_i'(\mu) = 1 \qquad i = 1, \ldots, n$$

and

$$\text{var}[\phi] = \sum_{i=1}^{n} \text{var}[x_i] + \sum_{i,\,j|i\neq j}^{k} \text{cov}[x_i, x_j]$$

which is simply the sum of the elements of the covariance matrix for all x_i.

D.5.4 A Variable Change

The Real and Ellner (1992) Example D.6 included the common problem of translating the variance of a random quantity to some function of that quantity. In that example we determined how the variability in allocation to reproduction affected the mean (expected) number of offspring. In many cases we need to know more than just the effect on the mean (or even the variance). If we know the full density of a random quantity x, $p(x)$, we can often use a *variable change* to determine the full density of a function $\phi(x)$,

$$f(\phi) = p(x)\left|\frac{dx}{d\phi}\right| \qquad\qquad \text{D.47}$$

In other words, the density $f(\phi)$ is determined by the density of x divided by the rate at which ϕ changes with respect to x. The derivative is termed the *Jacobian*. There is a direct extension for multivariate distributions. There are some restrictions on use of this method. The function ϕ must be differentiable (smooth), and it must be strictly increasing or decreasing.

Example D.8. Reexamine the Real and Ellner (1992) Example D.6 under the assumption that the random variates are normally distributed,

$$\varepsilon_i \sim N(0, \sigma_\varepsilon^2)$$

Based on the previous analysis the mean and variance for x is

$$E[x] = Ab$$

$$\text{var}[x] = A^2 \sigma_\varepsilon^2$$

If x_i is normally distributed, we expect its distribution to have these moments, that is,

$$p(x) = N(x|Ab, A^2 \sigma_\varepsilon^2)$$

Because the function is linear,

$$x_i = A(b + \varepsilon_i)$$

and the stochastic term is normal, then x_i is normally distributed. I use the variable change (Equation D.47) to show that this is indeed the case. Write Equation D.47 as

$$f(x) = p(\varepsilon)\left|\frac{d\varepsilon}{dx}\right|$$

The necessary quantities are $\frac{dx}{d\varepsilon} = A$ and $\varepsilon = \frac{x}{A} - b$. Given the normal distribution of ε, Equation D.47 becomes

$$f(x) = \frac{N(\varepsilon|0,\sigma_\varepsilon^2)}{A} = \frac{1}{A} N\left(\frac{x}{A} - b\Big|0, \sigma_\varepsilon^2\right)$$

We need to write this function in terms of x rather than $\frac{x}{A} - b$. Write out this normal distribution,

$$f(x) = \frac{1}{A} \times \frac{1}{\sqrt{2\pi}\sigma_\varepsilon}\exp\left[-\frac{\varepsilon^2}{2\sigma_\varepsilon^2}\right] = \frac{1}{A} \times \frac{1}{\sqrt{2\pi}\sigma_\varepsilon}\exp\left[-\frac{\left(\frac{x}{A} - b\right)^2}{2\sigma_\varepsilon^2}\right]$$

Now rearrange the exponent

$$f(x) = \frac{1}{\sqrt{2\pi}A\sigma_\varepsilon}\exp\left[-\frac{(x - Ab)^2}{2A^2\sigma_\varepsilon^2}\right]$$

$$= N(x|Ab, A^2 \sigma_\varepsilon^2)$$

to obtain the normal distribution with mean Ab and variance $A^2\sigma^2$ as expected.

Appendix E

Basic Life History Calculations

In this appendix I provide background on some of the standard life history calculations that link population growth to underlying demography. I include this material, because it can be relevant and confusing for ecologists applying statistical methods in this book. Material from this appendix is used in Chapters 2, 9, and 10.

E.1 Age Structure and Population Growth

E.1.1 Life Tables for Age Structure

Life tables contain demographic information, including reproduction and survivorship. Table E.1 contains the northern spotted owl (NSO) demographic rates up to ten years of age from Chapter 2. The table does not contain data—these are calculated values based on the model described in Section 2.8. For ages greater than two years, the table simply follows the model assumption that the fecundity schedule (Equation 3.1) and survivorship (Equation 3.6) remain constant. The assumption that adult survival rate is constant is a default assumption made in the absence of age-specific data. It results in the geometric tail of the survival function (Figure 2.7a). Likewise, the assumption of constant fecundity suggests that the model may overemphasize the contribution of old age classes to population growth rate (e.g., Lande 1987).

E.1.2 Net Reproductive Rate

Population growth can be determined from the elements of a structured population model as $n_{t+1} = n_t \lambda$ (Section 2.8). Before solving the problem involving λ, I address a simpler one that is equally valuable and will help with the more complex one. The simpler question is: "How many female offspring can we expect a female to produce in her lifetime?" This number of offspring is

TABLE E.1.
Life Table for the First Ten Years for Spotted Owl

Stage	x	s_x	l_x	m_x	$l_x m_x$
juvenile	1	0.159	1.0	0.0	0
subadult	2	0.868	0.1590	0.0	0
adult	3	0.868	0.1380	0.382	0.0527
	4	0.868	0.1197	0.382	0.0458
	5	0.868	0.1040	0.382	0.0397
	6	0.868	0.0903	0.382	0.0345
	7	0.868	0.0783	0.382	0.0299
	8	0.868	0.0680	0.382	0.0260
	9	0.868	0.0590	0.382	0.0225
	10	0.868	0.0512	0.382	0.0196

Note: Parameter estimates from McKelvey et al. (1993).

termed the *net reproductive rate*, R_0. Clearly, if the population is to replace itself, then each female must produce at least one female offspring. So if λ is at least one, then R_0 must also be at least one. If females of the population do not reproduce, then λ and R_0 *both* equal zero. Otherwise, λ and R_0 can differ in value, because they can refer to different time scales; R_0 is measured over a generation, whereas λ is defined for the implicit time step in the model.

To calculate R_0, first consider the effect of fecundity. A female that survives through age x produces $\sum_{i=1}^{x} m_i$ offspring. This summation is not the expected reproduction for all individuals, because many will not survive to age x. There is a probability of mortality associated with each age x that downweights the contribution of each age by survivorship. The expected lifetime reproductive output is

$$R_0 = \sum_{x=1}^{\infty} m_x l_x \qquad \text{E.1}$$

The fecundity of each age class is weighted by the fraction expected to survive to that age.[1] Potential offspring production that comes late in life makes a decreasing contribution to R_0, because the expectation of survival decreases with age (Figure 2.14b). For this initial estimate of net reproductive rate I used the assumption of a constant mortality rate for all ages beyond the juvenile stage (e.g., Lande 1988).

---

[1] From Appendix D.5.1 you may recognize R_0 to be the expectation of a function, that function being the fecundity schedule m_x.

The population can maintain itself if R_0 is equal to one. Then, on average, each female replaces herself with a reproductive output of one female offspring. The result for these parameter values, $R_0 = 0.399$ (Section E.2), indicates that the population is in decline. Reproductive output is far too low to ensure persistence.

E.1.3 Generation Time

To understand how rapidly the population will decline, we need a time scale. Thus far, we have a model for growth in terms of R_0 offspring per generation. We need to rescale growth from a generation to some standard unit of time. Over the course of a generation, the population density changes by a factor of R_0. For a generation of length T,

$$n_{t+T} = n_t R_0 \qquad\qquad \text{E.2}$$

A comparison with the growth equation

$$n_{t+1} = n_t \lambda$$

shows that λ and R_0 are equal if time is measured in units of generation T. Thus, the relationship is

$$\lambda = R_0^{1/T} \qquad\qquad \text{E.3}$$

For one time increment

$$n_{t+1} = n_t \lambda = n_t R_0^{1/T}$$

The length of a generation is measured in several ways. It is important to know these definitions, because they produce different estimates depending on how fast the population is growing (Leslie 1966; Cochran and Ellner 1992):

1. *Time required to change by a factor of* R_0—From Equation E.3, calculate the generation time to be

$$T = \frac{\ln R_0}{\ln \lambda} \qquad\qquad \text{E.4}$$

2. *Mean age at which a cohort produces offspring*—The $l_x m_x$ schedule can be normalized to produce a distribution. This is obtained by dividing the schedule by its sum over all classes:

$$\Pr\{\text{offspring has an age } x \text{ parent}\} = \frac{l_x m_x}{\displaystyle\sum_{x=1}^{\infty} l_x m_x} = \frac{l_x m_x}{R_0}$$

This is a discrete distribution (Appendix D.2.2). R_0 is the *normalization constant* for this distribution. The mean of this distribution is obtained as a *weighted average*, that is, by weighting each age x by its contribution, $l_x m_x$,

$$T = \frac{1}{R_0} \sum_{x=1}^{\infty} x l_x m_x \qquad \text{E.5}$$

This is the expected age of parents producing the offspring.[2]

3. *Mean age of parents in a population at stable age distribution*—This estimate is the expected age of parents producing offspring in a population having a stable age distribution,

$$T = \frac{\sum_{x=1}^{\infty} x \lambda^{-x} l_x m_x}{\sum_{x=1}^{\infty} \lambda^{-x} l_x m_x} = \sum_{x=1}^{\infty} x \lambda^{-x} l_x m_x \qquad \text{E.6}$$

As for the previous estimate, the denominator is a normalization constant that ensures that the distribution sums to one. In this case, the normalization constant is Lotka's equation and *is* equal to one (see below).

When the population is at constant density, $\lambda = R_0 = 1$, and E.5 and E.6 yield the same answer,

$$T = \sum_{x=1}^{\infty} x l_x m_x$$

Equation E.5 does not assume that the population has achieved a stable age distribution. The first method (Equation E.4) is sometimes used, but it may not be very meaningful. Consider that for the special case ecologists most like to analyze ($R_0 = \lambda = 1$), it yields no answer at all (as pointed out by Leslie [1966]).

E.2 Application: Lotka's Equation and Life History Calculations for NSO

E.2.1 Net Reproductive Rate for the NSO Example

The net reproductive rate can be solved in terms of life history parameters. This illustration provides some practice with treatment of series. The approach follows Lande (1987) for NSO. First simplify the survivor function. Equation E.1 is the general definition of R_0. For the NSO model described thus far, $m_x = 0$

---

[2] This is Equation D.15 from Appendix D.3.1, $T = E[x] = \Sigma x \, p(x)$, where $p(x) = l_x m_x / R_0$.

at values of $x \leq \alpha$, where $\alpha = 2$ is the maturation age and $m_x = b$ otherwise. The summation looks like this:

$$R_0 = m_1 l_1 + m_2 l_2 + m_3 l_3 + \ldots$$
$$= 0 \times l_1 + 0 \times l_2 + b \times l_3 + \ldots$$
$$= 0 + 0 + 0.382 \times 0.138 + \ldots$$

From Equation 2.13, survivorship to the beginning of the reproductive age at $\alpha = 3$ years is $l_\alpha = l_2 = s_0 s$. Survivorship beyond age $\alpha = 2$ can be written as

$$l_x = l_\alpha s^{x - \alpha} \qquad\qquad \text{E.7}$$

The first terms of the summation do not contribute to R_0, because $m_1 = m_2 = 0$, so we need only sum over ages at or above maturation age,

$$R_0 = b \sum_{x=\alpha}^{\infty} l_x$$

To solve this, substitute for the survivor function (Equation E.7),

$$R_0 = b \sum_{x=\alpha}^{\infty} l_x = b l_\alpha \sum_{x=\alpha}^{\infty} s^{x-\alpha} = \frac{b l_\alpha}{s^\alpha} \sum_{x=\alpha}^{\infty} s^x$$

The finite progression can be rewritten using the series,

$$\sum_{x=\alpha}^{\infty} s^x = \frac{s^\alpha}{1 - s} \qquad\qquad \text{E.8}$$

From Table E.1, the probability of surviving to maturation age is

$$l_3 = 0.159 \times 0.868 = 0.138$$

Together with Equation E.8 we have

$$R_0 = \frac{b l_\alpha}{1 - s} = \frac{0.382 \times 0.138}{1 - 0.868} = 0.399$$

This is the expected number of female offspring over the lifetime of the parent.

E.2.2 Generation Time

Each of the generation times (E.4, E.5, E.6) gives a somewhat different answer. For the NSO example, the generation time from Equation E.6 can be written as

$$T = \frac{\sum_{x=1}^{\infty} x \lambda^{-x} l_x m_x}{\sum_{x=1}^{\infty} \lambda^{-x} l_x m_x} = \frac{l_\alpha/_{s^\alpha} \sum_{x=1}^{\infty} x \lambda^{-x} s^x}{l_\alpha/_{s^\alpha} \sum_{x=1}^{\infty} \lambda^{-x} s^x} = \frac{\sum_{x=\alpha}^{\infty} x (s/\lambda)^x}{\sum_{x=\alpha}^{\infty} (s/\lambda)^x}$$

To solve this, let $q = (s/\lambda)^x$, and use the series

$$\sum_{x=\alpha}^{\infty} xq^x = \frac{q^\alpha}{1-q}\left(\alpha + \frac{q}{1-q}\right) \tag{E.9}$$

Using Equation E.9 for the numerator and Equation E.8 for the denominator we obtain

$$T = \frac{\sum_{x=\alpha}^{\infty} xq^x}{\sum_{x=\alpha}^{\infty} q^x} = \frac{\frac{q^\alpha}{1-q}\left(\alpha + \frac{q}{1-q}\right)}{\frac{q^\alpha}{1-q}} = \alpha + \frac{s}{\lambda - s} \tag{E.10}$$

We do not yet know the value of λ needed to calculate T, but if it is close to one, then, $T \sim 9.6$ years. We could iteratively converge on a closer approximation by reestimating λ using Equation E.3. This iterative estimate converges to $T = 15$ years and $\lambda = 0.94$. The generation time from Equation E.5 is the answer we obtained for $\lambda = 1$ in Equation E.10.

E.2.3 Population Growth from Lotka's Equation

The relationship between demographic rates and overall population growth is given implicitly by Lotka's equation, which says that

$$1 = \sum_{x=1}^{\infty} \lambda^{-x} l_x m_x \tag{E.11}$$

We can solve this expression for NSO. Assuming a value of $\alpha = 2$ years as the age of first reproduction, and that fecundity is constant thereafter, then Lotka's Equation (E.11) becomes

$$1 = b\sum_{x=\alpha}^{\infty} \lambda^{-x} s^{x-1} = bl_\alpha \sum_{x=0}^{\infty} \lambda^{-x} s^{x-\alpha} = \frac{bl_\alpha}{s^\alpha} \sum_{x=\alpha}^{\infty} (s/\lambda)^x$$

Using the progression from Equation E.9, where $q = s/\lambda$, we obtain

$$1 = \frac{bl_\alpha}{\lambda^2(1 - s/\lambda)}$$

For $\alpha = 2$, this equation is quadratic (terms in λ^2) and has roots given by the quadratic equation,

$$\lambda = \frac{s \pm \sqrt{s^2 + 4bl_\alpha}}{2}$$

There is a positive and real root at $\lambda = 0.93$. The generation time obtained from Equation E.4 is 11.8 years. Recall that the fractional annual rate of decline is

$$r = \ln\lambda = -0.074$$

or approximately 7 percent per year. If we truncate the model (assume that all owls die by age A), Lotka's equation becomes

$$1 = \frac{bl_\alpha[1 - (s/\lambda)^{A-\alpha+1}]}{\lambda^\alpha(1 - s/\lambda)}$$

Lande (1988) examined how the assumption of finite age affected demographic calculations. If owls have a maximum life span of only about $A = 10$ years, then we might approximate this effect by truncating the characteristic equation at, say, ten years (instead of summing out to infinity). This change has a substantial effect on the growth rate, with various parameter estimates predicting λ as low as 0.77.

E.2.4 Stable Age Structure

If a population grows at a constant rate, it eventually reaches a stable age structure, where the relative abundance of age classes remains constant. This age structure is given by

$$w_x = \frac{l_x}{\sum\limits_{x}^{\infty} l_x} \qquad\qquad \text{E.12}$$

The stage distribution is proportional to the survivor function (Figure.3.3a).

E.3 Life History Calculations from Stage Structure

Demographic analysis is often done using matrixes (Caswell 2001). The approach is most powerful for linear models, in which case the transition elements of the matrix are fixed constants, and analysis involves standard linear algebra (Appendix C). Linear models are most appropriate for short time intervals and for populations at a structural equilibrium. They are of less utility for models that include density dependence and environmental variability, considerations that become increasingly important at longer time scales (Grant and Benton 2000). In the linear case, a transition matrix can be subjected to eigenanalysis, which extracts summary information from the transition matrix that bears on population growth rate and population structure (Appendix C).

E.3.1 The Leslie Matrix and a Related Stage-Structured Model

A Leslie matrix is the transition matrix used to project age structure. The matrix summarizes population vital rates in a manner that facilitates analysis of population growth. As mentioned in Section 2.2.4, the elements of a_{ij} of Leslie matrix **A** are the per capita contributions of age j (columns) to age i (rows) in the next time step. The elements are mostly zeros, because transition can only occur by births and by aging. For instance, age 5 cannot contribute to age 4 or to age 7 in one time step. New births all enter in the first row of the matrix, the contributions of each class to the youngest class. This row is the fertility schedule, m_x. Age-specific survivorship, the probability of not dying in age interval x,

$$s_x = 1 - \rho_x$$

enters on the subdiagonal, being the contribution of age j to age $i = j + 1$ in the course of one time step. For three age classes, the matrix looks like this,

$$\mathbf{A} = \begin{bmatrix} m_1 & m_2 & m_3 \\ s_1 & 0 & 0 \\ 0 & s_2 & 0 \end{bmatrix} \qquad \text{E.13a}$$

The zero in the lower right-hand corner means that all individuals die after the third age class. This particular matrix can be easily modified for the NSO example by providing for the fraction s that remain within the adult stage:

$$\mathbf{A} = \begin{bmatrix} 0 & 0 & b \\ s_1 & 0 & 0 \\ 0 & s & s \end{bmatrix} \qquad \text{E.13b}$$

This matrix differs from Equation E.13a in having a nonzero element in the lower right-hand corner. The element $a_{3,3} = s$ says that individuals in class 3 can survive (remain in the adult class). For the NSO, the following matrix has been used in several analyses,

$$\mathbf{A} = \begin{bmatrix} 0 & 0 & 0.382 \\ 0.159 & 0 & 0 \\ 0 & 0.868 & 0.868 \end{bmatrix}$$

Equation E.13a is a Leslie matrix. Equation E.13b is a stage-structured model, because there is at least one stage within which individuals can reside for multiple time steps. A diagram of this life cycle is shown in Figure E.1.

E.3.2 Matrix Projection

The matrix framework helps us analyze growth. For instance, we can project the population forward using the relation

$$\mathbf{n}_{t+1} = \mathbf{A}\mathbf{n}_t$$

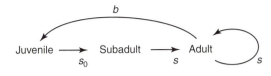

FIGURE E.1. The NSO model described by Equation 2.10.b. It is not a strict Leslie matrix, because individuals can remain in at least one stage (adult) for multiple time steps.

with rules of matrix multiplication (Appendix C). For the matrix E.13b the one-step ahead projection is

$$\begin{bmatrix} n_1 \\ n_2 \\ n_3 \end{bmatrix}_{t+1} = \begin{bmatrix} 0 & 0 & b \\ s_1 & 0 & 0 \\ 0 & s & s \end{bmatrix} \begin{bmatrix} n_1 \\ n_2 \\ n_3 \end{bmatrix}_t$$

$$= \begin{bmatrix} bn_3 \\ s_1 n_1 \\ s(n_2 + n_3) \end{bmatrix}_t$$

Because the model is linear, we can project in a manner analogous to that used for the nonstructured model

$$\mathbf{n}_{t+2} = \mathbf{A}^2 \mathbf{n}_t$$

and, continuing,

$$\mathbf{n}_t = \mathbf{A}^t \mathbf{n}_0$$

Like the nonstructured linear model of Equation 2.4, the result is exponential growth or decay. If the population grows exponentially, then each stage also grows exponentially, for example, $\lambda n_{x,t}$. In other words, there may be a constant λ that could apply to the population as a whole, which, in some sense, summarizes the impact of detailed, stage-specific information in $\mathbf{A}$. A comparison with the nonstructured model suggests the relation

$$\lambda \mathbf{n}_t = \mathbf{A} \mathbf{n}_t \qquad \text{E.14}$$

indicating that λ is an eigenvalue and $\mathbf{n}_t$ is the corresponding *right eigenvector* of $\mathbf{A}$. Note that the scalar λ has an effect like the matrix $\mathbf{A}$. Thus, it represents a useful summary of $\mathbf{A}$. Such relationships are valuable, because they submit to eigenanalysis (Appendix C). The relative contribution of different life history stages to λ can be determined from the elasticities (Box E.1), which come from the transition matrix.

The calculations for elasticities (Box E.1) apply to age-structured and stage-structured models. For the northern spotted owl, elasticities are highest for adult survival, because recruitment is low. Long life is often among the most important elements of growth for populations in decline. Here is the full elasticity matrix for the spotted owl:

	juvenile	subadult	adult
juvenile	0	0	0.0581
subadult	0.0581	0	0
adult	0	0.0581	0.826

Note how these proportionate effects on growth rate sum to one. A summary of elasticities is provided in Box E.1.

E.4 More Complex Stage Structure

The Leslie model used for analysis of age-structured data is a type of projection matrix having a first row that contains the fecundity schedule and a subdiagonal containing age-specific survival. Because classes are defined by age, the Leslie matrix is sparse; all other elements equal zero, as transitions can only occur by new births (contributions from a class j to class 1) and by aging (from class j to $j + 1$).

Age-structured models can be difficult to parameterize, because age may be unknown. If cohorts cannot be followed over time, horizontal data can be used to parameterize the transition matrix. In other words, examine short-term transitions for many stages simultaneously. For most plants and animals we are generally limited to a few recognizable classes, such as insect instars or plant life stages (seed, rosette, flowering adult, and so forth), or more arbitrary groupings, such as those based on size.

Stage-structured models are a generalization of the Leslie matrix where classes are defined by stage rather than age (Figure E.2). Like the Leslie matrix, a stage-structured model describes the probabilities of transitions in one time step. When I added a nonzero element to the bottom right-hand corner of the Leslie matrix for NSO, I introduced stage structure, because I allowed that adults could include individuals of any age greater than two years. The stage transition matrix contains fewer zeros than the age matrix. Diagonal elements result where stages span multiple time steps. The nonzero elements of a stage transition matrix **A** may include not only the first row (offspring produced by stage j) and the subdiagonal (the fraction of individuals in stage j that survive and grow to stage $j + 1$), but also the diagonal (individuals in stage j that survive but do not grow to stage $j + 1$), and potentially others (e.g., the probability that a plant in stage j will change to a rosette stage $i \neq j$).

The advantage of stage structure over the Leslie matrix is ease of identification (we do not need to know ages). A disadvantage is the greater degree of uncertainty associated with transitions among stages that are imprecisely defined.

The analysis of stage-structured models entails more thought than is needed for age-structured models. As with the Leslie matrix, we can use eigenanalysis to

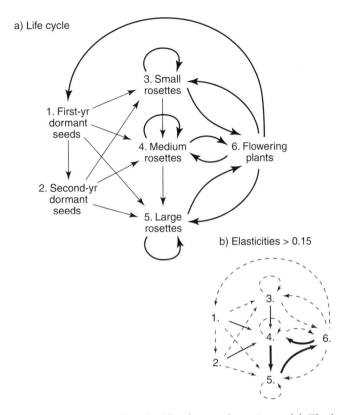

FIGURE.E.2. A life cycle diagram described by the teasel matrix model. The loop involving thick arrows in *b* contributes most to population growth rate (Box E.1).

determine growth rate and equilibrium stage structure. We may also desire information related to age. Despite the fact that stage-structured models do not explicitly follow population age, the stage transitions are defined over a common time scale; age is implicit.

Calculations of survival and fecundity schedules, age of first reproduction, and expected lifetimes are straightforward from age-structured models, making use of matrixes. This material has application to Chapters 2 and 8. In Chapters 9 and 10, I consider data modeling for capture-recapture data and population spread, respectively. The implicit information on aging contained in the stage-structured model can also be extracted with some linear algebra. The relationships here summarize Cochran and Ellner (1992) and especially Caswell (2001), the latter providing a more straightforward approach for many of the calculations. These references should be consulted for elaboration. Here I focus on a few useful relationships, beginning with two familiar data sets.

The standard teasel example (Werner and Caswell 1977) represents a contrasting life history, with a network of potential connections among stages (Figure. E.2a). The transition matrix is

A =

	1	2	3	4	5	6
1	0	0	0	0	0	322
2	0.966	0	0	0	0	0
3	0.013	0.01	0.125	0	0	3.450
4	0.007	0	0.125	0.238	0	30.2
5	0.008	0	0.038	0.245	0.167	0.862
6	0	0	0	0.023	0.750	0

with the dominant eigenvalue of $\lambda_1 = 2.33$ and an elasticity matrix of

	1	2	3	4	5	6
1	0	0	0	0	0	0.0675
2	0.000258	0	0	0	0	0
3	0.000808	0.000258	0.000146	0	0	0.0015
4	0.00767	0	0.00257	0.0275	0	0.231
5	0.0588	0	0	0.1900	0.0227	0.0443
6	0	0	0	0.0514	0.293	0

Caswell points out that the loop involving flowering plants, medium rosettes, and large rosettes accounts for 73 percent of λ (Figure E.2b). Thus, this portion of the life cycle makes the largest contribution to growth rate. Of course, other stages remain critical, as this vegetative reproduction would not allow for dispersal.

E.5 Application: Analysis of Age from Stage Structure

Methods for analysis of stage-structured linear models begin by decomposing the transition matrix **A** into contributions from fecundity **F** versus other types of transitions **S**,

$$\mathbf{A} = \mathbf{F} + \mathbf{S}$$

New individuals enter the population through **F**. Elements of the fecundity matrix **F** describe the number of stage-i offspring produced by a stage-j individual. Elements of the transition matrix **S** describe the probability that a stage-j individual moves to stage i in one time step. Both **F** and **S** have the same dimensions as **A**. This decomposition is done, because analysis of age-related events involves transitions that occur during the life cycle, that is, after birth and contained in **S**, whereas matrix **F** is used to derive fecundity schedules.

Example E.1. The decomposition of the projection matrix for northern spotted owl is

S =

	1	2	3
1	0	0	0
2	0.159	0	0
3	0	0.868	0.868

and

F =

	1	2	3
1	0	0	0.382
2	0	0	0
3	0	0	0

Note that the fertility matrix contains a single nonzero element.

Example E.2. The teasel life cycle has transitions not involving births in the matrix

S =

	1	2	3	4	5	6
1	0	0	0	0	0	0
2	0.966	0	0	0	0	0
3	0.013	0.01	0.125	0	0	0
4	0.007	0	0.125	0.238	0	0
5	0.008	0	0.038	0.245	0.167	0
6	0	0	0	0.023	0.750	0

Fecundity involves several types of offspring. Because each of these offspring types involves new individuals (as opposed to transitions of the same individuals from one stage to another), they are included in the F matrix:

F =

	1	2	3	4	5	6
1	0	0	0	0	0	322
2	0	0	0	0	0	0
3	0	0	0	0	0	3.45
4	0	0	0	0	0	30.2
5	0	0	0	0	0	0.862
6	0	0	0	0	0	0

The fecundity matrix is a column vector containing all of the different types of offspring that a flowering plant may beget.

Survival is assessed from the *fundamental matrix* $\mathbf{N}$, elements of which represent the expected duration in stage i given that an individual starts in stage j. It depends only on transitions $\mathbf{S}$ and is solved as

$$\mathbf{N} = (\mathbf{I} - \mathbf{S})^{-1}$$

The matrix of variances is

$$V_N = (2\mathbf{I} \circ \mathbf{N} - \mathbf{I})\mathbf{N} - \mathbf{N} \circ \mathbf{N}$$

The Hadamard product used here is simply the product of elements from the two matrixes that have the same position (Appendix C).

All spotted owls begin life in the same stage and pass through stages in sequence. The fundamental matrix,

$\mathbf{N} =$

	1	2	3
1	1	0	0
2	0.159	1	0
3	1.05	6.58	7.58

has a coefficients of variation

$cv(\mathbf{N}) =$

	1	2	3
1	0	—	—
2	2.30	0	—
3	3.54	1.07	0.93

Because all owls begin life in stage 1, we might focus on the first column of $\mathbf{N}$. Stage 1 has duration of one, because we assume they are already in that class, and the stage has a duration of one year. Whether or not they survive, by definition they reside here one time increment. Residence times in later stages reflect survivorship. The average owl might expect to reside in the adult stage slightly more than one year (1.05), as this is an average over the many that die before reaching this stage and the few that might live long. Note the coefficient of variation for this stage is 3.54 years.

By contrast, teasel has the fundamental matrix

$\mathbf{N} =$

	1	2	3	4	5	6
1	1	0	0	0	0	0
2	0.966	1	0	0	0	0
3	0.0259	0.0114	1.14	0	0	0
4	0.0134	0.00187	0.187	1.31	0	0
5	0.0147	0.00107	0.107	0.386	1.2	0
6	0.0114	0.000848	0.0848	0.32	0.9	1

The fundamental matrix has zeros for stage transitions that cannot be reached except by new births. Thus, for the individual that starts life in stage 3 (small rosette), the expected residence times in stages 1 and 2 (seeds) are zero; a stage 3 individual will not pass through these stages. Individuals starting life in later life history stages are expected to reside longer in the higher stages, because they do not experience the risks of earlier stages.

The expected time until death given that an individual begins life in stage j is given by column sums of $\mathbf{N}$,

$$E[T_j] = \sum_{i=1}^{m} \mathbf{N}_{i,j}$$

with variance

$$\mathrm{var}[T_j] = \sum_{i=1}^{m} (2\mathbf{N}^2 - \mathbf{N})_{i,j} - \sum_{i=1}^{m} \mathbf{N}_{i,j} \circ \sum_{i=1}^{m} \mathbf{N}_{i,j}$$

For spotted owl, the means and standard deviations for each stage are

E[T_j]	2.20	7.58	7.58
sd[T_j]	3.95	7.06	7.06

For teasel, we obtain

E[T_j]	2.03	1.02	1.52	2.02	2.10	1
sd[T_j]	0.29	0.182	1.01	1.21	0.575	0

The survivor function is also based on the matrix $\mathbf{S}$. For individuals beginning life in stage j we have

$$l_{x,j} = \sum_{i=1}^{m} \mathbf{S}_{i,j}^{x} \qquad \text{E.15}$$

Note that by summing over i we account for survival of individuals that have made transitions to other stages. This is the total probability of survival of individuals that began life in stage j regardless of where they reside by age x. The example for teasel demonstrates the different survival functions that apply to individuals that begin life in different stages (Figure E.3).

The fecundity schedule is represented by a matrix for each age. At age x, the number of stage i offspring produced by an individual that began life in stage j is $\mathbf{B}_x$. To obtain the fraction of individuals in stage i that began life in stage j determine $\mathbf{S}^x$ (the probability of surviving to i given that life began in j) and divide each column by its sum,

$$\mathbf{S}^x \left[Diag\left(\sum_{i=1}^{m} \mathbf{S}_{i,j}^x \right) \right]^{-1}$$

This notation looks more imposing than it is. First, determine the column sums of $\mathbf{S}^x$ (the summation) and create a diagonal matrix: these sums occupy the diagonal, with zeros for all off-diagonal elements. At this point we have

$$Diag\left(\sum_{i=1}^{m} \mathbf{S}_{i,j}^x \right) = \begin{bmatrix} \sum_{i=1}^{m} \mathbf{S}_{i,1}^x & 0 & \cdots & 0 \\ 0 & \sum_{i=1}^{m} \mathbf{S}_{i,2}^x & \cdots & 0 \\ \vdots & \vdots & \ddots & \vdots \\ 0 & 0 & \cdots & \sum_{i=1}^{m} \mathbf{S}_{i,m}^x \end{bmatrix}$$

Because the matrix is diagonal, inversion produces a new diagonal matrix, with each diagonal element replaced by its reciprocal.

The fecundity matrix for age x scales this quantity by the fecundity schedule

$$\mathbf{B}_x = \mathbf{F}\mathbf{S}^x \left[Diag\left(\sum_{i=1}^{m} \mathbf{S}_{i,j}^x \right) \right]^{-1}$$

For teasel, the fecundity schedule depends on the stage at which an individual begins life (Figure E.3b). First-year dormant seeds require some time to reach the flowering plant stage, whereas plants beginning life in later stages can be expected to reach the flowering stage sooner.

For a stage-structured model, estimating the *mean age of first reproduction* requires that we average over the different pathways individuals may take on the way to reproduction. This average time to first reproduction depends on the initial stage, so we will have one estimate for each initial stage. We condition on the event that individuals survive to reproductive age; we do not wish to average over individuals that never reproduce. This requires two sets of quantities. First is a fundamental matrix that does not allow individuals to leave the reproductive state. The new fundamental matrix requires a version of the transition matrix $\mathbf{S}$ with the exception that all elements in columns for reproductive

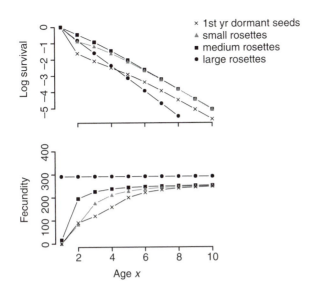

FIGURE. E.3. Survival and fecundity schedules for teasel beginning life in four different stages.

stages are zeros. This creates an absorbing state for reproduction; once an individual enters a reproductive state it does not leave. This new transition matrix is $\mathbf{S'}$, and it has a corresponding fundamental matrix

$$\mathbf{N'} = (\mathbf{I} - \mathbf{S'})^{-1}$$

This is the expected duration in stage i prior to reproduction for an individual that began life in stage j.

The second part of this analysis is a matrix $\mathbf{M}$ with elements that represent the probability of passing from nonreproductive to the two absorbing states, which are reproductive stages and mortality. The first row $\mathbf{m}_1$ is the probability of death in stage i

$$m_{1j} = 1 - \sum_{i=1}^{m} \mathbf{S'}_{i1}$$

The second row, $\mathbf{m}_2$ contains coefficients indicating whether or not reproduction occurs in each state, with ones for reproductive stages and zeros for nonreproductive stages. The probability of reproduction before death for individuals that began life in stage j is the product

$$\mathbf{b} = \mathbf{M}^{\mathrm{T}}\mathbf{N'}$$

The second row of $\mathbf{b}$ is the probability of reproduction before death.

From these quantities we create a new transition matrix that applies to individuals that reproduce before dying,

$$\mathbf{S''} = Diag(\mathbf{b}_2)^{-1} \, \mathbf{S'} \, Diag\,(\mathbf{b}_2)$$

The ij^{th} element of $\mathbf{S}''$ is

$$\mathbf{S}''_{ij} = \mathbf{S}'_{ij} \cdot \frac{\mathbf{b}_j}{\mathbf{b}_i}$$

Postreproductive stages will have $\mathbf{b}_j = 0$ and, thus, are ignored.

We now proceed with calculation of expected time to reproduction, with a new fundamental matrix,

$$\mathbf{N}'' = (\mathbf{I} - \mathbf{S}'')^{-1}$$

followed by summation over columns,

$$E[\alpha_j] = \sum_{i=1}^{m} \mathbf{N}''_{i,j} \qquad \text{E.16}$$

Example E.3. Expected reproductive age for the northern spotted owl model
The transition matrix for owls with an absorbing state at the reproductive stage is simply the matrix $\mathbf{S}$, with all zeros in the last column,

$\mathbf{S}' =$

	1	2	3
1	0	0	0
2	0.159	0	0
3	0	0.868	0

with matrix

$$\mathbf{M} = \begin{bmatrix} 0.841 & 0.132 & 0 \\ 0 & 0 & 1 \end{bmatrix}$$

The corresponding fundamental matrix is

$\mathbf{N}' =$

	1	2	3
1	1	0	0
2	0.159	1	0
3	0.138	0.868	1

with the probability that individuals reproduce before dying given by $\mathbf{b}_2$

0.138   0.868   1.000

The transition matrix for individuals that will reproduce before dying is

$S' =$

1	0	0	0
2	0.0253	0	0
3	0	0.753	0

and the expected time to reproduction, for those individuals that will survive to that age is

$$E[\alpha_1] = 1.04$$

Net Reproductive Rate—For an individual beginning life in stage j, net reproductive rate is determined as the expected time in a stage i, given by N_{ij}, multiplied by offspring production F_{ij}. Summed over all possible fates, we have

$$\mathbf{R} = \mathbf{FN}$$

The matrix of $\mathbf{R}$ contains the expected reproduction of stage i offspring by a mother that began life in stage j. The net reproductive rate for the population can be interpreted as the dominant eigenvalue of $\mathbf{R}$ (Caswell 2001).

Example E.4. Net reproductive rate of spotted owls
The multiplication $\mathbf{FN}$ yields the matrix

$\mathbf{R} =$

	1	2	3
1	0.399	2.51	2.89
2	0	0	0
3	0	0	0

having dominant eigenvalue in position 1,1. Note this to be the same result we obtained using Lotka's equation.

Generation Time—The three methods for calculating generation time are available for the stage-structured model. However, complex life cycles confuse the concept, because different types of offspring can have different reproductive values—all offspring are not equal. Cochran and Ellner (1992) suggest weighting reproduction by the reproductive value of offspring.

The first method (Equation E.4), is obtained as the log ratio of dominant eigenvalues for **R** (R_0) and **A** (λ). The second two methods are obtained from the survival and fecundity schedules,

$$T = \frac{\sum_{x=1}^{m} x\, B_{x,1} l_{x,1}}{\sum_{x=1}^{m} B_{x,1} l_{x,1}}$$

and

$$T = \frac{\sum_{x=1}^{m} x\, \lambda^{-x} B_{x,1} l_{x,1}}{\sum_{x=1}^{m} \lambda^{-x} B_{x,1} l_{x,1}}$$

Example E.5. Generation times for spotted owls
The dominant eigenvalues for matrixes **R** (R_0) and **A**(λ) are 0.399 and 0.929, respectively. By the first method, the generation time is thus $T\text{-ln}R_0/\text{ln}\lambda = 12.5$ years. By the second and third methods, we obtain $T = 8.6$ years (unweighted for population growth rate) and $T = 16.1$ years (weighted for population growth rate).

Example E.6. Analysis of the teasel model.
Caswell (2001) demonstrates the analysis of the **R** matrix for teasel

	1	2	3	4	5	6
1	3.66	0.273	27.3	103	290	322
2	0	0	0	0	0	0
3	0.0392	0.00292	0.292	1.1	3.11	3.45
4	0.343	0.0256	2.56	9.65	27.2	30.2
5	0.00979	0.000731	0.0731	0.276	0.776	0.862
6	0	0	0	0	0	0

The expected number of stage j offspring from an individual that begins life in stage 1 (seed) is 3.66, 0.039 small rosettes, and so on. The dominant eigenvalue is 14.4 and is an index for R_0.

E.6 Continuous Time

E.6.1 Continuous Survival

As for the discrete model, *survivorship* is defined as the probability of surviving to age a,

$$l(a) = \Pr\{a_i > a\}$$

where a_i is age of death. The subscript means that survival is treated as a stochastic event associated with individual i. $l(a)$ is the *survivor function*. It is a probability (it is dimensionless—Appendix D). The age-specific mortality rate $h(a)$ is the rate of mortality for an age a individual, and has units of 1/age. The probability of dying in the age interval $(a, a + da)$ given that one has survived that long is written as

$$h(a)da = \Pr\{a_i \in (a, a + da) \mid a_i > a\} \qquad \text{E.17}$$

The bar notation identifies this as a conditional probability. Left of the bar is the probability that age of death a_i occurs in the designated age interval. Right of the bar is the condition that the individual has survived at least as long as the beginning of the interval—age a (Appendix D). This conditional probability is not the same as the probability of dying at age a (see below). It refers only to the individuals still at risk of death at the beginning of the interval.

The relationship between age-specific mortality risk $h(a)$ and the probability of surviving to age a, $l(a)$, is derived as follows. The rate of change in the density of a cohort of age a is given by

$$\frac{dn(a)}{da} = -h(a)n(a)$$

This equation is separable (Appendix B), having the solution

$$n(a) = n_0 \exp\left[-\int_0^a h(x)dx \right]$$

The probability of surviving to age a is often approximated by the fraction of the initial density remaining at age a,

$$l(a) = \frac{n(a)}{n_0} = \exp\left[-\int_0^a h(x)dx \right] \qquad \text{E.18}$$

If age-specific mortality is constant at rate ρ for all ages,

$$h(a) = \rho$$

then survivorship is a simple exponential function,

$$l(a) = e^{-\rho a}$$

You may recognize this as the zero category of a Poisson process, $Pois(0|\rho a)$ (Appendix F), the probability of no events (an event being death) before a given a constant probability ρ. The probability that an individual survives one time increment,

$$l(1) = e^{-\rho}$$

is close to the result one would obtain in discrete time,

$$1 - \rho$$

This is the first two terms of a Taylor expansion of the survivor function (Appendix A). In other words, as the time increment gets small, the discrete result tends to the continuous survivor function

$$\lim_{da \to 0} (1 - \rho da)^{1/da} \to e^{-\rho}$$

The probability density of ages of deaths is the product of the survivor function (probability of surviving to an age a) and the age-specific mortality rate (the fraction of individuals that will die at that age),

$$f(a) = h(a)l(a) \qquad \text{E.19}$$

For a constant mortality rate, this density is exponential,

$$f(a) = h(a)l(a) = \rho e^{-\rho a}$$

The survivor function is also the complement of the distribution function of $f(a)$, because the distribution function describes the total probability of death before age a. Let $F(a)$ be the distribution function of $f(a)$, that is,

$$F(a) = \int_0^a f(x)dx \qquad \text{E.20}$$

(Appendix D). Then

$$l(a) = \int_a^\infty f(x)dx = 1 - F(a) \qquad \text{E.21}$$

Because $f(a)$ is the rate of change in $F(a)$, then

$$\frac{dF(a)}{da} = \frac{-dl(a)}{da} = f(a)$$

Figure E.4 shows examples of survival functions and densities of mortality that result from three different schedules of risk. The three examples are drawn from a Weibull distribution and have scale parameter ρ and shape parameter c (Appendix F). The exponential is a special case ($c = 1$).

Semilog plots (right-hand side of Figure E.4) are useful, because they show schedules on a proportionate basis. The derivative of this curve is per capita change. Constant risk means that $h(a)$ does not depend on age, and the density $f(a)$ is exponential (Figure E.4b). On a log scale $f(a)$ and $l(a)$ are plotted as straight lines, because per capita risk is constant. If risk decreases with age, then $f(a)$ and $l(a)$ are concave up on a semilog plot (Figure E.4a), and vice versa

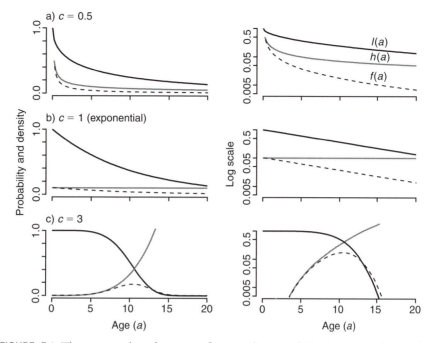

FIGURE. E.4. Three examples of age-specific mortality rate $h(a)$, density of deaths $f(a)$, and survivor functions $l(a)$ based on a Weibull distribution.

(Figure E.4c). The three shapes in Figure E.4 are sometimes termed Type I, Type II, and Type III (Deevey 1947).

The inferred shapes of mortality schedules depend in large part on availability of data for different aged organisms. For longitudinal data sets (following individuals over time), the tail of the distribution is especially difficult to estimate, because sample sizes decline with age. Estimating mortality for cross-sectional data is also difficult, as few populations include many large, old individuals. Exponential models are typically used as a default when information is limited. For many long-lived organisms (e.g., vertebrates and trees), the mortality schedules can resemble a combination of shapes in Figure E.5, with high risk at young and old ages, and low risk in-between. Lee and Carter (1992) show examples of age-specific mortality rates fitted to U.S. census data for different time periods and forecasts of future rates.

For a large cohort, we might view $l(a)$ as the fraction of the initial population that survives to age a. Alternatively, we can view $l(a)$ as the probability that an individual survives to age a. In this case, $h(a)da$ becomes the Bernoulli probability that an individual dies at age a. Estimation is discussed in Chapters 3 and 4.

Each function can be written in terms of the others. For the general case (i.e., mortality rate need not be constant) I provide in Table E.2 a summary of the relationships among three representations of survival in continuous time. For example, given the age-specific mortality rate $h(a)$, solve for the survivor function $l(a)$ by integrating and then exponentiating.

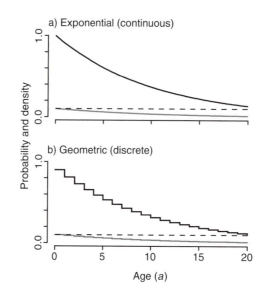

a) Exponential (continuous)

b) Geometric (discrete)

Probability and density

Age (a)

FIGURE. E.5. Comparison of continuous and discrete survival schedules. Symbolism follows Figure E.4.

TABLE E.2.
Relationships Involving the Survival Function

	$h(a)$	$l(a)$	$f(a)$
$h(a)$		$-\dfrac{d}{da}\ln l(a)$	$\dfrac{f(a)}{\displaystyle\int_a^\infty f(x)dx} = \dfrac{f(a)}{l(a)}$
$l(a)$	$\exp\left[-\displaystyle\int_0^a h(x)da\right]$		$\displaystyle\int_a^\infty f(x)dx$
$f(a)$	$h(a)\exp\left[-\displaystyle\int_0^a h(x)dx\right] = h(a)l(a)$	$-\dfrac{dl(a)}{da} = h(a)l(a)$	

We could construct a table analogous to this for discrete time survival models. As with continuous models you may encounter all three formulations in statistical models (e.g., Barry et al. 2003). For the constant risk case, the discrete model is geometric and converges to the exponential with diminishing age intervals (Figure E.5). To summarize survival relationships for discrete and continuous time models, define the following for a model of constant mortality risk ρ:

a age (continuous)
x age (discrete)

TABLE E.3.
Comparison of continuous and discrete time models for survival

Quantity	Continuous	Discrete
Age of mortality	$f(a) = \rho e^{-\rho a}$	$f_x = \rho(1 - \rho)^x$
Survivorship	$l(a) = e^{-\rho a}$	$l_x = (1 - \rho)^x$
Age-specific mortality	$h(a) = \rho$	$h_x = \rho$

The three functions of interest are compared in Table E.3.

E.6.2 Continuous Age Structure at Equilibrium

In the discrete case age structure was expressed as the fraction of the population in each age class (Equation E.12). In continuous time, age structure is expressed as a probability density. (Probability densities are described in Appendix D.) As with the discrete case, age structure of a population at equilibrium is simply proportional to the survivor function. For a population that is not changing in density, define the age structure as the probability density of age a individuals,

$$w(a) = \frac{n(a)}{\int_0^\infty n(a)\,da}$$

The denominator is a constant, so the distribution is simply proportional to the density of the cohort $n(a)$. We can express $n(a)$ in terms of initial density (at birth) and the survivor function

$$w(a) = \frac{n_0 l(a)}{n_0 \int_0^\infty l(a)\,da}$$

and cancel n_0

$$w(a) = \frac{l(a)}{\int_0^\infty l(a)\,da} \qquad\qquad \text{E.22}$$

to show that the age distribution has the same form as the survivor function, as demonstrated for the discrete case (Equation E.12). It is important to remember that this relationship holds for the special case of a population growing at a constant rate after the age structure has stabilized.

E.7 Life History to Population Growth in Continuous Models

This section discusses how demography (life history information) determines population growth rate. In continuous time models, I have used the notation $n(t)$ to refer to density at time t or $n(a)$ to indicate density at age a. Because we are now interested in age structure, I combine subscripts to denote the density of individuals of age a at time t. For continuous time I use the notation $n(a,t)$. The density of individuals falling in the age interval from a to $a + da$ is

$$\int_a^{a+da} n(x,t)dx \approx n(a,t)da$$

Think of the left-hand side as an integral over a rectangle that is approximately $n(x,t)$ tall and da wide. This rectangle approximation is reasonable if da is small. The right-hand side is the area of this rectangle.

E.7.1 Net Reproductive Rate and Generation Time

Life history contributes to demography and, thus, population dynamics and evolution. This link derives from the fact that population growth rate is determined by survivorship and reproduction (life history). As with discrete-state models, we have the important concepts of net reproductive rate, the contribution of a single female to the next generation, and generation time. For simplicity, I initially omit the t subscript. Recall, net reproductive rate is the lifetime production of female offspring scaled by the probability of surviving to each age,

$$R_0 = \int_0^\infty l(a)m(a)da \qquad \text{E.23}$$

(compare Equation E.1). If $R_0 > 1$, then population density must be increasing (each female is more than replacing herself), but how fast? As for discrete time, we require the generation time. The generation time involves an average age for reproduction (in the numerator) and the net reproductive rate (in the denominator). As in the discrete case, we have three definitions. The mean age at which a cohort produces offspring (compare Equation E.5) is

$$T = \frac{\int_0^\infty a l(a)m(a)da}{\int_0^\infty l(a)m(a)da} = \frac{\int_0^\infty a l(a)m(a)da}{R_0} \qquad \text{E.24}$$

Recognize this expression as the weighted average of the density $p(a) = l(a)m(a)/R_0$, where R_0 is the normalization constant (see Equation D.20). This normalization means that T is the weighted average of the $l(a)m(a)$ schedule. Comparable to Equation E.6 is a somewhat different definition,

$$T = \int_0^\infty ae^{-ra}l(a)m(a)da \qquad \text{E.25}$$

The continuous-time analog for the time required to change by a factor of R_0 is solved by remembering that

$$\frac{1}{n}\frac{dn}{dt} \approx \frac{\ln\left(n(t+T)/n(t)\right)}{T}$$

(assuming T is not large—Appendix A). Then for constant per capita growth rate r we can write

$$r = \frac{1}{n}\frac{dn}{dt} = \frac{\ln R_0}{T}$$

$$T = \frac{\ln R_0}{r},$$

and

$$n_{t+1} = n_t e^r$$

The following example shows how these quantities could be calculated for a life history where both fecundity schedule and survivorship follow simple parametric forms.

Example E.7. Consider a population with constant per capita birth rate b and mortality rate ρ. In other words survivorship is exponential $l(a) = e^{-\rho a}$ and fecundity schedule is $m(a) = b$. What is the net reproductive rate and approximate generation time? If the population density is constant, what is the relationship between b and ρ?

The equation for R_0 depends on fecundity and mortality schedules (Equation E.23),

$$R_0 = b\int_0^\infty e^{-\rho a}da = \left.\frac{be^{-\rho a}}{\rho}\right|_\infty^0 = \frac{b}{\rho}$$

or the ratio of births to deaths.

The generation time defined in Equation E.24 uses the net reproductive rate in the denominator, so we are half done. The numerator is a bit trickier.

Because I find the gamma function convenient, I use the substitution $u = \rho a$. Then $du/da = \rho$, and the integral is written as

$$T = \frac{b \int_0^\infty a e^{-\rho a} da}{R_0} = \frac{b \int_0^\infty \frac{u}{\rho} \times e^{-u} \times \frac{du}{\rho}}{b/\rho} = \frac{1}{\rho} \int_0^\infty u e^{-u} du$$

This integral is the gamma function of 2. Recall that $\Gamma(n) = (n-1)!$ (Box D.1). Thus, the numerator is $\Gamma(2) = 1$ and the generation time is $1/\rho$.

We could also calculate generation time using Equation E.25 (allowing the effect of population growth),

$$T = b \int_0^\infty a e^{-a(r+\rho)} da = \frac{b}{(r+\rho)^2} \int_0^\infty u e^{-u} du = \frac{b}{(r+\rho)^2}$$

If density is not changing, then R_0 must equal one; otherwise, the population would not remain at constant density. For this to be true, b and ρ must be equivalent (check the equation for R_0). This makes sense: density is constant when the per capita mortality rate equals the per capita reproductive rate and lifetime reproduction (R_0) is self-replacing. Under these conditions, both equations for T give the same answer, $\frac{1}{\rho}$. If there is net population growth, then $R_0 > 1$, $b > \rho$, and generation time for the second calculation is shorter than for the first.

E.7.2 Reproductive Value

The reproductive value has a representation analogous to the discrete case. For a population that is not growing,

$$\frac{v(a)}{v(0)} = \frac{\int_a^\infty l(x)m(x)dx}{l(a)}$$

Recognize this as the net reproductive rate in the numerator except we integrate from the current age rather than from age 0. Then divide by the probability of surviving to age a. If the population is growing at constant rate r, weight future births by change in population density,

$$\frac{v(a)}{v(0)} = \frac{\int_a^\infty e^{-r(x-a)}l(x)m(x)dx}{l(a)} \qquad \text{E.26}$$

Births that come much later than the current age a correspond to large values of x. If the population is growing rapidly, these births in the distant future have little impact on population growth rate, because they are diluted by the overall increase in density.

E.7.3 Lotka-Euler Equation

Information on vital rates is difficult to use unless we possess some way of linking it to behavior of the population as a whole. As for the discrete model, the Lotka or Euler equation provides such a link for the case of linear (density-independent) growth. There is no simple formulation for density-dependent growth. The linear result is nonetheless useful, both heuristically and for a number of applications to real populations. Here is how to derive it.

Let $B(t)$ be the number of newborn individuals (age $a = 0$) at time t,

$$B(t) = \int_0^\infty n(a,t)m(a)da \qquad \text{E.27}$$

Offspring production from age a individuals at time t is $n(a,t)m(a)$. Total offspring production is the integral over all ages. The density of age a individuals at time t is the offspring production from $t - a$ years ago that survive to age a (i.e., time t)

$$n(a,t) = B(t - a)l(a) \qquad \text{E.28}$$

Substituting Equation E.28 in Equation E.27 yields the renewal equation

$$B(t) = \int_0^\infty B(t - a)l(a)m(a)da \qquad \text{E.29}$$

Equation E.29 says that reproduction at time t is determined by survivors from a years ago multiplied by the birth rate of an a-year-old integrated over all age classes.

From expression E.29 derive Lotka's Equation as follows. Note that we have B on both sides of this expression. The simplest possible assumption about population growth is that it is constant. If the population grows at a constant rate, then so does the zero age class,

$$B(t) = B(0)e^{rt} \qquad \text{E.30}$$

for rate constant r. This is the linear assumption: the corresponding differential equation is linear, so the number of newborns is increasing exponentially. In terms of Equation E.29, then

$$B(t - a) = B(0)e^{r(t - a)} \qquad \text{E.31}$$

Substituting Equation E.31 in Equations E.29 obtain

$$B(0)e^{rt} = \int_0^\infty B(0)e^{r(t-a)}l(a)m(a)da$$

$$= B(0)e^{rt} \int_0^\infty e^{-ra}l(a)m(a)da$$

Canceling $B(0)e^{rt}$ from both sides we have Lotka's Equation

$$1 = \int_0^\infty e^{-ra}l(a)m(a)da \qquad \text{E.32}$$

This is an implicit solution for growth rate r in terms of the life history schedule, represented by $l(a)$ and $m(a)$. This population is growing exponentially (at constant rate) with constant age distribution. This equation thus relates growth of the entire population to the life history of the individuals that make up that population. This equation has one real root. A recent summary is provided by Kot (2001).

Example E.8. For the life history schedules in Example E.7, use Lotka's equation to calculate population growth rate.
Lotka's Equation is

$$1 = b\int_0^\infty e^{-ra}e^{-\rho a}da = b\int_0^\infty e^{-a(r+\rho)}da = \left.\frac{be^{-a(r+\rho)}}{r+\rho}\right|_\infty^0 = \frac{b}{r+\rho}$$

giving $r = b - \rho$. The per capita growth rate is the difference in per capita birth and death rates.

Example E.9. Patch demography
The relationships used to develop Lotka's equation can be applied to a range of life history calculations. Paine and Levin (1991) used demographic models to estimate the formation of open patches in mussel beds of the rocky intertidal zone that were created by disturbance and closure by colonization and encroachment from edges. If $M(t)$ is the fraction of area that is patch in time t, then

$$M(t) = \int_0^\infty B(t - a)l(a)da$$

where $B(t)$ is birth rate of patches at time t and $l(a)$ is patch survival until age a. Patch birth is assumed to be independent of current patch area. Patch survival constitutes the area opened a year-ago that is not yet recolonized. These two types of patch mortality constitute competing risks $h_1(a)$ and $h_2(a)$. If they operate independently, then

$$h(a) = h_1(a) + h_2(a)$$

and survival is

$$l(a) = \exp\left[-\int_0^a h_1(u) + h_2(u)du \right]$$

$$= l_1(a)l_2(a)$$

where the product of two survivor functions indicates the independent risks. Both sources of patch recolonization were estimated from field data.

E.7.4 Continuous Population Growth with Continuous Age Structure

Just as the discrete time model can include population structure, so too can a continuous model. This approach requires use of partial differential equations. Here I provide the simplest linear model for structured population growth.

We can write a continuous population growth model as an explicit function of age and time. Let $n(a,t)$ be the density of age a at time t. The McKendrick-von Foerster Equation describes how age structure changes over time. Ignore reproduction for the moment and focus on change in cohorts at ages $a > 0$. These cohorts change only due to mortality. The per capita mortality rate is $h(a)$, so the fraction that die during an interval of duration dt is $h(a)dt$. The total loss to the population is this fraction multiplied by density of the cohort, $h(a)n(a,t)dt$. The density of age class $a + dt$ at time $t + dt$ is given by

$$n(a + dt, t + dt) = n(a,t) - h(a)n(a,t)dt$$

The duration dt is arbitrary, so we need to eliminate it. Expanding the left-hand side in a Taylor series and ignoring higher-order terms,

$$n(a + dt, t + dt) \approx n(a,t) + \frac{\partial n}{\partial a}dt + \frac{\partial n}{\partial t}dt$$

Now, substitute the right-hand side for the left-hand side of the preceding Equation,

$$n(a,t) + \frac{\partial n}{\partial a}dt + \frac{\partial n}{\partial t}dt = n(a,t) - h(a)n(a,t)dt$$

Subtract $n(a,t)$ from both sides, and divide through by dt. This gives the McKendrick-von Foerster Equation

$$\frac{\partial n}{\partial a} + \frac{\partial n}{\partial t} = -h(a)n(a,t)$$

This partial differential equation describes mortality effects on the dynamics of age structure. We still require a boundary condition, a rule stating how new individuals enter the population. New individuals enter at age class $a = 0$, and they represent the product of fecundity × density integrated over all age classes,

$$n(0,t) = \int_0^\infty m(a,t)n(a,t)da$$

This equation says that the density of age 0 individuals at time t on the left-hand side is equal to the combined contribution of fecundity from all ages on the right-hand side.

E.8 Discrete Time and Continuous State

The integral projection model is discrete in time, but continuous in state (Easterling et al. 2000). This combination provides an alternative to matrix projection models for cases where stages are not clearly differentiated, but shift more-or-less continuously. Over a discrete time increment, the population structure, described by a continuous variable x, changes as

$$n_{t+1}(x) = \int_0^\infty k(x|y)n_t(y)dy$$

where the kernel $k(x|y)$ represents transitions from state y to state x. The kernel summarizes transitions that result from fecundity, survival, and growth. Let $f(x|y)$ be the fecundity function, describing the density of state y offspring produced by a state x parent. If newborns are defined as state $x = 0$, then the rate of offspring production by state y parents at time t could be represented as

$$f(0|y)n_t(y)$$

If there were no growth or mortality of a state y individual on the interval $(t, t + 1)$, then newborns would simply be

$$n_{t+1}(0) = \int_0^\infty f(0|y)n_t(y)dy$$

There typically will be growth and mortality, so the full kernel might be constructed as

$$k(x|y) = s(y)g(x|y) + f(x|y)$$

the first term representing the effects of survivorship $s(y)$ and growth $g(x|y)$ and the second being production of new offspring. The full integral equation is now the sum of two terms, growth/survival and newborns.

$$n_{t+1}(x) = \int_0^\infty s(y)g(x|y)n_t(y)dy + \int_0^\infty f(x|y)n_t(y)dy$$

Moreover, if newborns only enter state $x = 0$, then we can retain only the first term and use the second term as a boundary condition (because it is zero for all states other than $x = 0$). Note that the growth function is a density,

$$\int_0^\infty g(x|y)dx = 1$$

Like the matrix projection model, the integral projection model admits analysis of growth rate λ, continuous analogs of eigenvectors, and elasticities (Easterling et al. 2000). It avoids imposition of arbitrary boundaries and the artifacts that arise from such classifications.

Appendix F

Common Distributions

F.1 Discrete Univariate Distributions

F.1.1 Binomial, Bernoulli, and Beta-Binomial Distributions

The binomial distribution arises in any experiment described by sampling with replacement. It is the basis for games of chance, logistic regression, and survivor analysis, to name but a few. The distribution was derived by James Bernoulli in 1713 (Johnson and Kotz 1970).

The probability of y successes in n trials is described by the binomial distribution

$$Bin(y|n, \theta) = \binom{n}{y}\theta^y(1 - \theta)^{n-y}, \qquad y = 0, 1, \ldots, n; n = 1, 2, \ldots; 0 < \theta < 1,$$

where the binomial coefficient is

$$\binom{n}{y} = \frac{n!}{y!(n - y)!}$$

To see that the binomial sums to one, recall the *binomial series*

$$(a + b)^n = \sum_{y=0}^{n}\binom{n}{y}a^y b^{n-y}$$

using $a = \theta$ and $b = 1 - \theta$.

Moments are readily obtained using characteristic functions, which I do not review here. The mean is

$$E[y] = n\theta$$

and the variance is

$$\text{var}[y] = n\theta(1 - \theta)$$

The binomial tends to a normal distribution with large n (Laplace's theorem),

$$\lim_{n\to\infty} Bin(\theta, n) \to N(n\theta, n\theta(1 - \theta))$$

(Figure. D.4). A Poisson is obtained if θ tends to zero as n becomes large,

$$\lim_{\substack{n\to\infty \\ \theta\to 0}} Bin(\theta, n) \to Pois(n\theta)$$

The *Bernoulli* distribution is a special case of the binomial, where $n = 1$.

The *beta-binomial* distribution arises as a mixture. If θ is distributed as $Beta(\alpha, \beta)$, and y is distributed as $Bin(n, \theta)$, then y has the distribution $Betabin(n, \alpha, \beta)$.

F.1.2 Poisson Distribution

The Poisson describes the probability of obtaining a sample of y objects given that the mean is a parameter λ. It was derived by Poisson in 1837 as the limit of a binomial (Section F.1.1). Among the more famous early applications was the Prussian Army mule-kick data set, describing the distribution of deaths by mule kick per year (Johnson and Kotz 1970). The distribution is

$$Pois(y|\lambda) = \frac{e^{-\lambda}\lambda^y}{y!} \qquad y = 0, 1, 2, \ldots; \lambda > 0$$

It naturally arises as the limit of a sequence of binomials, as described in Section F.1.1, and as the number of events to occur in an interval of specified duration, length, area, or volume. Thus, it plays a central role in temporal and spatial processes. If the age-specific rate of death for a population is constant at ρ per unit time, then the distribution of deaths is exponential (Chapter 2), and the distribution of the number of deaths in an interval of duration t is Poisson with parameter $\lambda = \rho t$ In a similar vein, the distribution of objects to occur in a sample of area A could have Poisson parameter $\lambda = \rho A$, and in area A during an interval of duration t, we have $\lambda = \rho A t$.

A Poisson process is one for which all objects are independent, that is, the location (space) or failure (time) of any one does not depend on the location or timing of others. This property results in the variance being equal to the mean, both being equal to the Poisson parameter λ.

F.1.3 Negative Binomial and Geometric Distributions

There are two contexts in which the negative binomial arises. The first is the distribution of the number of trials needed to obtain a given number of occurrences (e.g., how many coin flips are needed to produce y tails). This context sometimes arises in ecological problems when the number of occurrences is one. This special case of the negative binomial is termed a *geometric density* (see below).

The second context typically applies to density data (counts) where the distribution is taken to be a spatial Poisson process but having a Poisson parameter that varies with a density of its own. Here we see how the negative binomial arises as a marginal distribution of Poisson process conditioned on a gamma-distributed Poisson parameter. The Poisson probability of y counts is

$$p(y|\lambda) = \frac{e^{-\lambda}\lambda^y}{y!}$$

Assume that the parameter of this Poisson process (λ) is described by a gamma density

$$p(\lambda) = \frac{\beta^\alpha \lambda^{\alpha-1} e^{-\beta\lambda}}{\Gamma(\alpha)}$$

The marginal distribution of counts is obtained by integrating over the variability in λ,

$$p(y) = \int_0^\infty p(y|\lambda)p(\lambda)d\lambda = \frac{\beta^\alpha}{y!\Gamma(\alpha)} \int_0^\infty \lambda^{y+\alpha-1} e^{-\lambda(1+\beta)} d\lambda$$

Using the substitution $u = \lambda(1 + \beta)$, write this as a gamma integral (Box D.1)

$$p(y) = \frac{\beta^\alpha}{y!\Gamma(\alpha)(1 + \beta)^{y+\alpha}} \int_0^\infty u^{y+\alpha-1} e^{-u} du = \frac{\Gamma(y + \alpha)\beta^\alpha}{y!\Gamma(\alpha)(1 + \beta)^{y+\alpha}}$$

This result is the negative binomial. It is written in several other ways. Using relationships involving combinatorials, factorials, and gamma functions from Box D.1, write this as

$$p(y) = \frac{\Gamma(y + \alpha)\beta^\alpha}{\Gamma(y + 1)\Gamma(\alpha)(1 + \beta)^{y+\alpha}} = \frac{(y + \alpha - 1)!}{y!(\alpha - 1)!} \times \frac{\beta^\alpha}{(1 + \beta)^{y+\lambda}}$$

$$= \binom{\alpha + y - 1}{\alpha - 1} \frac{\beta^\alpha}{(1 + \beta)^{y+\alpha}}$$

or

$$NB(y|\alpha, \beta) = \binom{\alpha + y - 1}{\alpha - 1}\left(\frac{1}{1 + \beta}\right)^y\left(\frac{\beta}{1 + \beta}\right)^\alpha \quad y = 0, 1, 2, \ldots; \alpha, \beta > 0$$

Although this is often written in terms of combinatorials, unlike the binomial distribution, α need not be an integer. The negative binomial is sometimes written in terms of the mean m,

$$NB(y|\alpha, m) = \binom{\alpha + y - 1}{\alpha - 1}\left(\frac{m}{\alpha}\right)^y\left(1 + \frac{m}{\alpha}\right)^{-(\alpha + y)}$$

The mean and variance of the distribution are

$$E[y] = \frac{\alpha}{\beta} = m$$

$$\text{var}[y] = \frac{\alpha}{\beta}\left(1 + \frac{1}{\beta}\right) = m\left(1 + \frac{m}{\alpha}\right)$$

The special case for $\alpha = 1$ yields the *geometric distribution*. Define the parameter

$$\theta = \frac{1}{1 + m}$$

Substitution gives the geometric distribution

$$p(y) = \theta(1 - \theta)^y$$

The geometric distribution is a discrete analog of the exponential, being the probability that the first occurrence occurs on the yth trial. Parameter θ is the probability of occurrence on any given trial.

F.2 Discrete Multivariate Distribution

F.2.1. Multinomial Distribution

Suppose we have n events that fall in any of $k = 1, \ldots, K$ ordinal classes. The probability of the kth class is θ_k. The multinomial distribution is

$$Multinom(\mathbf{y}|n, \theta_1, \ldots, \theta_K) = \binom{n}{y_1 \ldots y_K}\prod_{k=1}^K \theta_k^{y_k} \quad y_k = 0, 1, \ldots, n; \sum_{k=1}^K y_k = n;$$
$$0 < \theta_k < 1; \sum_{k=1}^K \theta_k = 1$$

The length-K vector $\mathbf{y}$ contains the number of events that fall in each class. There is a corresponding probability θ_k for each of these classes. The multinomial coefficient is a generalization of the binomial coefficient, used to partition events into only two disjoint sets of size y and and $n - y$. For K sets, the multinomial coefficient is

$$\binom{n}{y_1 \ldots y_K} = \frac{n!}{y_1! y_2! \ldots y_K!}$$

The marginal distribution of any one of the K classes is binomial. This situation occurs when we lump $\mathbf{y}$ into class k and class not k. The equivalent binomial is then

$$Bin(y_k|n, \theta_k) = \binom{n}{y_k} \theta_k^{y_k} (1 - \theta_k)^{n-y_k}$$

The mean for the kth element of vector $\mathbf{y}$ is

$$E[y_k] = n\theta_k$$

with variance

$$\text{var}[y_k] = n\theta_k (1 - \theta_k)$$

The covariance between any two elements of vector $\mathbf{y}$ is

$$\text{cov}[y_j, y_k] = -n\theta_j \theta_k$$

All covariances are negative, due to the sum-to-n constraint—increase in one class means a decrease in others. Box 7.1 describes sampling from a multinomial distribution.

F.3 Continuous Univariate Distributions

F.3.1 Normal Distribution

The important role of the normal distribution stems from the theoretical justification for its application to the sums of random variables and its tractability (Stigler 1986). The former motivation comes from Central Limit Theorems, which state that the sum of independent, identically distributed random variables $z = y_1 + y_2 + \cdots + y_n$ having finite mean and variance tends asymptotically to the normal density

$$N(nE[y], n \text{ var}[y])$$

It is often termed a *Gaussian distribution*, in recognition of Gauss's important contribution to its use as an error distribution, among other things. The density function is

$$N(y|\mu, \sigma^2) = \frac{1}{\sqrt{2\pi}\sigma} \exp\left[-\frac{(y - \mu)^2}{2\sigma^2}\right] \qquad \sigma > 0$$

The distribution function does not have a solution. It is noteworthy in being the only member of the exponential family having independent parameters for mean μ and variance σ^2. Thus we can write the density of a random variable in terms of the standardized unit normal,

$$y_i = \mu + \sigma\varepsilon_i$$

$$\varepsilon_i \sim N(0,1)$$

By extension, if y is normally distributed, then so is any linear function of y. The density is symmetric about μ, with inflexion points at $\mu \pm \sigma$.

F.3.2. Uniform Distribution

The uniform distribution is flat, with parameters designating the boundaries of the distribution. The density is

$$Unif(y|\alpha, \beta) = \frac{1}{\beta - \alpha} \quad \alpha < y < \beta$$

The distribution function increases linearly from zero at the lower bound to one at the upper bound,

$$\int_\alpha^y Unif(z|\alpha, \beta)dz = \frac{y - \alpha}{\beta - \alpha}$$

The mean lies halfway between the endpoints

$$E[y] = \frac{\beta - \alpha}{2}$$

and the variance is

$$var[y] = \frac{(\beta - \alpha)^2}{12}$$

The $Beta(1,1)$ density is $Unif(0,1)$.

F.3.3 Gamma and Inverse Gamma Distributions

The gamma density is defined as

$$Gam(y|\alpha, \beta) = \frac{\beta^\alpha y^{\alpha-1} e^{-\beta y}}{\Gamma(\alpha)} \quad y > 0; \alpha > 0, \beta > 0$$

The distribution function does not have a solution, but is sometimes termed an incomplete gamma function. Moments can be obtained from

$$\mu_m = \int_0^\infty x^m \cdot \frac{\beta^\alpha x^{\alpha-1} e^{-\beta x}}{\Gamma(\alpha)} dx = \frac{\beta^\alpha}{\Gamma(\alpha)} \int_0^\infty x^{m+\alpha-1} e^{-\beta x} dx$$

Now substitute $u = \beta x$ (and, thus, $x = u/\beta$, and $dx = du/\beta$) to obtain

$$= \frac{\beta^\alpha}{\beta \Gamma(\alpha)} \int_0^\infty \left(\frac{u}{\beta}\right)^{m+\alpha-1} e^{-u} du = \frac{1}{\beta^m \Gamma(\alpha)} \int_0^\infty u^{m+\alpha-1} e^{-u} du$$

The integral is now a gamma function, so

$$\mu_m = \frac{\Gamma(m+\alpha)}{\beta^m \Gamma(\alpha)}$$

Thus, the first moment (the mean) is

$$\mu_1 = \frac{\Gamma(1+\alpha)}{\beta \Gamma(\alpha)} = \frac{\alpha \Gamma(\alpha)}{\beta \Gamma(\alpha)} = \frac{\alpha}{\beta}$$

The second moment is

$$\mu_2 = \frac{\Gamma(2+\alpha)}{\beta^2 \Gamma(\alpha)} = \frac{(1+\alpha)\alpha \Gamma(\alpha)}{\beta^2 \Gamma(\alpha)} = \frac{(1+\alpha)\alpha}{\beta^2}$$

and the variance is

$$\bar{\mu}_2 = \mu_2 - \mu_1^2 = \frac{\alpha}{\beta^2}$$

Gamma distributions play important roles as priors and are sometimes used to describe failures in survival analysis.

If y^{-1} has a gamma distribution then y has an *inverse gamma distribution* (the Jacobian is unity). It is used as the conjugate prior for σ^2 where the likelihood is normal (Appendix G). If y is distributed as IG(a,b), then mean and variances are

$$\mu_1 = \frac{\beta}{\alpha - 1}$$

$$\bar{\mu}_2 = \frac{\beta^2}{(\alpha - 1)^2(\alpha - 2)}$$

Note that the mean is finite if $\alpha > 1$, and the variance is finite if $\alpha > 2$.

F.3.4. Exponential and Laplace Distributions

The exponential distribution arises naturally as the probability that an event will occur in elapsed time y, given a constant probability ρ with units of y^{-1}. It is a special case of the gamma with $\alpha = 1$,

$$Exp(y|\rho) = \rho e^{-\rho y} \qquad y > 0; \rho > 0$$

You should recognize $Exp(y|\rho) = Gam(y|1,\rho)$ from Section F.3.3.

Due to its common application to problems of survival, the distribution and survivor functions are frequently used. To obtain the distribution function, make the substitution $u = -\rho z$ (which means that $dz = -du/\rho$) to obtain

$$P(y|\rho) = \rho \int_0^y e^{-\rho z} dz = \rho \int_0^y e^{u} \frac{du}{-\rho}$$

Substituting back for u we obtain

$$P(y|\rho) = -e^{-\rho z}\Big|_0^y = 1 - e^{-\rho y}$$

The survivor function is

$$l(y|\rho) = 1 - P(y|\rho) = e^{-\rho y}$$

The mth moment is obtained from

$$\mu_m = \rho \int_0^\infty y^m e^{-\rho y} dy$$

With the substitution $u = \rho y$

$$\mu_m = \rho^{-m} \int_0^\infty u^m e^{-u} du = \frac{\Gamma(m + 1)}{\rho^m}$$

Using properties of the gamma function (Box D.1), we determine the mean and variance to be

$$E[y] = \frac{1}{\rho}$$

$$\text{var}[y] = \mu_2 - \mu_1^2 = \frac{2}{\rho^2} - \frac{1}{\rho^2} = \frac{1}{\rho^2}$$

Related to the exponential is the double exponential, or *Laplace distribution*.

$$p(y|\rho) = \frac{\rho}{2} e^{-\rho|y|} \qquad \rho > 0$$

This is a two-tailed distribution centered on zero (unless a location parameter is specified). Laplace originally conceived of this as an error distribution (Stigler 1986), a purpose for which it is still sometimes used. It is also used as a dispersal kernel for one-dimensional movement as in, say, rivers (Lutscher et al. 2005).

F.3.5 Beta Distribution

The beta density is defined as

$$Beta(y|\alpha, \beta) = \frac{y^{\alpha-1}(1 - y)^{\beta-1}}{B(\alpha, \beta)} \qquad 0 < y < 1; \alpha > 0, \beta > 0$$

The mth moment can be found by the integral

$$\mu_m = \frac{1}{B(\alpha, \beta)} \int_0^1 x^{\alpha-1+m}(1 - x)^{\beta-1}dx = \frac{B(\alpha + m, \beta)}{B(\alpha, \beta)}$$

To simplify, rewrite the beta functions as gamma functions (Box D.1)

$$\frac{B(\alpha + m, \beta)}{B(\alpha, \beta)} = \frac{\Gamma(\alpha + m)\Gamma(\alpha + \beta)}{\Gamma(\alpha + \beta + m)\Gamma(\alpha)}$$

Because moments involve integer values of m, these simplify using properties of the gamma function. The mean is

$$\mu_1 = \frac{\Gamma(\alpha + 1)\Gamma(\alpha + \beta)}{\Gamma(\alpha + \beta + 1)\Gamma(\alpha)} = \frac{\alpha\Gamma(\alpha)\Gamma(\alpha + \beta)}{(\alpha + \beta)\Gamma(\alpha + \beta)\Gamma(\alpha)} = \frac{\alpha}{\alpha + \beta}$$

The second moment is

$$\mu_2 = \frac{\Gamma(\alpha + 2)\Gamma(\alpha + \beta)}{\Gamma(\alpha + \beta + 2)\Gamma(\alpha)} = \frac{(\alpha + 1)\Gamma(\alpha + 1)\Gamma(\alpha + \beta)}{(\alpha + \beta + 1)\Gamma(\alpha + \beta + 1)\Gamma(\alpha)}$$

$$= \frac{\alpha(\alpha + 1)\Gamma(\alpha)\Gamma(\alpha + \beta)}{(\alpha + \beta + 1)(\alpha + \beta)\Gamma(\alpha + \beta)\Gamma(\alpha)}$$

$$= \frac{\alpha(\alpha + 1)}{(\alpha + \beta + 1)(\alpha + \beta)}$$

so the variance is

$$var_x = \frac{\alpha\beta}{(\alpha + \beta)^2(\alpha + \beta + 1)}$$

The beta distribution is most commonly used as a prior for the binomial.

F.4 Continuous Multivariate Distributions

F.4.1. Multivariate Normal

The multivariate normal distribution of dimension n is

$$N_n(\mathbf{y}|\mu, \Sigma) = \frac{1}{(2\pi)^{n/2}|\Sigma|^{1/2}} \exp\left[-\frac{1}{2}(\mathbf{y} - \mu)^T \Sigma^{-1}(\mathbf{y} - \mu) \right]$$

Σ symmetric, positive definite,

where μ is the vector of means (and modes) and Σ is the symmetric positive definite covariance matrix. If we integrate out any of the n variables, those remaining also have a multivariate normal distribution; hence, each of the n variables is marginally normal. As with the univariate normal, any linear function of vector $\mathbf{y}$ is multivariate normal.

F.4.2. Wishart

The Wishart distribution of dimension n is

$$W_n(\Sigma|\mathbf{R},\rho) = \frac{1}{2^{\rho n/2}\pi^{n(n-1)/4}\prod_{i=1}^{n} \Gamma\left(\frac{\rho + 1 - i}{2}\right)}|\mathbf{R}|^{-\rho/2}|\Sigma|^{(\rho-n-1)/2}\exp\left[-\frac{1}{2}tr(\mathbf{R}^{-1}\Sigma) \right]$$

$\mathbf{R}$ symmetric, positive definite

The expectation of Σ is $\rho\mathbf{R}$. The Wishart is used as the conjugate prior for the inverse covariance matrix and is a multivariate version of the gamma distribution. The density is finite if the degrees of freedom parameter $\rho > n + 1$.

F.4.3. Dirichlet

The Dirichlet distribution is a multivariate extension of the beta distribution. For data modeling, it is primarily used as a prior distribution that is conjugate with the multinomial. For a length-K random vector θ, the density is

$$Dir(\theta|\alpha_1,\ldots,\alpha_K) = \frac{\Gamma\left(\sum_{k=1}^{K}\alpha_k\right)}{\prod_{k=1}^{K}\Gamma(\alpha_k)}\prod_{k=1}^{K}\theta_k^{\alpha_k-1} \quad \theta_1,\ldots,\theta_K \geq 0, \sum_{k=1}^{K}\theta_k = 1$$

$$\alpha_1,\ldots,\alpha_K > 0$$

The parameters in vector α must be positive. Box 7.1 describes how to draw a random vector from the Dirichlet.

Appendix G

Common Conjugate Likelihood-Prior Pairs

This appendix contains some conjugate pairs often used in Bayesian models.

TABLE G.1

Likelihood	Prior	Posterior
	Discrete likelihoods	
Poisson/Gamma		
$Pois(\mathbf{y}|\lambda)$	$Gam(\lambda|\alpha,\beta)$	$Gam\left(\lambda \,\middle|\, \alpha + \sum_{i=1}^{n} y_i, \beta + n\right)$
$\propto \lambda^{\sum_{i=1}^{n} y_i} e^{-n\lambda}$	$\propto \lambda^{\alpha-1} e^{-\beta\lambda}$	$\propto^{\alpha + \Sigma y_{i-1}} e^{-(\beta+n)\lambda}$
Binomial/Beta		
$Bin(\mathbf{y}|\theta)$	$Beta(\theta|\alpha,\beta)$	$Beta(\theta|y + \alpha, n - y + \beta)$
$\propto \theta^{y}(1-\theta)^{n-y}$	$\propto \theta^{\alpha-1}(1-\theta)^{\beta-1}$	$\propto \theta^{y+\alpha-1}(1-\theta)^{n-y+\beta-1}$
Multinomial/Dirichlet		
$Multinom(\mathbf{y}|n,\theta_1,\ldots,\theta_k)$	$Dir(\theta|\alpha_1\ldots,\alpha_k)$	$Dir(\theta|y_1 + \alpha_1\ldots,y_k\alpha_k)$
$\propto \prod_{k=1}^{K}\theta_k^{y_k}$	$\propto \prod_{k=1}^{K}\theta_k^{a_k-1}$	$\propto \prod_{k=1}^{K}\theta_k^{y_k+a_k-1}$
	Continuous likelihoods	
Exponential/Gamma		
$Exp(\mathbf{y}|\lambda)$	$Gam(\lambda|\alpha,\beta)$	$Gam\left(\lambda \,\middle|\, \alpha + n, \beta + \sum_{i=1}^{n} y_i\right)$
$\lambda^n e^{-\lambda\Sigma_{i=1}^{n} y_i}$	$\propto \lambda^{\alpha-1} e^{-\beta\lambda}$	$\propto \lambda^{n+\alpha-1} e^{-(\beta+\Sigma_{i=1}^{n} y_i)\lambda}$
Normal/Normal for the mean, fixed variance		
$N(\mathbf{y}|\mu,\sigma^2)$	$N(\mu|\theta_0, \tau_0^2)$	$N(\mu|\theta_1, \tau_1^2)$, where $\tau_1^2 = \left(\dfrac{1}{\tau_0^2} + \dfrac{n}{\sigma^2}\right)^{-1}$,
		$\theta_1 = \tau_1^2\left(\dfrac{\theta_0}{\tau_0^2} + \dfrac{n\bar{y}}{\sigma^2}\right)$

(Continued)

TABLE G.1 (*continued*)

Likelihood	Prior	Posterior

$$\propto \exp\left[-\frac{\sum_{i=1}^{n}(y_i - \mu)^2}{2\sigma^2} \right]$$

$$\propto \exp\left[-\frac{(\mu - \theta_0)^2}{2\tau_0^2} \right]$$

$$\propto \exp\left[-\frac{1}{2}\left(\frac{(\mu - \theta_0)^2}{\tau_0^2} \right.\right.$$

$$\left.\left. + \frac{\sum_{i=1}^{n}(y_1 - \mu)^2}{2\sigma^2} \right) \right]$$

Normal/Inverse Gamma for the variance, fixed mean

$$N(\mathbf{y}|\mu,\sigma^2) \qquad\qquad IG(\sigma^2|\alpha,\beta) \qquad\qquad IG\left(\sigma^2 \left| \frac{n}{2} + \alpha + 1, \beta \right.\right.$$

$$\left. + \frac{1}{2}\sum_{i=1}^{n}(y_i - \mu)^2 \right)$$

$$\propto \sigma^{-n}\exp\left[\frac{\sum_{i=1}^{n}(y_i - \mu)^2}{2\sigma^2} \right]$$

$$\propto \sigma^{-2(\alpha+1)}\exp\left[-\frac{\beta}{\sigma^2} \right]$$

$$\propto \sigma^{-2(\frac{n}{2}+\alpha+1)}\exp\left[-\frac{1}{\sigma^2}\left(\beta \right.\right.$$

$$\left.\left. + \frac{1}{2}\sum_{i-1}^{n}(y_i - \mu)^2 \right) \right]$$

References

Agarwal, D. K., R. E. Dewar, A. E. Gelfand, J. G. Mickelson, and J. A. Silander. 2005. Tropical deforestation in Madagascar: Analysis using hierarchical, spatially explicit, Bayesian regression models. *Ecological Modeling*, in press.

Agren, G. I., and E. Bosatta. 1996. *Theoretical Ecosystem Ecology*. Cambridge University Press, Cambridge.

Aitkin, M. 1991. Posterior Bayes factors. *Journal of the Royal Statistical Society* B 53:111–142.

Akaike, H. 1973. Information theory and an extension of the maximum likelihood principle. Pages 267–281 in B. N. Petrov and F. Czaki, eds. *Information Theory*. Academia Kiado, Budapest.

Akaike, H. 1974. A new look at the statistical model identification. *IEEE Transaction on Automatic Control* AC-19:716–723.

Albert, J. H., and S. Chib. 1993. Inference via Gibbs sampling of autoregressive time series subject to Markov mean and variance shifts. *Journal of Business & Economic Statistics* 11:1–15.

Allee, W. C., A. E. Emerson, O. Park, T. Park, and K. P. Schmidt, 1949. *Principles of Animal Ecology*. W. B. Saunders, Philadelphia, Pennsylvania.

Andersen, M. 1991. Mechanistic models for the seed shadows of wind-dispersed plants. *American Naturalist* 137:476–497.

Anderson, D. R., and K. P. Burnham. 1990. Demographic analysis of northern spotted owl populations. Appendix to 1990 status review: Northern spotted owl (*Strix occidentalis caurina*). U.S. Fish and Wildlife Service, Washington, D.C.

Anderson, R. M., and R. M. May. 1991. *Infectious Diseases of Humans: Dynamics and Control*. Oxford University Press, Oxford.

Andow, D. A., P. M. Kareiva, S. Levin, and A. Okubo. 1990. Spread of invading organisms. *Landscape Ecology* 4:177–188.

Andrewartha, H. G., and L. C. Birch. 1954. *The Distribution and Abundance of Animals*. University of Chicago Press, Chicago, Illinois.

Armstrong, R. A., and R. McGehee. 1980. Competitive exclusion. *American Naturalist* 115:151–170.

Arnason, A. N. 1972. Parameter estimates from mark-recapture experiments on two populations subject to migration and death. *Researches on Population Ecology* 13:97–113.

Bafumi, J., and A. Gelman. 2006. Fitting multilevel models when predictors and group effects correlate, in review.

Bagnold, R. A. 1941. *The Physics of Blown Sand and Desert Dunes*. Methuen & Company, London.

Banerjee, S., B. P. Carlin, and A. E. Gelfand. 2004. *Hierarchical Modeling and Analysis for Spatial Data*. Chapman and Hall/CRC Press, Boca Raton, Florida.

Barnes, A., and G. J. E. Hill. 1989. Census and distribution of black noddy *Anous minutus* nests on Heron Island, November 1985. *The Emu* 89:129–134.

Barrowclough, G. F., and S. L. Coats. 1985. The demography and population genetics of owls, with special reference to the conservation of the spotted owl (*Strix occidentalis*). Pages 74–85 in R. J. Gutierrez and B. Carey, eds. *Ecology and Management of the Spotted Owl in the Pacific Northwest*. U.S. Forest Service, Pacific Northwest Forest and Range Experiment Station, Portland, Oregon.

Barrowman, N. J., R. A. Myers, R. Hilborn, D. G. Kehler, and C. A. Field. 2003. The variability among populations of coho salmon in the maximum reproductive rate and depensation. *Ecological Applications* 13:784–793.

Barry, S. C., S. P. Brooks, E. A. Catchpole, and B. J. T. Morgan. 2003. The analysis of ring-recovery data using random effects. *Biometrics* 59:54–65.

Bartlett, M. S. 1960. *Stochastic Population Models in Ecology and Epidemiology*. Methuen, London.

Bates, S. C., A. Cullen, and A. E. Raftery. 2003. Bayesian uncertainty assessment in multicompartment deterministic simulation models for environmental risk assessment. *Environmetrics* 14:355–371.

Batista, W. B., W. J. Platt, and R. E. Macchiavelli. 1998. Demography of a shade-tolerant tree (*Fagus grandifolia*) in a hurricane disturbed forest. *Ecology* 79:38–53.

Bayes, T. 1763. An essay towards solving a problem in the doctrine of changes. *Philosophical Transactions of the Royal Society*, 330–418. Reprinted in *Biometrika* (1958) 45:293–315.

Beckage, B., and J. S. Clark. 2003. Seedling survival and growth in southern Appalachian forests: Does spatial heterogeneity maintain species diversity? *Ecology* 84:1849–1861.

Beckage, B., and W. J. Platt. 2003. Predicting severe wildfire years in the Florida Everglades. *Frontiers in Ecology and the Environment* 1:235–239.

Bekker, P. A. 1986. Comment on identification in the linear errors in variables model. *Econometrica* 54:215–217.

Bengtsson, T., C. Snyder, and D. Nychka. 2003. A nonlinear filter that extends to high dimensional systems. *Journal of Geophysical Research* 108(D24), 8775, doi:10.1029/2002JD002900.

Benoit, H. P., E. McCauley, and J. R. Post. 1998. Testing the demographic consequences of cannibalism in *Tribolium confusum*. *Ecology* 79:2839–2851.

Berger, J. O. 1985. *Statistical Decision Theory and Bayesian Analysis*. Springer Verlag, New York.

Berger, J. O. 2000. Bayesian analysis: A look at today and thoughts of tomorrow. *Journal of the American Statistical Association* 95:1269–1276.

Berger, J. O. 2003. Could Fisher, Jeffreys, and Neyman have agreed on testing? (with discussion). *Statistical Science* 18:1–32.

Berger, J., and D. Berry. 1988. Analyzing data: Is objectivity possible? *American Scientist* 76:159–165.

Berger, J., V. De Oliveira, and B. Sanso. 2001. Objective Bayesian analysis of spatially correlated data. *Journal of the American Statistical Association* 96:1361–1374.

Berliner, L. M. 1996. Hierarchical Bayesian time series models. Pages 15–22 in K. M. Hanson and R. N. Silver, eds. *Maximum Entropy and Bayesian Methods.* Kluwer Academic, The Netherlands.

Berliner, L. M., A. J. Royle, C. K. Wikle, and R. F. Milliff. 1999. Bayesian methods in the atmospheric sciences. Pages 83–100 in J. M. Bernardo, J. O. Berger, A. P. Dawid, and A. F. M. Smith, eds. *Bayesian Statistics* 6. Oxford University Press, Oxford.

Berliner, L. M., C. K. Wikle, and N. Cressie. 2000. Long-lead prediction of Pacific SSTs via Bayesian dynamic modeling. *Journal of Climate* 13:3953–3958.

Berryman, A. A., and P. Turchin. 2001. Identifying the density-dependent structure underlying ecological time series. *Oikos* 92:265–270.

Besag, J. E., P. J. Green, D. M. Higdon, and K. L. Mengersen. 1995. Bayesian computation and stochastic systems (with discussion). *Statistical Science* 10:3–66.

Bickel, P. J., E. Hammel, and J. O'Connell. 1975. Sex bias in graduate admissions: Data from Berkeley. *Science* 187:398–404.

Biging, G. S., and L. C. Wensel. 1988. The effect of eccentricity on the estimation of basal area and basal area increment of coniferous trees. *Forest Science* 32:621–633.

Bjørnstad, O. N., J.-M. Fromentin, N. C. Stenseth, and J. Gjøsæter. 1999. Cycles and trends in cod populations. *Proceedings of the National Academy of Sciences USA* 96:5066–5071.

Bjørnstad, O. N., W. Falck, and N. C. Stenseth. 1995. A geographic gradient in small rodent density fluctuations: A statistical modelling approach. *Proceedings of the Royal Society of London* B 262:127–133.

Bjørnstad, O. N., B. F. Finkenstadt, and B. T. Grenfell. 2002. Dynamics of measles epidemics: Estimating scaling of transmission rates using a time series SIR model. *Ecological Monographs* 72:169–184.

Bjørnstad, O. N., and B. T. Grenfell. 2001. Noisy clockwork: Time series analysis of population fluctuations in animals. *Science* 293:638–643.

Bjørnstad, O. N., R. A. Ims, and X. Lambin. 2000. Spatial population dynamics: Causes and consequences of spatial synchrony in density fluctuations. *Trends in Ecology and Evolution* 14:427–431.

Bolker, B. M., and S. W. Pacala. 1999. Spatial moment equations for plant competition: Understanding spatial strategies and the advantage of short dispersal. *American Naturalist* 153:575–602.

Bolker, B. M., S. W. Pacala, and C. Neuhauser. 2003. Spatial dynamics in model plant communities: What do we really know? *American Naturalist* 162:135–148.

Bonsall, M. B., and A. Hastings. 2004. Demographic and environmental stochasticity in predator-prey metapopulation dynamics. *Journal of Animal Ecology* 73:1043–1055.

Bonsall, M. B., and M. Mangel. 2004. Life-history trade-offs and ecological dynamics in the evolution of longevity. *Proceedings of the Royal Society of London* B 271: 1143–1150.

Box, G. E. P. 1980. Sampling and Bayes inference in scientific modeling and robustness. *Journal of the Royal Statistical Society* A 143:383–430.

Box, G. E. P., and N. R. Draper. 1987. *Empirical model building and response surfaces.* Wiley, New York.

Boyce, M. S., J. S. Mao, E. H. Merrill, D. Fortin, M. G. Turner, J. M. Fryxell, and P. Turchin. 2003. Scale and heterogeneity in habitat selection by elk in Yellowstone National Park. *Écoscience* 10:21–332.

Braswell, B. H., W. J. Sacks, E. Linder, and D. S. Schimel. 2005. Estimating diurnal to annual ecosystem parameters by synthesis of a carbon flux model with eddy covariance net ecosystem exchange observations. *Global Change Biology* 11:335–355.

Brown, J. H. 1995. *Macroecology*. University of Chicago Press, Chicago, Illinois.

Brown, L. D. 2000. An essay on statistical decision theory. *Journal of the American Statistical Association* 95:1277–1281.

Brownie, C., J. E. Hines, J. D. Nichols, K. H. Pollock, and J. B. Hestbeck. 1993. Capture-recapture studies for multiple strata including non-Markovian transition probabilities. *Biometrics* 49:1173–1187.

BUGS documentation project. The project documentation. www.mrc-bsu.cam.ac.uk/bugs/documentation/bugs05/manual05.html.

Burnham, K. P., and D. R. Anderson. 1998. *Model Selection and Inference*. Springer, New York.

Burnham, K. P., and W. S. Overton. 1978. Estimation of the size of a closed population when capture probabilities vary among animals. *Biometrika* 65:625–633.

Burnham, K. P., and A. White. 2002. Evaluation of some random effects methodology applicable to ringing data. *Journal of Applied Statistics* 29:254–262.

Cain, M. L. 1990. Models of clonal growth in *Solidago altissima*. *Journal of Ecology* 78:27–46.

Cain, M. L., S. W. Pacala, J. A. Silander, and M. J. Fortin. 1995. Neighborhood models of clonal growth in the white clover *Trifolium repens*. *American Naturalist* 145:888–917.

Calder, C. A., C. H. Holloman, S. M. Bortnik, W. J. Strauss, and M. Morara, 2004. Relating ambient particulate matter concentration levels to mortality using an exposure simulator. Department of Statistics Preprint No. 725, Ohio State University. Submitted.

Calder, K., M. Lavine, P. Mueller, and J. S. Clark. 2003. Incorporating multiple sources of stochasticity in population dynamic models. *Ecology* 84:1395–1402.

Callaway, R. M., and J. L. Maron. 2006. What have exotic plant invasions taught us over the past twenty years? *Trends in Ecology and Evolution*, in press.

Cam, E., W. A. Link, E. G. Cooch, J.-Y. Monnat, and E. Danchin. 2002. Individual covariation in life-history traits: Seeing the trees despite the forest. *American Naturalist* 159:96–105.

Cappe, O., and C. P. Robert. 2000. Markov chain Monte Carlo: 10 years and still running! *Journal of the American Statistical Association* 95:1282–1286.

Carey, J. R. 2001. Insect biodemography. *Annual Review of Entomology* 46:79–100.

Carey, J. R., P. Liedo, H.-G. Müller, J.-L Wang, and J. W. Vaupel. 1998. A simple graphical technique for displaying individual fertility data and cohort survival: Case study of 1000 Mediterranean fruit fly females. *Functional Ecology* 12:359–363.

Carlin, B. P., and T. A. Louis. 2000. *Bayes and Empirical Bayes Methods for Data Analysis*. Chapman and Hall, Boca Raton, Florida.

Carlin, B. P., N. G. Polson, and D. S. Stoffer. 1992. A Monte Carlo approach to nonnormal and nonlinear state-space modeling. *Journal of the American Statistical Association* 87:493–500.

Carothers, A. D. 1973. The effects of unequal catchability on Jolly-Seber estimates. *Biometrics* 29:79–100.

Carothers, A. D. 1979. Quantifying unequal catchability and its effect on survival estimates in an actual population. *Journal of Animal Ecology* 48:863–869.

Carpenter, S. R. 1996. Microcosm experiments have limited relevance for community and ecosystem ecology. *Ecology* 77:677–680.

Carpenter, S. R. 2002. Ecological futures: Building an ecology of the long now. *Ecology* 83:2069–2083.

Carpenter, S. R., J. J. Cole, J. R. Hodgson, J. F. Kitchell, M. L. Pace, D. Bade, K. L. Cottingham, T. E. Essington, J. N. Houser, and D. E. Schindler. 2001. Trophic cascades, nutrients, and lake productivity: Whole-lake experiments. *Ecological Monographs* 71:163–186.

Carpenter, S. R., K. L. Cottingham, and C. A. Stow. 1994. Fitting predator-prey models to time series with observation errors. *Ecology* 75:1254–1264.

Case, T. J. 2000. *An Illustrated Guide to Theoretical Ecology*. Oxford University Press, Oxford.

Casella, G., and R. L. Berger. 1990. *Statistical Inference*. Duxbury Press, Belmont, California.

Caswell, H. 1988. Theory and models in ecology: A different perspective. *Ecological Modelling* 43:33–44.

Caswell, H. 2001. *Matrix Population Models*. Sinauer, Sunderland, Massachusetts.

Caswell, H., and J. E. Cohen. 1991. Disturbance, interspecific interaction and diversity in metapopulations. Pp. 193–218 in M. Gilpin, and I. Hanski, eds. *Metapopulation Dynamics: Empirical and Theoretical Investigations*. Academic Press, London.

Catchpole, E. A., B. J. T. Morgan, T. N. Coulson, S. N. Freeman, and S. D. Albon. 2000. Factors influencing Soay sheep survival. *Applied Statistics* 49:453–472.

Charnov, E. L. 1991. Evolution of life history variation among female mammals. *Proceedings of the National Academy of Sciences USA* 88:1134–1137.

Chase, J. M., and M. A. Leibold. 2002. Spatial scale dictates the productivity-biodiversity relationship. *Nature* 416:427–430.

Chen, L., M. Fuentes, and J. M. Davis. 2006. Spatial temporal statistical modeling and prediction of environmental processes. Pages 121–144 in J. S. Clark and A. Gelfand, eds. *Applications of Computational Statistics in the Environmental Sciences: Hierarchical Bayes and MCMC Methods*. Oxford University Press, Oxford.

Chen, Z., and D. B. Dunson. 2003. Random effects selection in linear mixed models. *Biometrics* 59:762–769.

Chesson, P. 2000a. General theory of competitive coexistence in spatially-varying environments. *Theoretical Population Biology* 58:211–237.

Chesson, P. 2000b. Mechanisms of maintenance of species diversity. *Annual Reviews of Ecology and Systematics* 31:343–366.

Chib, S. 1993. Bayes regression with autoregressive errors: A Gibbs sampling approach. *Journal of Econometrics* 58:275–294.

Chib, S., and B. P. Carlin. 1999. On MCMC sampling in hierarchical longitudinal models. *Statistics and Computing* 9:17–26.

Chib, S., and E. Greenberg. 1994. Bayes inference in regression models with ARMA(p,q) errors. *Journal of Econometrics* 64:183–206.

Clark, D. A., and D. B. Clark. 1999. Assessing the growth of tropical rain forest trees: Issues for forest modeling and management. *Ecological Applications* 9:981–997.

Clark, J. S. 1990. Fire and climate change during the last 750 years in northwestern Minnesota. *Ecological Monographs* 60:135–159.

Clark, J. S. 1991. Disturbance and tree life history on the shifting mosaic landscape. *Ecology* 72:1102–1118.

Clark, J. S. 1998. Why trees migrate so fast: Confronting theory with dispersal biology and the Paleo record. *American Naturalist* 152:204–224.

Clark, J. S. 2003. Uncertainty in population growth rates calculated from demography: The hierarchical approach. *Ecology* 84:1370–1381.

Clark, J. S. 2005. Why environmental scientists are becoming Bayesians. *Ecology Letters* 8:2–14.

Clark, J. S., B. Beckage, P. Camill, B. Cleveland, J. Hille Ris Lambers, J. Lichter, J. MacLachlan, J. Mohan, and P. Wyckoff. 1999b. Interpreting recruitment limitation in forests. *American Journal of Botany* 86:1–16.

Clark, J. S., and O. Bjørnstad. 2004. Population time series: Process variability, observation errors, missing values, lags, and hidden states. *Ecology* 85:3140–3150.

Clark, J. S., S. R. Carpenter, M. Barber, S. Collins, A. Dobson, J. Foley, D. Lodge, M. Pascual, R. Pielke Jr., W. Pizer, C. Pringle, W. V. Reid, K. A. Rose, O. Sala, W. H. Schlesinger, D. Wall, and D. Wear. 2001a. Ecological forecasts: An emerging imperative. *Science* 293:657–660.

Clark, J. S., M. Dietze, I. Ibanez, and J. Mohan. 2003a. Coexistence: How to identify trophic tradeoffs. *Ecology* 84:17–31.

Clark, J. S., G. Ferraz, and N. Oguge. 2005. Hierarchical Bayes for structured and variable populations: From capture-recapture data to life-history prediction. *Ecology* 86:2232–2244.

Clark, J. S., and A. E. Gelfand. 2006a. The emergence of integrated hierarchical models in ecology. *Trends in Ecology and Evolution*, in press.

Clark, J. S., and A. E. Gelfand, eds. 2006b. *Hierarchical Modelling for the Environmental Sciences.* Oxford University Press, Oxford.

Clark, J. S., E. C. Grimm, J. J. Donovan, S. C. Fritz, D. R. Engstrom, and J. E. Almendinger. 2002. Drought cycles and landscape responses to past aridity on prairies of the Northern Great Plains, USA. *Ecology* 83:595–601.

Clark, J. S., L. Horvath, and M. Lewis. 2001a. On the estimation of spread for a biological population. *Statistics and Probability Letters* 51:225–234.

Clark, J. S., and S. L. LaDeau. 2006. Synthesizing ecological experiments and observational data with hierarchical Bayes. Pages 41–58 in J. S. Clark and A. Gelfand, eds. *Applications of Computational Statistics in the Environmental Sciences: Hierarchical Bayes and MCMC Methods.* Oxford University Press, Oxford.

Clark, J. S., S. LaDeau, and I. Ibanez. 2004. Fecundity of trees and the colonization-competition hypothesis. *Ecological Monographs* 74:415–442.

Clark, J. S., and M. Lavine. 2001. Bayesian statistics in ecology. Pages 327–346 in S. M. Scheiner and J. Gurevitch, eds. *Design and Analysis of Ecological Experiments.* Oxford University Press, Oxford.

Clark, J. S., M. Lewis, and L. Horvath. 2001b. Invasion by extremes: Variation in dispersal and reproduction retards population spread. *American Naturalist* 157:537–554.

Clark, J. S., M. Lewis, J. S. McLachlan, and J. Hille Ris Lambers. 2003b. Estimating population spread: What can we forecast and how well? *Ecology* 84:1979–1988.

Clark, J. S., E. Macklin, and L. Wood. 1998. Stages and spatial scales of recruitment limitation in southern Appalachian forests. *Ecological Monographs* 68:213–235.

Clark, J. S., and J. S. McLachlan. 2003. Stability of forest diversity. *Nature* 423:635–638.

Clark, J. S., M. Silman, R. Kern, E. Macklin, and J. Hille Ris Lambers. 1999a. Seed dispersal near and far: Generalized patterns across temperate and tropical forests. *Ecology* 80:1475–1494.

Clark, J. S., M. Wolosin, M. Dietze, I. Ibanez, S. LaDeau, and M. Welsh. Tree growth inference and prediction from diameter censuses and ring widths, in review.

Cochran, M. E., and S. Ellner. 1992. Simple methods for calculating age-based life history parameters for stage-structured populations. *Ecological Monographs* 62:345–364.

Cohen, J. E. 1986. Population forecasts and confidence intervals for Sweden: A comparison of model-based and empirical approaches. *Demography* 23:105–126.

Comins, H. N., and I. R. Noble. 1985. Dispersal, variability, and transient niches: Species coexistence in a uniformly variable environment. *American Naturalist* 126:706–723.

Condit, R., S. P. Hubbell, and R. B. Foster. 1995. Mortality rates of 205 neotropical tree and shrub species and the impact of severe drought. *Ecological Monographs* 65:419–439.

Condit, R., R. Sukumar, S. P. Hubbell, and R. B. Foster. 1998. Predicting population trends from size distributions: A direct test in a tropical tree community. *American Naturalist* 152:495–509.

Congdon, P. 2001. *Bayesian Statistical Modelling*. Wiley, New York.

Costantino, R. F., R. A. Desharnais, J. M. Cushing, and B. Dennis. 1997. Chaotic dynamics in an insect population. *Science* 275:389–391.

Costanza, R., R. d'Arge, R. de Groot, S. Farber, M. Grasso, B. Hannon, K. Limburg, S. Naeem, R. O'Neill, J. Paruelo, R. G. Raskin, P. Sutton, and M. van den Belt. 1997. The value of the world's ecosystem services and natural capital. *Nature* 387:253–260.

Coulson, T., E. A. Catchpole, S. D. Albon, B. J. T. Morgan, J. M. Pemberton, T. H. Clutton-Brock, M. J. Crawley, and B. T. Grenfell. 2001. Age, sex, density, winter weather and population crashes in Soay sheep. *Science* and 292:1528–1531.

Cousins, R. D. 1995. Why isn't every physicist a Bayesian? *American Journal of Physics* 63:398–410.

Cox, D. R., and D. Oakes. 1984. *Analysis of Survival Data*. Chapman and Hall, London.

Crank, J. 1975. *The Mathematics of Diffusion*. Clarendon Press, Oxford.

Cressie, N. A. C. 1993. *Statistics for Spatial Data*. Wiley, New York.

Cressie, N., and H.-C. Huang. 1999. Classes of nonseparable, spatio-temporal stationary covariance functions. *Journal of the American Statistical Association* 94:1330–1340.

Crouse, D. T., L. B. Crowder, and H. Caswell. 1987. A stage-based population model for loggerhead sea turtles and implications for conservation. *Ecology* 68:1412–1423.

Dalling, J. W., H. C. Muller-Landau, S. J. Wright, and S. P. Hubbell. 2002. Role of dispersal in the recruitment limitation of neotropical pioneer species. *Journal of Ecology* 90:714–727.

D'Antonio, C. M, J. T. Tunison, and R. Loh. 2000. Variation in impact of exotic grass fueled fires on native plant composition in relation to fire across an elevation gradient in Hawai'i. *Austral Ecology* 25:507–522.

Darwin, C. 1859. *On the Origin of Species*. Murray, London.

Daszak, P., A. A. Cunningham, and A. D. Hyatt. 2000. Emerging infectious diseases of wildlife—threats to biodiversity and human health. *Science* 287:443–449.

Dawid, A. P. 2004. Probability, causality and the empirical world: A Bayes-de Finetti-Popper-Borel synthesis. *Statistical Science* 19:44–57.

DeAngelis, D. L. 1992. *Dynamics of Nutrient Cycling and Food Webs*. Chapman and Hall, New York.

de Kroon, H., A. Plaisier, J. van Groenendael, and H. Caswell. 1986. Elasticity: The relative contribution of demographic parameters to population growth rate. *Ecology* 67:1427–1431.

Deevey, E. S. 1947. Life tables for natural populations of animals. *Quarterly Review of Biology* 22:283–314.

DeLucia, E. H., and R. T. Thomas. 2000. Photosynthetic responses of four hardwood species in a forest understory to atmospheric $[CO_2]$ enrichment. *Oecologia* 122:11–19.

DeLucia, E., J. Hamilton, S. Naidu, R. Thomas, J. Andrews, A. Finzi, M. Lavine, R. Matamala, J. Mohan, G. Hendrey, and W. Schlesinger. 1999. Net primary production of a forest ecosystem with experimental CO_2 enrichment. *Science*, 284:1177–1179.

Dennis, B., R. A. Desharnais, J. M. Cushing, and R. F. Costantino. 1995. Nonlinear demographic dynamics: Mathematical models, statistical methods, and biological experiments. *Ecological Monographs* 65:261–281.

Dennis, B., R. A. Desharnais, J. M. Cushing, S. M. Henson, and R. F. Costantino. 2001. Estimating chaos and complex dynamics in an insect population. *Ecological Monographs* 71:277–303.

Dennis, B., and M. L. Taper. 1994. Density-dependence in time-series observations of natural-populations—estimation and testing. *Ecological Monographs* 4:205–224.

de Valpine, P. 2003. Better inferences from population-dynamics experiments using Monte Carlo state-space likelihood methods. *Ecology* 84:3064–3077.

de Valpine, P., and A. Hastings. 2002. Fitting population models incorporating process noise and observation error. *Ecological Monographs* 72:57–76.

Dial, R., and J. Roughgarden. 1995. Experimental removal of insectivores from rain forest canopy: Direct and indirect effects. *Ecology* 76:1821–1834.

Diamond, J. M. 1975. Assembly of species communities. Pages 342–444 in M. L. Cody and J. M. Diamond, eds. *Ecology and Evolution of Communities*. Harvard University Press, Cambridge, Massachusetts.

Dickey, D. A., and W. A. Fuller. 1979. Distribution of the estimators for autoregressive time series with a unit root. *Journal of the American Statistical Association* 74:427–433.

Diez-Roux, A. V. 1998. Bringing context back into epidemiology: Variables and fallacies in multilevel analysis. *American Journal of Public Health* 88:216–222.

Diggle, P. J., K.-Y. Liang, and S. L. Zeger. 1996. *Analysis of Longitudinal Data*. Oxford University Press, Oxford.

Diggle, P. J., J. A. Tawn, and R. A. Moyeed. 1998. Model-based geostatistics (with discussion). *Applied Statistics* 47:299–350.

Dixon, P., and A. M. Ellison. 1996. Introduction: Ecological applications of Bayesian inference. *Ecological Applications* 6:1034–1035.

Doak, D. F. 1989. Spotted owls and old growth logging. *Conservation Biology* 3: 389–396.

Doak, D. F., D. Bigger, E. Harding-Smith, M. A. Marvier, R. O'Malley, and D. Thomson. 1998. The statistical inevitability of stability-diversity relationships in community ecology. *American Naturalist* 151:264–276.

Doak, D., P. Kareiva, and B. Klepetka. 1994. Modeling population viability for the desert tortoise in the western Mojave Desert. *Ecological Applications* 4:446–460.

Doak, D. F., and W. Morris. 1999. Detecting population-level consequences of ongoing environmental change without long-term monitoring. *Ecology* 80:1537–1551.

Dorazio, R. M., and F. A. Johnson. 2003. Bayesian inference and decision theory—a framework for decision making in natural resource management. *Ecological Applications* 13:556–563.

Dorazio, R. M., and J. A. Royle. 2003. Mixture models for estimating the size of a closed population when capture rates vary among individuals. *Biometrics* 59:351–364.

Draper, D., A. Pereira, P. Prado, A. Saltelli, R. Cheal, S. Eguilior, B. Mendes, and S. Tarantola. 1999. Scenario and parametric uncertainty in GESAMAC: A methodological study in nuclear waste disposal risk assessment. *Computer Physics Communications* 117:142–155.

Drechsler, M., K. Frank, I. Hanski, R. B. O'Hara, and C. Wissel. 2003. Ranking metapopulation extinction risk: From patterns in data to conservation management decisions. *Ecological Applications* 13:990–998.

Dupuis, J. A. 1995. Bayesian estimation of movement and survival probabilities from capture-recapture data. *Biometrika* 82:761–772.

Earn, D. J. D., P. Rohani, B. M. Bolker, and B. T. Grenfell. 2000. A simple model for complex dynamical transitions in epidemics. *Science* 287:667–670.

Easterling, M. R., S. P. Ellner, and P. M. Dixon. 2000. Size-specific sensitivity: Applying a new structured population model. *Ecology* 81:694–708.

Edelstein-Keshet, L. 1988. *Mathematical Models in Biology*. Random House, New York.

Edwards, A. W. F. 1972. *Likelihood*. Cambridge University Press, Cambridge.

Efron, B., and R. J. Tibshirani. 1993. *An Introduction to the Bootstrap*. Chapman and Hall, New York.

Ellison, A. M. 2004. Bayesian inference in ecology. *Ecology Letters* 7:509–520.

Ellner, S. P., and J. Feiberg. 2003. Using PVA for management in light of uncertainty: Effects of habitat, hatcheries, and harvest on salmon viability. *Ecology* 84:1359–1369.

Ellner, S. and P. Turchin. 1995. Chaos in a noisy world: New methods and evidence from time series analysis. *American Naturalist* 145:343–375.

Elser, J. J., and R. W. Sterner. 2002. *Ecological Stoichiometry: The Biology of Elements from Molecules to the Biosphere*. Princeton University Press, Princeton, New Jersey.

Elton, C. S. 1924. Periodic fluctuations in the number of animals: Their causes and effects. *British Journal of Experimental Biology* 2:119–163.

Elton, C. S. 1927. *Animal Ecology*. Sedgwick and Jackson, London.

Engen, S., O. Blakke, and A. Islam. 1998. Demographic and environmental stochasticity—concepts and definitions. *Biometrics* 54:840–846.

Enright, N. J., M. Franco, and J. Silvertown. 1995. Comparing plant life histories using elasticity analysis: The importance of life span and the number of life-cycle stages. *Oecologia* 104:79–84.

Farnsworth, E., A. Ellison, and W. Gong. 1996. Elevated CO_2 alters anatomy, physiology, growth, and reproduction of red mangrove (*Rhizophora mangle L*). *Oecoloia*, 108: 599–609.

Feiberg, J., and S. P. Ellner. 2003. Stochastic matrix models for conservation and management: A comparative review of methods. *Ecology Letters* 4:244–266.

Ferguson, N. M., C. A. Donnelly, and R. M. Anderson. 2001. The foot-and-mouth epidemic in Great Britain: Pattern of spread and impact of interventions. *Science* 292:1155–1160.

Finzi, A. C., and C. D. Canham. 2000. Sapling growth in response to light and nitrogen availability in a southern New England forest. *Forest Ecology and Management* 131:153–165.

Fisher, R. A. 1934. Two new properties of mathematical likelihood. *Proceedings of the Royal Society* A 144:285–307.

Fisher, R. A. 1935. The logic of inductive inference. *Journal of the Royal Statistical Society* 98:39–54.

Fisher, R. A. 1937. The wave of advance of advantageous genes. *Annals of Eugenics* 7:355–369.

Fisher, R. A. 1959. *Statistical Methods and Scientific Inference*. Hafner, New York.

Flinn, K. M., M. Velland, and P. L. Marks. 2005. Environmental causes and consequences of forest clearance and agricultural abandonment in central New York, USA. *Journal of Biogeography* 32:439–452.

Fortin, D., H. L. Beyer, M. S. Boyce, D. W. Smith, T. Duchesne, T., and J. S. Mao. 2005. Wolves influence elk movements: Behavior shapes a trophic cascade in Yellowstone National Park. *Ecology* 86:1320–1330.

Fortin, M.-J., T. H. Keitt, B. A. Maurer, M. L. Taper, D. M. Kaufman, and T. M. Blackburn. 2005. Species ranges and distributional limits: Pattern analysis and statistical issues. *Oikos* 108:7–17.

Foster, D. R. 1988. Disturbance history, community organization and vegetation dynamics of the old-growth Pisgah forest, south-western New Hampshire, U.S.A. *Journal of Ecology* 76:105–134.

Fournier, D., and C. P. Archibald. 1982. A general theory for analyzing catch at age data. *Canadian Journal of Fisheries and Aquatic Sciences* 39:1195–1207.

Fox, G. A., and J. Gurevitch. 2000. Transient analysis of matrix models. *American Naturalist* 156:242–256.

Franklin, A. B. 1992. Population regulation in northern spotted owls: Theoretical implications for management. Pages 815–827 in D. R. McCullough and R. H. Barrett, eds. *Wildlife 2001: Populations*. Elsevier Applied Science, London.

Franklin, A. B., D. R. Anderson, R. J. Gutierrez, and K. P. Burham. 2000. Climate, habitat quality, and fitness in northern spotted owl populations in northwestern California. *Ecological Monographs* 70:539–590.

Fraser, D. F., J. F. Gilliam, M. J. Daley, A. N. Le, and G. T. Skalski. 2001. Explaining leptokurtic movement distributions: Intrapopulation variation in boldness and exploration. *American Naturalist* 158:124–135.

Friedman, F. 2004. Inferring cellular networks using probabilistic graphical models. *Science* 303:799–805.

Fuentes, M. 2001. High frequency kriging for nonstationary environmental processes. *Environmetrics* 12:469–483.

Fuentes, M. 2002. Spectral methods for nonstationary processes. *Biometrika* 89:197–210.

Fuentes, M., and A. E. Raftery. 2005. Model evaluation and spatial interpolation by Bayesian combination of observations with outputs from numerical models. *Biometrics* 61:36–45.

Fujiwara, M., and H. Caswell. 2002. Estimating population projection matrices from multi-stage mark-recapture data. *Ecology* 83:3257–3265.

Gaggiotti, O. E., S. P. Brooks, W. Amos, and J. Harwood. 2004. Combining demographic, environmental and genetic data to test hypotheses about colonization events in metapopulations. *Molecular Ecology* 13:811–825.

Galloway, J. N., F. J. Dentener, D. G. Capone, E. W. Boyer, R. W. Howarth, S. P. Seitzinger, G. P. Asner, C. Cleveland, P. Green, E. Holland, D. M. Karl, A. F. Michaels, J. H. Porter, A. Townsend, and C. Vorosmarty. 2004. Nitrogen cycles: Past, present, and future. *Biogeochemistry* 70:153–226.

Galton, F. 1889. *Natural Inheritance*. Macmillan, London.

Gamerman, D. 1997. *Markov Chain Monte Carlo: Stochastic Simulation for Bayesian Inference*. Chapman and Hall, London.

Gelfand, A. E., and D. K. Dey. 1994. Bayesian model choice: Asymptotics and exact calculation. *Journal of the Royal Statistical Society* B 56:501–514.

Gelfand, A. E., and S. K. Ghosh. 1998. Model choice: A minimum posterior predictive loss approach. *Biometrika* 85:1–11.

Gelfand, A. E., S. K. Ghosh, J. R. Knight, and C. F. Sirmans. 1998. Spatio-temporal modeling of residential sales data. *Journal of the American Statistical Association* 16:312–321.

Gelfand, A. E., S. E. Hills, A. Racine-Poon, and A. F. M. Smith. 1990. Illustration of Bayesian inference in normal data models using Gibbs sampling. *Journal of the American Statistical Association* 85:972–985.

Gelfand, A. E., H.-J. Kim, C. F. Sirmans, and S. Banerjee. 2003. Spatial modeling with spatially varying coefficient process. *Journal of the American Statistical Association* 98:387–396.

Gelfand, A. E., and S. K. Sahu. 1999. Identifiability, improper priors, and Gibbs sampling for generalized linear models. *Journal of the American Statistical Association* 94:247–253.

Gelfand, A. E., A. M. Schmidt, S. Wu, J. A. J. Silander, A. Latimer, and A. G. Rebelo. 2004. Explaining species diversity through species level hierarchical modeling. *Applied Statistics* 65:1–20.

Gelfand, A. E., and A. F. M. Smith. 1990. Sampling-based approaches to calculating marginal densities. *Journal of the American Statistical Association* 85:398–409.

Gelfand, A. E., L. Zhu, and B. P. Carlin. 2001. On the change of support problem for spatio-temporal data. *Biostatistics* 2:31–45.

Gelman, A., J. B. Carlin, H. S. Stern, and D. B. Rubin. 1995. *Bayesian Data Analysis*. Chapman and Hall, London.

Gelman, A., J. B. Carlin, H. S. Stern, and D. B. Rubin. 2006. *Bayesian Data Analysis*. 2nd ed. Chapman and Hall, London.

Gelman, A., and D. B. Rubin. 1992. Inference from iterative simulation using multiple sequences. *Statistical Science* 7:457–511.

George, E. I., U. E. Makov, and A. F. M. Smith. 1993. Conjugate likelihood distributions. *Scandinavian Journal of Statistics* 20:147–156.

George, E. I., and C. P. Robert. 1992. Capture-recapture estimation via Gibbs sampling. *Biometrika* 79:677–683.

Geweke, J., and H. Tanizaki. 2001. Bayesian estimation of state-space models using the Metropolis-Hastings algorithm within Gibbs sampling. *Computational Statistics and Data Analysis* 37:151–170.

Geweke, J., and N. Terui. 1993. Bayesian threshold autoregressive models for nonlinear time series. *Journal of Time Series Analysis* 14:441–454.

Ghosh, J. K., and R. Mukerjee. 1992. Non-informative priors. Pp. 195–203 in J. M. Bernardo, J. O. Berger, A. P. Dawid, and A. F. M. Smith, eds. *Bayesian Statistics*. Vol. 4. Oxford University Press, London.

Gilks, W. R., S. Richardson, and D. J. Spiegelhalter. 1996. *Markov Chain Monte Carlo in Practice*. Chapman & Hall, London.

Gilleland, E., D. Nychka, and U. Schneider. 2006. Spatial models for the distribution of extremes. Pages 170–184 in J. S. Clark and A. Gelfand, eds. *Hierarchical Models of the Environment*. Oxford University Press, in press.

Girosi, F., and G. King. 2006. *Demographic Forecasting*, in press.

Gneiting, T. 2002. Nonseparable, stationary covariance functions for space-time data. *Journal of the American Statistical Association* 97:590–600.

Govindarajan, S., M. Dietze, P. Agarwal, and J. S. Clark. 2004. A scalable algorithm for dispersing populations. *Journal of Intelligent Information Systems*, in press.

Govindarajan, S., M. Dietze, P. Agarwal, and J. S. Clark. 2004. A scalable model of forest dynamics. *Proceedings of the ACM Symposium on Computational Geometry* 106–115.

Gradshteyn, I. S., and I. M. Ryzhik. 1980. *Table of Integrals, Series, and Products*. Academic Press, New York.

Grant, A., and T. G. Benton. 2000. Elasticity analysis for density-dependent populations in stochastic environments. *Ecology* 81:680–693.

Grant, P. R., and B. R. Grant. 1992. Demography and the genetically effective sizes of two populations of Darwin's finches. *Ecology* 73:766–784.

Green, P. J. 1995. Reversible jump Markov chain Monte Carlo computation and Bayesian model determination. *Biometrika* 82:711–732.

Greenberg, C. H., and B. R. Parresol. 2002. Dynamics of acorn production by five species of southern Appalachian oaks. Pages 149–172 in W. J. McSchea and W. M. Healy, eds. *Oak Forest Ecosystems*. Johns Hopkins University Press, Baltimore, Maryland.

Greene, D. F., and E. A. Johnson. 1989. A model of wind dispersal of winged or plumed seeds. *Ecology* 70:339–347.

Greenland, S., and I. Robins. 1994. Ecologic studies—biases, counterexamples. *American Journal of Epidemiology* 139:747–760.

Gregoire, T. G., S. M. Zedaker, and N. S. Nicholas. 1990. Modeling relative error in stem basal area estimates. *Canadian Journal of Forest Research* 20:496–502.

Gregory, P. H. 1968. *The Microbiology of the Atmosphere*. Wiley, New York.

Grenfell, B. T., O. N. Bjørnstad, and B. Finkenstädt. 2002. Dynamics of measles epidemics: Scaling noise, determinism and predictability with the TSIR model. *Ecological Monographs* 72:185–202.

Grenfell B. T., O. N. Bjørnstad, and J. Kappey. 2001. Traveling waves and spatial hierarchies in measles epidemics. *Nature* 414:716–723.

Grenfell, B. T., K. Wilson, B. F. Finkenstadt, T. N. Coulson, S. Murray, S. D. Albon, J. M. Pemberton, T. H. Clutton-Brock, and M. J. Crawley. 1998. Noise and determinism in synchronized sheep dynamics. *Nature* 394:674–676.

Grenfell, B. T., B. F. Finkenstadt, K. Wilson, T. N. Coulson, and M. J. Crawley. 2000. Nonlinearity and the Moran effect. *Nature* 406:847.

Gross, K., A. R. Ives, and E. V. Nordheim. 2005. Estimating fluctuating vital rates from time series data: A case study of aphid biocontrol. *Ecology* 86: 740–752.

Grover, J. P. 1991. Resource competition in a variable environment: Phytoplankton growing according to the variable-internal-stores model. *American Naturalist* 138:811–835.

Grünbaum, D. 1998. Using spatially explicit models to characterize foraging performance in heterogeneous landscapes. *American Naturalist* 151:97–115.

Guevara, S., and J. Laborde. 1993. Monitoring seed dispersal at isolated standing trees in tropical pastures: Consequences for local species availability. *Vegetatio* 107/108:319–338.

Gurevitch, J., and S. T. Chester. 1986. Analysis of repeated measures experiments. *Ecology* 67:251–255.

Gutierrez, R. J., A. B. Franklin, W. Lehaye, V. J. Meretsky, and J. P. Ward. 1985. Juvenile spotted owl dispersal in northwestern California: Preliminary results. Pages 60–65 in R. J. Gutierrez and A. B. Carey, eds. *Ecology and Management of the Spotted Owl in the Pacific Northwest*. U.S. Forest Service Technical Report PNW 185.

Haddad, N. M. 1999. Corridor use predicted from behaviors at habitat boundaries. *American Naturalist* 153:215–227.

Hanski, I. 1991. Single-species metapopulation dynamics: Concepts, models and observations. *Biological Journal of the Linnean Society* 42:17–38.

Hanski I. 1998. Metapopulation dynamics. Nature 396:41–49.

Hanson, P. J., J. S. Amthor, S. D. Wullschleger, K. B. Wilson, R. F. Grant, A. Hartley, D. Hui, E. R. Hunt Jr., D. W. Johnson, J. S. Kimball, A. W. King, Y. Luo, S. G. McNulty, G. Sun, P. E. Thornton, S. Wang, M. Williams, D. D. Baldocchi, and R. M. Cushman. 2004. Oak forest carbon and water simulations: Model intercomparisons and evaluations against independent data. *Ecological Monographs* 74:443–489.

Harrison, S. 1989. Long-distance dispersal and colonization in the bay checkerspot butterfly, *Euphydryas editha bayensis*. *Ecology* 70:1236–1243.

Harrison, S., A. Stahl, and D. Doak. 1993. Spatial models and spotted owls: Exploring some biological issues behind recent events. *Conservation Biology* 7:950–953.

Hartemink, A., D. Gifford, T. Jaakkola, and R. Young. 2002. Bayesian methods for elucidating genetic regulatory networks. *IEEE Intelligent Systems in Biology* 17:37–43.

Harvey, A. C. 1989. *Forecasting, Structural Time Series Models and the Kalman Filter*. Cambridge University Press, Cambridge.

Harvey, C. A. 2000. Windbreaks enhance seed dispersal into agricultural landscapes in Monteverde, Costa Rica. *Ecological Applications* 10:155–173.

Hassell, M. P., J. H. Lawton, and R. M. May. 1976. Patterns of dynamical behaviour in single-species populations. *Journal of Animal Ecology* 45:471–486.

Hastings, A. 1980. Disturbance, coexistence, history, and competition for space. *Theoretical Population Biology* 18:363–373.

Hastings, A. 1997. *Population Biology: Concepts and Models*. Springer, New York.

Hastings, A. 2004. Transients: The key to long-term ecological understanding? *Trends in Ecology and Evolution* 19:39–45.

Hastings, A., C. L. Horn, S. Ellner, P. Turchin, and H. C. J. Godfray. 1993. Chaos in ecology: Is Mother Nature a strange attractor? *Annual Reviews of Ecology and Systematics* 24:1–33.

Hastings, A., and C. L. Wolin. 1989. Within-patch dynamics in a metapopulation. *Ecology* 70:1261–1266.

Hattenschwiler, S., and C. Korner. 1996. Effects of elevated CO_2 and increased nitrogen deposition on photosynthesis and growth of understory plants in spruce model ecosystems. *Oecologia* 106:172–180.

Havel, J. E., J. B. Shurin, and J. R. Jones. 2002. Estimating dispersal from patterns of spread: Spatial and local control of lake invasions. *Ecology* 83:3306–3318.

He, C. Z. 2003. Bayesian modeling of age-specific survival in bird nesting studies under irregular visits. *Biometrics* 59:962–973.

Henderson, R., S. Shimakura, and D. Gorst. 2002. Modeling spatial variation in leukemia survival data. *Journal of the American Statistical Association* 97:965–972.

Henson, S. M., R. F. Costantino, J. M. Cushing, R. A. Desharnais, B. Dennis, and A. A. King. 2001. Lattice effects observed in chaotic dynamics of experimental populations. *Science* 294:602–605.

Hestbeck, J. B., J. D. Nichols, and R. A. Malecki. 1991. Estimates of movement and site fidelity using mark–resight data of wintering Canada geese. *Ecology* 72:523–533.

Higdon, D. 2002. Space and space-time modeling using process convolutions. Pages 37–56 in C. Anderson, V. Barnett, P. C. Chatwin, and A. H. El-Shaarawi, eds. *Quantitative Methods for Current Environmental Issues*. Springer Verlag, New York.

Higgens, K., A. Hastings, J. N. Sarvela, and L. W. Botsford. 1997. Stochastic dynamics and deterministic skeletons: Population behavior of dungeness crab. *Science* 276:1431–1435.

Higgins, S. I., and D. M. Richardson. 1999. Long-distance dispersal. *American Naturalist* 153:464–475.

Hilborn, R., and M. Mangel. 1997. *The Ecological Detective*. Princeton University Press, Princeton, New Jersey.

Hille Ris Lambers, J., and J. S. Clark. 2003. Effects of dispersal, shrubs, and density-dependent mortality on seed and seedling distributions in temperate forests. *Canadian Journal of Forest Research* 33:783–795.

Hille Ris Lambers, J., J. S. Clark, and M. Lavine. 2005. Seed banking in temperate forests: Implications for recruitment limitation. *Ecology* 86:85–95.

Hixon, M. A., S. W. Pacala, and S. A. Sandin. 2002. Population regulation: Historical context and contemporary challenges of open vs. closed systems. *Ecology* 83:1490–1508.

Hoeting, J. A., D. Madigan, A. E. Raftery, and C. T. Volinsky. 1999. Bayesian model averaging: A tutorial (with discussion). *Statistical Science* 14:382–417.

Holt, R. D. 1977. Predation, apparent competition, and the structure of prey communities. *Theoretical Population Biology* 12:197–229.

Holthuijzen, A. M. A., and T. L. Sharik. 1985. The red cedar (*Juniperus virginiana* L.) seed shadow along a fenceline. *American Midland Naturalist* 113:200–202.

Hubbell, S. P. 2001. *The Unified Neutral Theory of Biodiversity and Biogeography.* Princeton University Press, Princeton, New Jersey.

Hubbell, S. P., and R. B. Foster. 1986. Biology, chance, and history and the structure of tropical rain forest tree communities. Pages 314–330 in J. Diamond and T. J. Case, eds. *Community Ecology.* Harper and Row, New York.

Hudgens, B. 2000. When does migration matter to aphid metapopulations? Ph.D. thesis, Duke University, Durham, North Carolina.

Hudson, P. J., A. P. Dobson, and D. Newborn. 1998. Prevention of population cycles by parasite removal. *Science* 282:2256–2258.

Huerta, G., and M. West. 1999. Priors and component structures in autoregressive time series models. *Journal of the Royal Statistical Society* B 61:881–899.

Huitu, O., M. Koivula, E. Korpimäki, T. Klemola, and N. Norrdahl. 2003. Winter food supply limits growth of northern vole populations in the absence of predation. *Ecology* 84:2108–2118.

Hurlbert, S. H. 1984. Pseudoreplication and the design of ecological field experiments. *Ecological Monographs* 54:187–211.

Huston, M. A. 1994. *Biological Diversity.* Cambridge University Press, Cambridge.

Huston, M., and T. Smith. 1987. Plant succession: Life history and competition. *American Naturalist* 130:168–198.

Hutchinson, G. E. 1959. Homage to Santa Rosalia; or, Why are there so many kinds of animals. *American Naturalist* 93:145–159.

Hutchinson, G. E. 1961. The paradox of the plankton. *American Naturalist* 95:137–145.

Ibáñez, I., J. S. Clark, M. C. Dietze, K. Feeley, M. Hersh, S. LaDeau, A. McBride, N. E. Welch, and M. S. Wolosin. 2006. Predicting biodiversity change: Outside the climate envelope, beyond the species-area curve. *Ecology*, in press.

Ibrahim, J. G., and M.-H. Chen. 2000. Power prior distributions for regression models. *Statistical Science* 15:46–60.

Inchausti, P., and J. Halley. 2001. Investigating long-term ecological variability using the global population dynamics database. *Science* 293:655–657.

Ives, A. R., S. R. Carpenter, and B. Dennis. 1999. Community interaction webs and the response of a zooplankton community to experimental manipulations of planktivory. *Ecology* 80:1405–1421.

Ives, A. R., B. Dennis, K. L. Cottingham, and S. R. Carpenter. 2003. Estimating community stability and ecological interactions from time-series data. *Ecological Monographs* 73:301–330.

Jackson, R. B., and W. H. Schlesinger. 2004. Curbing the U.S. carbon deficit. *Proceedings of the National Academy of Sciences* 101:15827–15829.

Jacquier, E., N. Polson, and P. Rossi. 1994. Bayesian analysis of stochastic volatility models. *Journal of Business and Economic Statistics* 12:371–417.

Jarvis, P. G., J. M. Massheder, S. E. Hale, J. B. Moncrieff, M. Rayment, and S. L. Scott. 1997. Seasonal variation of carbon dioxide, water vapor and energy exchanges of a boreal black spruce forest. *Journal of Geophysical Research* 102:28953–28966.

Jedrzejewska, B., W. Jedrzejewski, A. N. Bunevich, L. Milkowsi, and Z. A. Krasinski. 1997. Factors shaping population densities and increase rates of ungulates in

Bialowieza primeval forest (Poland and Belarus) in the 19th and 20th centuries. *Acta Theriologica* 42:399–451.

Jeffreys, H. 1961. *Theory of Probability*. Oxford University Press, London.

Johnson, D. S., and J. A. Hoeting. 2003. Autoregressive models for capture-recapture data: A Bayesian approach. *Biometrics* 59:341–350.

Johnson, W. C., and T. Webb III. 1989. The role of blue jays in the postglacial dispersal of fagaceous trees in eastern North America. *Journal of Biogeography* 16:561–571.

Johnson, M. E. 1987. *Multivariate Statistical Simulation*. Wiley, New York.

Johnson, N. L., and S. Kotz. 1970. *Distributions in Statistics*. Wiley, New York.

Johnson, V. 1996. On Bayesian analysis of multirater ordinal data: An application to automated essay grading. *Journal of the American Statistical Association* 91:42–51.

Jolly, G. M. 1965. Explicit estimates from capture-recapture data with both death and dilution-stochastic model. *Biometrika* 52:225–247.

Jones, R. H., and L. M. Ackerson. 1990. Serial correlation in unequally spaced longitudinal data. *Biometrika* 77:721–731.

Jonsen, I. D., R. A. Myers, and J. Mills Flemming. 2003. Meta-analysis of animal movement using state–space models. *Ecology* 84:3055–3063.

Kalbfleisch, J. D., and R. L. Prentice. 1980. *The Statistical Analysis of Failure Time Data*. Wiley, New York.

Kareiva, P. M. 1982. Experimental and mathematical analyses of herbivore movement: Quantifying the influence of plant spacing and quality on foraging discrimination. *Ecological Monographs* 52:261–282.

Kareiva, P. M. 1983. Local movement in herbivorous insects: Applying a passive diffusion model to mark-recapture field experiments. *Oecologia* 57:322–324.

Kareiva, P. M., and N. Shigesada. 1983. Analyzing insect movement as a correlated random walk. *Oecologia* 56:234–238.

Katul, G. G., A. Porporato, R. Nathan, M. Siqueira, M. B. Soons, D. Poggi, H. S. Horn, and S. A. Levin. 2005. Mechanistic analytical models for long-distance seed dispersal by wind. *American Naturalist* 166:368–381.

Kauffman, M. J., W. F. Frick, and J. Linthicum. 2003. Estimation of habitat-specific demography and population growth for peregrine falcons in California. *Ecological Applications* 13:1802–1816.

Keeling, M. J., M. E. J. Woolhouse, D. J. Shaw, L. Matthews, M. Chase-Topping, D. T. Haydon, S. J. Cornell, J. Kappey, J. Wilesmith, and B. T. Grenfell. 2001. Dynamics of the 2001 UK foot and mouth epidemic: Stochastic dispersal in a heterogeneous landscape. *Science* 294:813–817.

Kelly, D, and V. Sork. 2002. Mast seeding: Patterns, causes, and consequences. *Annual Review of Ecology and Systematics* 33:427–447.

Kelsall, J. E., and J. C. Wakefield. 2002. Modeling spatial variation in disease risk: A geostatistical approach. *Journal of the American Statistical Association* 97: 692–701.

Kendall, B. E., C. J. Briggs, W. W. Murdoch, P. Turchin, S. P. Ellner, E. Mccauley, R. M. Nisbet, and S. N. Wood 1999. Why do populations cycle? A synthesis of statistical and mechanistic modeling approaches. *Ecology* 80:1789–1805.

Kendall, B. E., S. P. Ellner, E. McCauley, S. N. Wood, C. J. Briggs, W. W. Murdoch, and P. Turchin. 2005. Population cycles in the pine looper moth (*Bupalus piniarius*): Dynamical tests of mechanical hypotheses. *Ecological Monographs* 75:259–276.

Kendall, W. L., and J. D. Nichols. 2002. Estimating state-transition probabilities for unobservable states using capture-recapture/resighting data. *Ecology* 83: 3276–3284.

Kierstead, H., and L. B. Slobotkin. 1953. The size of water masses containing plankton blooms. *Journal of Marine Research* 12:141–147.

King, R., and S. P. Brooks. 2002. Bayesian model discrimination for multiple strata capture-recapture data. *Biometrika* 89:785–806.

Kingsland, S. E. 1985. *Modeling Nature*. University of Chicago Press, Chicago, Illinois.

Klein, J. P., and M. L. Moeschberger. 1997. *Survival Analysis*. Springer Verlag, Berlin.

Kobe, R. K. 1999. Light gradient partitioning among tropical tree species through differential seedling mortality and growth. *Ecology* 80:187–201.

Koenig, W. D., J. M. H. Knops, W. J. Carmen, and M. T. Stanback. 1999. Spatial dynamics in the absence of dispersal: Acorn production by oaks in central coastal California. *Ecography* 22:499–506.

Kolar, C. S., and D. M. Lodge. 2002. Ecological predictions and risk assessments for alien species. *Science* 298:1233–1236.

Kolmogorov, A., I. Petrovsky, and N. Piscounov. 1937. Étude de l'equation de la diffusion avec croissance de la quantité de matière et son application à un problèma biologigue. *Bulletin de l'Université d'Etat a Moscou se Compose des Sections Suivantes* A 1:1–25.

Korpimaki, E., and K. Norrdahl. 1991. Do breeding nomadic avian predators dampen population fluctuations of small mammals? *Oikos* 62:195–208.

Korpimaki, E., K. Norrdahl, and T. Rinta-Jaskari. 1991. Responses of stoats and least weasels to fluctuating vole abundances: Is the low phase of the vole cycle due to mustelid predation? *Oecologia* 88:552–561.

Kot, M. 2001. *Elements of Mathematical Ecology*. Cambridge University Press, Cambridge.

Kot, M., M. A. Lewis, and P. van den Driessche. 1996. Dispersal data and the spread of invading organisms. *Ecology* 77:2027–2042.

Kot, M., J. Medlock, T. Reluga, and D. B. Walton. 2004. Stochasticity, invasions, and branching random walks. *Theoretical Population Biology* 66:175–184.

Krebs, C. J., R. Boonstra, S. Boutin, and A. R. E. Sinclair. 2001. What drives the 10-year cycle of snowshoe hares? *BioScience* 51:25–35.

Krebs, C. J., S. Boutin, R. Boonstra, A. R. E. Sinclair, J. N. M. Smith, M. R. T. Dale, K. Martin, and R. Turkington. 1995. Impact of food and predation on the snowshoe hare cycle. *Science* 269:1112–1115.

Lack, D. 1954. *The natural regulation of animal numbers*. Oxford University Press, Oxford.

LaDeau, S., and J. S. Clark. 2001. Rising CO_2 and the fecundity of forest trees. *Science* 292:95–98.

LaDeau, S. L., and J. S. Clark. 2006. Elevated CO_2 and tree fecundity: The role of tree size, interannual variability, and population heterogeneity. *Global Change Biology*, in press.

Lairde, N. M., and J. H. Ware. 1982. Random effects models for longitudinal data. *Biometrics* 38:963–974.

Lamberson, R. H., R. McKelvey, B. R. Noon, and C. Voss. 1992. A dynamic analysis of northern spotted owl viability in a fragmented landscape. *Conservation Biology* 6:505–512.

Lande, R. 1987. Extinction thresholds in demographic models of territorial populations. *American Naturalist* 130:624–635.

Lande, R. 1988. Demographic models of the northern spotted owl (*Strix occidentalis caurina*). *Oecologia* 75:601–607.

Lande, R., S. Engen, B.-E. Saether, F. Filli, E. Matthysen, and H. Weimerskirch. 2002. Estimating density dependence from population time series: Using demographic theory and life-history data. *American Naturalist* 159:321–337.

Lande, R., S. Engen, and B.-E. Saether. 2003. *Stochastic Population Dynamics in Ecology and Conservation*. Oxford University Press, Oxford.

Lange, N., B. P. Carlin, and A. E. Gelfand. 1992. Hierarchical Bayes models for the progression of HIV infection using longitudinal CD4 T-cell numbers. *Journal of the American Statistical Association* 87:615–631.

Langtimm, C. A., and C. A. Beck. 2003. Lower survival probabilities for adult Florida manatees in years with intense coastal storms. *Ecological Applications* 13:257–268.

Lavine, M., B. Beckage, and J. S. Clark. 2002. Statistical modeling of seedling mortality. *Journal of Agricultural, Biological, and Environmental Statistics* 7:21–41.

Layton, D. F., and R. A. Levine. 2003. How much does the far future matter? A hierarchical Bayesian analysis of the public's willingness to mitigate ecological impacts of climate change. *Journal of the American Statistical Association* 98:533–544.

Lebreton, J.-D., K. P. Burnham, J. Clobert, and D. R. Anderson. 1992. Modeling survival and testing biological hypotheses using marked animals: A unified approach with case studies. *Ecological Monographs* 62:67–118.

Lee, R. D., and L. R. Carter. 1992. Modeling and forecasting U.S. mortality. *Journal of the American Statistical Association* 87:659–675.

Legendre, P. 1993a. Spatial autocorrelation: Trouble or new paradigm? *Ecology* 74:1659–1673.

Legendre, P. 1993b. Real data are messy. *Statistics and Computing* 3:197–199.

Leirs, H., N. C. Stenseth, J. D. Nichols, J. E. Hines, R. Verhagen, and W. Verheyen. 1997. Stochastic seasonality and nonlinear density-dependent factors regulate population size in an African rodent. *Nature* 389:176–180.

Leslie, P. H. 1945. On the use of matrices in certain population mathematics. *Biometrika* 33:14–212.

Leslie, P. H. 1966. The intrinsic rate of increase and the overlap of successive generations in a population of guillemots (*Uria aalge Pont.*). *Journal of Animal Ecology* 35:291–301.

Levin, S. A. 1986. Random walk models of movement and their implications. Pages 149–154 in T. G. Hallam and S. A. Levin, eds. *Mathematical Ecology*. Springer Verlag, Berlin.

Levin, S. A. 1992. The problem of pattern and scale in ecology. *Ecology* 73:1943–1967.

Levin, S. A. 1998. Ecosystems and the biosphere as complex adaptive systems. *Ecosystems* 1:431–436.

Levins, R. 1979. Coexistence in a variable environment. *American Naturalist* 114:765–783.

Levins, R. 1969. Some demographic and genetic consequences of environmental heterogeneity for biological control. *Bulletin of the Entomological Society of America* 15:237–240.

Lewis, M. A. 1997. Variability, patchiness, and jump dispersal in the spread of an invading population. Pages 46–69 in D. Tilman and P. Kareiva, eds. *Spatial Ecology*. Princeton University Press, Princeton, New Jersey.

Lewis, M. A. 2000. Spread rate for a nonlinear stochastic invasion. *Journal of Mathematical Biology* 41:430–454.

Lewis, M. A., and P. Kareiva. 1993. Allee dynamics and the spread of invading organisms. *Theoretical Population Biology* 43:141–158.

Lewis, M. A., and J. D. Murray. 1993. Modeling territoriality and wolf–deer interactions. *Nature* 366:738–740.

Lewis, M. A., and S. Pacala. 2000. Modeling and analysis of stochastic invasion processes. *Journal of Mathematical Biology* 41:387–429.

Liebhold, A. M., and J. Bascompte. 2003. The allee effect, stochastic dynamics and the eradication of alien species. *Ecology Letters* 6:133–140.

Lima, M., P. A. Marquet, and F. M. Jaksic. 1999. El Niño events, precipitation patterns and rodent outbreaks are statistically associated in semiarid Chile. *Ecography* 22:213–218.

Lindsey, J. K. 1999. *Models for Repeated Measurements*. Oxford University Press, Oxford.

Lindley, D. V., and L. D. Phillips. 1976. Inference for a Bernoulli process (a Bayesian view). *American Statistician* 30:112–119.

Lindstrom, J., H. Kokko, E. Ranta, and H. Linden. 1999. Density dependence and the response surface methodology. *Oikos* 85:40–52.

Lindstrom, M., and D. Bates. 1990. Nonlinear mixed effects models for repeated measures data. *Biometrics* 46:673–687.

Link, W. A. 2002. Scaling in sensitivity analysis. *Ecology* 83:3299–3305.

Link, W. A. 2003. Nonidentifiability of population size from capture-recapture data with heterogeneous detection probabilities. *Biometrics* 59:1123–1130.

Link, W. A., and J. R. Sauer. 1998. Estimating population change from count data: Application to the North American Breeding Bird Survey. *Ecological Applications* 8:258–268.

Liu, J. S., and R. Chen. 1998. Sequential Monte Carlo methods for dynamic systems. *Journal of the American Statistical Association* 93:1032–1044.

Lorenz, E. N. 1963. Deterministic nonperiodic flow. *Journal of Atmospheric Sciences* 20:130–141.

Loreau, M., S. Naeem, P. Inchaustl, J. Bengtsseon, J. P. Grime, A. Hector, D. U. Hooper, M. A. Huston, D. Raffaelli, B. Schmid, D. Tilman, and D. A. Wardle. 2001. Biodiversity and ecosystem functioning: Current knowledge and future challenges. *Science* 294:804–808.

Lubina, J. A., and S. A. Levin. 1988. The spread of a reinvading species: Range expansion in the California sea otter. *American Naturalist* 131:526–543.

Ludwig, D. 1999. Is it meaningful to estimate an extinction probability? *Ecology* 10:298–310.

Lunn, D. J., J. Wakefield, and A. Racine-Poon. 2001. Cumulative logit models for ordinal data. A case study involving allergic rhinitis severity scores. *Statistics in Medicine* 20:2261–2285.

Lutscher, F., E. Pachepsky, and M. A. Lewis. 2006. The effect of dispersal patterns on stream populations. *SIAM Journal of Applied Mathematics*, in press.

MacArthur, R. H., and E. O. Wilson. 1967. *The Theory of Island Biogeography*. Princeton University Press, Princeton, New Jersey.

MacGibbon, B., and T. J. Tomberlin. 1989. Small area estimates of proportions via empirical Bayes techniques. *Survey Methodology* 15:237–252.

MacArthur, R. H. 1972. *Geographical Ecology*. Princeton University Press, Princeton, New Jersey.

Mack, R. N., D. Simberloff, W. M. Lonsdale, H. Evans, M. Clout, and F. A. Bazzaz. 2000. Biotic invasions: Causes, epidemiology, global consequences, and control. *Ecological Applications* 10:689–710.

Madigan, D., and J. York. 1995. Bayesian graphical models for discrete data. *International Statistical Review* 63:215–232.

Mangel, M. 2003. Environment and longevity: The demography of the growth rate. *Population and Development Review* 29 (Suppl.):57–70.

Manne, L. L., S. L. Pimm, J. M. Diamond, and T. M. Reed. 1998. The form of the curves: A direct evaluation of MacArthur and Wilson's classic theory. *Journal of Animal Ecology* 67:784–794.

Martin, T. G., P. M. Kuhnert, K. Mengersen, and H. P. Possingham 2005. The power of expert opinion in ecological models using Bayesian methods: Impact of grazing on birds. *Ecological Applications* 15:266–280.

May, R. M. 1973. *Stability and Complexity in Model Ecosystems*, Monographs in Population Biology 6, Princeton University Press, Princeton, New Jersey.

May, R. M. 1974. Biological populations with nonoverlapping generations: Stable points, stable cycles, and chaos. *Science* 186:645–647.

May, R. M. 1981. Models for single populations. Pages 5–29 in R. M. May, ed. *Theoretical Ecology*. Blackwell, Oxford.

May, R. M., and G. F. Oster. 1976. Bifurcations and dynamic complexity in simple ecological models. *American Naturalist* 110:573–599.

McCauley, E., R. M. Nisbet, W. W. Murdoch, A. M. De Roos, and W. S. C. Gurney 1999. Large-amplitude cycles of *Daphnia* and its algal prey in enriched environments. *Nature* 402:653–656.

McCullagh, P., and J. A. Nelder. 1989. *Generalized Linear Models*. Chapman and Hall, London.

McCulloch, C. E. 2003. *Generalized Linear Mixed Models*. NSF-CBMS Regional Conference Series in Probability and Statistics, Institute of Mathematical Sciences 7.

McCulloch, C. E., and S. R. Searle. 2001. *Generalized, Linear, and Mixed Models*. Wiley, New York.

McCulloch, R. E., and R. S. Tsay. 1993. Bayesian inference and prediction for mean and variance shifts in autoregressive time series. *Journal of the American Statistical Association* 88:968–978.

McKelvey, K., B. R. Noon, and R. H. Lamberson. 1993. Conservation planning for species occupying fragmented landscapes: The case of the northern spotted owl. Pages 424–450 in P. M. Kareiva, J. G. Kingsolver, and R.B. Huey, eds,. *Biotic Interactions and Global Change*. Sinauer, Sunderland, Massachusetts.

Metz, J. A. J., and O. Diekmann. 1986. *The Dynamics of Physiologically Structured Populations*. Springer Verlag, Berlin.

Meyer, R. and R. B. Millar. 1999. Bugs in Bayesian stock assessments. *Canadian Journal of Fisheries and Aquatic Sciences* 56:1078–1086.

Millar, R. B., and R. Meyer. 2000. State-space modeling of nonlinear fisheries biomass dynamics using the Gibbs sampler. *Applied Statistics* 49:327–342.

Milliff, R. F., P. Niler, J. Morzel, A. Sybrandy, D. Nychka, and W. Large, 2003. Mesoscale correlation length scales from NSCAT and minimet surface wind retrievals in the Labrador Sea. *Journal of Atmospheric and Oceanic Technology* 20:513–533.

Mitchell, T. 1997. *Machine Learning*. McGraw Hill, New York.

Mohan, J., J. S. Clark, and W. H. Schlesinger. 2006. Long-term CO_2 enrichment of an intact forest ecosystem: Implications for temperate forest regeneration and succession. *Ecological Applications*, in press.

Mollison, D. 1972. The rate of spatial propagation of simple epidemics. *Proceedings of the Sixth Berkeley Symposium on Mathematics, Statistics, and Probability* 3:579–614.

Mollison, D. 1977. Spatial contact models for ecological and epidemic spread. *Journal of the Royal Statistical Society* B 39:283–326.

Mollison, D. 1991. Dependence of epidemic and population velocities on basic parameters. *Mathematical Biosciences* 107:255–287.

Molofsky, J., J. D. Bever, et al. 2001. Coexistence under positive frequency dependence. *Proceedings of the Royal Society of London* B 268:273–277.

Moloney, K. A. 1986. A generalized algorithm for determining category size. *Oecologia* 69:176–180.

Moorcroft, P. R., and M. A. Lewis. 2006. *Mechanistic Home Range Analysis*. Princeton Monographs in Population Biology, Princeton University Press, Princeton, New Jersey.

Moorcroft, P. R., M. A. Lewis, and R. L. Crabtree. 1999. Home range analysis using a mechanistic home range model. *Ecology* 80:1656–1665.

Morales, J. M., and S. P. Ellner. 2002. Scaling up movements in heterogeneous landscapes: The importance of behavior. *Ecology* 83:2240–2247.

Morales, J. M., D. T. Haydon, J. Frair, K. E. Holsinger, and J. M. Fryxell. 2004. Extracting more out of relocation data: Building movement models as mixtures of random walks. *Ecology* 85:2436–2445.

Morris, C. 1983. Parametric empirical Bayes inference: Theory and applications. *Journal of the American Statistical Association* 78:47–65.

Morris, W. F., and D. F. Doak. 2003. *Quantitative Conservation Biology: Theory and Practice of Population Viability Analysis*. Sinauer Associates, Sunderland, Massachusetts.

Murdoch, W. W. 1994. Population regulation in theory and practice. *Ecology* 75:271–287.

Murdoch, W. W., C. J. Briggs, and R. M. Nisbet. 2003. *Consumer-Resource Dynamics*. Princeton University Press, Princeton, New Jersey.

Murdoch, W. W., R. M. Nisbet, E. McCauley, A. M. deRoos, and W. S. C. Gurney. 1998. Plankton abundance and dynamics across nutrient levels: Tests of hypotheses. *Ecology* 79:1339–1356.

Murdoch, W. W., B. E. Kendall, R. M. Nisbet, C. J. Briggs, E. McCauley, and R. Bolser. 2002. Single-species models for many-species food webs. *Nature* 417:541–543.

Murray, J. D. 1989. *Mathematical Biology*. Springer Verlag, Berlin.

Mwalili, S. M., E. Lesaggre, and D. Declerck. 2005. A Bayesian ordinal logistic regression model to correct for interobserver measurement error in a geographical oral health study. *Applied Statistics* 54:77–93.

Myers, R. A., and N. G. Cadigan. 1993. Density-dependent juvenile mortality in marine demersal fish. *Canadian Journal of Fisheries and Aquatic Sciences* 50:1576–1590.

Nathan, R., G. G. Katul, H. S. Horn, S. M. Thomas, R. Oren, R. Avissar, S. W. Pacala, and S. A. Levin. 2002. Mechanisms of long-distance dispersal of seeds by wind. *Nature* 418:409–413.

Neubert, M. G., and H. Caswell. 2000. Demography and dispersal: Calculation and sensitivity analysis of invasion speed for structured populations. *Ecology* 81:1613–1628.

Neuhauser, C. 2001. Mathematical challenges in spatial ecology. *Notices of the American Mathematical Society* 48:1304–1314.

Neyman, J., and E. S. Pearson. 1933. On the problem of the most efficient tests of statistical hypotheses. *Philosophical Transactions of the Royal Society of London* A 231:289–337.

Nichols, J. D., J. R. Sauer, K. H. Pollock, and J. B. Hestbeck. 1992. Estimating transition probabilities for stage-based population projection matrices using capture-recapture data. *Ecology* 73:306–312.

Niklas, K. J., and B. J. Enquist. 2002. On the vegetative biomass partitioning of seed plant leaves, stems, and roots. *American Naturalist* 159:482–497.

Nisbet, R. M., and W. S. C. Gurney. 1982. *Modelling Fluctuating Populations*. Wiley, Chichester.

Nychka, D., C. Wikle, and J. A. Royle. 2002. Multiresolution models for nonstationary spatial covariance functions. *Statistical Modeling* 2:315–332.

Ogden, J. 1993. Population increase and nesting patterns of the black noddy *Anous minutus* in Pisonia forest on Heron Island: Observations from 1978, 1979, and 1992. *Australian Journal of Ecology* 18:395–403.

Ogle, K., M. Uriarte, J. Thompson, J. Johnstone, A. Jones, Y. Lin, E. J. B. McIntire, and J. K. Zimmerman. 2006. Implications of vulnerability to hurricane damage for long-term survival of tropical tree species: A Bayesian hierarchical analysis. Pages 98–118 in J. S. Clark and A. E. Gelfand, eds. *Hierarchical Modelling for the Environmental Sciences*. Oxford University Press, Oxford.

Ogle, K., R. L. Wolpert, and J. F. Reynolds. 2004. Reconstructing plant root area and water uptake profiles. *Ecology* 85:1967–1978.

O'Hagan, A. 1994. *Kendall's Advanced Theory of Statistics*: vol. 2B, *Bayesian Inference*. Edward Arnold, London.

Okubo, A. 1971. Oceanic diffusion diagrams. *Deep-Sea Research* 18:789–802.

Okubo, A. 1980. *Diffusion and Ecological Problems: Mathematical Models*. Springer Verlag, Berlin.

Okubo, A., and S. A. Levin. 1989. A theoretical framework for data analysis of wind dispersal of seeds and pollen. *Ecology* 70:329–338.

Okuyama, T., and B. M. Bolker. 2005. Combining genetic and ecological data to estimate sea turtle origins. *Ecological Applications* 15:315–325.

Openshaw, S., and P. Taylor. 1979. A million or so correlation coefficients: Three experiments on the modifiable area unit problem. Pp. 127–144 in *Statistical Applications in the Spatial Sciences*. Pion, London.

Oreskes, N. A., K. Shrader-Frechette, and K. Belitz. 1994. Verification, validation, and confirmation of numerical models in the earth sciences. *Science* 263:641–646.

Ostfeld, R. S., and C. G. Jones. 1996. Of mice and mast. *BioScience* 46:323–330.

Ovaskainen, O. 2004. Habitat-specific movement parameters estimated using mark–recapture data and a diffusion model. *Ecology* 85:242–257.

Ovaskainen, O., and S. J. Cornell. 2003. Biased movement at boundary and conditional occupancy times for diffusion processes. *Journal of Applied Probability* 40:557–580.

Pacala, S. W., C. D. Canham, J. Saponara, J. A. Silander, R. K. Kobe, and E. Ribbens. 1996. Forest models defined by field measurements: Estimation, error analysis, and dynamics. *Ecological Monographs* 66:1–44.

Pacala, S. W., and M. Rees. 1998. Field experiments that test alternative hypotheses explaining successional diversity. *American Naturalist* 152:729–737.

Pacala, S. W., and J. A. Silander. 1990. Field tests of neighborhood population dynamic models of two annual weed species. *Ecological Monographs* 60:113–134.

Paine, R. T. 1988. Road maps of interactions or grist for theoretical development? *Ecology* 69:1648–1654.

Paine, R. T., and S. A. Levin. 1991. Intertidal landscapes: Disturbance and the dynamics of pattern. *Ecological Monographs* 51:145–178.

Palmer, T. N., G. J. Shutts, R. Hagedorn, F. J. Doblas-Reyes, T. Jung, and M. Leutbecher. 2005. Representing weather uncertainty in weather and climate prediction. *Annual Review of Earth and Planetary Sciences* 33:163–193.

Parmesan, C., and G. Yohe. 2003. A globally coherent fingerprint of climate change impacts across natural systems. *Nature* 421:37–42.

Pascual, M. A., and P. Kareiva. 1996. Predicting the outcome of competition using experimental data: Maximum likelihood and Bayesian approaches. *Ecology* 77:337–349.

Patil, D. J., B. Hunt, E. Kalnay, J. A. Yorke, and E. Ott. 2001. Using bred vectors to establish local low dimensionality of atmospheric dynamics. *Physical Review Letters* 86:5878–5881.

Pearl, J. 2000. *Causality: Models, Reasoning, and Inference.* Cambridge University Press, Cambridge.

Pearl, R. 1925. *The Biology of Population Growth.* A. A. Knopf, New York.

Pearson, K. 1920. The fundamental problem of practical statistics. *Biometrika* 13:1–16.

Pearson, K., and L. N. G. Filon. 1898. Mathematical contributions to the theory of evolution IV. On the probable errors of frequency constants and on the influence of random selection on variation and correlation. *Philosophical Transactions of the Royal Society* A 191:229–311.

Peltonen, A., and I. Hanski. 1991. Patterns of island occupancy explained by colonization and extinction rates in shrews. *Ecology* 72:1698–1708.

Peterson, G. D., S. R. Carpenter, and W. A. Brock. 2003. Uncertainty and the management of multistate ecosystems: An apparently rational route to collapse. *Ecology* 84:1403–1411.

Pfister, C. A. 1995. Estimating competition coefficients from census data: A test with field manipulations of tidepool fishes. *American Naturalist* 146:271–291.

Pielke, R. A., and R. T. Conant. 2003. Best practices in prediction for decision-making: Lessons from the atmospheric and earth sciences. *Ecology* 84:1351–1358.

Pielou, E. C. 1981. The usefulness of ecological models: A stock-taking. *Quarterly Review of Biology* 56:17–31.

Pitelka, L. F., J. Ash, S. Berry, R. H. W. Bradshaw, L. Brubaker, J. S. Clark, M. B. Davis, J. M. Dyer, R. H. Gardner, H. Gitay, G. Hope, R. Hengeveld,

B. Huntley, G. A. King, S. Lavorel, R. N. Mack, G. P. Malanson, M. McGlone, I. R. Noble, I. C. Prentice, M. Rejmanek, A. Saunders, A. M. Solomon, S. Sugita, and M. T. Sykes. 1997. Plant migration and climate change. *American Scientist* 85:464–473.

Poole, D. J., and A. E. Raftery. 2000. Inference for deterministic simulation models: The Bayesian melding approach. *Journal of the American Statistical Association* 95:1244–1255.

Portnoy, S., and M. F. Willson. 1993. Seed dispersal curves: Behavior of the tail of the distribution. *Evolutionary Biology* 7:25–44.

Pulliam, H. R. 1988. Sources, sinks, and population regulation. *American Naturalist* 132(5):652–661.

Raftery, A. E., F. Balabdaoui, T. Gneiting, and M. Polakowski. 2003. Using Bayesian model averaging to calibrate forecast ensembles. Technical Report no. 440, Department of Statistics, University of Washington.

Raftery, A. E., T. Gneiting, F. Balabdaoui, and M. Polakowski. 2005. Using Bayesian model averaging to calibrate forecast ensembles. *Monthly Weather Review* 133:1155–1174.

Rannala, B., and J. L. Mountain. 1997. Detecting immigration by using multilocus genotypes. *Proceedings of the National Academy of Sciences* 94:9197–9201.

Ranta, E., V. Kaitala, J. Lindström, and H. Lindén. 1995. Synchrony in population dynamics. *Proceedings of the Royal Society of London* B 262:113–118.

Rasmussen, P. W., D. M. Heisey, E. V. Nordheim, and T. M. Frost. 2001. Time series intervention analysis: Unreplicated large-scale experiments. Pages 158–177 in S. M. Scheiner and J. Gurevitch, eds. *Design and Analysis of Ecological Experiments*. Oxford University Press, Oxford.

Real, L. A., and S. Ellner. 1992. Life history evolution in stochastic environments: A graphical mean-variance approach. *Ecology* 73:1227–1236.

Real, L. A., and S. A. Levin. 1991. The role of theory in the rise of modern ecology. Pages 177–191 in L. A. Real and J. H. Brown, eds. *Foundations of Ecology*. University of Chicago Press, Chicago, Illinois.

Reckhow, K. H. 1994a. Importance of scientific uncertainty in decision making. *Environmental Management* 18:161–166.

Reckhow, K. H. 1994b. A decision analytic framework for environmental analysis and simulation modeling. *Environmental Toxicology and Chemistry* 13:1901–1906.

Rees, M., R. Condit, M. Crawley, S. Pacala, and D. Tilman. 2001. Long-term studies of vegetation dynamics. *Science* 293:650–655.

Rees, M., D. Kelly, and O. N. Bjørnstad. 2002. Snow grass, chaos and the evolution of mast seeding. *American Naturalist* 160:44–59.

Reich, P. B., C. Buschena, M. Tjoelker, K. Wrage, J. Knops, D. Tilman, and J.-L. Machado. 2003. Variation in growth rate and ecophysiology among 34 grassland and savanna species under contrasting N supply: A test of functional group differences. *New Phytologist* 157:617–631.

Ribbens, E., J. A. Silander, and S. W. Pacala. 1994. Seedling recruitment in forests: Calibrating models to predict patterns of tree seedling dispersion. *Ecology* 75:1794–1806.

Ricciardi, A., R. J. Neves, and J. B. Rasmussen. 1998. Impending extinctions of North American freshwater mussels (*Unionidae*) following the zebra mussel (*Dreissena polymorpha*) invasion. *Journal of Animal Ecology* 67:613–619.

Ricciardi, L. M. 1986. Stochastic population theory: Birth and death processes. Pages 155–190 in T. G. Hallam and S. A. Levin, eds. *Mathematical Ecology*. Springer Verlag, Berlin.

Ricker, W. 1954. Stock and recruitment. *Journal of the Fisheries Research Board of Canada* 11:559–623.

Ries, L., and D. M. Debinski. 2001. Butterfly responses to habitat edges in the highly fragmented prairies of Central Iowa. *Journal of Animal Ecology* 70:840–852.

Ripley, B. D. 1987. *Stochastic Simulation*. Wiley, New York.

Robert, C. P., and G. Casella. 1999. *Monte Carlo Statistical Methods*. Springer Verlag, New York.

Robertson, G. P. 1987. Geostatistics in ecology: Interpolating with known variance. *Ecology* 68:744–748.

Rohani, P., D. J. D. Earn, and B. T. Grenfell. 1999. Opposite patterns of synchrony in sympatric disease metapopulations. *Science* 286:968–971.

Root, T. L., J. T. Price, K. R. Hall, S. H. Schneider, C. Rosenzweig, and J. A. Pounds. 2003. Fingerprints of global warming on wild animals and plants. *Nature* 421:57–60.

Rosenzweig, M. L. 1971. Paradox of enrichment: Destabilization of exploitation ecosystems in ecological time. *Science* 171:385–387.

Royama, T. 1992. *Analytical Population Dynamics*. Chapman and Hall, London.

Royle, A. J., and W. Link, 2002. Random effects and shrinkage estimation in capture-recapture models. *Journal of Applied Statistics* 29:329–351.

Sabo, J. L., and M. E. Power. 2003. Numerical response of lizards to aquatic insects and short-term consequences for terrestrial prey. *Ecology* 83:3023–3036.

Saether, B.-E., J. Tufto, S. Engen, K. Jerstad, O. W. Rostad, and J. E. Skatan. 2000. Population dynamical consequences of climate change for a small temperate songbird. *Science* 287:854–856.

Sampson, P. D., and P. Guttorp. 1992. Nonparametric estimation of nonstationary spatial covariance structure. *Journal of the American Statistical Association* 87:108–119.

Sarewitz, D., and R. A. Pielke Jr. 2000. *Prediction: Science, Decision Making and the Future of Nature*. Island Press, Washington, D.C.

Saucy, F. 1994. Density dependence in time series of the fossorial form of the water vole (*Arvicola terrestris*). *Oikos* 71:381–392.

Sauer, J. R., and W. A. Link. 2002. Hierarchical modeling of population stability and species group attributes from survey data. *Ecology* 83:1743–1751.

Schabenberger, O., and C. A. Gotway. 2005. *Statistical Methods for Spatial Data Analysis*. Chapman and Hall, Boca Raton, Florida.

Schaffer, W. M. 1981. Ecological abstraction: The consequences of reduced dimensionality in ecological models. *Ecological Monographs* 51:383–401.

Scheffer, M., and S. R. Carpenter. 2003. Catastrophic regime shifts in ecosystems: Linking theory to observation. *Trends in Ecology and Evolution* 12:648–656.

Scheffer, M., S. R. Carpenter, J. A. Foley, C. Folke, and B. Walker. 2001. Catastrophic shifts in ecosystems. *Nature* 413:591–596.

Scheiner, S. M., and J. Gurevitch, eds. 2002. *Design and Analysis of Ecological Experiments*. Oxford University Press, Oxford.

Schlesinger, W. H. 2004. Better living through biogeochemistry. *Ecology* 85:2402–2407.

Schwartz, C. J., J. F. Schweigert, and A. N. Aronson. 1993. Estimating migration rates using tag-recovery data. *Biometrics* 49:177–193.

Schwartz, G. 1978. Estimating the dimension of a model. *Annals of Statistics* 6:461–464.

Schwartz, S. 1994. The fallacy of the ecological fallacy: The potential misuse of a concept and the consequences. *American Journal of Public Health* 84:819–824.

Seber, C. A. F. 1965. A note on the multiple-recapture census. *Biometrika* 52:249–259.

Seitz, R. D., R. N. Lipcius, A. H. Hines, and D. B. Eggleston. 2002. Density-dependent predation, habitat variation, and the persistence of marine bivalve prey. *Ecology* 82:2435–2451.

Sharov, A. A., and A. M. Leibold. 1998. Model of slowing the spread of gypsy moth (Lepidoptera: Lymantriidae) with a barrier zone. *Ecological Applications* 8:1170–1179.

Sharpe, D. M., and D. E. Fields. 1982. Integrating the effects of climate and seed fall velocities on seed dispersal by wind: A model and application. *Ecological Modelling* 17:297–310.

Shea, K., and H. P. Possingham. 2000. Optimal release strategies for biological control agents: An application of stochastic dynamic programming to population management. *Journal of Animal Ecology* 37:77–86.

Shigesada, N., K. Kawasaki, and Y. Takeda. 1995. Modeling stratified diffusion in biological invasions. *American Naturalist* 146:229–251.

Shipley, B. 2000. *Cause and Correlation in Biology: A User's Guide to Path Analysis, Structural Equations and Causal Inference*. Cambridge University Press, New York.

Simberloff, D. 1980. A succession of paradigms in ecology: Essentialism to materialism and probabilism. *Synthese* 43:3–39.

Simberloff, D. 1983. Competition theory, hypothesis testing, and other community-ecological buzzwords. *American Naturalist* 122:626–635.

Skellam, J. G. 1951. Random dispersal in theoretical populations. *Biometrika* 38:196–218.

Skelly, D. 2002. Experimental venue and estimation of interaction strength. *Ecology* 83:2097–2101.

Soons, M. B., G. W. Heil, R. Nathan, and G. G. Katul. 2004. Determinants of long-distance seed dispersal by wind in grasslands. *Ecology* 85:3069–3079.

Spiegelhalter, D. J., N. G. Best, B. P. Carlin, and A. van der Linde. 2002. Bayesian measures of model complexity and fit (with discussion). *Journal of the Royal Statistical Society* B 64:583–639.

Spiegelhalter, D. J., K. R. Abrams, and J. P. Myles. 2003. *Bayesian Approaches to Clinical Trials and Health-Care Evaluation*. Wiley, New York.

Staples, D. F., M. L. Taper, and B. Dennis. 2004. Estimating population trend and process variation for PVA in the presence of sampling error. *Ecology* 85:923–929.

Stenseth, N. C., O. N. Bjørnstad, and T. Saitoh. 1996. A gradient from stable to cyclic populations of *Clethrionomys rufocanus* in Hokkaido, Japan. *Proceedings of the Royal Society of London* B 263:1117–1126.

Stenseth, N.C., K.-S. Chan, H. Tong, R. Boonstra, S. Boutin, C. J. Krebs, E. Post, M. O'Donoghue, N. G. Yoccoz, M. C. Forchhammer, and J. W. Hurrell. 1999. Common dynamic structure of Canada lynx populations within three climatic regions. *Science* 285:1071–1073.

Stenseth, N. C., W. Falck, O. N. Bjørnstad, and C. J. Krebs. 1997. Population regulation in snowshoe hare and Canadian lynx: Asymmetric food web configurations between hare and lynx. *Proceedings of the National Academy* USA 94:5147–5152.

Stenseth, N. C., W. Falck, K.-S. Chand, O. N. Bjørnstad, M. O'Donoghuee, H. Tong, R. Boonstraa, of Sciences H. S. Boutini, C. J. Krebs, and N. G. Yoccoz. 1998. From patterns to processes: Phase and density dependencies in the Canadian lynx cycle. *Proceedings of the National Academy of Sciences* 95:15430–15435.

Stenseth, N. C., H. Viljugrein, T. Saitoh, T. F. Hansen, M. O. Kittilsen, E. Bolviken, and F. Glockner. 2003. Seasonality, density dependence, and population cycles in Hokkaido voles. *PNAS* 100:11478–11483.

Stewart-Oaten, A., W. W. Murdoch, and K. R. Parker. 1986. Environmental impact assessment: "Pseudoreplication" in time? *Ecology* 67:929–940.

Stigler, S. M. 1986. *The History of Statistics*. Harvard University Press, Cambridge, Massachusetts.

Stock, J. H., and M. W. Watson. 1995. *Introduction to Econometrics*. Addison Wesley, Upper Saddle River, New Jersey.

Strong, D. R. 1986. Density vagueness: Abiding the variance in the demography of real populations. Pages 257–268 in J. Diamond and T. J. Case, eds. *Community Ecology*. Harper and Row, Cambridge, England.

Stroud, J. R., P. Mueller, and N. G. Polson. 2003. Nonlinear state-space models with state-dependent variances. *Journal of the American Statistical Association* 98:377–386.

Stroud, T. 1994. Bayesian analysis of binary survey data. *Canadian Journal of Statistics* 22:33–45.

Stuart, A., and J. K. Ord. 1994. *Kendall's Advanced Theory of Statistics*. Edward Arnold, London.

Suarez, A. V., D. A. Holway, and T. J. Case. 2001. Patterns of spread in biological invasions dominated by long-distance jump dispersal: Insights from Argentine ants. *Proceedings of the National Academy of Sciences* 98:1095–1100.

Sutton, O. G. 1953. *Micrometeorology*. McGraw Hill, New York.

Swedlow, B. 2003. Scientists, judges, and the spotted owls: Policy makers in the Pacific Northwest. *Duke Environmental Law and Policy Forum*. Duke University School of Law, Durham, NC.

Tanizaki, H. 2003. On regression models with autocorrelated error: Small sample properties. *International Journal of Pure and Applied Mathematics* 5:161–175.

Tanner, M. A. 1996. *Tools for Statistical Inference*. Springer Verlag, Berlin.

Taylor, H. M., and S. Karlin. 1994. *An Introduction to Stochastic Modeling*. Academic Press, Boston, Massachusetts.

Thogmartin, W. E., J. R. Sauer, and M. G. Knutson. 2004. A hierarchical spatial model of avian abundance with application to cerulean warblers. *Ecological Applications* 14:N1766–1779.

Tilman, D. 1982. *Resource Competition and Community Structure*. Princeton University Press, Princeton, New Jersey.

Tilman, D. 1988. *Plant Strategies and the Dynamics and Structure of Plant Communities*. Princeton University Press, Princeton, New Jersey.

Tilman, D. 1994. Competition and biodiversity in spatially structured habitats. *Ecology* 75:2–16.

Tilman, D., and P. Kareiva, eds. 1997. *Spatial Ecology*. Princeton University Press, Princeton, New Jersey.

Tilman, D., D. Wedin, and J. Knops. 1996. Productivity and sustainability influenced by biodiversity in grassland ecosystems. *Nature* 379:718–720.

Tong, H. 1990. *Non-linear Time Series: A Dynamical System Approach*. Oxford University Press., Oxford.

Turchin, P. 1995. Population regulation: Old arguments and a new synthesis. Pages 19–40 in *Population Dynamics*. Academic Press, New York.

Turchin, P. 1996. Nonlinear time-series modeling of vole population fluctuations. *Researches in Population Ecology* 38:121–132.

Turchin, P. 1998. *Quantitative Analysis of Movement*. Sinauer, Sunderland, Massachussetts.

Turchin, P. 2003. *Complex Population Dynamics*. Princeton University Press, Princeton, New Jersey.

Turchin, P., and S. Ellner. 2000. Living on the edge of chaos: Population dynamics of Fennoscandian voles. *Ecology* 81:3099–3116.

Turchin, P., F. J. Odendaal, and M. D. Rausher. 1991. Quantifying insect movement in the field. *Environmental Entomology* 20:955–963.

Turchin, P., and A. D. Taylor. 1992. Complex dynamics in ecological time series. *Ecology* 73:289–305.

Turchin, P., A. D. Taylor, and J. D. Reeve. 1999. Dynamical role of predators in population cycles of a forest insect: An experimental test. *Science* 285:1068–1071.

Turcotte, D. L., B. D. Malamud, et al. 2002. Self-organization, the cascade model, and natural hazards. *Proceedings of the National Academy of Sciences* USA 99 Suppl. 1:2530–2537.

Turner, M. G., R. H. Gardner, and R. V. O'Neill. 2001. *Landscape Ecology in Theory and Practice: Pattern and Process*. Springer Verlag, New York.

Urban, D. L. 2000. Using model analysis to design monitoring programs for landscape management and impact assessment. *Ecological Applications* 10:1820–1832.

Van den Bosch, F., J. A. J. Metz, and O. Diekmann. 1990. The velocity of spatial population expansion. *Journal of Mathematical Biology* 28:529–565.

Vandermeer, J. 1978. Choosing category size in a stage projection matrix. *Oecologia* 32:79–84.

Van Groenendael, J., H. de Kroon, and H. Caswell. 1988. Projection matrices in population biology. *TREE* 3:264–269.

Vaupel, J. W., K. G. Manton, and E. Stallard. 1979. The impact of heterogeneity in individual frailty on the dynamics of mortality. *Demography* 16:439–454.

Veit, R. R., and M. A. Lewis. 1996. Dispersal, population growth, and the allee effect: Dynamics of the house finch invasion of eastern North America. *American Naturalist* 148:255–274.

Ver Hoef, J. M. 1996. Parametric empirical Bayes methods for ecological applications. *Ecological Applications* 6:1047–1055.

Ver Hoef, J. 2002. Sampling and geostatistics for spatial data. *Ecoscience* 9:152–161.

Ver Hoef, J. M., N. Cressie, R. N. Fisher, and T. J. Case. 2001. Uncertainty in spatial linear models for ecological data. Pages 214–237 in C. T. Hunsaker, M. F. Goodchild, M. A. Friedl, and T. J. Case, eds. *Spatial Uncertainty for Ecology: Implications for Remote Sensing and GIS Applications*. Springer Verlag, New York.

Ver Hoef, J. M., and K. J. Frost. 2003. A Bayesian hierarchical model for monitoring harbor seal changes in Prince William Sound, Alaska. *Environmental and Ecological Statistics* 10:201–219.

Vitousek, P. M., C. M. D'Antonio, L. L. Loope, and R. Westbrooks. 1996. Biological invasions as global environmental change. *American Scientist* 84:468–478.

Wakefield, I., and G. Shaddick. 2005. Health-exposure modeling and the ecological fallacy. *Biostatistics* 1:1–19.

Walter, E. 1982. Identifiability of state space models. *Lecture Notes Biomath* 46.

Warner, R. R., and P. L. Chesson. 1985. Coexistence mediated by recruitment fluctuations: A field guide to the storage effect. *American Naturalist* 125:769–787.

Wedin, D. A., and D. Tilman. 1996. Influence of nitrogen loading and species composition on the carbon balance of grasslands. *Science* 274:1720–1723.

Weinberger, H. F. 1982. Long-time behavior of a class of biological models. *SIAM Journal of Mathematical Analysis* 13:353–396.

Weitzman, M. L. 2001. Gamma discounting. *American Economic Review* 91:260–271.

Werner, P. A., and H. Caswell. 1977. Population growth rates and age versus stage-distribution models for teasel (*Dipsacus sylvestris* Huds.). *Ecology* 58:1103–1111.

West, M., and J. Harrison. 1997. *Bayesian Forecasting and Dynamic Models.* Springer Verlag, New York.

White, G. C., and R. E. Bennett. 1997. Analysis of frequency count data using the negative binomial distribution. *Ecology* 77:2549–2557.

White, G. C., and K. P. Burnham. 1999. Program MARK: Survival rate estimation for both live and dead encounters. *Bird Study* 46 (Suppl.): S120–S139.

Whittaker, J. 1990. *Graphical Models in Applied Multivariate Statistics.* Wiley, New York.

Wikle, C. K. 2003a. Hierarchical Bayesian models for predicting the spread of ecological processes. *Ecology* 84:1382–1394.

Wikle, C. K. 2003b. Hierarchical models in environmental science. *International Statistical Review* 71:181–199.

Wikle, C. K., and C. J. Anderson. 2003. Climatological analysis of tornado report counts using a hierarchical Bayesian spatio-temporal model. *Journal of Geophysical Research—Atmospheres* (D24), 9005, doi:10.1029/2002JD002806.

Wikle, C. K., and L. M. Berliner. 2005. Combining information across spatial scales. *Technometrics* 47:80–91.

Wikle, C. K., Milliff, R. F., Nychka, D., and L. M. Berliner. 2001. Spatiotemporal hierarchical Bayesian modeling: Tropical ocean surface winds. *Journal of the American Statistical Association* 96:382–397.

Wikle, C. K., and J. A. Royle. 2005. Dynamic design of ecological monitoring networks for non-Gaussian spatio-temporal data. *Environmetrics* 16:507–522.

Wikle, C. K., and J. A. Royle, 2005. Predicting migratory bird settling patterns with hierarchical Bayesian spatio-temporal models, in review.

Williams, B. K., J. D. Nichols, and M. J. Conroy. 2001. *Analysis and Management of Animal Populations.* Academic Press, New York.

Williams, C. K., A. R. Ives, and R. D. Applegate. 2003. Population dynamics across geographical ranges: Time-series analyses of three small game species. *Ecology* 84:2654–2667.

Willson, M. F. 1993. Dispersal mode, seed shadows, and colonization patterns. *Vegetatio* 107/108:261–280.

Winkler, R. L. 1981. Combining probability distributions from dependent information sources. *Management Science* 27:479–488.

Wolpert, R. L. 1995. Comment on "Inference from a deterministic population dynamics model for bowhead whales" by A. E. Raftery, G. H. Givens and J. E. Zeh. *Journal of the American Statistical Association* 90:426–427.

Wood, S. N. 2001. Partially specified ecological models. *Ecological Monographs* 71:1–25.

Woodard, R., C. Z. He, and D. Sun. 2003. Bayesian estimation of hunting success rate and harvest for spatially correlated post-stratified data. *Biometrical Journal* 45:985–1005.

Wootton, J. T. 1993. Indirect effects and habitat use in an intertidal community: Interaction chains and interaction modifications. *American Naturalist* 141:71–89.

Wright, S. 1921. Correlation and causation. *Journal of Agricultural Research* 20:557–585.

Wright, S. J., C. Carrasco, O. Calderon, and S. Paton. 1999. The El Niño Southern Oscillation, variable fruit production, and famine in a tropical forest. *Ecology* 80:1632–1647.

Wyckoff, P. H., and J. S. Clark. 2002. Growth and mortality for seven co-occurring tree species in the southern Appalachian Mountains: Implications for future forest composition. *Journal of Ecology* 90:604–615.

Wyckoff, P., and J. S. Clark. 2005. Comparing predictors of tree growth: The case for exposed canopy area. *Canadian Journal of Forest Research* 35:13–20.

Yuan, Z., and Y. Yang. 2004. Combining linear regression models: When and how? *Journal of the American Statistical Association* 100:1202–1204.

Zeger, S. L., and M. R. Karim. 1991. Generalized linear models with random effects: A Gibbs sampling approach. *Journal of the American Statistical Association* 86:79–86.

INDEX